Chemometrics

Chemometrics

Data Driven Extraction for Science

Richard G. Brereton
University of Bristol (Emeritus)
UK

Second Edition

Registered Office(s)
John Wiley & Sons, Inc., 111 River Street, Hoboken, NJ 07030, USA
John Wiley & Sons Ltd, The Atrium, Southern Gate, Chichester, West Sussex, PO19 8SQ, UK

Editorial Office
The Atrium, Southern Gate, Chichester, West Sussex, PO19 8SQ, UK

For details of our global editorial offices, customer services, and more information about Wiley products visit us at www.wiley.com.

Wiley also publishes its books in a variety of electronic formats and by print-on-demand. Some content that appears in standard print versions of this book may not be available in other formats.

Library of Congress Cataloging-in-Publication Data:

Names: Brereton, Richard G., author.
Title: Chemometrics : data driven extraction for science / Richard G.
  Brereton.
Description: Second edition. | Hoboken, NJ : John Wiley & Sons, 2018. |
  Originally published in 2003 as: Chemometrics : data analysis for the
  laboratory and chemical plant. |
Identifiers: LCCN 2017054468 (print) | LCCN 2017059486 (ebook) | ISBN
  9781118904688 (epub) | ISBN 9781118904671 (pdf) | ISBN 9781118904664 (pbk.)
Subjects: LCSH: Chemometrics–Data processing. | Chemical
  processes–Statistical methods–Data processing.
Classification: LCC QD75.4.C45 (ebook) | LCC QD75.4.C45 B74 2018 (print) |
  DDC 543.01/5195–dc23
LC record available at https://lccn.loc.gov/2017054468

Cover design by Wiley
Cover images: (Background) © LiliKo/Gettyimages; (Diagram) Courtesy of Richard G. Brereton

Set in 10/12pt WarnockPro by SPi Global, Chennai, India

Printed and bound by CPI Group (UK) Ltd, Croydon, CR0 4YY

10 9 8 7 6 5 4 3 2 1

Contents

Preface to Second Edition

The first edition of this book has been well received, with a special emphasis on numerical illustration of a wide range of chemometric methods. Of particular importance were the problems at the end of each chapter that readers could work through in their own favourite environment, such as Excel or Matlab, but also R or Python or Fortran or any number of languages or computational packages if desired. I have performed calculations in both Matlab and Excel, but readers should not feel restricted if they have an alternative.

The reader of this book is likely to be an applied scientist or statistician who wishes to understand the basis and motivation of many of the main methods used in chemometrics.

Since the first edition, chemometrics has become much more widespread, including outside mainstream chemistry. In the early 2000s, the major applications were quantitative laboratory analytical science and chemical engineering including process control. Over the past few years, application areas have broadened, as large analytical laboratory-generated data sets become more widely available, for example, in metabolomics, heritage science and food science, reflecting a larger emphasis on pattern recognition in the second edition including some practical case studies from metabolomics in the form of worked problem sets.

Despite this, many of the original building blocks of the subject remain unchanged. A factorial design and a principal component is still the same, so parts of the text only involve small changes from the first edition. Nevertheless, feedback both from students and co-workers of mine and also from comments via the Internet have provided valuable guidance as to what changes are desirable for a second edition. Important structural changes such as multiple choice questions throughout the book and colour printing update the original edition as a modern day textbook.

Some major updates are as follows.
- Short multiple choice questions at the end of every section of the main text.
- Colour printing involving redrawing many figures.
- New chapter on supervised pattern recognition (classification) involving enhanced discussions of SIMCA, PLS-DA, LDA, QDA, EDC, kNN as well as validation.
- New case studies on NIR for distinguishing edible oils, and properties of elements, to illustrate unsupervised pattern recognition methods.
- New case studies in metabolomics, including Arabidopsis genotyping by MS, Raman of cancerous lymph nodes and NMR for diagnosing diabetes, as new problem sets.
- Additional description of MCR and ITTFA.
- New and expanded discussions of wavelets and of Bayesian methods in signal analysis.
- Updated description of Matlab R2016a under Windows 10, and Excel 2016 under Windows 10, in the context of the needs of the chemometrician.
- Enhanced discussion of the main statistical distributions.
- Enhanced discussions on validation and optimisation, including description of the bootstrap and of performance indicators.

To supplement this book, all data sets in this book, both from the main text and the problems at the end of each chapter, are downloadable. In addition, there is a downloadable Excel add-in to perform most of the common multivariate methods and a macro for labelling graphs. Matlab routines corresponding to many of the main methods are also available. The answers to the problems at the end of each chapter can also be found. These are available on the Wiley website associated with this book.

It is hoped that this text will be useful for students wishing to obtain a fundamental understanding of many chemometric methods. It will also be useful for any practicing chemometrician who needs to work through methods they may have only recently encountered, using numerical examples: as a researcher, when I encounter an unfamiliar approach, I usually like to reproduce numerical data from published case studies to check how it works before I am confident to use the method. For people encountering chemometrics for the first time, for example, in metabolomics and heritage science, this book presents many of the most widespread methods and so will serve as a good reference. And as a refresher, the multiple choice questions test the basic understanding. The worked case studies can be collected together and are helpful for courses.

Finally, I thank the publishers who have encouraged the development of this rather complex project, especially Jenny Cossham, through many stages and also colleagues who have provided data as listed in the acknowledgements.

Bristol, May 2017   *Richard G. Brereton*

Preface to First Edition

This book is a product of several years' activities from myself. First and foremost, the task of educating graduate students in my research group from a large variety of backgrounds over the past 10 years has been a significant formative experience, and this has allowed me to develop a large series of problems which we set every 3 weeks and present answers in seminars. From my experience, this is the best way to learn chemometrics! In addition, I have had the privilege to organise international quality courses mainly for industrialists with the participation of many representatives as tutors of the best organisations and institutes around the world, and I have learnt from them. Different approaches are normally taken while teaching industrialists who may be encountering chemometrics for the first time in mid-career and have a limited period of a few days to attend a condensed course, and university students that have several months or even years to practice and improve. However, it is hoped that this book represents a symbiosis of both needs.

In addition, it has been a great inspiration for me to write a regular fortnightly column for Chemweb (available to all registered users on www.chemweb.com) and some of the material in this book is based on articles first available in this format. Chemweb brings a large reader base to chemometrics, and feedback via e-mails or even travels around the world have helped me formulate my ideas. There is a very wide interest in this subject, but it is somewhat fragmented. For example, there is a strong group of Near Infrared Spectroscopists, primarily in the USA, that has led to the application of advanced ideas in process monitoring who see chemometrics as a quite technical industrially oriented subject. There are other groups of mainstream chemists that see chemometrics as applicable to almost all branches of research, ranging from kinetics to titrations to synthesis optimisation. Satisfying all these diverse people is not an easy task.

This book relies mainly on numerical examples: many in the body of the text come from my favourite research interests that are primarily in analytical chromatography and spectroscopy, to expand the text more to produce a huge book of twice the size, so I ask the indulgence of readers if your area of application differs. Certain chapters such as those on calibration could be approached from widely different viewpoints, but the methodological principles are the most important, and if you understand how the ideas can be applied in one area, you will be able to translate to your own favourite application. In the problems at the end of each chapter, I cover a wider range of applications to illustrate the broad basis of these methods. The emphasis of this book is on understanding ideas, which can then be applied to a wide variety of problems in chemistry, chemical engineering and allied disciplines.

It is difficult to select what material to include in this book without making it too long. Every expert I have shown this book to has made suggestions for new material. Some I have taken into account and I am most grateful for every proposal, and others I have mentioned briefly or not at all, mainly for the reason of length and also to ensure that this book sees the light of day rather than constantly expands without an end. There are many outstanding specialist books for the enthusiast. It is my experience, although, that if you understand the main principles (which are quite a few in number), and constantly apply them to a variety of problems, you will soon pick up the more advanced techniques, so it is the building blocks that are most important.

In a book of this nature, it is very difficult to decide on what detail is required for the various algorithms, some readers will have no real interest in the algorithms, whereas others will feel the text is incomplete without comprehensive descriptions. The main algorithms for common chemometric methods are presented in Appendix A.2. Step by step descriptions of methods, rather than algorithms, are presented in the text. A few approaches that will interest some readers such as cross-validation in PLS are described in the problems at the end of appropriate chapters which supplement the text. It is expected that readers will approach this book with different levels of knowledge and expectations, so it is possible to gain a great deal without having an in-depth appreciation of computational algorithms, but for interested readers, the information is nevertheless available. People rarely read texts in a linear fashion, they often dip in and out of parts of it according to their background and aspirations, and chemometrics is a subject which people approach with very different previous knowledge and skills, so it is possible to gain from this book without covering every topic in full. Many readers will simply use add-ins or Matlab commands and be able to produce all the results in this text.

Chemometrics uses a very large variety of software. In this book, we recommend two main environments, Excel and Matlab, the examples have been tried using both environments, and you should be able to get the same answers in both cases. Users of this book will vary from people that simply want to plug the data into existing packages to those that are curious and want to reproduce the methods in their own favourite language such as Matlab, VBA or even C. In some cases, instructors may use the information available with this book to tailor examples for problem classes. Extra software supplements are available via the publishers' website www. SpectroscopyNOW.com, together with all the data sets in this book.

The problems at the end of each chapter form an important part of the text, the examples being a mixture of simulations (which have an important role in chemometrics) and real case studies from a wide variety of sources. For each problem, the relevant sections of the text that provide further information are referenced. However, a few problems build on the existing material and take the reader further: a good chemometrician should be able to use the basic building blocks to understand and use new methods. The problems are of various types; thus, not every reader will to solve all the problems. In addition, instructors can use the data sets to construct workshops or course material that goes further than the book.

I am very grateful for the tremendous support I have had from many people when asking for information and help with data sets and permission where required. I thank Chemweb for agreement to present material modified from articles originally published in their e-zine, *The Alchemist*, and the RSC for permission to base the text of Chapter 5 on material originally published in the *Analyst* (**125**, 2125–2154 (2000)). A full list of acknowledgements for the data sets used in this text is presented after this foreword.

I thank Tom Thurston and Les Erskine for a superb job on the Excel add-in, and Hailin Shen for outstanding help in Matlab. Numerous people have tested the answers to the problems. Special mention should be given to Christian Airiau, Kostas Zissis, Tom Thurston, Conrad Bessant and Cevdet Demir for access to a comprehensive set of answers on disc for a large number of exercises so I can check mine. In addition, several people have read chapters and made detailed comments particularly checking numerical examples; in particular, I thank Hailin Shen for suggestions about improving Chapter 6 and Mohammed Wasim for careful checking of errors. In some ways, the best critics are the students and postdocs working with me because they are the people that have to read and understand a book of this nature, and it gives me great confidence that my co-workers in Bristol have found this approach useful and have been able to learn from the examples.

Finally, I thank the publishers for taking a germ on an idea and making valuable suggestions as to how this could be expanded and improved to produce what I hope is a successful textbook and having faith and patience over a protracted period.

Bristol, February 2002   *Richard G. Brereton*

Acknowledgements

The following have provided me with sources of data for this text. All other case studies are simulations.

Data set	Source
Problem 2.2	A. Nordin, L. Eriksson, M. Öhman, *Fuel*, 74, 128–135 (1995)
Problem 2.6	G. Drava, University of Genova
Problem 2.7	I.B. Rubin, T.J. Mitchell, G. Goldstein, *Anal Chem*, 43, 717–721 (1971)
Problem 2.10	G. Drava, University of Genova
Problem 2.11	Y. Yifeng, S. Dianpeng, H. Xuebing, W. Shulan, *Bull Chem Soc Japan*, 68, 1115–1118 (1995)
Problem 2.12	D.V. McCalley, University of West of England, Bristol
Problem 2.15	D. Vojnovic, B. Campisi, A. Mattei, L. Favreto, *Chemometrics Intell Lab Systems*, 27, 205–219 (1995)
Problem 2.16	L.E. Garcia-Ayuso, M.D. Luque de Castro, *Anal Chim Acta*, 382, 309–316 (1999)
Problem 3.8	K.D. Zissis, University of Bristol
Problem 3.9	C. Airiau, University of Bristol
Table 4.1	S. Dunkerley, University of Bristol
Table 4.2	S. Goswami and K. Olafsson, Camo ASA
Table 4.3	A. Javey, Chemometrics On-line
Problem 4.3	D. Duewer, National Institute of Standards Technology, US
Problem 4.5	S. Dunkerley, University of Bristol
Problem 5.3	S. Wold, University of Umeå (based on R. Cole and K. Phelps, *J Sci Food Agric*, 30, 669–676 (1979)
Problem 5.4	P. Bruno, M. Caselli, M.L. Curri, A. Genga, R. Striccoli, A. Traini, *Anal Chim Acta*, 410, 193–202 (2000)
Problem 5.5	R. Vendrame, R.S. Braga, Y. Takahata, D.S. Galvão, *J Chem Inf Comp Sci*, 39, 1094–1104 (1999)
Problem 5.7	R. Goodacre, University of Manchester (based on M. Kusano, A. Fukushima, M. Arita, P. Jonsson, T. Moritz, M. Kobayashi, M., et al., *BMC System Biology*, 1, 53 (2007) – Metabolights accession MTBLS40)
Problem 5.8	R. Goodacre, University of Manchester (based on R.M. Salek, M.L. Maguire, E. Bentley, D.V. Rubtsov, T. Hough, M. Cheeseman, et al., *Physiol Genomics*, 29, 99–10 (2007) – Metabolights accession MTBLS1)
Problem 5.9	G.R. Lloyd (based on G.R. Lloyd, L.E. Orr, J. Christie-Brown et al., *Analyst*, 138, 3900–3908 (2013))
Table 6.1	S.D. Wilkes, University of Bristol
Table 6.20	S.D. Wilkes, University of Bristol
Problem 6.1	M.C. Pietrogrande, F. Dondi, P.A. Borea, C. Bighi, *Chemometrics Intell Lab Systems*, 5, 257–262 (1989)
Problem 6.3	H. Martens, M. Martens, *Multivariate Analysis of Quality*, Wiley, Chichester, 2001, p. 14
Problem 6.6	P.M. Vacas, University of Bristol
Problem 6.9	K.D. Zissis, University of Bristol
Problem 7.1	S. Dunkerley, University of Bristol
Problem 7.3	S. Dunkerley, University of Bristol
Problem 7.5	R. Tauler, University of Barcelona (results published in R. Gargallo, R. Tauler, A. Izquierdo-Ridorsa, *Quimica Analitica*, 18, 117–120)
Problem 7.6	S.P. Gurden, University of Bristol

About the Companion Website

Do not forget to visit the companion website for this book:

http://booksupport.wiley.com

The accompanying website for this text, http://booksupport.wiley.com, provides valuable material designed to enhance your learning, including:

- Answers to problems at the end of each chapter
- Software
- Associated data sets
- Figures in PPT

1

Introduction

1.1 Historical Parentage

There are many opinions about the origin of chemometrics. Until quite recently, the birth of chemometrics was considered to have happened in the 1970s. Its name first appeared in 1972 in an article by Svante Wold [1]: in fact, the topic of this article was not one that we would recognise as being core to chemometrics, being relevant to neither multivariate analysis nor experimental design. For over a decade, the word chemometrics was considered to be of very low profile, and it developed a recognisable presence only in the 1980s, as described below.

However, if an explorer describes a new species in a forest, the species was there long before the explorer. Thus, the naming of the discipline just recognises that it had reached some level of visibility and maturity. As people re-evaluate the origins of chemometrics, the birth can be traced many years back.

Chemometrics burst into the world due to three fundamental factors, applied statistics (multivariate and experimental design), statistics in analytical and physical chemistry, and scientific computing.

1.1.1 Applied Statistics

The ideas of multivariate statistics have been around a long time. R.A. Fisher and colleagues working in Rothamsted, UK, formalised many of our modern ideas while applying primarily to agriculture. In the UK, before the First World War, many of the upper classes owned extensive land and relied on their income from tenant farmers and agricultural labourers. After the First World War, the cost of labour became higher, with many moving to the cities, and there was stronger competition of food from global imports. This meant that historic agricultural practices were seen to be inefficient and it was hard for landowners (or companies that took over large estates) to be economic and competitive, hence a huge emphasis on agricultural research, including statistics to improve these. R.A. Fisher and co-workers published some of the first major books and papers that we would regard as defining modern statistical thinking [2, 3], introducing ideas ranging from the null hypothesis to discriminant analysis to ANOVA. Some of the work of Fisher followed from the pioneering work of Karl Pearson in the University College London who had founded the world's first statistics department previously and had first formulated ideas such as p values and correlation coefficients.

During the 1920s and 1930s, a number of important pioneers of multivariate statistics published their work, many strongly influenced or having worked with Fisher, including Harold Hotelling, credited by many as defining principal components analysis (PCA) [4], although Pearson had independently described this method some 30 years ago, but under a different guise. As so often ideas are reported several times over in science, it is the person that names it and popularises it that often gets the credit: in the early twentieth century, libraries were often localised and there were very few international journals (Hotelling working mainly in the US) and certainly no internet; therefore, parallel work was often reported.

The principles of statistical experimental design were also formulated at around this period. There had been early reports on what we regard as modern approaches to formal designs before that, for example James Lind's work on scurvy in the eighteenth century and Charles Pierce's discussion on randomised trials in the nineteenth century, but Fisher's classic work of the 1930s put all the concepts together in a rigorous statistical format [5].

Much non-Bayesian, applied statistical thinking has been based on principles established in the 1920s and 1930s, for nearly a century. Early applications include agriculture, psychology, finance and genetics. After the Second World War, the chemical industry took an interest. In the 1920s, an important need was to improve agricultural practice, but by the 1950s, a major need was to improve processes in manufacturing, especially chemical engineering; hence, many more statisticians were employed within the industry. O.L. Davies edited an important book on experimental design with contributions from colleagues in ICI [6]. Foremost was G.E.P. Box, son-in-law of Fisher, whose book with colleagues is one of the most important post-war classics in experimental design and multi-linear regression [7].

These statistical building blocks were already mature by the time people started calling themselves chemometricians and have changed only a little during the intervening period.

1.1.2 Statistics in Analytical and Physical Chemistry

Statistical methods, for example, to estimate accuracy and precision of measurements or to determine a best-fit linear relationship between two variables, have been available to analytical and physical chemists for over a century. Almost every general analytical textbook includes chapters on univariate statistics and has done for decades. Although theoretically we could view this as applied statistics, on the whole, the people who advanced statistics in analytical chemistry did not class themselves as applied statisticians and specialist terminology has developed over time.

Most quantitative analytical and physical chemistry until the 1970s was viewed as a univariate field; that is, only one independent variable was measured in an experiment. Usually, all other external factors were kept constant. This approach worked well in mechanics or fundamental physics, the so-called 'One Factor at a Time' (OFAT) approach. Hence, statistical methods were primarily used for univariate analysis of data. By the late 1940s, some analytical chemists were aware of ANOVA, F-tests and linear regression [8], although the term *chemometrics* had not been invented, but multivariate data came along much later.

There would have been very limited cross-fertilisation between applied statisticians, working in mathematics departments, and analytical chemists in chemistry departments, during these early days. Different departments often had different buildings, different libraries and different textbooks. A chemist, however numerate, would feel a stranger walking into a maths building and would probably cocoon him or herself in their own library. There was no such thing as the Internet or Web or Knowledge or electronic journals. Maths journals published papers for mathematicians and vice versa for chemistry journals. Although in areas such as agriculture and psychology there was a tradition of consulting statisticians, chemists were numerate and tended to talk to each other – an experimental chemist wanting to fit a straight line would talk to a physical chemist in the tea room if need be. Hence, ideas did not travel in academia. Industry was somewhat more pragmatic, but even there, the main statistical innovations were in chemical engineering and process chemistry and often classed as industrial chemistry. The top Universities often did not teach or research industrial chemistry, although they did teach Newtonian physics and relativity. In fact, the treatment of variables and errors by physicists trying, for example, to measure gravitational effects or the distance of a star is quite different to multivariate statistics: the former try to design experiments so that only one factor is studied and to make sure any errors are minimised and from one source, whereas a multivariate statistician might accept and expect data to be multifactorial.

Hence, statistics in analytical chemistry diverged from applied statistics for many decades. Caulcutt and Body's book first published in 1983 contains nothing on multivariate statistics [9] and in Miller and Miller's book of 1993 just one out of six main chapters is devoted to experimental design, optimisation and pattern recognition (including PCA) [10].

Even now, there are numerous useful books aimed at analytical and physical chemists that omit multivariate statistics. An elaborate vocabulary has developed for the needs of analytical chemists, with specialist concepts that are rarely encountered in other areas. Some analytical chemists in the 1960s to 1980s were aware that multivariate approaches existed and did venture into chemometrics, but good multivariate data were limited. Most are aware of ANOVA and experimental design. However, statistics for analytical chemistry tends to lead a separate existence from chemometrics, although multivariate methods derived from chemometrics do have a small foothold within most graduate-level courses and books in general analytical chemistry, and certainly quantitative analytical (and physical) chemistry was an important building block for modern chemometrics.

Over the last two decades, however, applications of chemometrics have moved far beyond traditional quantitative analytical chemistry, for example, into the areas of metabolomics, environment, cultural heritage or food, where the outcome is not necessarily to measure accurately the concentration of an analyte or how many compounds are in the spectra of a series of mixtures. This means that the aim of some chemometric analysis has changed. We often do not always have, for example, well-established reference samples and, in many cases, we cannot judge a method by how efficiently it predicts properties of these reference samples. We may not know whether the spectra of some extracts of urine samples can contain enough information to tell whether our donors are diseased or not: it may depend on how the disease has progressed, how good the diagnosis is, what the genetics of the donor and so on. Hence, we may never have a model that perfectly distinguishes two groups of samples. In classical physical or analytical chemistry, the answer is usually known to a greater accuracy than we can predict, in advance, so we can always tell which methods are best. This gradual change in culture distinguishes much of modern chemometrics from traditional statistics in analytical chemistry, although analytical chemistry is definitely one of the ancestors of chemometrics, and the two are symbiotic.

1.1.3 Scientific Computing

Another revolution happened as from the 1960s, the use of computers in scientific research.

Many of the original statistical computations required complex matrix operations that may have taken days or even weeks to solve using manual calculations even with calculators or slide rules. This limited the applicability of many statistical methods. Many early statistical papers were intensely theoretical and some methods were applied only to important and economically significant problems: an agricultural experiment that took several years deserved a couple of weeks manually computing the trends in the data. However, they were not widespread, especially in scientific laboratories.

With the 1960s, scientists in the best resourced laboratories gained access to mainframe computers. Usually, they had to be programmed in languages such as FORTRAN and used punch cards, paper tape and line printers. However, they allowed a rapid adoption of computers by applied scientists, which became the third revolution that led to chemometrics.

Resolution and rank analysis of spectroscopy of mixtures had its vintage in the 1960s, with a small number of pioneering papers [11, 12] taking advantage of newly available computer power: in earlier papers, such methods were reported but were applied to very small problems, for example, of four mixtures and four wavelengths due to the difficulty of manual calculation. Multivariate spectroscopic resolution developed quite separately to multivariate statistics, primarily via physical chemistry. The original terminology differed quite considerably from statistics and was primarily that of physics.

Over the 1960s and 1970s, there were many papers about spectroscopic resolution in both the physical chemistry and the analytical chemistry literature, but Ed Malinowski, whose remarkable publication career stretches from 1955 to 2011, is best recognised to having put these concepts together with multivariate statistics. He published what many regard as the first book that covered one important area of chemometrics [13], which he called *Factor Analysis*, involving determining the number of components in spectroscopic mixtures together with their characteristics.

Meanwhile, a separate development in scientific computing emerged in the 1960s – partly catalysed by NASA's trip to the moon – to use AI to identify compounds spectroscopically [14], a project that involved Nobel Prize winners and spawned the whole new area of expert systems. This in turn led to the field of pattern recognition and the award of several competitive grants in scientific computing, particularly in the USA.

Isenhour, Jurs and Kowalski were early pioneers of computerised learning in chemistry, primarily using pattern recognition [15], the early group founded by Isenhour, who left in 1969. Kowalski took over the reins in 1974, initially with an interest in chemical pattern recognition.

Hence, computational chemistry arrived via both physical chemistry of spectroscopic mixtures and organic chemistry for pattern recognition and had important elements in the formative mix in the 1960s and 1970s. This allowed the application to comparatively large problems and wider access to algorithms that had previously been rather theoretical.

1.2 Developments since the 1970s

Chemometrics slowly gained an identity from the mid-1970s, after Wold first named it. However, some of the recognised pioneers were slow to identify with it. For example, both Wold and Kowalski published far more papers using 'chemical pattern recognition' than 'chemometrics' as a keyword in the 1970s.

The first symposia with chemometrics in the name, in the USA, took place in the late 1970s. The first analytical chemistry review entitled 'Chemometrics' was published in 1980 [16]. The International Chemometrics Society was founded by Wold and Kowalski in the 1970s.

By this stage, although still relatively few workers identified themselves with chemometrics, small groups of enthusiasts were promoting the name and idea. In those days, most of those that identified themselves as chemometricians were quite expert programmers who cut their teeth on a mainframe or, latterly, primitive micros. Some even started their scientific careers before scientists had ready access to computers and may have had to learn programming via Assembly language so were in practice extremely good programmers. If a method was reported in a paper, the authors would typically program it in itself rather than using a package.

A NATO sponsored workshop in Cosenza, Italy, in 1983, brought together many of the early experts of the time [17] and events moved fast after that. The first journals dedicated to chemometrics, *Chemometrics and Intelligent Laboratory Systems* (Elsevier) and *Journal of Chemometrics* (Wiley), were founded in 1986 and 1987. Kowalski and co-workers produced the first comprehensive book in 1986 [18] followed by Massart and co-workers in 1988 [19]. Software packages such as Arthur, Unscrambler and Simca emerged during this period.

By the 1990s, well-established books, journals, courses and software were available, although still only quite a small number of dedicated groups worldwide. However, this changed when laboratory-based data started to become more readily available – the size of data sets and complexity of problems increased massively. In the 1980s, the emphasis was primarily on small problems such as the resolution of a cluster of HPLC peaks or deconvolution of uv/vis spectra. Economically important problems in process control and NIR spectroscopy posed new challenges to chemometricians and gradually moved the subject from a rather theoretical application of quantitative analytical chemistry to a more applied subject. There was a special interest in the interface between chemical engineering and chemometrics.

A further revolution has happened in the last 15 years when complex real-world data have become available. This has allowed looking at applications ranging from metabolomics to heritage studies to forensics and so on where large data sets are available. It has resulted in chemometrics tools becoming very widely used, although the core community of experts is probably no bigger than a few decades ago. The widespread applicability of common chemometric methods such as PCA, classification, calibration, and so on, leads to an urgent need to understand these methods. This book is primarily aimed at potential users who want to understand the underlying mathematical approaches, rather than just use packages.

1.3 Software and Calculations

The key to chemometrics is to understand how to perform meaningful calculations on data. In most cases, these calculations are too complex to do by hand or using a calculator; hence, it is necessary to use some software.

The approach taken in this book, which differs from many books on chemometrics, is to understand the methods using numeric examples. Some excellent books and reviews are more descriptive, listing the methods available together with the literature references and possibly some examples. Others have a big emphasis on equations and output from packages. This book, however, is primarily based on how I personally learn and understand new methods, and how I have found it most effective to help students working with me. Data analysis is not really a knowledge-based subject but is more a skill-based subject. A good organic chemist may have encyclopaedic knowledge of reactions in their own area. The best supervisor will be able to list to his or her students thousands of reactions, or papers, or conditions that will aid their students, and with experience this knowledge base grows. In chemometrics, although there are quite a number of named methods, the key is not to learn hundreds of equations but to understand a few basic principles. These ideas, such as multiple linear regression (MLR), occur again and again but in different contexts. To become skilled in chemometric data analysis, practice to manipulating numbers is required, not an enormous knowledge base. Although equations are necessary for the formal description of methods, and cannot easily be avoided, it is easiest to understand the methods in this book by looking at numbers. Hence, the methods described in this book are illustrated using numerical examples, which are available for the reader to reproduce. The data sets employed in this book are available on the publisher's website. In addition to the main book, there are extensive problems at the end of each main chapter. All numerical examples are quite small and are designed in such a manner that you can check all the numbers yourselves. Some are reduced versions of larger data sets, such as spectra recorded at 5 nm rather than 1 nm intervals. Many real examples, especially in chromatography and spectroscopy, simply differ in size to those in this book. In addition, the examples are chosen so that they are feasible to analyse fairly simply.

One of the difficulties is to decide the software to be employed in order to analyse the data. This book is not restrictive and you can use any approach you like. Some readers may like to program their own methods, for example, in C or Visual Basic. Others may like to use statistical packages such as SAS or SPSS. There is a significant statistical community that uses R. Some groups use ready packaged chemometrics software such as Pirouette, Simca, Unscrambler, PLS Toolbox and several others in the market. One problem with using packages is that they are often very focussed in their facilities. What they do, they do excellently, but if they cannot do what you want, you may be stuck, even for relatively simple calculations. If you have an excellent multivariate package but want to use a Kalman filter, where do you turn? Perhaps you have the budget to buy another package, but if you just want to explore the method, the simplest implementation takes only an hour or less for an experienced Matlab programmer to implement. In addition, there are no universally agreed definitions, so a 'factor' or 'eigenvector' might denote something quite different according to the software used. Some software has limitations making it unsuitable for many applications of chemometrics, a very simple example being the automatic use of column centring in PCA in many general statistical packages, whereas some chemometric methods involve using uncentred PCA.

Nevertheless, many of the results from the examples in this book can quite successfully be obtained using commercial packages, but be aware of the limitations and also understand the output of any software you use. It is important to recognise that the definitions used in this book may differ from that employed by any specific package. As a huge number of often incompatible definitions are available, even for fairly common parameters, in order not to confuse the reader, we have had to adopt one single definition for each parameter; thus, it is important to carefully check with your favourite package or book or paper whether the results appear to differ from those presented in this book. It is not the aim of this book to replace an international committee that defines chemometric terms. Indeed, it is quite unlikely that such a committee would be formed because of the very diverse backgrounds of those interested in chemical data analysis.

However, in this book, we recommend that the readers use one of the two environments.

The first is Excel. Almost everyone has some familiarity with Excel, and in Appendix A.4, specific features that might be useful for chemometrics are described. Most calculations can be performed quite simply using normal spreadsheet functions. The exception is PCA for which a small program must be written. For instructors and users of VBA (a programming language associated with Excel), a small macro that can be edited is available, downloadable from the publisher's website. However, some calculations such as cross-validation and partial least squares (PLS), while possible to program using Excel, can be quite tedious. It is strongly recommended that readers do reproduce these methods step by step when first encountered, but after a few times, one does not learn much from setting up the spreadsheet each time. Hence, we also provide an Excel add-in to perform PCA, PLS, MLR and PCR (principal components regression). The software also contains facilities for validation. Readers of this book should choose what approach they wish to take.

A second environment, which many chemical engineers and statisticians enjoy, is Matlab described in Appendix A.5. Historically, the first significant libraries of programs in chemometrics became first available in the late 1980s. Quantum chemistry, originating in the 1960s, is still very much based on Fortran because this was the major scientific programming environment of the time, and over the years, large libraries have been developed and maintained; hence, a modern quantum chemist will probably learn to use Fortran. The vintage of chemometrics is such that a more recent environment to scientific programming has been adopted by the majority, and many chemometricians swap software using Matlab. The advantage is that Matlab is very matrix oriented and it is most convenient to think in terms of matrices, especially as most data are multivariate. In addition, there are special facilities for performing singular value decomposition (or PCA) and the pseudo-inverse used in regression, which means it is not necessary to program these basic functions. There have been a number of recent enhancements, including links to Excel that allow easy interchange of data which enables simple programs to be written that transfer data to and from Excel. There is no doubt at all that matrix manipulation, especially for complex algorithms, is tedious in VBA and Excel. Matlab is an excellent environment for learning the nuts and bolts of chemometrics. A slight problem with Matlab is that it is possible to avoid looking at the raw numbers, whereas most users of Excel will be forced to look at the raw numeric data in detail, and I have come across experienced Matlab users that are otherwise very good at chemometrics but who sometimes miss quite basic information because they are not constantly examining the numbers; hence, if you are a dedicated Matlab programmer, look at the numerical information from time to time!

An ideal situation would probably involve using both Excel and Matlab simultaneously. Excel provides a good interface and allows flexible examination of the data, whereas Matlab is best for developing matrix-based algorithms. The problems in this book have been tested both in Matlab and in Excel and identical answers were obtained. Where there are quirks of either package, the reader is guided.

Two final words of caution are needed. The first is that some answers in this book have been rounded to a few significant figures. Where intermediate results of a calculation have been presented, putting these intermediate results back may not necessarily result in exactly the same numerical results as retaining them to higher accuracy and continuing the calculations. A second issue that often perplexes new users of multivariate methods is that it is impossible to control the sign of a principal component (see Chapter 4 for a description of PCA). This is because PCs involve calculating square roots that may give negative as well as positive answers. Therefore, using different packages, or even the same package but with different starting points, can result in reflected graphs, with scores and loadings that are opposite in sign. It is therefore unlikely to be a mistake if you obtain PCs that are opposite in sign to those in this book.

1.4 Further Reading

A large number of books and review articles have been written, covering differing aspects of chemometrics, often aimed at a variety of audiences. In Sections 1.1 and 1.2, we list some of the more historic books and papers. This chapter summarises some of the most widespread and recent works. In most cases, these works will allow the reader to delve further into the methods introduced within this book. In each category, only a few main books will be mentioned, but most have extensive bibliographies, allowing the reader to access information especially from the primary literature. Although there are also internet resources and numerous tutorial and review papers, in order to restrict the bibliography, we only list books.

1.4.1 General

Largest authored book in chemometrics is published by Massart and co-workers, part of two volumes [20, 21]. These volumes provide an in-depth summary of many modern chemometric methods, involving a wide range of techniques, and many references to the literature. The first volume, although, is quite strongly oriented towards analytical chemists but contains an excellent grounding in basic statistics for measurement science. The books are especially useful as springboards for the primary literature. This is a complete rewrite of the original book published in 1988 [19], which is still cited as a classic in the analytical chemistry literature. *Comprehensive Chemometrics* [22] is a follow-on from the same publisher, an encyclopaedic collection of edited articles in four volumes covering much of the knowledge base of chemometrics in 2009 and is probably the most comprehensive detailed summary of the subject.

Otto's book on chemometrics [23] is a well-regarded book now in its third edition covering quite a range of topics but at a fairly introductory level. The book looks at computing, in general, in analytical chemistry including databases and instrumental data acquisition. It is a very clearly written introduction for the analytical chemist, by an outstanding educator.

Beebe and co-workers at Dow Chemicals have produced a book [24] that is useful for many practitioners and contains very clear descriptions especially of multivariate calibration in spectroscopy and although some years old is still recommended for those working in this area. However, there is a strong 'American School' originating in part from the pioneering work of Kowalski in NIR spectroscopy and process control, and while covering the techniques required in this area in an outstanding way, and is well recommended as a next step for readers of this book working in this application area, it lacks a little in generality, probably because of the very close association between NIR and chemometrics in the minds of some.

Kramer has produced a somewhat more introductory book [25]. He is well known for his consultancy company and highly regarded courses, and his approach is less mathematical. This will suit some people very well, but may not be presented in a way that suits statisticians and chemical engineers.

This current author published a book on chemometrics at an early stage of the development of the subject [26], which has an emphasis on signal resolution and minimises matrix algebra, and is an introductory tutorial book especially for the laboratory-based chemist. This author also published a later book based on web articles that covers a range of applications as well as simple descriptions of methods [27]. This author has a series of ongoing short tutorial articles covering aspects of chemometrics as a column in *Journal of Chemometrics*, starting in 2014: these look more into the statistical principles of the subject.

The journal *Chemometrics and Intelligent Laboratory Systems* published regular tutorial review articles over its first decade or more of existence. Some of the earlier articles are good introductions to general subjects such as PCA, Fourier transforms and Matlab. They are collected together as two volumes [28, 29]. They also contain some valuable articles on expert systems.

Varmuza and Filtzmoser have published a very well-regarded book involving using R with clear descriptions within a statistical context [30]. Gemperline edited a multi-author book, which is currently in its second edition [31]. Pomerantsev has published a book oriented towards users of Excel [32]. Mark and Workman have written a comprehensive book aimed at spectroscopists [33]: it is very good at analytical instrumental chemistry and the authors are well regarded.

Meloun and Militky published a large book based on extensive course work [34]. This covers many topics in chemometrics and has a special feature of 1250 numerical problems and data sets.

Martens and Martens have produced a book that is quite a detailed discussion of how multivariate methods can be used in quality control [35] and covers several aspects of modern chemometrics, and so could be classed as a general book on chemometrics.

Although this list is not comprehensive, it lists most general books on chemometrics. There are also several books in different application areas such as food, environment, various types of spectroscopy and so on.

1.4.2 Specific Areas

There are a large number of books and review articles dealing with specific aspects of chemometrics, interesting as a next step after this book, and for a comprehensive chemometrics library. We will list just a few.

1.4.2.1 Experimental Design

In the area of experimental design, there are innumerable books, many written by statisticians. Specifically aimed at chemists, Deming and Morgan have produced a highly regarded book [36], which is well recommended as a next step after this book. Bayne and Rubin have written a clear and thorough book [37]. An introductory book mainly discussing factorial designs was written by Morgan as part of the Analytical Chemistry by Open Learning Series [38]. For mixture designs, involving compositional data, the classic statistical book by Cornell is much cited and recommended [39] but is quite mathematical. More historical books such as those by Fisher [5] and by Box and co-workers [7] have already been described above but are still relevant today.

1.4.2.2 Pattern Recognition and Principal Component Analysis

There are several books on pattern recognition and PCA. An introduction to several of the main techniques is provided in an edited book [40]. For more statistical in-depth descriptions of *Principal Components Analysis*, read books by Joliffe [41] and Mardia and co-authors [42]. An early but still valuable book by Massart and Kaufmann covers more than just its title 'cluster analysis' [43] and provides clear introductory material. Varmuza [44] and Strouf [45] wrote early books in the area when much of the rest of chemometrics was focussed on calibration and signal resolution.

A more up-to-date book focussed on pattern recognition was recently published by this author [46] that illustrated using several case studies. Over the past decade, there has been a much more interest in pattern recognition compared with a few decades ago, with increased application to areas such as metabolomics.

1.4.2.3 Multivariate Signal Analysis

Multivariate curve resolution (MCR) is the main topic of Malinowski's book [47], which is the third edition of his original book [13]. The author is a physical chemist and so the book is oriented towards that particular audience and especially relates to the spectroscopy of mixtures. Although there have been notable advances in the area, especially in alternating least squares (ALS), these are primarily published in the form of papers. Malinowski's book is still the classic book in the area. For more up-to-date reading, search for papers on MCR and ALS. However, the third edition of this book covers ALS well, and most of the pioneering papers were published some 15–20 years ago.

1.4.2.4 Multivariate Calibration

Multivariate calibration is a very popular area, and the much reprinted classic by Martens and Næs [48] is one of the most cited books in chemometrics. Much of the book is based around NIR spectroscopy, which was one of the major success stories in applied chemometrics in the 1980s and 1990s, but the clear mathematical descriptions of algorithms are particularly useful for a wider audience. The book by Beebe and co-workers [24] also has good in-depth discussion about calibration. A more recent book by Naes *et al.* is somewhat less theoretical and is mainly about multivariate calibration [49].

1.4.2.5 Statistical Methods

There are a number of books on general statistical methods in chemistry, mainly oriented towards analytical and physical chemists. Miller and Miller's book [10] has gone through several editions and takes the reader through many of the basic significance tests, distributions and so on. There is a small amount on chemometrics in the final chapter. The Royal Society of Chemistry published quite a nice introductory tutorial book by Gardiner [50]. Caulcutt and Boddy's book [9] is also a much reprinted and useful reference. There are several other competing books, most of which are very thorough, for example, in describing applications of the *t*-test, *F*-test and ANOVA but which do not progress much into modern chemometrics. If you are a physical chemist, Gans' viewpoint on deconvolution and curve fitting may suit you more [51], covering many regression methods. Meier and Zund published a book in 2000 [52] with a very thorough discussion of univariate methods especially in industrial practice and a little introduction to multivariate methods. Ellison and co-workers published a book based on the UK Valid Analytical Measurement initiative [53].

Several other books about statistical approaches (mainly univariate) in analytical chemistry and a number of international initiatives that regularly produce reports are regularly being developed.

1.4.2.6 Digital Signal Processing and Time Series

There are numerous books on digital signal processing (DSP) and Fourier transforms (FTs). Unfortunately, many of the chemically based books are fairly technical in nature and oriented towards specific techniques such as NMR; however, books written primarily by and for engineers and statisticians are often quite understandable. A recommended reference to DSP contains many of the main principles [54], but several similar books are available. A couple of recent general books on FTs are recommended [55, 56]. For non-linear deconvolution, Jansson's book is well known [57]. Methods for time series analysis are described in more depth in an outstanding and much reprinted book written by Chatfield [58].

1.4.2.7 Multi-way Methods

For chemometricians, the best book available is by Smilde *et al.* [59], which is a thorough description and illustration of the algorithms. There was much development in this area in the 1990s, which was a very exciting era for new algorithms, and the three authors were pioneers of some of the original papers in the chemometrics literature. This book is the best comprehensive summary of the application of such methods in chemistry.

References

1 Wold, S. (1972) Spline functions, a new tool in data-analysis. *Kemisk Tidskrift*, **3**, 34–37.
2 Fisher, R.A. (1925) *Statistical Methods for Research Workers*, Oliver and Boyd, Edinburgh.
3 Fisher, R.A. (1936) The use of multiple measurements in taxonomic problems. *Ann. Eugen.*, 7, 179–188.
4 Hotelling, H. (1933) Analysis of a complex of statistical variables into principal components. *J. Educ. Psychol.*, **24**, 417–441.
5 Fisher, R.A. (1935) *The Design of Experiments*, Hafner, New York.
6 Davies, O.L. (ed.) (1956) *Statistical Methods in Research and Production*, Longman, London.
7 Box, G.E.P., Hunter, W.G. and Hunter, J.S. (1978) *Statistics for Experimenters*, John Wiley & Sons, Inc., New York.
8 Mandel, J. (1949) Statistical Methods in Analytical Chemistry. *J. Chem. Educ.*, **26**, 534–539.
9 Caulcutt, R. and Boddy, R. (1983) *Statistics for Analytical Chemists*, Chapman and Hall, London.
10 Miller, J.C. and Miller, J.N. (1993) *Statistics for Analytical Chemistry*, 2nd edn, Prentice-Hall, Hemel Hempstead.
11 Wallace, R.M. and Katz, S.M. (1964) A method for determination of rank in analysis of absorption spectra of multi-component systems. *J. Phys. Chem.*, **68**, 3890–3892.
12 Katakis, D. (1965) Matrix rank analysis of spectral data. *Anal. Chem.*, **37**, 876–878.
13 Malinowski, E.R. and Howery, D.G. (1980) *Factor Analysis in Chemistry*, John Wiley & Sons, Inc., New York.
14 Lindsay, R.K., Buchanan, B.G., Feigenbaum, E.A. and Lederberg, J. (1980) *Applications of Artificial Intelligence for Organic Chemistry: The DENDRAL Project*, McGraw-Hill, New York.
15 Kowalski, B.R., Jurs, P.C., Isenhour, T.L. and Reilly, C.N. (1969) Computerized learning machines applied to chemical problems: interpretation of infrared spectrometry data. *Anal. Chem.*, **41**, 1945–1949.
16 Kowalski, B.R. (1980) Chemometrics. *Anal. Chem.*, **52**, R112–R122.
17 Kowalski, B.R. (ed.) (1984) *Chemometrics: Mathematics and Statistics in Chemistry*, Reidel, Dordrecht.
18 Sharaf, M.A., Illman, D.L. and Kowalski, B.R. (1986) *Chemometrics*, John Wiley & Sons, Inc., New York.
19 Massart, D.L., Vandeginste, B.G.M., Deming, S.N. *et al.* (1988) *Chemometrics: A Textbook*, Elsevier, Amsterdam.
20 Massart, D.L., Vandeginste, B.G.M., Buydens, L.M.C. *et al.* (1997) *Handbook of Chemometrics and Qualimetrics Part A*, Elsevier, Amsterdam.
21 Vandeginste, B.G.M., Massart, D.L., Buydens, L.M.C. *et al.* (1997) *Handbook of Chemometrics and Qualimetrics Part B*, Elsevier, Amsterdam.
22 Tauler, R., Walczak, B. and Brown, S.D. (eds) (2009) *Comprehensive Chemometrics*, Elsevier, Amsterdam.
23 Otto, M. (2016) *Chemometrics: Statistics and Computer Applications in Analytical Chemistry*, 3rd edn, Wiley-VCH Verlag GmbH, Weinheim.
24 Beebe, K.R., Pell, R.J. and Seasholtz, M.B. (1998) *Chemometrics: A Practical Guide*, John Wiley & Sons, Inc., New York.
25 Kramer, R. (1998) *Chemometrics Techniques for Quantitative Analysis*, Marcel Dekker, New York.

26 Brereton, R.G. (1990) *Chemometrics: Applications of Mathematics and Statistics to Laboratory Systems*, Ellis Horwood, Chichester.

27 Brereton, R.G. (2007) *Applied Chemometrics for Scientists*, John Wiley & Sons, Ltd, Chichester.

28 Massart, D.L., Brereton, R.G., Dessy, R.E. *et al.* (eds) (1990) *Chemometrics Tutorials*, Elsevier, Amsterdam.

29 Brereton, R.G., Scott, D.R., Massart, D.L. *et al.* (eds) (1992) *Chemometrics Tutorials II*, Elsevier, Amsterdam.

30 Varmuza, K. and Filzmoser, P. (2009) *Introduction to Multivariate Statistical Analysis in Chemometrics*, CRC Press, Boca Raton.

31 Gemperline, P.J. (ed.) (2006) *Chemometrics: A Practical Guide*, CRC Press, Boca Raton.

32 Pomerantsev, A.L. (2014) *Chemometrics in Excel*, John Wiley & Sons, Ltd, Chichester.

33 Mark, H. and Workman, J. (2007) *Chemometrics in Spectroscopy*, Academic Press, London.

34 Meloun, M. and Militky, J. (2011) *Statistical Data Analysis: A Practical Guide*, Woodhead, New Delhi.

35 Martens, H. and Martens, M. (2000) *Multivariate Analysis of Quality*, John Wiley & Sons, Ltd, Chichester.

36 Deming, S.N. and Morgan, S.L. (1994) *Experimental Design: A Chemometric Approach*, Elsevier, Amsterdam.

37 Bayne, C.K. and Rubin, I.B. (1986) *Practical Experimental Designs and Optimisation Methods for Chemists*, Wiley-VCH Verlag GmbH, Deerfield Beach.

38 Morgan, E. (1995) *Chemometrics: Experimental Design*, John Wiley & Sons, Ltd, Chichester.

39 Cornell, J.A. (1990) *Experiments with Mixtures: Design, Models, and the Analysis of Mixture Data*, 2nd edn, John Wiley & Sons, Inc., New York.

40 Brereton, R.G. (ed.) (1992) *Multivariate Pattern Recognition in Chemometrics, Illustrated by Case Studies*, Elsevier, Amsterdam.

41 Joliffe, I.T. (1987) *Principal Components Analysis*, Springer-Verlag, New York.

42 Mardia, K.V., Kent, J.T. and Bibby, J.M. (1979) *Multivariate Analysis*, Academic Press, London.

43 Massart, D.L. and Kaufmann, L. (1983) *The Interpretation of Analytical Chemical Data by the Use of Cluster Analysis*, John Wiley & Sons, Inc., New York.

44 Varmuza, K. (1980) *Pattern Recognition in Chemistry*, Springer, Berlin.

45 Strouf, O. (1986) *Chemical Pattern Recognition*, Research Studies Press, Letchworth.

46 Brereton, R.G. (2009) *Pattern Recognition for Chemometrics*, John Wiley & Sons, Ltd, Chichester.

47 Malinowski, E.R. (2002) *Factor Analysis in Chemistry*, 3rd edn, John Wiley & Sons, Inc., New York.

48 Martens, H. and Næs, T. (1989) *Multivariate Calibration*, John Wiley & Sons, Ltd, Chichester.

49 Naes, T., Isaksson, T., Fearn, T. and Davies, T. (2002) *A User Friendly guide to Multivariate Calibration and Classification*, NIR Publications, Chichester.

50 Gardiner, W.P. (1997) *Statistical Analysis Methods for Chemists: A Software-Based Approach*, Royal Society of Chemistry, Cambridge.

51 Gans, P. (1992) *Data Fitting in the Chemical Sciences: By the Method of Least Squares*, John Wiley & Sons, Ltd, Chichester.

52 Meier, P.C. and Zund, R.E. (2000) *Statistical Methods in Analytical Chemistry*, 2nd edn, John Wiley & Sons, Inc., New York.

53 Ellison, S.L.R., Barwick, V.J. and Duguid Farrant, T.J. (2009) *Practical Statistics for the Analytical Scientist: A Bench Guide*, 2nd edn, Royal Society of Chemistry, Cambridge.

54 Lynn, P.A. and Fuerst, W. (1998) *Introductory Digital Signal Processing with Computer Applications*, 2nd edn, John Wiley & Sons, Ltd, Chichester.

55 James, J.F. (2011) *A Student's Guide to Fourier Transforms*, 3rd edn, Cambridge University Press, Cambridge.

56 Bracewell, R.N. (2000) *Fourier Transform and Its Applications*, McGraw-Hill, Boston.

57 Jansson, P.A. (ed.) (1984) *Deconvolution: with Applications in Spectroscopy*, Academic Press, New York.

58 Chatfield, C. (2003) *Analysis of Time Series: An Introduction*, 6th edn, Chapman and Hall/CRC, Boca Raton.

59 Smilde, A., Bro, R. and Geladi, P. (2004) *Multi-way Analysis*, John Wiley & Sons, Ltd, Chichester.

2

Experimental Design

2.1 Introduction

Although all chemists acknowledge the need to be able to design laboratory-based experiments, formal statistical (or chemometric) rules are rarely developed as part of mainstream chemistry. In contrast, a biologist or a psychologist will often spend weeks in carefully constructing a formal statistical design before investing months or years in time-consuming and often unrepeatable experiments and surveys. The simplest of experiments in chemistry are relatively quick and can be repeated, if necessary, under slightly different conditions; hence, not all chemists observe the need for formalised experimental design early in their career. For example, there is little point in spending a week for constructing a set of experiments that take a few hours to perform. This lack of expertise in formal design permeates all levels from management to professors and students. However, in contrast, some real-world experiments are expensive; for example, optimising conditions for a synthesis, testing compounds in a quantitative structure–activity relationships (QSAR) study or improving the chromatographic separation of isomers, and can take days or months of people's time, and it is essential to, under such circumstances, to have a good appreciation of the fundamentals of design.

There are several key reasons why the chemist can be more productive if he or she understands the basis of design, including the following four main areas.

- *Screening*. These types of experiments involve considering factors that are important for the success of a process. An example may be the study of a chemical reaction, dependent on the proportion of the solvent, catalyst concentration, temperature, pH, stirring rate and so on. Typically, 10 or more factors might be relevant. Which can be eliminated, and which should be studied in detail? Approaches such as factorial and Plackett–Burman designs (Sections 2.3.1–2.3.3) are useful in this context.
- *Optimisation*. This is one of the commonest applications in chemistry. How to improve a synthetic yield or a chromatographic separation? Systematic methods can result in a better optimum, found more rapidly. Simplex is a classical method for optimisation (Section 2.6), although several designs such as mixture designs (Section 2.5) and central composite designs (Section 2.4) can also be employed to find optima.
- *Saving time*. In industry, this is possibly the major motivation for experimental design. There are obvious examples in optimisation and screening, but even more radical cases, as in the area of quantitative structure–property relationships. From structural data of existing molecules, it is possible to predict a small number of compounds for further testing, representative of a larger set of molecules. This allows saving of enormous time. Fractional factorial, Taguchi and Plackett–Burman designs (Sections 2.3.2 and 2.3.3) are good examples, although almost all experimental designs have this aspect in mind.
- *Quantitative modelling*. Almost all experiments, ranging from simple linear calibration in analytical chemistry to complex physical processes, where a series of observations are required to obtain a mathematical model of the system, benefit from good experimental design. Many such designs are based around the central composite design (Section 2.4), although calibration designs (Section 2.3.4) are also useful.

An example of where systematic experimental design is valuable is the optimisation of the yield of a reaction as a function of reagent concentration and pH. A representation is given in Figure 2.1. In reality, this relationship is unknown in advance, but the experimenter wishes to determine that the pH and concentration (in mM) provide the best reaction conditions. Within 0.2 of a pH and concentration unit, this optimum happens to be pH 4.4 and 1.0 mM. Many experimentalists will start by guessing one of the factors, say concentration, and then finding the best pH at that concentration.

Chemometrics: Data Driven Extraction for Science, Second Edition. Richard G. Brereton.
© 2018 John Wiley & Sons Ltd. Published 2018 by John Wiley & Sons Ltd.
Companion website: http://booksupport.wiley.com

Figure 2.1 Yield of a reaction as a function of pH and catalyst concentration.

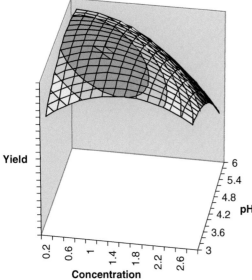

Figure 2.1 Yield of a reaction as a function of pH and catalyst concentration.

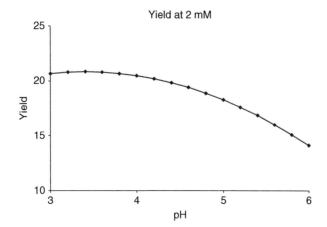

Figure 2.2 Cross-section through surface in Figure 2.1 at 2 mM catalyst concentration.

Consider an experimenter who chooses to start the experiment at 2 mM and wants to find the best pH. Figure 2.2 shows the yield at 2.0 mM. The best pH is undoubtedly a low one, in fact pH 3.4. Hence, the next stage is to perform the experiments at pH 3.4 and improve the concentration, as shown in Figure 2.3. The best concentration is 1.4 mM. These answers, pH 3.4 and 1.4 mM, are quite far from the true ones.

The reason for this problem is that the influence of pH and temperature is not independent. In chemometric terms, they 'interact'. In many cases, interactions are common sense. For example, the optimum pH in one solvent may be different to that in another solvent. Chemistry is complex, but how to find the true optimum, by a quick and efficient manner, and be confident in the result? Experimental design provides the chemists with a series of rules to guide the optimisation process, which will be explored later.

A rather different example relates to choosing compounds for biological tests. Consider the case where it is important to determine what type of compounds in a group are harmful, often involving biological experiments. Say there are 50 potential compounds in the group. Running comprehensive and expensive tests on each compound is prohibitive. However, it is likely that certain structural features will relate to toxicity. The trick of experimental design is to choose a selection of the compounds and then decide to perform tests only on this subset.

Chemometrics can be employed to develop a mathematical relationship between chemical property descriptors (e.g. bond lengths, polarity, steric properties, reactivities and functionalities) and biological functions, via a computational model such as principal components analysis. The question asked is whether it is really necessary to test all the 50 compounds for this model? The answer is no. Choosing a set of 8 or 16 compounds may provide adequate information

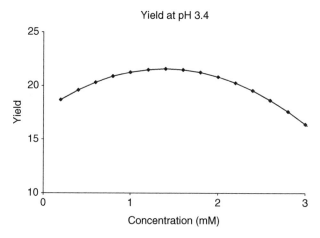

Figure 2.3 Cross-section through surface in Figure 2.1 at pH 3.4.

Figure 2.4 Choice of nine molecules based on two properties.

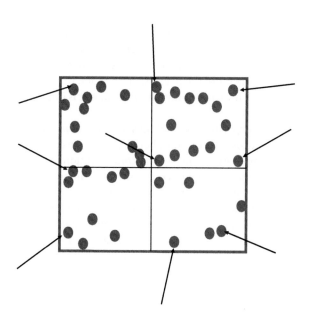

to predict the influence of not only the remaining compounds (and this can be tested) but also any unknown compound in the group.

Figure 2.4 illustrates a simple example. An experimenter is interested in studying the influence of hydrophobicity and dipoles on a set of candidate compounds, for example, in chromatography. He or she finds out these values simply by reading the literature and plotting them in a simple graph. Each red circle in the figure represents a compound. How to narrow down the test compounds? One simple design involves selecting nine candidates, those at the edges, corners and centre of the square, indicated by arrows in the diagram. These candidates are then tested experimentally and represent a typical range of compounds. In reality, there are vastly more chemical descriptors, but similar approaches can be employed using, instead of raw properties, statistical functions of these to reduce the number of axes, typically to about 3, and then choose a good and manageable selection of compounds.

The potential uses of rational experimental design throughout chemistry are large, and some of the most popular designs will be described below. Only certain selective, and generic, classes of design are discussed in this chapter, but it is important to recognise that the large number of methods reported in the literature is based on a small number of fundamental principles. Most important is to appreciate the motivations behind using statistical experimental design rather than any particular named method. The material in this chapter should permit the generation of a variety of common designs. If very specialist designs are employed, there must be correspondingly specialist reasons for such choice; hence, the techniques described in this chapter should be applicable to most common situations. Applying a design without appreciating the underlying motivation is dangerous.

For introductory purposes, multiple linear regression (MLR) analysis is used to relate the experimental response to the values of the factors, as is common to most texts in this area, but it is important to realise that other regression methods such as partial least squares (PLS) are applicable in many cases, as discussed in Chapter 6. Certain designs such as those discussed in Section 2.3.4 have direct relevance to multivariate calibration. In some cases, multivariate methods such as PLS can be modified by inclusion of squared and interaction terms as described below for MLR. It is important to remember, however, that in many areas of chemistry, quite a lot of information is available about a data set, and conceptually simple approaches based on MLR are often adequate.

2.2 Basic Principles

2.2.1 Degrees of Freedom

Fundamental to the understanding of experimental designs is the idea of degrees of freedom. An important outcome of many experiments is the measurement of errors. This can tell us how confidently a phenomenon can be predicted; for example, are we really sure that we can estimate the activity of an unknown compound from its molecular descriptors, or are we happy with the accuracy with which a concentration can be determined using spectroscopy? In addition, what is the weak link in a series of experiments? Is it the performance of a spectrometer or the quality of the volumetric flasks? Each experiment involves making a series of observations, which allow us to try to answer some of these questions, the number of degrees of freedom relating to the amount of information available for each answer. Of course, the more the degrees of freedom, the more certain we can be of our answers, but the more the effort and work required. If we have only a limited time, it is important to provide some information to allow us to answer all the desired questions.

Most experiments result in some sort of *model*, which is a mathematical way of relating an experimental *response* to the value or state of a number of *factors*. An example of a response is the yield of a synthetic reaction; the factors may be pH, temperature and catalyst concentration. An experimenter wishes to run a reaction under a given set of conditions and predict the yield. How many experiments should be performed in order to provide confident predictions of the yield at any combination of the three factors? 5, 10 or 20? Usually, the more the experiments, the more certain the predictions, but the greater the time, effort and expense. Hence, there is a balance, and experimental design helps to guide the chemist as to how many and what type of experiments should be performed.

Consider a linear calibration experiment, for example, measuring the peak height using electronic absorption spectroscopy as a function of concentration, at five different concentrations, illustrated in Figure 2.5. A chemist may wish to fit a straight-line model to the experiment of the form

$$y = b_0 + b_1 x$$

where y is the response (in this case the peak height), x is the value of the factor (in this case concentration) and b_0 and b_1 are the *coefficients* of the model. There are two coefficients in this equation, but five experiments have been

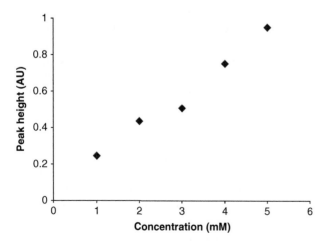

Figure 2.5 Graph of spectroscopic peak height against concentration at five concentrations.

performed. More than enough experiments have been performed to give an equation for a straight line, the remaining experiments help answer the question 'how well is the linear relationship obeyed?' This could be quite important to the experimenter. For example, there may be unknown interferences, or the instrument might be very irreproducible, or there may be non-linearities at high concentrations. Hence, the experiments must be used not only to determine the equation relating peak height to concentration but also to answer whether the relationship is truly linear and reproducible.

The ability to determine how well the data fits a linear model depends on the number of degrees of freedom, which are given, in this case, by

$$D = N - P$$

where N is the number of experiments and P the number of coefficients in the model. In this example,

- $N = 5$
- $P = 2$ (the number of coefficients in the model $y = b_0 + b_1 x$)

so that

- $D = 3$

There are three degrees of freedom allowing us to determine the ability to predict the model, often referred to as the *lack-of-fit*.

From this, we can obtain a value that relates to how well the experiment obeys the underlying linear model (or mathematical relationship between response and the values of the independent factors), often referred to as an *error*. It is important to understand that the formal statistical definition of an error is different from the colloquial definition (which really means a mistake): errors arise because nothing can be measured precisely – there are, of course, several contributions to the overall observed error, which will be discussed below. The error can be reported as a number, which, in the case discussed, will probably be expressed in AUs (absorbance units). Physical interpretation is not so easy. Consider an error that is reported as 100 mAU: this looks large, but then express it as AU and it becomes 0.1. Is it now a large error? The value of this error must be compared with something, and here the importance of *replication* comes into play. It is useful to repeat the experiment a few times under, as far as possible, identical conditions: this gives an idea of the reproducibility of the experimental error, sometimes called the *analytical* or *experimental* or *replicate* error. The larger the error, the harder it is to make good predictions. Figure 2.6 is of a linear calibration experiment with relatively large experimental errors: these may be due to many reasons, for example, instrumental performance, quality of volumetric flasks and precision of weighing. It is hard to see visually whether the results can be adequately described by a linear equation or not. The reading resulting from the experiment at the top right-hand corner of the graph might be a 'rogue' experiment, often called an *outlier*. Consider a similar experiment, but with lower experimental error (Figure 2.7). Now it looks as if a linear model is unlikely to be suitable, but only because the experimental error is small compared with the deviation from linearity. In Figures 2.6 and 2.7, an extra five degrees of freedom (the five replicates)

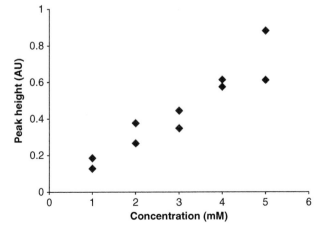

Figure 2.6 Experiment with high instrumental errors.

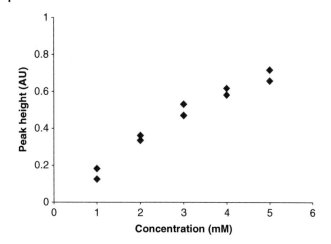

Figure 2.7 Experiment with low instrumental errors.

have been added to provide information on experimental error. The degrees of freedom available to test for lack-of-fit to a linear model are now given by

$$D = N - P - R$$

where R equals the number of replicates so that

$$D = 10 - 2 - 5 = 3$$

Although this number remains the same as in Figure 2.5, five extra experiments have been performed to give an idea of the experimental error.

In many designs, it is important to balance the number of unique experiments against the number of replicates. Each replicate provides a degree of freedom towards measuring experimental error. Some investigators use a degree-of-freedom tree that represents this information, a simplified version illustrated in Figure 2.8. A good rule of thumb is that the number of replicates (R) should be similar to the number of degrees of freedom for the lack-of-fit (D), unless there is an overriding reason for studying one aspect of the system in preference to another. Consider three

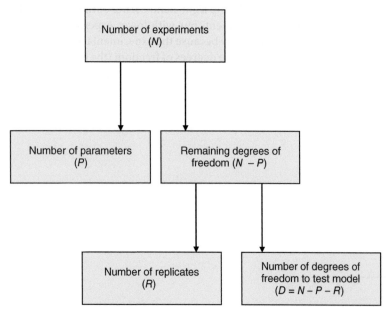

Figure 2.8 Degree-of-freedom tree.

Table 2.1 Three experimental designs.

Experiment number	Design 1		Design 2		Design 3	
	[A]	[B]	[A]	[B]	[A]	[B]
1	1	1	1	2	1	3
2	2	1	2	1	1	1
3	3	1	2	2	3	3
4	1	2	2	3	3	1
5	2	2	3	2	1	3
6	3	2	2	2	1	1
7	1	3	2	2	3	3
8	2	3	2	2	3	1
9	3	3				

experimental designs in Table 2.1. The aim is to produce a linear model of the form

$$y = b_0 + b_1 x_1 + b_2 x_2$$

The response y may represent the absorbance in a spectrum, the two x's the concentrations of two compounds. The value of P is equal to 3 in all cases.

- *Design 1.* This has a value of R equal to 0 and D of 6. There is no information about experimental error and all effort has taken into determining the model. If the relationship between the response and concentration is known with certainty (or this information is not of interest), this experiment may be a good one, but otherwise too little effort is taken in measuring replicates. Although this design may appear to provide an even distribution over the experimental domain, the lack of replication could, in some cases, lose crucial information.
- *Design 2.* This has a value of R equal to 3 and D of 2. There is a reasonable balance between taking replicates and examining the model. If nothing much is known about the certainty of the system, this is a good design taking into account the need to economise on experiments.
- *Design 3.* This has a value of R equal to 4 and D of 1. The number of replicates is rather large compared with the number of unique experiments. However, if the main aim is simply to investigate experimental reproducibility over a range of concentrations, this approach might be useful.

It is always possible to break down a set of planned experiments in this manner, and it is recommended as a first step before modelling.

1. An experiment is performed to study the relationship of the yield of a reaction as a function of pH, all other conditions are kept constant. Five pH values are studied, seven experiments are performed, three of which are at pH 6. The number of replicates is

    (a) 2
    (b) 3
    (c) 4

2. The number of degrees of freedom for estimating the lack-of-fit is always more than the number of coefficients in the model.

    (a) True
    (b) False

2.2.2 Analysis of Variance

A key aim of modelling is to ask how significant a factor is. In Section 2.2.1, we discussed about designing an experiment that allows sufficient degrees of freedom to determine the significance of a given factor; below we will introduce an important way of providing numerical information about this significance.

There are many situations where this information is useful, some examples are being listed.

- In an enzyme-catalysed extraction, many possible factors, such as incubation temperature, extraction time, extraction pH, stirring rates and so on, could have an influence over the extraction efficiency. Often 10 or more possible factors can be identified. Which factors are significant and should be studied or optimised further?
- In linear calibration, is the baseline important? Are there curved terms, is the concentration too high so that the Beer–Lambert law is no longer obeyed?
- In the study of a simple reaction dependent on temperature, pH, reaction time and catalyst concentration, are the interactions between these factors important? In particular, are higher order interactions (between more than two factors) significant?

A conventional approach is to set up a mathematical model linking the response to coefficients of the various factors. Consider the simple linear calibration experiment, discussed in Section 2.2.1, where the response and concentration are linked by the equation

$$y = b_0 + b_1 x$$

The term b_0 represents to an intercept term, which might be a consequence of the baseline of the spectrometer, the nature of a reference sample (for a double beam instrument) or the solvent absorption. Is this term significant? Extra terms in an equation will *always* improve the fit to the model; hence, simply determining how well a straight line fits the data does not provide the full picture.

The way to study this is to determine a model of the form

$$y = b_1 x$$

and ask how much worse the fit to the data is when the intercept term is removed. If it is not much worse, then the extra (intercept) term is not very important. The overall lack-of-fit to the model excluding the intercept term can be compared with the replicate error. Often, mean squared errors are calculated and are called *variances*; hence, the statistical term *analysis of variance* is abbreviated as *ANOVA*. If the lack-of-fit error is much larger than the replicate error, it may be significant; hence, the intercept term must be taken into account (and the experimenter may wish to check carefully how the baseline, solvent background and reference sample influence the measurements).

Above, we discussed how an experimental design is divided into different types of degrees of freedom, and we need to use this information in order to obtain a measure of significance.

Two data sets, A and B, are illustrated in Figures 2.9 and 2.10: the question asked is whether there is a significant intercept term. The numerical data are given in Table 2.2. These provide an indication as to how serious a baseline error is in a series of instrumental measurements. The first step is to determine the number of degrees of freedom. For each experiment,

- N (the total number of experiments) equals 10,
- R (the number of replicates) equals 4, measured at concentrations 1, 3, 4 and 6 mM. Remember if three experiments are performed at one concentration, this only adds 2 to the value of R.

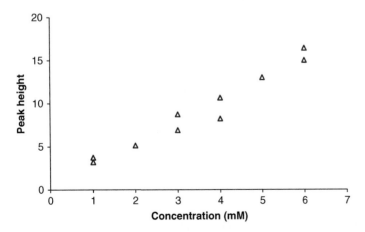

Figure 2.9 Graph of peak height against concentration for ANOVA example, data set A.

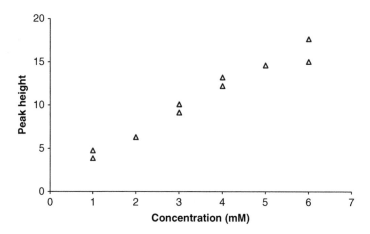

Figure 2.10 Graph of peak height against concentration for ANOVA example, data set B.

Table 2.2 Numerical information for data sets A and B.

Concentration (mM)	A	B
1	3.803	4.797
1	3.276	3.878
2	5.181	6.342
3	6.948	9.186
3	8.762	10.136
4	10.672	12.257
4	8.266	13.252
5	13.032	14.656
6	15.021	17.681
6	16.426	15.071

Two models can be determined, the first without an intercept of the form $y = b_1 x$ and the second with an intercept of the form $y = b_0 + b_1 x$. In the former case,

- $D = N - R - 1 = 5$

and in the latter case,

- $D = N - R - 2 = 4$.

The tricky part comes in determining the size of the errors.

- The total replicate error can be obtained by observing the difference between the responses under identical experimental conditions. For the data in Table 2.2, replication is performed at 1, 3, 4 and 6 mM. A simple way of determining this error is as follows:
  - Take the average reading at each replicated level or concentration.
  - Determine the differences between this average and the true reading for each replicated measurement.
  - Then, calculate the sum of squares of these differences (note that if the sums are not squared, they will add up to zero).

This procedure is illustrated in Table 2.3(a) for the data set A and the total replicate sum of squares equals 5.665 in this case.

- Algebraically, this sum of squares is defined as

$$S_{rep} = \sum_{n=1}^{N} (\bar{y}_n - y_n)^2$$

Table 2.3 Calculation of errors for data set A, model including intercept.

(a) Replicate error

Concentration	Absorbance	Replicate average	Differences	Squared differences
1	3.803		0.263	0.069
1	3.276	3.540	−0.263	0.069
2	5.181			
3	6.948		−0.907	0.822
3	8.762	7.855	0.907	0.822
4	10.672		1.203	1.448
4	8.266	9.469	−1.203	1.448
5	13.032			
6	15.021		−0.702	0.493
6	16.426	15.724	0.702	0.493
Sum of square replicate error				5.665

(b) Overall error (data fitted using univariate calibration and an intercept term)

Concentration	Absorbance	Fitted data	Differences	Squared differences
1	3.803	3.048	0.755	0.570
1	3.276	3.048	0.229	0.052
2	5.181	5.484	−0.304	0.092
3	6.948	7.921	−0.972	0.945
3	8.762	7.921	0.841	0.708
4	10.672	10.357	0.315	0.100
4	8.266	10.357	−2.091	4.372
5	13.032	12.793	0.238	0.057
6	15.021	15.230	−0.209	0.044
6	16.426	15.230	1.196	1.431
Total squared error				8.370

where $\bar{y}_n$ is the mean response at each unique experimental condition. If, for example, only one experiment is performed under a specified condition, it equals the response, whereas if three experiments are performed under identical conditions, it is their average. There are R degrees of freedom associated with this parameter.

- The total residual error sum of squares is simply the difference between the sum of squares of the observed readings and those predicted using a best-fit model (e.g. obtained using standard regression procedures in Excel). How to determine the best-fit model using MLR analysis will be described in more detail in Section 2.4. For a model with an intercept, $y = b_0 + b_1 x$, the calculation is presented in Table 2.3(b), where the predicted model is of the form $y = 0.6113 + 2.4364x$, giving a residual sum of square error of $S_{resid} = 8.370$. The sum of squares of the predicted or estimated response is always less than the observed response.
- Algebraically, this can be defined by

$$S_{resid} = \sum_{n=1}^{N} \left(y_n - \hat{y}_n \right)^2$$

and has $(N - P)$ degrees of freedom associated with it. The ˆ symbol means estimated (using the corresponding model). It is also equal to the difference between the total sum of squares for the raw data set given by

$$S_{total} = \sum_{n=1}^{N} y_n^2 = 1024.587$$

and the sum of squares for the predicted data

$$S_{reg} = \sum_{n=1}^{N} \hat{y}_n^2 = 1016.207$$

so that

$$S_{resid} = S_{total} - S_{reg} = 1024.587 - 1016.207 = 8.370$$

- The lack-of-fit sum of square error is simply the difference between these two numbers or 2.705 and may be defined by

$$S_{lof} = S_{resid} - S_{rep} = 8.370 - 5.665$$

or

$$S_{lof} = \sum_{n=1}^{N} (\bar{y}_n - \hat{y}_n)^2 = S_{mean} - S_{reg}$$

where

$$S_{mean} = \sum_{n=1}^{N} \bar{y}_n^2$$

and has $D = (N - P - R)$ degrees of freedom associated with it.

Note that there are several numerically equivalent ways of calculating these errors.

There are, of course, two ways in which a straight line can be fitted, one with and one without the intercept. Each generates different sum of square errors according to the model. The values of the coefficients and the errors are given in Table 2.4 for both data sets. Note that although the size of the term for the intercept for data set B is larger than data set A, this does not in itself indicate significance, unless the replicate error is taken into account.

Errors are often presented either as mean square or as root mean square. The root mean square error is given by

$$s = \sqrt{(S/d)}$$

where d is the number of degrees of freedom associated with a particular sum of squares. Note that the calculation of residual error for the overall data set differs according to authors. Strictly speaking, this sum of squares should be divided by $(N - P)$ or, for example, with the intercept, 8 ($=10 - 2$). The reason for this is that if there are no degrees of freedom for determining the residual error, the apparent error will be equal to exactly 0, but this does not mean too much. Hence, the root mean square residual error for data set A using the model with the intercept is, strictly speaking, equal to $\sqrt{(8.370/8)}$ or 1.0228. This error can also be converted to a percentage of the mean reading for the entire data set (which is 9.139), resulting in a mean residual of 11.19% by this criterion. However, it is also possible, provided the number of parameters is significantly less than the number of experiments, simply to divide by N for the residual error, giving a percentage of 10.01% in this example. In many areas of modelling such as principal components analysis and PLS regression (see Section 6.5), it is not always easy to determine the number of degrees of freedom in a

Table 2.4 Error analysis for data sets A and B.

		A	B
Model without intercept		$y = 2.576x$	$y = 2.948x$
Total sum of square error	S_{resid}	9.115	15.469
Sum of square replicate error (*d.f.* = 4)	S_{rep}	5.665 (mean = 1.416)	4.776 (mean = 1.194)
Difference between sum of squares (*d.f.* = 5): lack-of-fit	S_{lof}	3.450 (mean = 0.690)	10.693 (mean = 2.139)
Model with intercept		$y = 0.611 + 2.436x$	$y = 2.032 + 2.484x$
Total sum of square error	S_{resid}	8.370	7.240
Sum of square replicate error (*d.f.* = 4)	S_{rep}	5.665 (mean = 1.416)	4.776 (mean = 1.194)
Difference between sum of squares (*d.f.* = 4): lack-of-fit	S_{lof}	2.705 (mean = 0.676)	2.464 (mean = 0.616)

straightforward manner, and sometimes acceptable, if, for example, there are 40 objects in a data set, to simply obtain the mean residual error dividing by the number of objects. Many mathematicians debate the meaning of probabilities and errors: is there an inherent physical (or natural) significance to an error, in which case the difference between 10% and 11% could mean something or do errors primarily provide general guidance as to how good and useful as set of results is? For chemists, it is often more important to get a ballpark figure for an error rather than debate the ultimate meaning of the number. The degrees of freedom would have to take into account the number of principal components in the model, as well as data pre-processing such as normalisation and standardisation as discussed in Chapter 4. In this book, we adopt the convention of dividing by the total number of degrees of freedom to get a root mean square residual error, unless there are specific difficulties determining this number.

Several conclusions can be drawn from Table 2.4.

- The replicate sum of squares is obviously the same regardless of the model employed for a given experiment, but differs for each data set. The two data sets result in roughly similar replicate errors, suggesting that the experimental procedure (e.g. dilutions and instrumental method) is similar in both cases. Only four degrees of freedom are used to measure this error; hence, it is unlikely that these two measured replicate errors will be exactly equal. A series of measurements is a sample from a larger population, and it is necessary to have a large sample size to obtain very close agreement to the overall population variance. Obtaining a high degree of agreement may involve several hundred repeat measurements, which clearly overkills for such a comparatively straightforward series of experiments.
- The total error reduces when an intercept term is added in both cases. This is inevitable and does not necessarily imply that the intercept is significant.
- The difference between the total error and the experimental error equals the 'lack-of-fit'. The bigger this is, the worse the model.
- The lack-of-fit error is slightly smaller than the experimental error, in all cases, except when the intercept is removed from the model for the data set B when it is large, 10.693. This suggests that adding the intercept term to the second data set makes a big difference to the quality of the model and so the intercept is significant for this data set.

Conventionally, these numbers are often compared using ANOVA. In order for this to be meaningful, each sum of squares should be divided by the number of degrees of freedom to give the 'mean' sum of squares (Table 2.5) because the more the measurements, the greater the underlying sum of squares is likely to be. These mean squares are often called *variances*, and it is simply necessary to compare their sizes, by taking ratios. The larger the ratio to the mean replicate error, the greater the significance. In all cases apart from the model without the intercept arising from data set B, the mean lack-of-fit error is considerably less than the mean replicate error. Often the results are presented in a tabular form, a typical example given for the two-parameter model of data set B, and is presented in Table 2.5: the five sums of squares S_{total}, S_{reg}, S_{resid}, S_{rep}, and S_{lof}, together with the relevant degrees of freedom, mean square and variance ratio being presented. The number 0.516 is the key to assess how well the model describes the data and is often called the *F-ratio* between the mean lack-of-fit error and the mean replicate error, which will be discussed in more detail in Section 2.2.4.4. Suffice it to say, the higher this number, the more significant the error. A lack-of-fit error that is much lesser than the replicate error is not significant, within the constraints of the experiment. Assuming an underlying normal distribution of errors, the value of F can be converted to a probability (see Section A.3.5). Usually this is cited as a p value that the null hypothesis that the model is not significant is rejected: the widespread idea of a null hypothesis was introduced by the statistician R.A. Fisher about 100 years ago. A p value of 0.01 suggests a low chance that the lack-of-fit is not significant. Some chemists prefer to talk about confidence; a p value of 0.01 represents a 99% confidence that the model is significant.

Table 2.5 ANOVA table: two-parameter model, data set B.

Source of variation	Sum of squares	Degrees of freedom	Mean sum of squares	Variance ratio
Total	1345.755	10	134.576	
Regression	1338.515	2	669.258	
Residual	7.240	8	0.905	
Replicate	4.776	4	1.194	
Lack-of-fit	2.464	4	0.616	0.516

Most statistical packages produce ANOVA tables if required, and it is not always necessary to determine these errors manually, although it is important to appreciate the principles behind such calculations. However, for simple examples, a manual calculation is often quite quick and a good alternative to the interpretation of the output of complex statistical packages.

The use of ANOVA is widespread and is based on these simple ideas. Normally, two mean errors are compared, for example, one due to replication and the other due to lack-of-fit, although any two errors or variances may be compared. As an example, if there are 10 possible factors that might have an influence over the yield in a synthetic reaction, try modelling the reaction removing one factor at a time and see how much the lack-of-fit error increases. If not much relative to the replicates, the factor is probably not significant. It is important to recognise that reproducibility of the reaction has an influence over apparent significance also, as does the accuracy of measurement of the yield. If there is a large replicate error, then some significant factors might be missed out.

1. The sum of squares for predicted data is always less than the sum of squares for the observed data.
   (a) True
   (b) False

2. The mean sum of square lack-of-fit error is always less than the mean sum of square replicate error.
   (a) True
   (b) False

3. Nine experiments are performed as follows: condition 1 (1 experiment), condition 2 (2 experiments), condition 3 (3 experiments), condition 4 (2 experiments) and condition 5 (1 experiment). The total sum of square replicate error is 0.736. The mean square replicate error is
   (a) 0.105
   (b) 0.184
   (c) 0.245

2.2.3 Design Matrices and Modelling

The design matrix is a key concept. A design may consist of a series of experiments performed under different conditions, for example, a reaction at differing pHs, temperatures and concentrations. Table 2.6 illustrates a typical experimental set-up, together with an experimental response, for example, the rate constant of a reaction. Note the replicates in the final five experiments: in Section 2.4, we will discuss such an experimental design commonly called a *central composite* design.

2.2.3.1 Models

It is normal to describe experimental data by forming a mathematical relationship between the factors or independent variables such as temperature and a response or dependent variable such as a synthetic yield, a reaction time or a percentage impurity. A typical equation involving three factors might be of the form

$\hat{y} =$ (response)

b_0+ (an intercept or average)

$b_1x_1 + b_2x_2 + b_3x_3+$ (linear terms depending on each of the three factors)

$b_{11}x_1{}^2 + b_{22}x_2{}^2 + b_{33}x_3{}^2+$ (quadratic terms depending on each of the three factors)

$b_{12}x_1x_2 + b_{13}x_1x_3 + b_{23}x_2x_3$ (interaction terms between the factors).

Notice the 'hat' on top of y, this is because the model estimates its value and is unlikely to give an exact value that agrees with the observed value because of error.

The explanation for these terms is as follows:

- The intercept is the estimated response when the values of all factors is 0: if the factors are coded to be centred at 0 (see Section 2.2.4.1), it is the estimated response in the centre of the design usually corresponding to the average values of the factors. It is an important term because a response of 0 is not normally achieved when the factors are at their average values. Only in certain circumstances (e.g. spectroscopy: if it is known, there are no baseline problems or interferences), this term can be safely ignored.

Table 2.6 Typical experimental design.

pH	Temperature (°C)	Concentration (mM)	Response (y)
6	60	4	34.841
6	60	2	16.567
6	20	4	45.396
6	20	2	27.939
4	60	4	19.825
4	60	2	1.444
4	20	4	37.673
4	20	2	23.131
6	40	3	23.088
4	40	3	12.325
5	60	3	16.461
5	20	3	33.489
5	40	4	26.189
5	40	2	8.337
5	40	3	19.192
5	40	3	16.579
5	40	3	17.794
5	40	3	16.650
5	40	3	16.799
5	40	3	16.635

- The linear terms allow for a direct relationship between the response and a given factor. For some experimental data, there are only linear terms. If pH increases, does the yield increase or decrease, and if so by how much?
- In many situations, quadratic terms are important. This allows curvature and is one way of obtaining a maximum or minimum. Most chemical reactions have an optimum performance at a particular pH, for example. Almost all enzymic reactions work in this way. Quadratic terms balance out the linear terms.
- In Section 2.1, we discussed the need for interaction terms. These arise because the influence of two factors on the response is rarely independent. For example, the optimum pH at one temperature may differ from that at a different temperature.

Some of these terms may not be very significant or relevant, but it is up to the experimenter to check using approaches such as ANOVA (Section 2.2.2) and the related significance tests (Section 2.2.4). In advance of modelling, it is often hard to predict which factors are important.

1. If we are interested in the relationship between a response y and two factors x_1 and x_2 and obtain a model with all possible intercept, linear, quadratic and interaction terms, how many terms are there in the model?
    (a) 4
    (b) 5
    (c) 6

2.2.3.2 Matrices
There are 10 terms or parameters in the above equation. Many chemometricians find it convenient to work using matrices. Although a significant proportion of traditional statistical texts often shy away from matrix-based notation, with modern computer packages and spreadsheets, it is easy and rational to employ matrices. In later chapters, we will use matrix notation more frequently. The design matrix is simply one in which

- the *rows* refer to experiments and
- the *columns* refer to individual terms in the mathematical model or equation linking the response to the values of the individual factors.

Note that the experimental matrix in Table 2.6 is not, in statistical terms, a design matrix, as the columns relate only to the experimental conditions and not to the terms of the model.

Using the model of Section 2.2.3.1, the design matrix consists of

- 20 rows as there are 20 experiments and
- 10 columns as there are 10 terms in the model,

as is illustrated symbolically in Figure 2.11. Note that for the experimental matrix in Table 2.6, there can be several different design matrices, according to the model used. If we did not want quadratic terms but still wanted all linear and two-factor interaction terms, the design matrix would have 7 rather than 10 columns, for example.

For the experiment and model discussed above, the design matrix is given in Table 2.7. Note the first column of 1's: this corresponds to the intercept term b_0, which is can be regarded as multiplied by the number 1 in the equation. The figures in the table can be checked numerically. For example, the interaction between pH and temperature for the first experiment is 360, which equals 6×60, and appears in the eighth column of the first row corresponding to the term b_{12}.

Two considerations are required when computing a design matrix, namely

- the number and arrangement of the experiments, including replication and
- the mathematical model to be tested.

It is easy to see that, in our case,

- the 20 responses form a vector with 1 column and 20 rows, called y,
- the design matrix has 10 columns and 20 rows as illustrated in Table 2.7, called D and
- the 10 coefficients of the model form a vector with 1 row by 10 columns, called b.

1. The design matrix can be obtained once we know the arrangements of our experiments.
   (a) True
   (b) False

2. We obtain 15 unique observations and fit the results to a model with seven parameters. The vector b has the dimensions.
   (a) 1 row by 7 columns
   (b) 7 rows by 1 column
   (c) 15 rows by 1 column
   (d) 15 rows by 7 columns

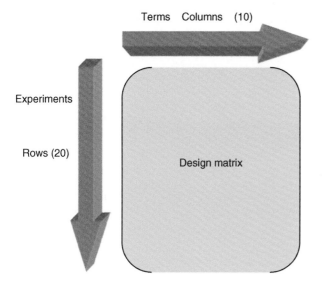

Terms   Columns   (10)

Experiments

Rows (20)

Design matrix

Figure 2.11 Design matrix.

Table 2.7 Design matrix for the experiment in Table 2.6 using the model discussed in Section 2.2.3.1.

Intercept	Linear terms			Quadratic terms			Interaction terms		
b_0	b_1	b_2	b_3	b_{11}	b_{22}	b_{33}	b_{12}	b_{13}	b_{23}
Intercept	pH	Temp.	Conc.	pH^2	$Temp.^2$	$Conc.^2$	pH × temp.	pH × conc.	Temp. × conc.
1	6	60	4	36	3600	16	360	24	240
1	6	60	2	36	3600	4	360	12	120
1	6	20	4	36	400	16	120	24	80
1	6	20	2	36	400	4	120	12	40
1	4	60	4	16	3600	16	240	16	240
1	4	60	2	16	3600	4	240	8	120
1	4	20	4	16	400	16	80	16	80
1	4	20	2	16	400	4	80	8	40
1	6	40	3	36	1600	9	240	18	120
1	4	40	3	16	1600	9	160	12	120
1	5	60	3	25	3600	9	300	15	180
1	5	20	3	25	400	9	100	15	60
1	5	40	4	25	1600	16	200	20	160
1	5	40	2	25	1600	4	200	10	80
1	5	40	3	25	1600	9	200	15	120
1	5	40	3	25	1600	9	200	15	120
1	5	40	3	25	1600	9	200	15	120
1	5	40	3	25	1600	9	200	15	120
1	5	40	3	25	1600	9	200	15	120
1	5	40	3	25	1600	9	200	15	120

2.2.3.3 Determining the Model

The relationship between the response, the coefficients and the experimental conditions can be expressed in matrix form by

$$\hat{y} = D\,b$$

as illustrated in Figure 2.12. Readers unfamiliar with matrix notation should read Section A.1.

It is simple to show that this is the matrix equivalent to the equation introduced in Section 2.2.3.1. It is surprisingly easy to calculate b (or the coefficients in the model) knowing D and y using MLR. This approach will be discussed in greater detail in Chapter 6.

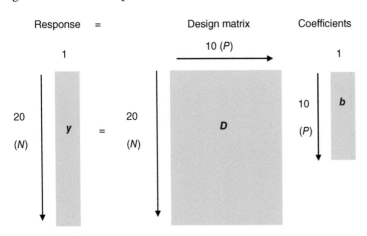

Figure 2.12 Relationship between response, design matrix and coefficients.

- If D is a square matrix, then there are exactly the same number of experiments as coefficients in the model and

$$b = D^{-1}y$$

- If D is not a square matrix (as in the case in this section), then use the pseudo-inverse, an easy calculation in Excel, Matlab and almost all matrix-based software, as follows:

$$b = (D'D)^{-1}D'y$$

The idea of the pseudo-inverse is used in several places in this text, for example, see Sections 6.2 and 6.3 for a general treatment of regression. A simple derivation is as follows:

$$y \approx Db \quad \text{so } D'y \approx D'Db \text{ or } (D'D)^{-1}D'y \approx (D'D)^{-1}(D'D)b \approx b$$

In practice, we obtain estimates of b from regression; hence, strictly speaking, there should be an approximation sign in all the equations, or a hat on top of the b; however, in order to simplify the text, we ignore the hat and so the approximation sign becomes an equals sign. There is quite a significant statistical literature about estimation, which is beyond the scope of this introductory text.

It is important to recognise that for some designs, there are several alternative methods for calculating these regression coefficients, which will be described in the relevant sections, but the method of regression described above will *always* work, provided the experiments are designed appropriately. A limitation before the computer age was the inability to determine matrix inverses easily; hence, classical statisticians often got around this by devising methods often for summing functions of the response and, in some cases, designed experiments specifically to overcome the difficulty of computing inverses and for ease of calculation. The dimensions of the square matrix ($D'D$) equal the number of parameters in a model; hence, if there are 10 parameters, it would not have been easy to compute the relevant inverse manually, although this is now a simple operation using modern computer-based packages, and it is usually no longer necessary to understand computational formulae for the inverse of a matrix.

There are a number of important consequences.

- If the matrix D is a square matrix, the estimated values of $\hat{y}$ are identical to the observed values y. The model provides an exact fit to the data, and there are no degrees of freedom remaining to determine the lack-of-fit. Under such circumstances, there will not be any replicate information, but nevertheless, the values of b can provide valuable information about the size of different effects. Such a situation might occur, for example, in factorial designs (Section 2.3). The residual error between the observed and fitted data will be zero. This does not imply that the predicted model exactly represents the underlying (or population) data, simply that the number of degrees of freedom is insufficient for determination of prediction errors. In all other circumstances, there is likely to be an error as the predicted and observed response will differ.
- The matrix D – or $D'D$ (if the number of experiments is more than the number of terms) – must have an inverse. If it does not, it is impossible to calculate the coefficients b. This is a consequence of poor design and may occur if two terms or factors are correlated with each other. For well-designed experiments, this problem will not occur. Notice that a design in which the number of experiments is less than the number of terms in the model has no solution; in order to solve this, either perform more experiments or reduce the size of the model. If we are, for example, performing a four-factor experiment, higher order interactions may not be relevant and so can be removed from the model.

1. The matrix D is square if

    (a) The number of factors in the model equals the number of experiments
    (b) The number of terms in the model equals the number of experiments

2. The matrix $D'D$ is the same as DD'.

    (a) True
    (b) False

Table 2.8 The vectors b and $\hat{y}$ for data in Table 2.6.

Parameters		Predicted y
b_0	58.807	35.106
b_1	−6.092	15.938
b_2	−2.603	45.238
b_3	4.808	28.399
b_{11}	0.598	19.315
b_{22}	0.020	1.552
b_{33}	0.154	38.251
b_{12}	0.110	22.816
b_{13}	0.351	23.150
b_{23}	0.029	12.463
		17.226
		32.924
		26.013
		8.712
		17.208
		17.208
		17.208
		17.208
		17.208
		17.208

2.2.3.4 Predictions

Once b is determined, it is then possible to predict or estimate y and to calculate the sums of squares and other statistics as outlined in Sections 2.2.2 and 2.2.4. For the data in Table 2.6, the results are presented in Table 2.8, using the pseudo-inverse to obtain b and then predict $\hat{y}$. Note that the *size* of the parameters does not necessarily relate to significance, in this example. It is a common misconception that the larger the coefficient, the more important it is. For example, it may appear that b_{22} is small (0.020) relative to the b_{11} parameter (0.598), but this depends on the physical measurement units.

- The pH range is between 4 and 6; hence, the square of pH varies between 16 and 36 or by 20 units overall.
- The temperature range is between 20 and 60 °C, the squared range varying between 400 and 3600 or by 3200 units overall, which is a 160-fold difference in range compared with pH.
- Therefore, to be of equal importance, b_{22} would need to be 160 times smaller than b_{11}.
- As the ratio between of $b_{11}:b_{22}$ is 29.95, b_{22} is in fact considerably more significant than b_{11}.

In Section 2.2.4, we discuss in more detail how to tell whether a given parameter is significant, but it is indeed very dangerous to rely on visual inspection of tables of regression parameters and make deductions from these without understanding carefully how the data are scaled.

If carefully interpreted, three types of information can come from the model.

- The size of the coefficients can help the experimenter decide how significant the coefficient is, although the absolute value may be misleading in some cases. For example, does a change in pH significantly change the yield of a reaction? Or is the interaction between pH and temperature significant? In other words, does the temperature at which the reaction has a maximum yield differ substantially at pH 5 and at pH 7?
- The coefficients can be used to construct a model of the response, for example, the yield of a reaction as a function of pH and temperature, and to establish the optimum conditions for obtaining the best yield. In this case, the experimenter is not so interested in the precise equation for the yield but is very interested in the best pH and temperature.
- Finally, a quantitative model may be interesting. Predicting the concentration of a compound from the absorption in a spectrum requires an accurate knowledge of the relationship between the variables. Under such circumstances,

Table 2.9 Coding of data.

Variable	Units	−1	+1
pH	−log[H^+]	4	6
Temperature	°C	20	60
Concentration	mM	2	4

the value of the coefficients is important. In some cases, it is known that there is a certain kind of model, and the task is primarily to obtain a regression or calibration equation.

Although the emphasis in this chapter is on using MLR techniques, it is important to recognise that the analysis of designed experiments is not restricted to such approaches, and it is legitimate to employ multivariate methods such as principal components regression and PLS as described in detail in Chapter 6.

1. The term b_1 in a model is estimated at 0.736 and b_2 at 5.129. Does this imply b_1 is less significant than b_2?

   (a) Yes
   (b) Cannot tell unless the design matrix is known
   (c) Cannot tell unless the replicate error is known.

2.2.4 Assessment of Significance

In many traditional books on statistics and analytical chemistry, large sections are devoted to significance testing. Indeed, an entire and very long book could easily be written about the use of significance tests in chemistry. However, much of the work on significance testing goes back nearly a 100 years, to the work of 'Student' (a pseudonym for W.S. Gossett), and slightly later to R.A. Fisher. Although their methods based primarily on the t-test and F-test have had a huge influence in applied statistics, they were developed before the modern computer age. A typical statistical calculation using pen and paper and perhaps a book of logarithm or statistical tables might take several days, compared with a few seconds on a modern microcomputer. Ingenious and elaborate approaches were developed, including special types of graph papers and named methods for calculating the significance of various effects.

These early methods were developed primarily for use by specialised statisticians, mainly trained as mathematicians, in an environment where user-friendly graphics or easy analysis of data was inconceivable. A mathematical statistician will have a good feeling for the data and so is unlikely to perform calculations or compute statistics from a data set unless satisfied that the quality of data is appropriate. In the modern age, everyone can have access to these tools without a great deal of mathematical expertise, but correspondingly, it is possible to misuse these methods in an inappropriate manner. The practicing chemist needs to have a numerical and graphical feel for the significance of his or her data, and traditional statistical tests are only one of a battery of approaches to determine the significance of a factor or effect in an experiment.

This section provides an introduction to a variety of approaches for assessing significance. For historic reasons, some methods such as cross-validation and independent testing of models are best described in the chapters on multivariate methods (see Chapters 4–6), although the chemometrician should have a broad appreciation of all such approaches and not be restricted to any one set of methods.

2.2.4.1 Coding

In Section 2.2.3, we introduced an example of a three-factor design, given in Table 2.6, described by 10 terms in the corresponding model. Our comment was that the significance of the coefficients cannot easily be assessed by inspection because the physical scale for each variable is different. In order to have a better idea of the significance, it is useful to put each variable on a comparable scale. It is common to *code* experimental data. Each variable is placed on a common scale, often with the highest coded value of each variable equal to +1 and the lowest to −1. Table 2.9 represents a possible way to scale the data; hence, for factor 1 (pH), a coded value (or level) or −1 corresponds to a true pH of 4. Note that coding does not need to be linear: in fact, pH is actually measured on a logarithmic scale; hence, we are not coding [H^+] using a linear scale.

Table 2.10 Coded design matrix together with estimated values of coded coefficients.

1	x_1	x_2	x_3	x_1^2	x_2^2	x_3^2	x_1x_2	x_1x_3	x_2x_3
1	1	1	1	1	1	1	1	1	1
1	1	1	−1	1	1	1	1	−1	−1
1	1	−1	1	1	1	1	−1	1	−1
1	1	−1	−1	1	1	1	−1	−1	1
1	−1	1	1	1	1	1	−1	−1	1
1	−1	1	−1	1	1	1	−1	1	−1
1	−1	−1	1	1	1	1	1	−1	−1
1	−1	−1	−1	1	1	1	1	1	1
1	1	0	0	1	0	0	0	0	0
1	−1	0	0	1	0	0	0	0	0
1	0	1	0	0	1	0	0	0	0
1	0	−1	0	0	1	0	0	0	0
1	0	0	1	0	0	1	0	0	0
1	0	0	−1	0	0	1	0	0	0
1	0	0	0	0	0	0	0	0	0
1	0	0	0	0	0	0	0	0	0
1	0	0	0	0	0	0	0	0	0
1	0	0	0	0	0	0	0	0	0
1	0	0	0	0	0	0	0	0	0
1	0	0	0	0	0	0	0	0	0

Estimated values

b_0	b_1	b_2	b_3	b_{11}	b_{22}	b_{33}	b_{12}	b_{13}	b_{23}
17.208	5.343	−7.849	8.651	0.598	7.867	0.154	2.201	0.351	0.582

The design matrix simplifies considerably and, together with the corresponding regression coefficients, is presented in Table 2.10. Now the coefficients are approximately on the same scale, and it appears that there are radical differences between these new numbers and the coefficients in Table 2.8. Some of the differences and their interpretation are listed below.

- The coefficient b_0 is very different. In the current calculation, it represents the predicted response in the centre of the design, where the coded levels of the three factors are (0,0,0). In the calculation of Section 2.2.3, it represents the estimated response at 0 pH units, 0 °C and 0 mM, conditions that will not normally be reached experimentally. Note also that this approximates to the mean of the entire data set (21.518) and is close to the average over the six replicates in the central point (17.275). For a perfect fit, with no error, it will equal the mean of the entire data set, as it will for designs centred on the point (0,0,0) in which the number of experiments equals the number of terms in the model such as a factorial designs discussed in Section 2.3, where no degrees of freedom are available for the lack-of-fit error.
- The relative size of the coefficients b_{11} and b_{22} changes dramatically compared with that in Table 2.8, the latter increasing hugely in apparent size when the coded data set is employed. Provided the experimenter chooses appropriate physical conditions, it is the coded values that are most helpful for interpretation of significance. A change in pH of 1 unit is more important than a change in temperature of 1 °C. A temperature range of 40 °C is quite small, whereas a pH range of 40 units would be almost unconceivable. Therefore, it is important to be able to directly compare the size of parameters in the coded scale.
- Another very important observation is that the *sign* of significant parameters can also change as the coding of the data is changed. For example, the sign of the parameter for b_1 is negative (−6.092) in Table 2.8 but positive (+5.343) in Table 2.10, yet the size and sign of the b_{11} term does not change. The difference between the highest and the lowest true pH (2 units) is the same as the difference between the highest and lowest coded values of pH, also 2 units. In Table 2.8 and Table 2.10, the value of b_1 is approximately 10 times greater in magnitude than b_{11}, so might

appear much more significant. Furthermore, it is one of the largest terms apart from the intercept. What has gone wrong with the calculation? Does the value of y increase with increasing pH or does it decrease? There can be only one physical answer. The clue to change of sign comes from the mathematical transformation. Consider a simple equation of the form

$$y = 10 + 50x - 5x^2$$

and a new transformation from a range of raw values between 9 and 11 to coded values between -1 and $+1$ so that

$$c = x - 10$$

where c is the coded value. Then,

$$y = 10 + 50(c + 10) - 5(c + 10)^2$$
$$= 10 + 50c + 500 - 5c^2 - 100c - 500$$
$$= 10 - 50c - 5c^2$$

an apparent change in sign. Using raw data, we might conclude that the response increases with increasing x_1, whereas with the coded data, the opposite conclusion might be drawn. Which is correct? Returning to our example, although the graph of the response depends on interaction effects, and the relationship between y and pH is different at each temperature and concentration, at central point of the design, it is given in Figure 2.13, monotonically increasing over the experimental region. Indeed, the average value of the response when the pH is equal to 6 is higher than the average value when it is equal to 4. Hence, it is correct to conclude that the response increases with pH in the central region of the design, and the negative coefficient in Table 2.8 is misleading. Using coded data provides correct conclusions about the trends, whereas using coefficients obtained from the raw data may lead to incorrect deductions.

Therefore, without taking great care, misleading conclusions can be obtained about the significance and influence of the different factors. It is essential that the user of simple chemometric software is fully aware of this and always interprets numbers in terms of physical meaning.

1. For a symmetrical design, if variables are coded, the term b_0
   (a) Always equals the average of the observed responses for a given design
   (b) Always equals the predicted response in the centre of the design
   (c) Always equals the mean observed response in the centre of the design averaged over replicates.

2. A coefficient that has a negative value for an uncoded design matrix will also have a negative value for the corresponding coded design matrix, provided the low values of the factors are coded as more negative than high values.
   (a) True
   (b) False

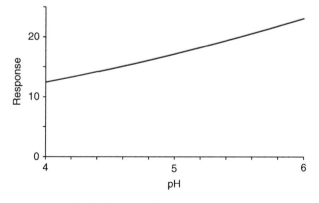

Figure 2.13 Graph of estimated response versus pH at the central temperature and concentration of the design in Table 4.6.

2.2.4.2 Size of Coefficients

The simplest approach to determining significance is to simply look at the magnitude of the terms (or coefficients) in the model. Provided the data is coded correctly, the larger the term, the greater its significance. This depends on each coded factor varying over approximately the same range (between +1 and −1 in this case). Clearly, small differences in range are not important, often the aim is to say whether a particular term has a significant influence or not, rather than a detailed interpretation of their size. A value of 5.343 for b_1 implies that on average, in the centre of the design, the response is higher by 5.343 if the value of b_1 is increased by one coded pH unit. This is easy to verify and provides an alternative, classical approach to the calculation of the coefficients.

- Consider the 10 experiments in Table 2.6 at which b_1 is at a coded level of either +1 or −1, namely the first 10 experiments.
- Then, group these in five pairs, each of which the levels of the other two main factors are identical. These pairs are {1,5}, {2,6}, {3,7}, {4,8} and {9,10}.
- Take the difference between the responses at the levels and average them

$$[(34.841 − 19.825) + (16.567 − 1.444) + (45.396 − 37.673) + (27.939 − 23.131) + (23.088 − 12.325)]/5,$$

  which gives an answer of 10.687, representing the average change in value of the response when the pH is increased from a coded value of −1 to one of +1, half of which equals the coefficient 5.343.
- Note that this calculation is only valid because b_1 is at its central value (i.e. a coded value of 0) for the remaining 10 experiments. It will not be valid for the b_2 and b_3 unless different experiments are chosen.

It is useful to make practical deductions from the data, which will guide the experimenter.

- The response varies over a range of 43.953 units between the lowest and highest observation in the experimental range.
- Hence, the linear effect of pH, on average, is to increase the response by twice the coded coefficient or 10.687 units over this range, approximately 25% of the variation, probably quite significant. The effect of the interaction between pH and concentration (b_{13}), however, is only 0.702 units or a relatively small contribution, rather less than the replicate error; hence, this factor is unlikely to be significant.
- The squared terms must be interpreted slightly differently. The lowest possible coded value for the squared terms is 0, not −1; hence, we do not double these values to obtain an indication of significance, the range of variation of the squared terms being between 0 and +1, or half that of the other terms.

It is not, of course, necessary to have replicates to perform this type of analysis. If the yield of a reaction varies between 50% and 90% over a range of experimental conditions, then a factor that contributes, on average, to only 1% of this increase is unlikely to be too important. However, it is vital that the factors are coded for meaningful comparison. In addition, certain important properties of the design (namely orthogonality), which will be discussed in detail in later sections, are also important.

Provided the factors are coded correctly, it is quite easy to make qualitative comparisons of significance simply by examining the size of the coefficients either numerically and graphically. In some cases, the range of variation of each individual factor might differ slightly (e.g. squared and linear terms above), but provided this is not dramatic, for rough indications, the size of the factors can be legitimately compared. In the case of two-level factorial designs (described in Sections 2.3.1–2.3.3), each factor is normally scaled between −1 and +1; hence, all coefficients are on the same scale.

1. A design is coded so that a temperature of 15 °C is −1 and 55 °C +1. Temperature is donated as factor 1 and $b_1 = +8\%$ of the range of yields of a reaction over the design. This implies, on average, in the centre of the design

    (a) We expect the yield to increase at the rate of 8% every 10 °C
    (b) We expect the yield to increase at the rate of 4% every 10 °C
    (c) We expect the yield to increase at the rate of 16% every 10 °C.

2.2.4.3 Student's *t*-Test

An alternative, statistical indicator based on Student's *t*-test can be used, provided more experiments are performed than the terms in the model. Although this and related statistical indicators have a long and venerated history, it is always important to back up the statistics by simple graphs and considerations about the data. There are many diverse applications of a *t*-test, but in the context of analysing the significance of factors on designed experiments, the following main steps are used and are illustrated in Table 2.11 for the example described above using the coded values in Table 2.9.

Table 2.11 Calculation of *t*-statistic.

(a) Matrix $(D'D)^{-1}$

	b_0	b_1	b_2	b_3	b_{11}	b_{22}	b_{33}	b_{12}	b_{13}	b_{23}
b_0	**0.118**	0.000	0.000	0.000	−0.045	−0.045	−0.045	0.000	0.000	0.000
b_1	0.000	**0.100**	0.000	0.000	0.000	0.000	0.000	0.000	0.000	0.000
b_2	0.000	0.000	**0.100**	0.000	0.000	0.000	0.000	0.000	0.000	0.000
b_3	0.000	0.000	0.000	**0.100**	0.000	0.000	0.000	0.000	0.000	0.000
b_{11}	−0.045	0.000	0.000	0.000	**0.364**	−0.136	−0.136	0.000	0.000	0.000
b_{22}	−0.045	0.000	0.000	0.000	−0.136	**0.364**	−0.136	0.000	0.000	0.000
b_{33}	−0.045	0.000	0.000	0.000	−0.136	−0.136	**0.364**	0.000	0.000	0.000
b_{12}	0.000	0.000	0.000	0.000	0.000	0.000	0.000	**0.125**	0.000	0.000
b_{13}	0.000	0.000	0.000	0.000	0.000	0.000	0.000	0.000	**0.125**	0.000
b_{23}	0.000	0.000	0.000	0.000	0.000	0.000	0.000	0.000	0.000	**0.125**

(b) Values of t and significance

	s	0.79869			
	v	$\sqrt{sv}$	b	t	% Probability
b_0	0.118	0.307	17.208	56.01	>99.9
b_1	0.100	0.283	5.343	18.91	>99.9
b_2	0.100	0.283	−7.849	−27.77	>99.9
b_3	0.100	0.283	8.651	30.61	>99.9
b_{11}	0.364	0.539	0.598	1.11	70.7
b_{22}	0.364	0.539	7.867	14.60	>99.9
b_{33}	0.364	0.539	0.154	0.29	22.2
b_{12}	0.125	0.316	2.201	6.97	>99.9
b_{13}	0.125	0.316	0.351	1.11	70.7
b_{23}	0.125	0.316	0.582	1.84	90.4

- Calculate the matrix $(D'D)^{-1}$. This will be a square matrix with dimensions equal to the number of terms in the model.
- Calculate the error sum of squares between the predicted and observed data (compare the actual response in Table 2.6 with the predictions in Table 2.8), $S_{resid} = \sum_{i=1}^{I} (y_i - \hat{y}_i)^2 = 7.987$.
- Take the mean of the sum of square error (divided by the number of degrees of freedom available for testing for regression), $s = S_{resid}/(N - P) = 7.987/(20 - 10) = 0.799$. Note that the t-test is not applicable to data where the number of experiments equals the number of terms in the model, such as full factorial designs discussed in Section 2.3.1 where all possible terms are included in the model.
- For each of the P parameters (=10 in this case), take the appropriate number from the diagonal of the matrix in Table 2.11(a) obtained in step 1 above. This is called the *variance* for each parameter, so that, for example, $v_{11} = 0.364$ (the variance of b_{11}).
- For each coefficient, b, calculate $b/\sqrt{sv}$. The higher this ratio, the more significant the coefficient. This ratio is used for the t-test.
- The statistical significance can then be obtained from a two-tailed t-distribution, described in detail in Section A.3.4, or most packages such as Excel have simple functions for the t-test. Note that the reason why it is two tailed is that the coefficient could deviate in both positive and negative directions from 0 and that some tables are for the one-tailed distribution, so check this first. Take the absolute value of the ratio calculated above. If you use a table, along the left-hand column of a t-distribution table are tabulated degrees of freedom, which equal the number available to test for regression, or $N - P$ or 10 in this case. The percentage probability is located along the columns (often the higher

the significance, the smaller the percentage, so simply subtract from 100). The higher this probability, the greater the confidence that the factor or term is significant. Hence, using Table A.3, we see that a critical value of 4.1437 indicates 99.9% certainty and that a parameter is significant for 10 degrees of freedom; hence, any value above this is highly significant. A 95% significance results in a value of 1.8125, so b_{23} is just above this level. In fact, the numbers in Table 2.11 were calculated using the Excel function TDIST, which provides probabilities for any value of t and any number of degrees of freedom. Normally, quite high probabilities are expected if a factor is significant, often in excess of 95%.

- In some articles, p values are used instead of percentage confidence. Analytical chemists tend to like the idea of confidence but statisticians like the idea of p values, for historic reasons. A p value of 0.05 corresponds to a percentage confidence of 95%. Strictly speaking, we talk about the probability of rejecting the null hypothesis. The null hypothesis test whether any given term has no significance in the model. If a p value is calculated to be 0.01, this means that there is a 1% chance of rejecting the null hypothesis, or that it is very unlikely the term has no significance, but chemists often like to say that there is 99% confidence that the term is significant. Additionally, this confidence or p value assumes that the underlying sample errors are normally distributed, an assumption that is often only roughly obeyed in practice; however, low p values or high confidence will usually suggest that a term or factor should be retained.

1. The p value for a term is calculated to be 0.02. This implies that it is not significant.
    - (a) True
    - (b) False

2. A design consists of 15 experiments in total, and a 10-term model is formed. How many degrees of freedom are remaining for determining the confidence in the regression model?
    - (a) 10
    - (b) 5
    - (c) 1

3. The t-test can be employed to determine the significance of terms in a full factorial model for a full factorial design.
    - (a) True
    - (b) False

2.2.4.4 F-Test

The F-test is another alternative. A common use of the F-test is together with ANOVA and determines the significance of one variance (or mean sum of squares) in relation to another. Typically, the lack-of-fit is compared with the replicate error. Simply determine the ratio S_{lof}:S_{rep} (e.g. see Table 2.4) and check the size of this number. F-distribution tables are commonly presented at various probability levels, which are often called *confidence levels*. We use a one-tailed F-test in this case as the aim is to see whether one variance is significantly bigger than another, not whether it differs significantly; this differs from the t-test that is two tailed in the application described in Section 2.2.4.3. The columns correspond to the number of degrees of freedom for S_{lof} and the rows to S_{rep} (in the case discussed in here). The table allows to determine the significance of the error (or variance) represented along the columns relative to that represented along the rows. Consider the proposed models for data sets A and B, both excluding the intercept. Locate the relevant number (for a 95% confidence, the lack-of-fit is significant – or for a p value of 0.05, the null hypothesis is rejected, with five degrees of freedom for the lack-of-fit and four degrees of freedom for the replicate error, this number is 6.26, see Table A.5 (given by a distribution often called $F_{(5,4)}$ see Section A.3.5) – hence, an F-ratio must be greater than this value for this level of confidence). Returning to Table 2.4, it is possible to show that the chances of lack-of-fit to a model without an intercept are not very high for the data in Figure 2.9 (ratio = 0.49); however, there is some doubt about the data arising from Figure 2.10 (ratio = 1.79); using the FDIST function in Excel, we can see that the probability is 70.4%, below the 95% confidence the intercept is significant (or in other words with a $p = 0.396$, the null hypothesis is correct, so over one time in three if there were no underlying significance in the intercept, we will still expect to obtain this value of F or more), but still high enough to give us some doubts. Nevertheless, the evidence is not entirely conclusive because the intercept term (2.032) is approximately the same order of magnitude as the replicate error (1.194); for this level of experimental variability, it will never be possible to predict and model the presence of an intercept of this size with a high degree of confidence.

Table 2.12 *F*-ratio for experiment with low experimental error.

Concentration	Absorbance		Model with intercept	Model without intercept
1	3.500	b_0	0.854	n/a
1	3.398	b_1	2.611	2.807
2	6.055			
3	8.691	S_{reg}	0.0307	1.4847
3	8.721	S_{rep}	0.0201	0.0201
4	11.249	S_{lof}	0.0107	1.4646
4	11.389			
5	13.978			
6	16.431			
6	16.527	*F*-ratio	0.531	58.409

The solution is to perform new experiments, perhaps on a different instrument, in which the reproducibility is much greater. Table 2.12 is of such a data set, with essential statistics indicated. Now the *F*-ratio for the lack-of-fit without the intercept becomes 58.409, which is significant at the >99% level (critical value from Table A.4), whereas the lack-of-fit with the intercept included is less than the experimental error.

1. $F_{(6,3)}$ is equal to $F_{(3,6)}$.
   (a) True
   (b) False

2. It is essential to have replicate measurements for both an *F*-test and a *t*-test to determine the significance of terms in a model using the methods described in this chapter.
   (a) Yes.
   (b) No, only for an *F*-test.
   (c) No, only for a *t*-test.
   (d) No, for neither.

2.2.4.5 Normal Probability Plots

For designs where there are no replicates (essential for most uses of the *F*-test) and also where there no degrees of freedom available to assess the lack-of-fit to the data (essential for a *t*-test), other approaches can be employed to examine the significance of coefficients.

As will be discussed in Section 2.3, two-level factorial designs are common, and providing the data are appropriately coded, the size of the coefficients directly relates to their significance. Normally, several terms are calculated, and an aim of modelling is to determine which have significance, the next step, possibly being to then perform another, a more detailed design for quantitative modelling of the significant effects. Often it is convenient to present the coefficients graphically, and a classical approach is to plot them on a normal probability paper. Before the computer age, a large number of different types of statistical graph paper would be available, assisting data interpretation. However, nowadays, it is quite easy to obtain relevant graphs using simple computer packages.

The principle of normal probability plots is that in the absence of any systematic trend, a randomly generated set of numerical values will fall into a normal distribution (see Section A.3.2). Hence, if, for example, we look at the size of seven effects, for example, as assessed by their values of b (provided the data are properly coded and the experiment is well designed, of course), and the effects are simply random, on average, we would expect the probable size of each effect to be normally distributed. In Figure 2.14, seven lines are indicated on the normal distribution curve (the horizontal axis representing the number of standard deviations from the mean) so that the areas between each line equal 1/7 the total area (the areas at the extremes adding up to 1/7 in total). If, however, an effect is very large, it will fall at a very high or low value, so large that it is unlikely to be the consequence of a random process, and is significant. Normal probability plots can be used to rank the coefficients in size (the most negative being the lowest, the most positive being the highest), from the rank determining the expected position in the normal probability plot and then

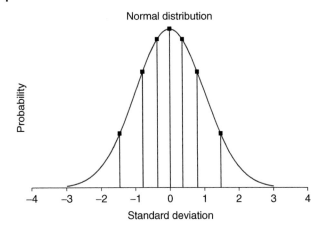

Figure 2.14 Seven lines, equally spaced in area, dividing the normal distribution into eight regions, including six central regions and two extreme regions whose summed area equals those of the central regions.

producing a graph of the coefficients against this likely position. The less significant effects should form approximately on a straight line in the centre of the graph; significant effects will deviate from this line because they are so far from the centre.

Table 2.13 illustrates the calculation for a typical model of the form $y = b_0 + b_1 x_1 + b_2 x_2 + b_3 x_3 + b_{12} x_1 x_2 + b_{13} x_1 x_3 + b_{23} x_2 x_3 + b_{123} x_1 x_2 x_3$.

- Seven possible coefficients are to be assessed for significance. Note that the b_0 coefficient cannot be analysed in this way.
- They are ranked from 1 to 7, where r is the rank.
- Then, the values of $(r - 0.5)/7$ are calculated. This indicates approximately where in the normal distribution each effect is likely to fall. For example, the value for the fourth (or middle ranking) coefficient is 0.5, which means the coefficient might be expected in the centre of the distribution, corresponding to a standard deviation from the mean of 0, as illustrated in Figure 2.14.
- Then, work out how many standard deviations corresponding to the area under the normal curve are calculated as above, using normal distribution tables or standard functions in most data analysis packages. For example, we expect the coefficient of rank 7 to fall approximately at a probability 0.9286 equivalent to 1.465 standard deviations above the mean. See Table A.1 in which 1.46 standard deviations correspond to a probability of 0.92785 (slightly less as the table is not cited to three decimal point accuracy), or use the NORMINV function in Excel.
- Finally, plot the size of the effects against the values obtained in the previous step, to give, for the case discussed, the graph in Figure 2.15. The four central values fall roughly on a straight line, suggesting that only coefficients b_1, b_2 and b_{12}, which deviate from the straight line, are significant.

Like many classical methods of data analysis, the normal probability plot has limitations. It is only useful if there are several factors and clearly will not be much use in the case of two or three factors. It also assumes that a large number of factors are not significant and will not give good results if there are too many significant effects. However,

Table 2.13 Normal probability calculation.

Effect	Coefficient	$(r - 0.5)/7$	Standard deviations
b_1	−6.34	0.0714	−1.465
b_{23}	−0.97	0.2143	−0.792
b_{13}	0.6	0.3571	−0.366
b_{123}	1.36	0.5	0
b_3	2.28	0.6429	0.366
b_{12}	5.89	0.7858	0.792
b_2	13.2	0.9286	1.465

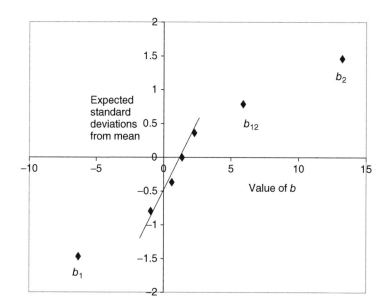

Figure 2.15 Normal probability plot for data in Table 2.13 with significant factors marked.

in certain cases, it can provide useful preliminary graphical information, although probably is not much used in modern computer-based chemometrics.

1. Normal probability plots can be used where there are no replicates and no degrees of freedom to determine lack of fit in a model.
   (a) True
   (b) False

2.2.4.6 Dummy Factors

Another very simple approach is to include one or more dummy factors. These can be built into a design and might, for example, be the colour of shoes worn by the experimenter, some factor that is not likely to have a real effect on the experiment: level −1 might correspond to black shoes and level +1 to brown shoes. Mathematical models can be built including this factor, and effects smaller than this are ignored (remembering as ever to ensure that the coding of the data is sensible).

1. The size of the term arising from the dummy factor is same size as the replicate error.
   (a) True
   (b) False

2.2.4.7 Limitations of Statistical Tests

Although many traditionalists often enjoy the security statistical significance tests give, it is important to recognise that these tests do depend on assumptions about the underlying data that may not be correct, and a chemist should beware of making decisions based only on a probability obtained from a computerised statistical software package, without looking at the data, often graphically. Some typical drawbacks are as follows.

- Most statistical tests assume that the underlying samples and experimental errors are normally distributed. In some cases, this is not so; for example, when analysing some analytical signals, it is unlikely that the noise distribution will be normal; it is often determined by electronics and sometimes even data pre-processing such as the common logarithmic transform used in electronic absorption and infrared spectroscopy. In such circumstances, it may be possible to transform the data first before performing statistical analysis.

- The tests assume that the measurements arise from the same underlying population. Sometimes this is not the case, and systematic factors will come into play. A typical example involves calibration curves. It is well known that the performance of an instrument can vary from day to day. Hence, an absorption coefficient measured on Monday morning is not necessarily the same as the measured coefficient on Tuesday morning; yet, all the coefficients calculated on Monday morning might fall into the same class. If a calibration experiment is performed over several days or even hours, the performance of the instrument may vary and one solution may be to make a very large number of measurements over a long timescale, which may be impractical. Another trick is to randomise the sequence of experiments, but then it may be important to test additional effects such as the day the samples were recorded, just to work out whether this has a significant influence on the experimental data.
- The precision of an instrument must be considered. Many typical measurements, for example, in atomic spectroscopy, are recorded to only two significant figures. Consider a data set in which about 95% of the readings were recorded between 0.10 and 0.30 absorbance units; yet, a statistically designed experiment is used to estimate 64 effects. The t-test provides information on the significance of each effect. However, statistical tests assume that the data are recorded to a high degree of accuracy and will not take this lack of numerical precision into account. For the obvious effects, chemometrics will not be necessary, but for less obvious effects, the statistical conclusions may be invalidated because of the low numerical accuracy in the raw data.

Often it is sufficient simply to look at the size of factors, the significance of the lack-of-fit statistics, perform quite simple ANOVA or produce a few graphs to make quite valid scientific deductions. In most cases, significance testing is used primarily for a preliminary modelling of the data and detailed modelling should be performed after eliminating those factors that are deemed unimportant. It is not necessary to have a very detailed theoretical understanding of statistical significance tests before the design and analysis of chemical experiments, although a conceptual appreciation of, for example, the importance of coding is essential.

1. For a series of measurements, the time they were performed can be included as a factor in the experimental design.
   - **(a)** True
   - **(b)** False

2.2.5 Leverage and Confidence in Models

An important experimental question relates to how well quantitative information can be predicted after a series of experiments has been carried out. For example, if observations have been made between 40 and 80 °C, what can we predict about the experiment at 90 °C? It is traditional to sharply cut off the model outside the experimental region, so that the model is used to predict only within the experimental limits. However, this approach misses much information. The ability to make a prediction often reduces smoothly from the centre of the experiments, being best at 60 °C and worse the farther away from the centre as described in the example above. This does not imply that it is impossible to make any statement about the response at 90 °C, simply that there is less confidence in the prediction than at 80 °C, which, in turn, is predicted less well than at 60 °C. It is important to be able to visualise how the ability to predict a response (e.g. a synthetic yield or a concentration) varies as the independent factors (e.g. pH and temperature) are changed.

When only one factor is involved in the experiment, the predictive ability is often visualised by confidence bands. The 'size' of these confidence bands depends on the magnitude of the experimental error. The 'shape', however, depends on the experimental design and can be obtained from the design matrix (Section 2.2.3) and is influenced by the arrangement of experiments, replication procedure and mathematical model. The concept of *leverage* is used as a measure of such confidence. The mathematical definition is given by

$$H = D(D'D)^{-1}D'$$

where D is the design matrix. This new matrix is sometimes called the *hat matrix* and is a symmetric square matrix with the number of rows and columns equal to the number of experiments. Each experimental point n has a corresponding value of leverage h_{nn} (the diagonal element of the hat matrix) associated with it. Alternatively, the value of leverage for experiment n can be calculated as follows:

$$h_{nn} = d_n(D'D)^{-1}d'_n$$

Table 2.14 Leverage values for a two-factor design and a model of the form.

$$y = b_0 + b_1x_1 + b_2x_2 + b_{11}x_1^2 + b_{22}x_2^2 + b_{12}x_1x_2$$

x_1	x_2	h
1	−1	0.597
1	1	0.597
−1	−1	0.597
−1	1	0.597
1.5	0	0.655
−1.5	0	0.655
0	1.5	0.655
0	−1.5	0.655
0	0	0.248
0	0	0.248
0	0	0.248
0	0	0.248

where d_n is the row of the design matrix corresponding to an individual experiment. The values of leverage for a simple experiment are presented in Table 2.14 for a two-factor design consisting of 12 experiments and a model of the form

$$y = b_0 + b_1x_1 + b_2x_2 + b_{11}x_1^2 + b_{22}x_2^2 + b_{12}x_1x_2$$

The steps in the calculation are as follows:

- Set up the 12×6 design matrix.
- Calculate $(D'D)^{-1}$. Note that this matrix is also used in the t-test as discussed in Section 2.2.4.3.
- Calculate the hat matrix and determine the diagonal values.
- These diagonal values are the values of leverage for each experiment.

This numerical value of leverage has certain properties.

- The value is always ≥ 0.
- The lower the value, the higher the confidence in the prediction. A value of 1 indicates very poor prediction. A value of 0 indicates perfect prediction and will not be achieved.

If there are P coefficients in the model, the sum of the values for leverage over all experimental points adds up to P. Hence, the sum of the values of leverage for the 12 experiments in Table 2.14 is equal to 6.

In the design in Table 2.14, the leverage is lowest at the centre as expected. However, the value of leverage for the first four points is slightly lower than that for the second four points. As discussed in Section 2.4, this design is a form of central composite design, with the points 1–4 corresponding to a factorial design and points 5–8 corresponding to a star design.

Leverage can also be converted to an equation form quite simply by substituting the algebraic expression for the coefficients in the equation

$$h = d(D'D)^{-1}d'$$

where, in the case in Table 2.14

$$d = \begin{pmatrix} 1 & x_1 & x_2 & x_1^2 & x_2^2 & x_1x_2 \end{pmatrix}$$

to give an equation, in this example, of the form

$$h = 0.248 - 0.115(x_1^2 + x_2^2) + 0.132(x_1^4 + x_2^4) + 0.316x_1^2x_2^2$$

The equation can be obtained by summing the appropriate terms in the matrix $(D'D)^{-1}$. This is illustrated graphically in Figure 2.16. Label each row and column by the corresponding terms in the model and then find the combinations of

	1	x_1	x_2	x_1^2	x_2^2	$x_1 x_2$
1	0.248	0	0	−0.116	−0.116	0
x_1	0	0.118	0	0	0	0
x_2	0	0	0.118	0	0	0
x_1^2	−0.116	0	0	0.132	0.033	0
x_2^2	−0.116	0	0	0.033	0.132	0
$x_1 x_2$	0	0	0	0	0	0.25

Figure 2.16 Method of calculating equation for leverage term for the coefficient of $x_1^2 x_2^2$, sum the shaded areas.

terms in the matrix that result in the coefficients of the leverage equation; for $x_1^2 x_2^2$, there are three such combinations so that the term $0.316 = 0.250 + 0.033 + 0.033$.

This equation can also be visualised graphically and used to predict the confidence at any point, not just where experiments were performed. Leverage can also be used to predict the confidence in the prediction under any conditions, which is given by

$$y_\pm = s\sqrt{(F_{(1,N-P)}[1 + h])}$$

where s is the mean squared residual error given by $S_{resid}/(N - P)$ as described in Section 2.2.2 and the F-statistic as introduced in Section 2.2.4.4, which can be obtained at any desired level of confidence but most usually at 95% limits and is one sided. Note that this equation technically refers to the confidence in the individual prediction and there is a slightly different equation for the mean response after replicates have been averaged:

$$y_\pm = s\sqrt{\left(F_{(1,N-P)}\left[\frac{1}{r} + h\right]\right)}$$

See Section 2.2.1 for definitions of N and P; r is the number of times a measurement is obtained at a given point; for example, if we repeat the experiment at 10 mM five times, r equals 5. If measured only once, then this equation equals the first one. Note that, strictly speaking, if a measurement is performed five times at a specific point, the number of replicates is 4; hence, r equals the number of replicates plus 1.

Although the details of these equations may seem esoteric, there are two important considerations.

- The shape of the confidence bands depends entirely on leverage and
- The size depends on the experimental error.

These and most other equations assume that the experimental error is the same over the entire response surface: there is no satisfactory agreement for how to incorporate heteroscedastic errors. Notice that there are several different equations in the literature according to the specific aims of the confidence interval calculations, but for brevity, we introduce only two that can be generally applied to most situations.

To show how leverage can help, consider the example of univariate calibration, three designs A to C (Table 2.15) will be analysed. Each experiment involves performing 11 experiments at five different concentration levels, the only difference being the arrangement of the replicates. The aim is simply to perform linear calibration to produce a model of the form $y = b_0 + b_1 x$, where x is the concentration, and to compare how each design predicts confidence. The leverage can be calculated using the design matrix D that consists of 11 rows (corresponding to each experiment) and two columns (corresponding to each term). The hat matrix consists of 11 rows and 11 columns, the numbers on the diagonal being the values of leverage for each experimental point. The leverage for each experimental point is given in Table 2.15. It is also possible to obtain a graphical representation of the equation as shown in Figure 2.17 for designs A to C, respectively.

What does this tell us?

- Design A contains more replicates at the periphery of the modelling to design B, thus results in a flatter graph. This design will provide predictions that are fairly even throughout the area of interest.

Table 2.15 Leverage for three possible single-variable designs using a two-parameter linear model.

Concentrations			Leverage		
Design A	Design B	Design C	Design A	Design B	Design C
1	1	1	0.234	0.291	0.180
1	1	1	0.234	0.291	0.180
1	2	1	0.234	0.141	0.180
2	2	1	0.127	0.141	0.180
2	3	2	0.127	0.091	0.095
3	3	2	0.091	0.091	0.095
4	3	2	0.127	0.091	0.095
4	4	3	0.127	0.141	0.120
5	4	3	0.234	0.141	0.120
5	5	4	0.234	0.291	0.255
5	5	5	0.234	0.291	0.500

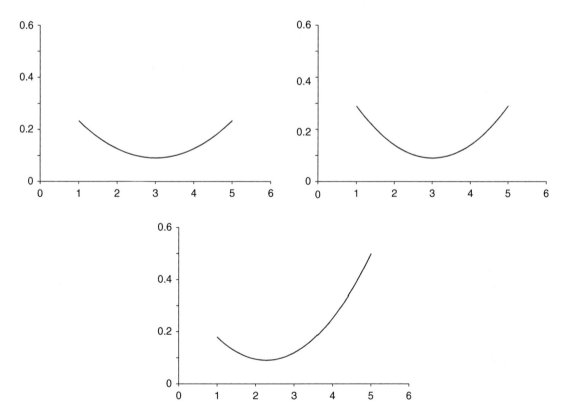

Figure 2.17 Graph of leverage for designs in Table 2.15, from top to bottom, designs A, B and C.

- Design C shows how replication can result in a major change in the shape of the curve for leverage. The asymmetric graph is a result of replication regime. In fact, the best predictions are no longer in the centre of modelling.

This approach can be used for univariate calibration experiments more generally. How many experiments are necessary to produce a given degree of confidence in the prediction? How many replicates are sensible? How good is the prediction outside the region of modelling? How do different experimental arrangements relate? In order to get a value for the confidence of predictions, it is also necessary, of course, to determine the experimental error, but this together with the leverage, which is a direct consequence of the design and model, is sufficient information. Note that leverage will change if the model changes.

Leverage is most powerful as a tool when several factors are to be studied. There is no general agreement to define an experimental space under such circumstances. Consider the simple design in Figure 2.18, consisting of five experiments. Where does the experimental boundary stop? The range of concentrations for the first compound is 0.5–0.9 mM and for the second compound 0.2–0.4 mM. Does this mean we can predict the response well when the concentrations of the two compounds are at 0.9 and 0.4 mM, respectively? Probably not, some people would argue that the experimental region is a circle, not a square. For this nice symmetric arrangement of experiments, it is possible to envisage an experimental region, but imagine telling the laboratory worker that if the concentration of the second compound is 0.34 mM and if the concentration of the first compound is 0.77 mM, the experiment is within the region, whereas if it is 0.80 mM, it is outside the region. There will be confusion as to where the model starts and stops. For some supposedly simple designs such as a full factorial design, the definition of the experimental region is even harder to conceive.

So the best solution is to produce a simple graph as to how confidence in the prediction varies over the experimental region. Consider the two designs in Figure 2.19. Using a very simple linear model, of the form $y = b_1 x_1 + b_2 x_2$, the leverage for both designs is illustrated in Figure 2.19. The consequence of the different experimental arrangements is now quite obvious, and the result of changing the design on the confidence in predictions can be seen. Although a two-factor example is fairly straightforward, for multi-factor designs (e.g. mixtures of several compounds), it is quite hard to produce an arrangement of samples in which there is symmetric confidence in the results over the experimental domain.

Leverage can show the effect of changing an experimental parameter such as the number of replicates, or, in the case of central composite design, the position of the axial points (see Section 2.4). Some interesting features emerge from this analysis. For example, confidence is not always highest in the centre of modelling but also depends on the number of replicates. The method in this section is an important tool for visualising how changing design relates to the ability to make quantitative predictions.

1. A design consists of 18 experiments and is modelled using 10 terms. The leverage is calculated over all experimental points and is summed. This sum equals
   (a) 18
   (b) 10
   (c) 8

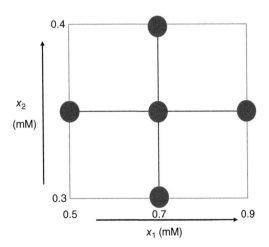

Figure 2.18 Two-factor design consisting of five experiments.

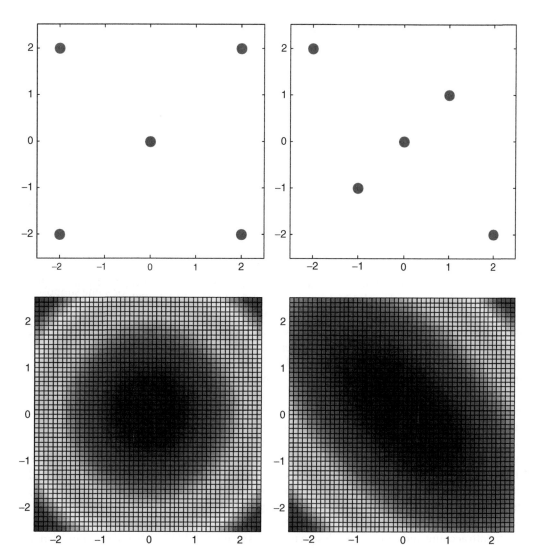

Figure 2.19 Two experimental arrangements together with the corresponding leverage for a linear model.

2. The hat matrix is always a diagonal matrix
   (a) True
   (b) False

3. If we know how far an experiment is from the centre of a design and the mathematical model used, we can then calculate leverage.
   (a) True
   (b) False

2.3 Factorial Designs

In this and the remaining sections of this chapter, we will introduce a number of possible designs, which can be understood using the building blocks introduced in Section 2.2. Factorial designs are some of the simplest, often used for screening or when there are a large number of possible factors. As will be seen, they have limitations but are the easiest to understand. Many designs are presented as a set of rules that provide the experimenter with a list of conditions, below we will present the rules for many common methods.

2.3.1 Full Factorial Designs

Full factorial designs at two levels are mainly used for screening, which is to determine the influence of a number of effects on a response and to eliminate those that are not significant, the next stage being to undertake a more detailed study. Sometimes where detailed predictions are not required, the information from factorial designs is adequate, at least in situations where the aim is fairly qualitative (e.g. to improve the yield of a reaction rather than obtain a highly accurate rate dependence that is then interpreted in fundamental molecular terms).

Consider a chemical reaction whose performance is known to depend on pH and temperature, including their interaction. A set of experiments can be proposed to study these two factors, each at two levels (e.g. two temperatures), using a two-level, two-factor experimental design. The number of experiments is given by $N = l^k$, where l is the number of levels (=2) and k the number of factors (=2); hence, in this case, N equals 4. For three factors, the number of experiments will equal 8 and so on, provided the design is performed only at two levels. The following stages are used to construct the design and interpret the results.

- The first step is to choose a high and low level for each factor; for example, if temperature is factor 1, use 30 and 60 °C, and if pH is factor 2, use 4 and 6. This choice will require the experimenter's judgment to determine sensible ranges.
- The next step is to use a standard design. The value of each factor is usually coded (see Section 2.2.4.1) as '−' (low) or '+' (high). Note that some authors use −1 and +1 or even 1 and 2 for low and high. When reading different texts, do not get confused; always first understand what notation has been employed. There is no universally agreed convention for coding; however, design matrices that are symmetric around 0 are almost always easier to handle computationally. There are four possible unique sets of experimental conditions that can be represented as a table analogous to four binary numbers 00 (−−), 01 (−+), 10 (+−) and 11 (++), which can be related to a set of physical conditions. For example, the set of experiments represented by −+ involve performing an experiment at 30 °C and pH 6 in our case.
- Next, perform the experiments and obtain the response under each set of experimental conditions. Table 2.16 illustrates the coded and true set of experimental conditions plus the corresponding response, which might, for example, be the percentage of a by-product, the lower the better. Something immediately appears strange from these results. Although it is obvious that the higher the temperature, the higher the percentage by-product, there does not at first seem to be any consistent trend as far as pH is concerned. Provided the experimental results were recorded correctly, this suggests that there must be an interaction between temperature and pH. At a lower temperature, the percentage decreases with an increase in pH, but the opposite is observed at a higher temperature. How can we interpret this?
- The next step, of course, is to analyse the data, by constructing a design matrix (Section 2.2.3). We know that an interaction term must be taken into account and can set up a design matrix as given in Table 2.17 based on a model of the

Table 2.16 Coding of a simple two factor, two level design and corresponding responses.

Experiment number	Factor 1	Factor 2	Temperature	pH	Response
1	−	−	30	4	12
2	−	+	30	6	10
3	+	−	60	4	24
4	+	+	60	6	25

Table 2.17 Design matrix.

Intercept	Temperature	pH	Temp. × pH
1	30	4	120
1	30	6	180
1	60	4	240
1	60	6	360

1	x_1	x_2	$x_1 x_2$
+	−	−	−
+	−	+	−
+	+	−	+
+	+	+	+

form $y = b_0 + b_1 x_1 + b_2 x_2 + b_{11} x_1 x_2$. This can be expressed either as a function of the true or coded concentrations, but, as discussed in Section 2.2.4.1, is probably best as coded values. Note that four possible coefficients can be obtained from the four experiments. Note also that each of the columns in Table 2.17 is different. This is an important and crucial property and allows each of the four possible terms to be distinguished uniquely from one another and is called *orthogonality*. Observe, also, that quadratic terms in the model are impossible because four experiments can be used to obtain only a maximum of four terms, and also the experiments are performed at only two levels: ways of introducing such terms will be described in Section 2.4.

- Calculate the coefficients. It is not necessary to employ specialist statistical software for this. In matrix terms, the response can be given by $y = D\,b$ where b is a vector of the four coefficients and D is presented in Table 2.17. Simply use the matrix inverse so that $b = D^{-1} y$, as the design matrix is a square matrix. Note that there are no replicates and the model will exactly fit the data. The parameters are presented below.
  - For raw values
    Intercept $= 10$; temperature coefficient $= 0.2$; pH coefficient $= -2.5$; interaction coefficient $= 0.05$.
  - For coded values
    Intercept $= 17.5$; temperature coefficient $= 6.75$; pH coefficient $= -0.25$; interaction coefficient $= 0.75$,
- Finally, interpret the coefficients. Note that for the raw values, it appears that pH is much more influential than temperature, also that the interaction is very small. In addition, the intercept term is not the average of the four readings. The reason why this happens is that the intercept is the predicted response at pH 0 and 0 °C, conditions unlikely to be reached experimentally. The interaction term appears very small because units used for temperature correspond to a range of 30 degrees as opposed to a pH range of 2. A better measure of significance comes from the coded coefficients. The effect of temperature is overwhelming. Changing pH has a very small influence, which is less than the interaction between the two factors, explaining why the response is higher at pH 4 when the reaction is studied at 30 °C, but the opposite is true at 60 °C.

Two-level full factorial designs (also sometimes called *saturated factorial designs*), as presented in this section, take into account all linear terms and all possible k way interactions. The number of different types of terms can be calculated by the binomial theorem (given by $k!/[(k - m)!m!]$ for mth order interactions and k factors; for example, there are 6 two-factor ($=m$) interactions for a full four-factor design ($=k$)): for interested readers, these numbers can be calculated by the binomial theorem or Pascal's triangle. Hence, for a four-factor, two-level design, there will be 16 experiments, the response being described by an equation with a total of 16 terms of which

- there is one interaction term,
- four linear terms such as b_1,
- six two-factor interaction terms such as $b_1 b_2$,
- four three-factor interactions terms such as $b_1 b_2\, b_3$ and
- one four-factor interaction term $b_1 b_2 b_3 b_4$.

The coded experimental conditions are given in Table 2.18(a) and the corresponding design matrix in Table 2.18(b). In common with the generally accepted convention, a '+' symbol is employed for a high level and a '−' symbol for a low level. The values of the interactions are obtained simply by multiplying the levels of the individual factors together. For example, the value of $x_1 x_2 x_4$ for experiment 2 is '+' as it is a product of '−' × '−' × '+'. Several important features should be noted.

- Every column (apart from the intercept) contains exactly eight high and eight low levels. This property is called *balance*.
- Apart from the first column, each of the other possible pairs of columns have the property that for each experiment at level '+' for one column, there are equal number of experiments for all the other columns at levels '+' and '−'. Figure 2.20 is the graph of the level of any one column (apart from the first) plotted against the level of any other column. For any combination of columns 2–16, this graph will be identical and is a key feature of the design. It relates to *orthogonality*, which is discussed in other contexts throughout this book. Some chemometricians regard each column as a vector in space, so that any two vectors are at right angles to each other. Algebraically, the *correlation coefficient* between each pair of columns equals 0. Why is this so important? Consider a case in which the values of two factors (or indeed any two columns) are related as in Table 2.19. In this case, every time the first factor is at a high level, the second is at a low level and *vice versa*. So, for example, every time a reaction is performed at pH 4, it is also performed at 60 °C and every time it is performed at pH 6, it is also performed at 30 °C, how can an effect due

Table 2.18 Four-factor, two-level full factorial design.

(a) Coded experimental conditions

Experiment	Factor 1	Factor 2	Factor 3	Factor 4
1	−	−	−	−
2	−	−	−	+
3	−	−	+	−
4	−	−	+	+
5	−	+	−	−
6	−	+	−	+
7	−	+	+	−
8	−	+	+	+
9	+	−	−	−
10	+	−	−	+
11	+	−	+	−
12	+	−	+	+
13	+	+	−	−
14	+	+	−	+
15	+	+	+	−
16	+	+	+	+

(b) Design matrix

x_0	x_1	x_2	x_3	x_4	x_1x_2	x_1x_3	x_1x_4	x_2x_3	x_2x_4	x_3x_4	$x_1x_2x_3$	$x_1x_2x_4$	$x_1x_3x_4$	$x_2x_3x_4$	$x_1x_2x_3x_4$
+	−	−	−	−	+	+	+	+	+	+	−	−	−	−	+
+	−	−	−	+	+	+	−	+	−	−	−	+	+	+	−
+	−	−	+	−	+	−	+	−	+	−	+	−	+	+	−
+	−	−	+	+	+	−	−	−	−	+	+	+	−	−	+
+	−	+	−	−	−	+	+	−	−	+	+	+	−	+	−
+	−	+	−	+	−	+	−	−	+	−	+	−	+	−	+
+	−	+	+	−	−	−	+	+	−	−	−	+	+	−	+
+	−	+	+	+	−	−	−	+	+	+	−	−	−	+	−
+	+	−	−	−	−	−	−	+	+	+	+	+	+	−	−
+	+	−	−	+	−	−	+	+	−	−	+	−	−	+	+
+	+	−	+	−	−	+	−	−	+	−	−	+	−	+	+
+	+	−	+	+	−	+	+	−	−	+	−	−	+	−	−
+	+	+	−	−	+	−	−	−	−	+	−	−	+	+	+
+	+	+	−	+	+	−	+	−	+	−	−	+	−	−	−
+	+	+	+	−	+	+	−	+	−	−	+	−	−	−	−
+	+	+	+	+	+	+	+	+	+	+	+	+	+	+	+

to increase in temperature be distinguished from an effect due to decrease in pH? It is impossible. The two factors are correlated. The only way to be completely sure that the influence of each effect is completely independent is to ensure that the columns are orthogonal, which is not correlated.

- The other remarkable property is that the inverse of the design matrix is related to the transpose by

$$\boldsymbol{D}^{-1} = (1/N)\boldsymbol{D'}$$

where there are N experiments. This a general feature of all full two-level designs (also called *saturated designs*) and results in an interesting classical approach for determining the size of each effect. Using modern matrix notation,

Figure 2.20 Graph of levels of one term against another in the design in Table 2.18.

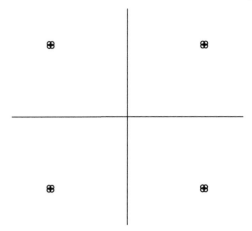

Table 2.19 Correlated factors.

−	+
+	−
−	+
+	−
−	+
+	−
−	+
+	−
−	+
+	−
−	+
+	−
−	+
+	−
−	+
+	−

the simplest method is simply to calculate

$$b = D^{-1}y$$

but some classical texts use an algorithm involving multiplying the response by each column (of coded coefficients) and dividing by the number of experiments to get the value of the size of each factor. For the example in Table 2.16, the value of the effect due to temperature can be given by

$$b_1 = (-1 \times 12 - 1 \times 10 + 1 \times 24 + 1 \times 25)/4 = 6.75$$

identical to that obtained by simple matrix manipulations. Such method for determining the size of the effects was extremely useful to statisticians before matrix-oriented software, and still occasionally used, but is limited only to certain very specific designs. It is also important to recognise that some texts divide the expression above by $N/2$ rather than N, making the classical numerical value of the effects equal to twice those obtained by regression. As long as all the effects are on the same scale, it does not matter which method is employed when comparing the size of each factor.

An important additional advantage of two-level factorial designs is that some factors can be 'categorical' in nature; that is, they do not need to refer to a quantitative parameter. For example, one factor may be whether a reaction mixture is stirred ('+' level) or not ('−' level), another factor may be whether it is carried out under nitrogen or not. Thus, these designs can be used to ask qualitative questions. The values of the b parameters relate to the significance or importance of these factors and their interactions.

Two-level factorial designs can be used very effectively for screening, but also have pitfalls.

- They only provide an approximation within the experimental range. Note that for the model above, it is possible to obtain nonsensical predictions of negative percentage yields outside the experimental region.
- They cannot take quadratic terms into account, as the experiments are performed only at two levels.
- There is no replicate information.
- If all possible interaction terms are taken into account, no error can be estimated, the F-test and t-test not being applicable. However, if it is known that some interactions are unlikely or irrelevant, it is possible to model only the most important factors. For example, in the case of the design in Table 2.18, it might be decided to model only the intercept, four single-factor and six two-factor interaction terms, making 11 terms in total, and to ignore the five remaining higher order interactions. Hence,
  - $N = 16$
  - $P = 11$
  - $(N - P) = 5$ degrees of freedom remain to determine the fit to the model.

Some valuable information about the importance of each term can be obtained under such circumstances. Note, however, the design matrix is no longer square, and it is not possible to use the simple approaches mentioned above to calculate the effects, regression using the pseudo-inverse is necessary. This concept is discussed in more detail in Section A.1.2.5. Instead of calculating $\boldsymbol{b} = \boldsymbol{D}^{-1}\boldsymbol{y}$, because $\boldsymbol{D}$ is no longer a square matrix, it cannot have an inverse; hence, the pseudo-inverse $\boldsymbol{D}^{+}$ is defined so that $\boldsymbol{b} = \boldsymbol{D}^{+}\boldsymbol{y} = (\boldsymbol{D}'\boldsymbol{D})^{-1}\boldsymbol{D}'\boldsymbol{y}$.

Two-level factorial designs remain largely popular because they are extremely easy to set up and understand; in addition, calculation of the coefficients is very straightforward. One of the problems is that once there are a significant number of factors involved, it is necessary to perform a large number of experiments: for six factors, $2^6 = 64$ experiments are required. The 'extra experiments' really only provide information about the higher order interactions. It is debatable whether a six-factor, or even four-factor, interaction is really meaningful or even observable. For example, if an extraction procedure is to be studied as a function of (a) whether an enzyme is present or not, (b) incubation time, (c) incubation temperature, (d) type of filter, (e) pH and (f) concentration, what meaning will be attached to the higher order interactions, and even if they are present can they be measured with any form of confidence? And is it economically practicable or sensible to spend such a huge effort studying these interactions? Information such as squared terms are not available; hence, neither detailed models of the extraction behaviour are not available, nor is any replicate information being gathered. Two possible enhancements are as follows. If it is desired to reduce the number of experiments by neglecting some of the higher order terms, use designs discussed in Sections 2.3.2 and 2.3.3. If it is desired to study squared or higher order terms whilst reducing the number of experiments, use the designs discussed in Section 2.4.

Sometimes it is not sufficient to study an experiment at two levels. For example, is it really sufficient to study only two levels? A more detailed model will be obtained using three temperatures; in addition, such designs will allow either the use of squared terms or, if only linear terms used in the model, there will be some degrees of freedom to assess goodness-of-fit. Three- and four-level designs for two factors are presented in Figure 2.21 with the values of the coded experimental conditions in Table 2.20. Note that the levels are coded to be symmetrical around 0 so that each level differs by 1 from the next, although there are no universally agreed conventions. These designs are called *multi-level factorial designs*. The number of experiments could become very large if there are several factors; for example, a five-factor design at three levels involves 3^5 or 243 experiments. In Section 2.3.4, we will discuss how to reduce the size safely and in a systematic manner.

1. A full two-level factorial design involving four factors consists of the following number of experiments.
    (a) 8
    (b) 16
    (c) 32

2. A full two-level, five factor design can be modelled using the following number of two-factor interaction terms.
    (a) 5
    (b) 6
    (c) 10

Figure 2.21 Three- and four-level full factorial designs.

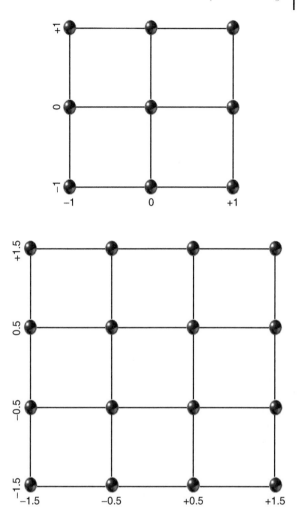

3. It is desired to form a two-level, five-factor full factorial design and model the data using just the intercept, linear and two-factor interaction terms. How many degrees of freedom are available to determine fit to the model?

   (a) 32
   (b) 16
   (c) 10

2.3.2 Fractional Factorial Designs

A weakness of full factorial designs is the large number of experiments that must be performed. For example, for a 10-factor design at two levels, 1024 experiments are required, which may be impracticable. The large number of experiments does not always result in useful or interesting extra information; hence, performing all these result in waste of time and resources. Especially in the case of screening, where a large number of factors may be of potential interest, it is inefficient to run so many experiments in the first instance. There are numerous tricks to reduce the number of experiments.

Consider a three-factor, two-level design. Eight experiments are listed in Table 2.21, the conditions being coded as usual. Figure 2.22 is a symbolic representation of the experiments, often presented on the corners of a cube, whose axes correspond to each factor. The design matrix for all the possible coefficients that can be set up as is also presented in Table 2.21 and consists of eight possible columns, equal to the number of experiments. Some columns represent interactions, such as the three-factor interaction, that may not be very significant. At first screening, we may primarily wish to say whether the three factors have any real influence on the response, not to study the model in detail. In a more complex situation, we may wish to screen 10 possible factors; hence, reducing the number of factors to be studied in

Table 2.20 Full factorial designs corresponding to Figure 2.21.

(a) Three levels

−1	−1
−1	0
−1	+1
0	−1
0	0
0	+1
+1	−1
+1	0
+1	+1

(b) Four levels

−1.5	−1.5
−1.5	−0.5
−1.5	+0.5
−1.5	+1.5
−0.5	−1.5
−0.5	−0.5
−0.5	+0.5
−0.5	+1.5
+0.5	−1.5
+0.5	−0.5
+0.5	+0.5
+0.5	+1.5
+1.5	−1.5
+1.5	−0.5
+1.5	+0.5
+1.5	+1.5

Table 2.21 Full factorial design for three factors together with the design matrix.

	Experiments			Design matrix							
	Factor 1	Factor 2	Factor 3	x_0	x_1	x_2	x_3	x_1x_2	x_1x_3	x_2x_3	$x_1x_2x_3$
1	+	+	+	+	+	+	+	+	+	+	+
2	+	+	−	+	+	+	−	+	−	−	−
3	+	−	+	+	+	−	+	−	+	−	−
4	+	−	−	+	+	−	−	−	−	+	+
5	−	+	+	+	−	+	+	−	−	+	−
6	−	+	−	+	−	+	−	−	+	−	+
7	−	−	+	+	−	−	+	+	−	−	+
8	−	−	−	+	−	−	−	+	+	+	−

Figure 2.22 Representation of a three-factor, two-level design.

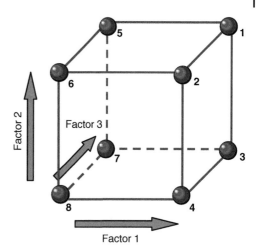

detail to three or four makes the next stage of modelling easier. If one factor clearly has little or no influence on the outcome of a process, there is little point on investing a lot of effort studying its effect, and we want at a first stage simply to eliminate it to save us time.

How can we reduce the number of experiments safely and systematically? Two-level fractional factorial designs are used to reduce the number of experiments by ½, ¼, ⅛ and so on. Can we halve the number of experiments? At first glance, a simple approach might be to take the first four experiments in Table 2.21. However, these would leave the level of the first factor at +1 throughout. A problem is that we now no longer study the variation of this factor; hence, we do not obtain any information on how factor 1 influences the response and are studying the wrong type of variation. In fact, such design would remove all the four terms from the model including the first factor, leaving the intercept, two single-factor terms and the interaction between factors 2 and 3, not the hoped for information, unless we know that factor 1 and its interactions are not significant.

Can a subset of four experiments be selected that allows us to study all three factors? Rules have been developed to produce these fractional factorial designs obtained by taking the correct subset of the original experiments. Table 2.22 illustrates a possible fractional factorial design that enables all factors to be studied. There are a number of important features.

- Every column in the experimental matrix is different.
- In each column, there are an equal number of '−' and '+' levels.
- For each experiment at level '+' for factor 1, there are equal number of experiments for factors 2 and 3, which are at levels '+' and '−', and the columns are orthogonal.

The properties of this design can be better understood by visualisation (Figure 2.23): half the experiments have been removed. For the remainder, each face of the cube now corresponds to two rather than four experiments, and every alternate corner corresponds to an experiment. From the figure, we can see that there are in fact two possible ways (in this case) for reducing the number of experiments from eight to four, whilst still studying each factor.

The matrix of effects in Table 2.22 is also interesting. The first four columns can be used to represent the new design matrix, whereas the last four columns represent the interactions. Although the first four columns are all different, the last four columns each correspond to one of the first four columns. For example, the x_1x_2 column exactly equals the

Table 2.22 Fractional factorial design.

	Experiments			Matrix of effects							
	Factor 1	Factor 2	Factor 3	x_0	x_1	x_2	x_3	x_1x_2	x_1x_3	x_2x_3	$x_1x_2x_3$
1	+	+	+	+	+	+	+	+	+	+	+
2	+	−	−	+	+	−	−	−	−	+	+
3	−	−	+	+	−	−	+	+	−	−	+
4	−	+	−	+	−	+	−	−	+	−	+

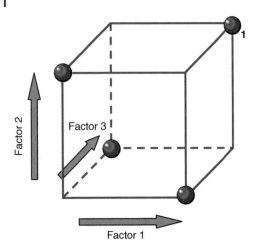

Factor 2

Factor 3

Factor 1

Figure 2.23 Fractional factorial design.

x_3 column. What does this imply in reality? As the number of experiments is reduced, the amount of information is correspondingly reduced. As only four experiments are now performed, it is only possible to measure four unique factors. The interaction between factors 1 and 2 is said to be *confounded* with factor 3. This might mean, for example, that using this design, the interaction between temperature and pH is indistinguishable from the influence of concentration alone. However, not all interactions will be significant, and the purpose of a preliminary experiment is often simply to sort out which main factors should be studied in detail later. When calculating the effects, it is important to use only four unique columns in the design matrix, rather than all eight columns, as otherwise the design matrix will not have a (pseudo)inverse. In simple terms, if we perform four experiments and measure one response, we cannot use more than four coefficients in a model.

Note that two-level fractional factorial designs can only be constructed when the number of experiments equals a power of 2. In order to determine the minimum number of experiments, do as follows.

- Determine how many terms are interesting.
- Then, construct a design whose size is the next greatest power of 2.

Setting up a fractional factorial design and determining which terms are confounded are relatively straightforward and will be illustrated with reference to a five-factor design.

A half factorial design involves reducing the experiments from 2^k to 2^{k-1}, or, in this case, from 32 to 16.

- In most cases, the aim is to
    - confound k factor interactions with the intercept,
    - $(k-1)$ factor interactions with single-factor terms,
    - up to $(k-1)/2$ factor interactions with $(k-1)/2+1$ factor interactions if the number of factors is odd, or $k/2$ factor interactions with themselves if the number of factors is even and
    - that is, for five factors, confound zero factor interactions (intercept) with 5, one-factor terms (pure variables) with 4 and two-factor interactions with three-factor interactions, and for six factors, confound 0 with five-factor, 1 with five-factor, 2 with four-factor interactions, and three-factor interactions with themselves.
- Set up a $k-1$ factor design for the first $k-1$ factors, that is, a four-factor design consisting of 16 experiments in our case.
- Confound the kth (or final) factor with the product of the other factors by setting the final column as either $-$ or $+$ the product of the other factors. A simple notation is often used to define these designs, whereby the final column is given by $k = +1*2* \ldots * (k-1)$ or $k = -1*2* \ldots * (k-1)$. The case where $5 = +1*2*3*4$ is illustrated in Table 2.23, where **1** is the value ($+1$ or -1) of factor 1 and so on. This means that a four-factor interaction (most unlikely to have any physical meaning) is confounded with the fifth factor. There are, in fact, only two different types of half factorial design with the properties mentioned above. Each design is defined by how the intercept (**I**) is confounded, and it is easy to show that this design is of the type $\mathbf{I} = +1*2*3*4*5$, the other possible design being of type $\mathbf{I} = -1*2*3*4*5$. Table 2.23 is, therefore, one possible half factorial design for five factors at two levels.
- It is possible to work out which of the other (interaction) terms are confounded with each other, either by multiplying the columns of the design together or from first principles as follows. Every column multiplied by itself will result in a column of $+$'s. or **I**, as the square of either -1 or $+1$ is always $+1$. Each term will be confounded with another term in

Table 2.23 Confounding factor 5 with the product of factors 1–4.

Factor 1	Factor 2	Factor 3	Factor 4	Factor 5
1	2	3	4	+1*2*3*4
−	−	−	−	+
−	−	−	+	−
−	−	+	−	−
−	−	+	+	+
−	+	−	−	−
−	+	−	+	+
−	+	+	−	+
−	+	+	+	−
+	−	−	−	−
+	−	−	+	+
+	−	+	−	+
+	−	+	+	−
+	+	−	−	+
+	+	−	+	−
+	+	+	−	−
+	+	+	+	+

Table 2.24 Confounding interaction terms in design in Table 2.23.

I	+1*2*3*4*5
1	+2*3*4*5
2	+1*3*4*5
3	+1*2*4*5
4	+1*2*3*5
5	+1*2*3*4
1*2	+3*4*5
1*3	+2*4*5
1*4	+2*3*5
1*5	+2*3*4
2*3	+1*4*5
2*4	+1*3*5
2*5	+1*3*4
3*4	+1*2*5
3*5	+1*2*4
4*5	+1*2*3

this particular design. To demonstrate which term **1*2*3** is confounded with, simply multiply **5** by **4** as **5 = 1*2*3*4**, hence **5*4 = 1*2*3*4*4 = 1*2*3** as **4*4** equals **I**. These interactions for the design in Table 2.23 are presented in Table 2.24.

- In the case of negative numbers, ignore the negative sign. If two terms are correlated, regardless if the correlation coefficient is positive or negative, they cannot be distinguished. In practical terms, this implies that if one term increases, the other decreases.
- From the table, we can see that it is possible to model the intercept, one-factor and two-factor terms whilst confounding with the three-, four- and five-factor terms.

A smaller factorial design can be constructed as follows.

- For a 2^{-f} fractional factorial, first set up a full factorial design consisting of 2^{k-f} experiments for the first k-f factors; that is, for a quarter ($f = 2$) of a 5 ($=k$) factorial experiment, set up a design consisting of eight experiments for the first three factors.
- Determine the lowest order interaction that must be confounded. For a quarter of a five factorial design, second-order interactions must be confounded. Then, almost arbitrarily (unless there are good reasons for specific interactions to be confounded) set up the last two columns as products (times '−' or '+') of combinations of the other columns, with the provision that the products must include at least as many terms as the lowest order interaction to be confounded. Therefore, for our example, any two-factor (or higher) interaction is entirely valid. In Table 2.25, a quarter factorial design where **4 = −1*2** and **5 = 1*2*3** is presented.
- Confounding can be analysed as above, but now each term will be confounded with three other terms for a quarter factorial design (or seven other terms for an eighth factorial design).

In more complex situations, such as 10-factor experiments, it is unlikely that there will be any physical interpretation of higher order interactions or at least that these interactions are not measurable. Therefore, it is possible to select specific interactions that are unlikely to be of interest and consciously reduce the number of experiments in a systematic manner by confounding these with lower order interactions.

There are obvious advantages in two-level fractional factorial designs, but these do have some drawbacks.

- There are no quadratic terms, as the experiments are performed only at two levels.
- There are no replicates.
- The number of experiments must be a power of 2.

Nevertheless, this approach is very popular in many exploratory situations and has the additional advantage that the data are easy to analyse. It is important to recognise, however, that the use of statistical experimental designs has a long history, and a major influence on the minds of early experimentalists and statisticians has always been ease of calculation. Sometimes extra experiments are performed simply to produce a design that could be readily analysed using pencil and paper. It cannot be over-stressed that inverse matrices were very difficult to calculate manually, but modern computers now remove this difficulty.

1. A quarter factorial, six-factor, two-level design consists of the following number of experiments.
   (a) 32
   (b) 16
   (c) 8

2. For a half factorial, seven-factor, two-level design, two-factor interactions are confounded with the following.
   (a) six-factor interactions
   (b) five-factor interactions
   (c) four-factor interactions

Table 2.25 Quarter factorial design.

Factor 1	Factor 2	Factor 3	Factor 4	Factor 5
1	2	3	−1*2	1*2*3
−	−	−	−	−
−	−	+	−	+
−	+	−	+	+
−	+	+	+	−
+	−	−	+	+
+	−	+	+	−
+	+	−	−	−
+	+	+	−	+

2.3.3 Plackett–Burman and Taguchi Designs

Where the number of factors is quite large, the constraint that the number of experiments must be equal to a power of 2 can be rather restrictive. As the number of experiments must always exceed the number of factors by at least one (to take into account the intercept), this would mean, for example, that 32 experiments are required for the study of 19 factors and 64 experiments for the study of 43 factors. In order to overcome this problem and reduce the number of experiments, other approaches are needed.

Plackett and Burman published their classic paper in 1946, which has been much cited by chemists. Their work originated from the need for wartime testing of components in equipment manufacture. A large number of factors influenced the quality of these components; thus, efficient procedures were required for screening. They proposed a number of two-level factorial designs, where the number of experiments is a multiple of 4. Hence, designs exist for $N = 4$, 8, 12, 16, 20, 24 and so on experiments. The number of experiments exceeds the number of factors, k, by 1, hence available for factors 3, 7, 11 and so on.

One such design is given in Table 2.26 for 11 factors and 12 experiments and has various features.

- In the first row, all factors are at the same level, in this case at the low (−) level.
- The first column from rows 2 to k is called a *generator*. The key to the design is that there are only certain allowed generators that can be obtained from tables. Note that the number of factors will always be an odd number equal to $k = 4m - 1$ (or 11 in this case), where m is any integer. If the first row consists of '−', the generator will consist of $2m$ (=6 in this case) experiments at '+' level and $2m - 1$ (=5 in this case) at '−' level, the reverse being true if the first row is at the '+' level. In Table 2.26, the generator is $+ + - + + + + - - - + -$, as outlined by a box.
- The next $4m - 2$ (=10) columns are generated from the first column simply by shifting the down cells by one row. This is indicated by diagonal arrows in the table. Notice that experiment 1 is not included in this procedure.
- The level of factor j in experiment (or row) 2 equals to the level of this factor in the row k for factor $j - 1$. For example, the level of factor 2 in experiment 2 equals the level of factor 1 in experiment 12.

There are as many high as low levels of each factor over the 12 experiments, as would be expected. The most important property of the design, however, is *orthogonality*. Consider the relationship between factors 1 and 2.

- There are six instances in which factor 1 is at a high level, and factor 6 at a low level.
- For each of the six instances at which factor 1 is at a high level, factor 2 is at a high level in three cases and it is at a low level in the other three cases. A similar relationship exists where factor 1 is at a low level. This implies that the factors are orthogonal or uncorrelated, an important condition for a good design.
- Any combination of two factors is related in a similar way.

Only certain generators possess all these properties; hence, it is important to use only known generators.

Standard Plackett–Burman designs exist for 7, 11, 15, 19 and 23 factors; generators are given in Table 2.27. A few of the designs for more than 23 factors are slightly more complicated to construct, for which readers should refer to their paper of 1946. Note that for 7 and 15 factors, it is also possible to use fractional factorial designs as discussed in Section 2.3.2. However, in the old adage, 'all roads lead to Rome'; in fact, fractional factorial and Plackett–Burman designs are equivalent, the difference simply being in the way the experiments and factors are organised in the data

Table 2.26 A Plackett–Burman design for 11 factors, generator outlined by a box.

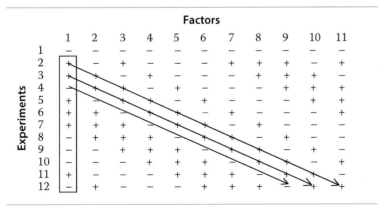

Table 2.27 Generators for Plackett–Burman design, first row is at − level.

Factors	Generator
7	+ + + − + − −
11	+ + − + + + − − − + −
15	+ + + + − + − + + − − + − − −
19	+ + − + + + + − + − + − − − − + + −
23	+ + + + + − + − + + − − + + − − + − + − − − −

Table 2.28 Equivalence of Plackett–Burman and fractional factorial designs for seven factors, the arrows showing how the rows are related.

Plackett-Burman design								1	2	3	4 = −1*3	5 = 1*2*3	6 = −1*2	7 = −2*3
−	−	−	−	−	−	−		−	−	−	−	−	−	−
+	+	+	−	+	−	−		−	−	+	+	+	−	+
−	+	+	+	−	+	−		−	+	−	−	+	+	+
−	−	+	+	+	−	+		−	+	+	+	−	+	−
+	−	−	+	+	+	−		+	−	−	+	+	+	−
−	+	−	−	+	+	+		+	−	+	−	−	+	+
+	−	+	−	−	+	+		+	+	−	+	−	−	+
+	+	−	+	−	−	+		+	+	+	−	+	−	−

table. In reality, it should make no difference in which order the experiments are performed (in fact, it is best that the experiments are run in a randomised order) and the factors can be represented in any order along the columns. Table 2.28 shows that for seven factors, a Plackett–Burman design is the same as a sixteenth factorial ($=2^{7-4}=8$ experiments), after rearranging the rows, as indicated by the arrows. The confounding of the factorial terms is also indicated. It does not really matter which approach is employed.

If the number of experimental factors is less than that of a standard design (a multiple of 4 minus 1), the additional factors can be set up as *dummy* ones. Hence, if there are only 10 real factors, use an 11-factor design, the final factor being a dummy one: this may be a variable that has no effect on the experiment, such as the technician that handed out the glassware or the colour of laboratory furniture.

If the intercept term is included, the design matrix is a square matrix; hence, the coefficients for each factor are given by

$$b = D^{-1}y$$

Provided coded values are used throughout, as there are no interactions or squared terms, the size of the coefficients are directly related to their importance. An alternative method of calculation is to multiply the response by each column, dividing by the number of experiments as in normal full factorial designs:

$$b_j = \sum_{i=1}^{N} x_{ij} y_i / N$$

where x_{ij} is a number equal to +1 or −1 according to the value in the experimental matrix. If one or more dummy factor is included, it is easy to compare the size of the real factors to that of the dummy factor, and factors that are demonstrably larger in magnitude have significance.

An alternative approach comes from the work of Genichi Taguchi. His method of quality control was much used by Japanese industry, and only fairly recently it was recognised that certain aspects of the theory are very similar to

Western practices. His philosophy was that consumers desire products that have constant properties within narrow limits. For example, a consumer panel may taste the sweetness of a product, rating it from 1 to 10. A good marketable product may result in a taste panel score of 8: above this, the product is too sickly, and below, the consumer expects the product to be sweeter. There will be a huge number of factors in the manufacturing process that might cause deviation from the norm, including suppliers of raw materials, storage and preservation of the food and so on. Which factors are significant? Taguchi developed designs for screening large number of potential factors.

His designs are presented in the form of table similar to that of Plackett and Burman, but with a '1' for a low and '2' for a high level. Superficially, Taguchi's designs might appear different, but by changing the notation, and swapping rows and columns around, it is possible to show that both types of design are identical, and, indeed, the simpler designs are the same as the well-known fractional factorial designs. There is a great deal of controversy surrounding Taguchi's work; although many statisticians feel that he has reinvented the wheel, he was an engineer, and his way of thinking had a major and positive effect on Japanese industrial productivity. Before globalisation and the Internet, there was less exchange of ideas between different cultures. His designs are part of a more comprehensive approach to quality control in industry.

Taguchi's designs can be extended to three or more levels, but construction becomes fairly complicated. Some texts do provide tables of multi-level screening designs, and it is also possible to mix the number of levels, for example, having one factor at two levels and another at three levels. This could be useful, for example, if there are three alternative sources of one raw material and two of another raw material. Remember that the factors can fall into discrete categories and do not have to be numerical values such as temperature or concentrations. A large number of designs have been developed from Taguchi's work, but most are quite specialist, and it is not easy to generalise. The interested reader is advised to consult the source literature.

1. For seven factors, a quarter factorial and a Plackett–Burman design are the same apart from rearrangement of rows and/or columns.

   (a) True
   (b) False

2. How many experiments are necessary for the study of 10 factors, using a Plackett–Burman design?

   (a) 12
   (b) 11
   (c) 10

3. If we want to study the effects of 13 factors using a Plackett–Burman design, how many dummy factors are necessary?

   (a) 0
   (b) 1
   (c) 2

2.3.4 Partial Factorials at Several Levels: Calibration Designs

Two-level designs are useful for exploratory purposes and can sometimes result in quite useful models, but in many areas of chemistry, such as calibration (see Chapter 5 for more details), it is desirable to have several levels, especially when studying mixtures. Much of chemometrics is concerned primarily with linearly additive models of the expected form $X = C\,S$, where X is an observed matrix, such as a set of spectra, each row consisting of a spectrum and each column of a wavelength, C is a matrix of, for example, concentrations, each row consisting of the concentration of a number of compounds in a spectrum, and S could consist of the corresponding spectra of each compound. There are innumerable variations on this theme; in some cases, all the concentrations of all the components in a mixture are known, whereas in other cases, the concentrations of only a few components in a mixture are known. In many situations, it is possible to control the experiments by mixing up components in the laboratory, but in other cases, this is not practicable, samples are being taken from the field. A typical laboratory-based experiment might involve recording a series of four component mixtures in which each component is studied at five concentration levels.

A recommended strategy is as follows:

- Perform a calibration experiment, by producing a set of mixtures of a series of compounds of known concentrations to give a 'training set'.

- Then, test this model on an independent set of mixtures called a *test set*.
- Finally, use the model on real data to produce predictions.

More detail is described in Section 6.6. Many brush aside the design of formalised training sets, often employing empirical or random approaches for choosing samples. Some chemometricians recommend huge training sets of several hundred samples in order to get a representative distribution of compounds, especially if there are known to be half a dozen or more significant components in a mixture. In large industrial calibration models, such a procedure is often considered important for robust predictions. However, this approach is expensive in time and resources and rarely possible in routine laboratory studies. In addition, many instrumental calibration models are unstable; hence, calibration on Monday might vary significantly to calibration on Tuesday; hence, if calibrations are to be repeated at regular intervals, the number of spectra in the training set must be limited. Finally, very ambitious calibrations can take months or even years to establish, by which time the instruments and often the detection methods are replaced.

For the most effective calibration models, the nature of the training set must be carefully considered using rational experimental design. Provided the spectra are linearly additive and there are no serious baseline problems or interactions, standard designs can be employed to obtain training sets. It is important to recognise that the majority of chemometric techniques for regression and calibration assume linear additivity. If this may not be so, either the experimental conditions can be modified (e.g. if the concentration of a compound is too high such that the absorbance does not obey the Beer–Lambert law, the solution is simply diluted) or various approaches for multi-linear modelling are required. It is important to recognise that there is a big difference between the application of chemometrics to primarily analytical or physical chemistry where it is usual to be able to attain conditions of linearity, and to organic or biological chemistry (e.g. QSAR) where this is often not possible. The designs in this section are most applicable in the former case.

In calibration, it is normal to use several concentration levels to form a model. Indeed for information on lack-of-fit and predictive ability, this is essential. Hence, two-level factorial designs are inadequate and typically four or five concentration levels may be required for each compound in the mixture (or factor). Consider an experiment carried out using various mixtures of methanol and acetone. What happens if the concentrations of acetone and methanol in a training set are completely correlated? What happens if the concentration of acetone increases so does that of methanol and similarly with a decrease. Such an experimental arrangement is shown in Figure 2.24. A more satisfactory design is given in Figure 2.25, in which the two concentrations are completely uncorrelated or orthogonal. In the former design, there is no way of knowing whether a change in the, for example, spectral characteristic is a result of change in the concentration of acetone or methanol. If this feature is consciously built into the training set and expected in all future samples, there is no problem, but if there is a future sample with a high acetone and low methanol concentration, calibration models will give a wrong answer for the concentration of each component and will have no way of distinguishing changes in acetone from those in methanol. This is potentially very serious especially when the result of chemometric analysis of spectral data is used to make decisions, such as about the quality of a batch of pharmaceuticals, based on the concentration of each constituent as predicted by computational analysis of spectra. Some packages do include elaborate diagnostics for so-called outliers, which may be perfectly good samples in many cases but whose correlation structure differs from that of the training set; however, assuming that outlier tests are not used or available, we have to develop models using a good training set. In this chapter, we will emphasise the importance

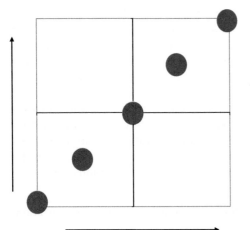

Figure 2.24 Poorly designed calibration experiment.

Figure 2.25 Well-designed calibration experiment.

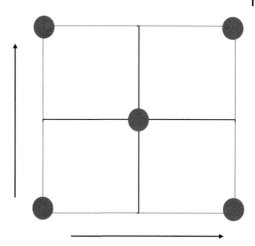

of good design. In the absence of any certain knowledge (e.g. in all conceivable future samples, the concentrations of acetone and methanol will be correlated), it is safest to design the calibration set so that the concentrations of as many compounds as possible in a calibration set are orthogonal.

A guideline to designing a series of multi-component mixtures for calibration is described below.

- Determine how many components in the mixture (or factors) ($=k$) and the maximum and minimum concentration of each component. Remember that, if studied by spectroscopy or chromatography, the overall absorbance when each component is at a maximum should be within the Beer–Lambert limit (about 1.2 AU for safety).
- Decide how many concentration levels are required for each compound ($=l$), typically 4 or 5. Mutually orthogonal designs are only possible if the number of concentration levels is a prime number or a power of a prime number, which means they are possible for 3, 4, 5, 7, 8 and 9 levels but not 6 or 10 levels.
- Decide how many mixtures to be produced or experiments to be performed. Designs exist involving $N = ml^p$ mixtures, where l equals the number of concentration levels, p is an integer at least equal to 2 and m an integer at least equal to 1. Setting both m and p at their minimum values, at least 25 experiments, are required to study a mixture (of more than one component) at five concentration levels, or l^2 at l levels.
- The maximum number of mutually orthogonal compound concentrations in a mixture design where $m = 1$ is 4 for a three-level design, 5 for a four-level design and 12 for a five-level design; hence, using five levels can dramatically increase the number of compounds we can study using calibration designs. We will discuss how to extend the number of mutually orthogonal concentrations below. Hence, choose the design and number of levels with the number of compounds of interest in mind.

The method for setting up a calibration design will be illustrated by a five-level, eight-compound (or factor), 25 experiments, design. The theory is rather complicated; hence, the design will be presented as a series of steps.

- The first step is to number the levels, typically coded −2 (lowest) to +2 (highest), corresponding to coded concentrations, for example, the level −2 = 0.7 mM and level +2 = 1.1 mM; note that the concentration levels can be coded differently for each component in a mixture.
- Next, choose a *repeater* level, recommended to be the middle level, 0. For a five-level design, and 7–12 factors (=components in a mixture), it is essential that this is 0. The first experiment is at this level for all factors.
- Third, select a *cyclical permuter* for the remaining ($l − 1$) levels. This relates each of these four levels, as will be illustrated below; only certain cyclic generators can be used, namely $−2 \rightarrow −1 \rightarrow 2 \rightarrow 1 \rightarrow −2$ and $−2 \rightarrow 1 \rightarrow 2 \rightarrow −1 \rightarrow −2$, which have the property that factors j and $j + l + 1$ are orthogonal (these are listed in Table 2.30 as discussed below). For less than $l + 2$ (=7 in our case) factors, any permuter can be used as long as it includes all the four levels. One such permuter is illustrated in Figure 2.26 and is used in the example below.
- Finally, select a *difference vector*; this consists of $l − 1$ numbers from 0 to $l − 2$, arranged in a particular sequence (or 4 numbers from 0 to 3 in this example). Only a very restricted set of such vectors as tabulated are acceptable of which {0 2 3 1} is one. The use of the difference vector will be described below.
- Then, generate the first column of the design consisting of l^2 (=25) levels in this case, each level corresponding to the concentration of the first compound in the mixture in each of 25 experiments.
  - The first experiment is at the repeater level for each factor.

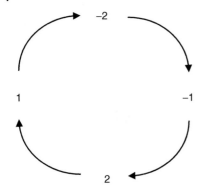

Figure 2.26 Cyclic permuter.

- The $l - 1$ (=4) experiments 2, 8, 14 and 20 are at the repeater level (=0 in this case). In general, the experiments 2, $2 + l + 1$, $2 + 2(l + 1)$ up to $2 + (l - 1) \times (l + 1)$ are at this level. These divide the columns into 'blocks' of five (=l) experiments.
- Now determine the levels for the first block, from experiments 3–7 (or in general, experiments 3 to $2 + l$). Experiment 3 can be at any level apart from the repeater. In the example below, we use level −2. The key to determining the levels for the next four experiments is the difference vector. The conditions for the fourth experiment are obtained from the difference vector and cyclic generator. The difference vector is {0 2 3 1} and implies that the second experiment of the block is 0 cyclical differences away from the third experiment or −2 using the cyclic permuter in Figure 2.26. The next number in the difference vector is 2, making the fifth experiment at level 2, which is two cyclic differences from −2. Continuously, the sixth experiment is three cyclic differences from the fifth experiment or at level −1, and the final experiment of the block is at level 2.
- For the second block (experiments 9–13), simply shift the first block by one cyclic difference using the permuter in Figure 2.26 and continue until the last (or fourth) block is generated.
- Then, generate the next column of the design as follows:
  - The concentration level for the first experiment is always at the repeater level.
  - The concentration level for the second experiment is at the same level as the third experiment of the previous column, up to the 24th – in our case – or ($l^2 - 1$)th – experiment.
  - The final experiment is at the same level as the second experiment for the previous column.
- Finally, generate successive columns using the principle in the step mentioned above.

The development of the design is illustrated in Table 2.29. Note that a full five-level factorial design for eight compounds would require 5^8 or 390 625 experiments; hence, there has been a dramatic reduction in the number of experiments required.

A number of important features have to be noted about the design in Table 2.29.

- In each column, there are an equal number of −2, −1, 0, +1 and +2 levels.
- Each column is orthogonal to every other column, that is the correlation coefficient is 0.
- A graph of the levels of any two factors against each other is given in Figure 2.27(a) for each combination of factors except factors 1 and 7 and factors 2 and 8, whose graph is given in Figure 2.27(b). In most cases, the levels of any two factors are related exactly as they would be for a full factorial design, which would require almost half a million experiments. The nature of the difference vector is crucial to this important property. Some compromise is required between factors differing by $l + 1$ (or 6) columns, such as factors 1 and 7. This is unavoidable unless more experiments are performed.

Table 2.30 summarises information required to generate some possible common designs, including the difference vectors and cyclic permuters, following the general rules mentioned above for different designs. According to the five level design, {0 2 3 1} is one possible difference vector, and also the permuter used above is one of the two possibilities. Obviously with different combinations of difference vectors and cyclic permuters, it is possible to generate a number of designs, especially if the levels are large. However, it is suggested that in most practical cases, no more than five levels would be required.

It is possible to expand the number of factors using a simple trick of matrix algebra. If a matrix A is orthogonal, then the matrix $\begin{bmatrix} A & A \\ A & -A \end{bmatrix}$ is also orthogonal. Therefore, new matrices can be generated from the original orthogonal designs

Table 2.29 Development of a multi-level partial factorial design.

	Factor 1	Factor 2	Factor 3	Factor 4	Factor 5	Factor 6	Factor 7	Factor 8
	0	0	0	0	0	0	0	0
Repeater	0	−2	−2	2	−1	2	0	−1
	−2	−2	2	−1	2	0	−1	−1
	−2	2	−1	2	0	−1	−1	1
Block 1	2	−1	2	0	−1	−1	1	2
	−1	2	0	−1	−1	1	2	1
	2	0	−1	−1	1	2	1	0
Repeater	0	−1	−1	1	2	1	0	2
	−1	−1	1	2	1	0	2	2
	−1	1	2	1	0	2	2	−2
Block 2	1	2	1	0	2	2	−2	1
	2	1	0	2	2	−2	1	−2
	1	0	2	2	−2	1	−2	0
Repeater	0	2	2	−2	1	−2	0	1
	2	2	−2	1	−2	0	1	1
	2	−2	1	−2	0	1	1	−1
Block 3	−2	1	−2	0	1	1	−1	−2
	1	−2	0	1	1	−1	−2	−1
	−2	0	1	1	−1	−2	−1	0
Repeater	0	1	1	−1	−2	−1	0	−2
	1	1	−1	−2	−1	0	−2	−2
	1	−1	−2	−1	0	−2	−2	2
Block 4	−1	−2	−1	0	−2	−2	2	−1
	−2	−1	0	−2	−2	2	−1	2
	−1	0	−2	−2	2	−1	2	0

Experiments

Figure 2.27 Graph of factor levels for design in Table 2.29: top factors 1 versus 2, bottom factors 1 versus 7.

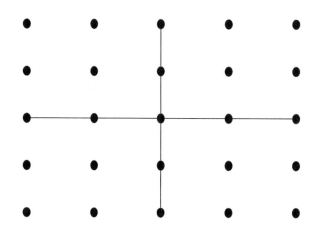

Table 2.30 Parameters for construction of a multi-level calibration design.

Levels	Experiments	Max number Orthogonal factors	Repeater	Difference vectors	Cyclic permuters
3	9	4	Any	{01}, {10}	
4	16	5	Any	{021}, {120}	
5	25	12	0	{0231}, {1320}, {2013}, {3102}	$-2 \rightarrow -1 \rightarrow 2 \rightarrow 1 \rightarrow -2,$ $-2 \rightarrow 1 \rightarrow 2 \rightarrow -1 \rightarrow -2$
7	49	16	0	{241035}, {514302}, {451023}, {124350}, {530142}, {203415}, {320154}, {053421}	$-3 \rightarrow 2 \rightarrow 3 \rightarrow -1 \rightarrow 1 \rightarrow -2 \rightarrow -3,$ $-3 \rightarrow 1 \rightarrow -1 \rightarrow 2 \rightarrow 3 \rightarrow -2 \rightarrow -3,$ $-3 \rightarrow -2 \rightarrow 3 \rightarrow 2 \rightarrow -1 \rightarrow 1 \rightarrow -3,$ $-3 \rightarrow -2 \rightarrow 1 \rightarrow -1 \rightarrow 3 \rightarrow 2 \rightarrow -3$

to expand the number of compounds in the mixture, involving twice the number of experiments for a doubling of the number of factors.

1. It is possible to set up a calibration design for three levels involving 18 experiments.
   (a) True
   (b) False

2. For each factor, only one experiment is performed at the repeater level for all factors in a calibration design.
   (a) True
   (b) False

3. It is possible to set up a calibration design using six levels
   (a) True
   (b) False

2.4 Central Composite or Response Surface Designs

Two-level factorial designs are primarily useful for exploratory purposes, and calibration designs have special uses in areas such as multivariate calibration, where we often expect an independent linear response from each component in a mixture. It is often, though, important to provide a more detailed model of a system. There are two prime reasons. The first is for optimisation – to find the conditions that result in a maximum or minimum as appropriate. An example is when improving the yield of synthetic reaction or a chromatographic resolution. The second is to produce a detailed quantitative model: to predict mathematically how a response relates to the values of various factors. An example may be to know how the near-infrared spectrum of a manufactured product relates to the nature of the material and process employed in manufacturing.

Most exploratory designs do not involve recording replicates, nor do they provide information on squared terms; some such as Plackett–Burman and highly fractional factorials do not even provide details of interactions. When we want to form a detailed model, it is often desirable at a first stage to reduce the number of factors via exploratory designs, described in Section 2.3, to a small number of main factors (perhaps 3 or 4) that are to be studied in detail, for which both squared and interaction terms in the model are of interest.

2.4.1 Setting up the Design

Many designs for use in chemistry for modelling are based on the central composite design (sometimes called a *response surface design*), the main principles of which will be illustrated via a three-factor example (Figure 2.28 and Table 2.31).

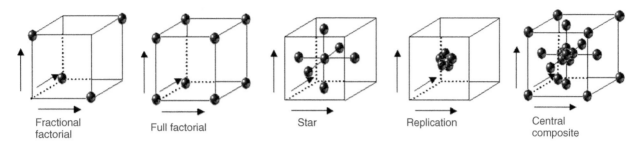

Figure 2.28 Elements of a central composite design: each axis represents a factor.

Table 2.31 Construction of a central composite design.

Fractional factorial

1	1	1
1	−1	−1
−1	−1	1
−1	1	−1

Full factorial

1	1	1
1	1	−1
1	−1	1
1	−1	−1
−1	1	1
−1	1	−1
−1	−1	1
−1	−1	−1

Star

0	0	−1
0	0	1
0	1	0
0	−1	0
1	0	0
−1	0	0
0	0	0

Replication in centre

0	0	0
0	0	0
0	0	0
0	0	0
0	0	0

(Continued)

Table 2.31 (Continued)

Central composite

1	1	1
1	1	−1
1	−1	1
1	−1	−1
−1	1	1
−1	1	−1
−1	−1	1
−1	−1	−1
0	0	−1
0	0	1
0	1	0
0	−1	0
1	0	0
−1	0	0
0	0	0
0	0	0
0	0	0
0	0	0
0	0	0
0	0	0

The first step, of course, is to code the factors, and it is always important to choose sensible physical values for each of the factors. It is assumed that the central point for each factor is coded by 0, and the design is symmetric around this. We will illustrate the design for three factors, which can be represented by points on a cube, each axis corresponding to a factor. A central composite design can be considered as several superimposed designs.

- The smallest possible *fractional factorial*, three-factor design consists of four experiments, used to estimate the three linear terms and the intercept. Such as design will not provide estimates of the interactions, replicates or squared terms.
- Extending this to eight experiments provides estimates of all interaction terms. When represented by a cube, these experiments are placed on the eight corners and consist of a *full factorial design*. All possible combinations of +1 and −1 for the three factors are observed. They can be used to estimate the intercept, linear and interaction terms. This involves performing 2^k experiments where there are k factors.
- Another type of design, often designated a *star design*, can be employed to estimate the squared terms. In order to do this, at least three levels are required for each factor, often denoted by $+a$, 0 and $−a$, with level 0 being in the centre because there must be at least three levels to fit a quadratic model. Points where one factor is at level $+a$ are called *axial* points. Each axial point involves setting one factor at level $\pm a$ and the remaining factors at level 0. One simple design sets a equal to 1, although, as discussed below, this value for the axial point is not always recommended. For three factors, a star design consists of the centre point, and six in the centre (or above) each of the six faces of the cube. This involves performing $2k + 1$ experiments.
- Finally, it is often useful to be able estimate the error (as discussed in Section 2.2.2), and one method is to perform *replicates* (typically five) in the centre. Obviously, other approaches to replication are possible, but it is usual to replicate in the centre and assume that the error is the same throughout the response surface. If there are any overriding reasons to assume that heteroscedasticity of errors has an important role, replication could be performed at the star or factorial points. However, much of experimental design is based on classical statistics where there is no real detailed information about error distributions over an experimental domain, or at least, obtaining such information would be very laborious.

- Performing a full factorial design, a star design and five replicates, results in 20 experiments. This design is a type of central composite design. When the axial or star points are situated at $a = \pm 1$, the design is sometimes also called a *face-centred cube design*, see Table 2.31. Note that the number of replicates is one less than the number of experiments performed in the centre.

1. How many experimental points are required for a four-factor star design?
   (a) 7
   (b) 4
   (c) 8
   (d) 9

2. How many experiments are required for a five-factor central composite design with five replicates in the centre?
   (a) 47
   (b) 48
   (c) 31
   (d) 32

2.4.2 Degrees of Freedom

In this section, we analyse the features of such designs in detail. However, many factors are used in most cases, only two-factor interactions are computed; hence, higher order interactions are ignored, although, of course, these provide sufficient degrees of freedom to estimate the lack-of-fit.

- The first step is to set up a model. A full model including all two-factor interactions consists of $1 + 2k + [k(k-1)]/2 = 1 + 6 + 3$ or 10 coefficients in the case of a three-factor design relating to
  - 1 intercept term (of the form b_0)
  - 3 ($=k$) linear terms (of the form b_1)
  - 3 ($=k$) squared terms (of the form b_{11})
  - and 3 ($=[k(k-1)]/2$) interaction terms (of the form b_{12})
  or in equation form

$$\hat{y} = b_0 + b_1 x_1 + b_2 x_2 + b_3 x_3 + b_{11}x_1{}^2 + b_{22}x_2{}^2 + b_{33}x_3{}^2 + b_{12}x_1x_2 + b_{13}x_1x_3 + b_{23}x_2x_3$$

- A degree-of-freedom tree can be drawn up as illustrated in Figure 2.29. We can see that
  - there are 20 ($=N$) experiments overall,
  - 10 ($=P$) parameters in the model,
  - 5 ($=R$) degrees of freedom to determine replication error and
  - 5 ($=N - P - R$) degrees of freedom for the lack-of-fit.
  Note that the number of degrees of freedom for the lack-of-fit equals that for replication in this case, suggesting quite a good design.

The total number of experiments N ($=20$) equals the sum of

- 2^k ($=8$) factorial points, often represented as the corners of the cube,
- $2k + 1$ ($=7$) star points, often represented as axial points on (or above) the faces of the cube plus one in the centre and
- R ($=5$) replicate points, in the centre.

A large number of variations are present on this theme, but each design can be defined by four parameters, namely,

- The number of factorial or cubic points (N_f),
- the number of axial points (N_a), usually one less than the number of points in the star design,
- the number of central points (N_c), usually one more than the number of replicates and
- the position of the axial points a.

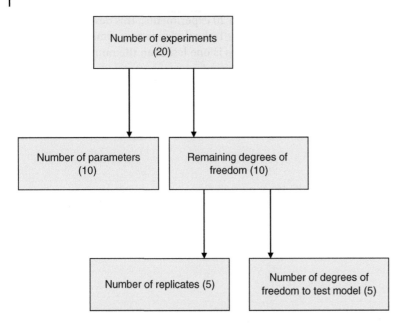

Figure 2.29 Degrees of freedom for central composite design.

In most cases, it is best to use a full factorial design for the factorial points; however, if the number of factors is large, it is legitimate to reduce this and use a partial factorial design. There are always $2k$ axial points.

The number of central points is often chosen according to the number of degrees of freedom required to assess errors via ANOVA and the F-test (see Sections 2.2.2 and 2.2.4.4) and should be approximately equal to the number of degrees of freedom for the lack-of-fit, with a minimum of about 4 unless there are special reasons for reducing this.

1. For a central composite design for four factors, using a model with the intercept, all possible linear, quadratic and two-factor interaction terms, how many degrees of freedom are there for determining the lack-of-fit error?

    (a) 6
    (b) 10
    (c) 15

2.4.3 Axial Points

The choice of the position of the axial (or star) points and how this relates to the number of replicates in the centre is an interesting consideration. Although many chemists use these designs fairly empirically, it is worth noting two statistical properties that influence the property of these designs. It is essential to recognise, although, that there is no single perfect design, indeed many of the desirable properties of a design are incompatible with each other.

- *Rotatability* implies that the confidence in the predictions depends only on the distance from the centre of the design. For a two-factor design, this means that all experimental points in a circle of a given radius will be predicted equally well. This has useful practical consequences; for example, if the two factors correspond to concentrations of acetone and methanol, we know that the farther the concentrations are from the central point (in coded values), the lower the confidence. Methods for visualising this were described in Section 2.2.5. If a design is rotatable, the confidence does not depend on the number of replicates in the centre, but only on the value of a, which should equal $\sqrt[4]{N_f}$, where N_f is the number of factorial points, equal to 2^k if a full factorial is used, for this property. Note that the position of the axial points will differ if a fractional factorial design is used.

- *Orthogonality* implies that all the terms (linear, squared and two factor interactions) are orthogonal to each other in the design matrix; that is, the correlation coefficient between any two terms (apart from the zero order term where it is not defined) equals 0. For linear and interaction terms, this will always be so, but squared terms are not so simple, and in the majority of central composite designs, they are not orthogonal. The rather complicated condition is

$a = \sqrt{\left(\sqrt{N \times N_f} - N_f\right)/2}$, which depends on the number of replicates, as a term for the overall number of experiments is included in the equation. A small lack of orthogonality in the squared terms can sometimes be tolerated, but it is often worth checking any particular design for this property.

Interestingly, these two conditions are usually not compatible, resulting in considerable dilemmas. Although in practical situations, the differences of a for the two different properties are not so large, and in some cases, it is not experimentally very meaningful to get too concerned about small differences in the axial points of the design. Table 2.32 analyses the properties of three, two-factor designs with a model of the form $y = b_0 + b_1 x_1 + b_2 x_2 + b_{11} x_1^2 + b_{22} x_2^2 + b_{12} x_1 x_2$ ($P = 6$). Design A is rotatable, Design B is orthogonal and Design C has both properties. However, the third is inefficient in which seven replicates are required in the centre; indeed, half the design points are in the centre, which makes little practical sense, although this design is both rotatable and orthogonal. Table 2.33 lists the values of a for rotatability and orthogonality for different numbers of factors and replicates. For the five-factor design, a half factorial design is also tabulated; in all other cases, the factorial part is full. It is interesting to note that for a two-factor design with one central point (i.e. no replication), the value of a for orthogonality is 1, making it identical to a two-factor, three-level design (see Table 2.20(a)), being four factorial and five star points or 3^2 experiments in total.

Terminology varies according to authors, some calling only the rotatable designs as *true central composite designs*. It is very important to recognise that the use of statistics is very widespread throughout science, especially in experimental areas such as biology, medicine and chemistry, and check carefully an author's precise terminology. It is important not to get locked in a single textbook (even this one!), a single software package or course provider. In many cases, to simplify, a single terminology is employed from any one source. As there are no universally accepted conventions, in which chemometrics differs from, for example, organic chemistry, and most historic attempts to set up committees have come to grief or been dominated by one specific strand of opinion, every major group has its own philosophy.

The true experimental conditions can be easily calculated from a coded design. For example, if coded levels +1, 0 and −1 for a rotatable design correspond to temperatures of 30°, 40° and 50° for a two-factor design, the axial points correspond to temperatures of 25.9° and 54.1°, whereas for a four-factor design, these points are 20° and 60°. Note that these designs are only practicable where factors can be numerically defined and cannot normally be employed if some data are categorical, unlike factorial designs. However, it is sometimes possible to set the axial points at values such as ±1 or ±2 under some circumstance to allow for factors that can take discrete values, for example, the number of cycles in an extraction procedure, although this does restrict the properties of the design.

A rotatable four-factor design consists of 30 experiments, namely

- 16 factorial points at all possible combinations of ±1,
- nine star points, including a central point of (0,0,0,0) and eight points of the form (±2,0,0,0) and so on and
- typically five further replicates in the centre; note that a very large number of replicates (11) would be required to satisfy orthogonality with the axial points at 2 units, and this is probably overkilled in many real experimental situations. Indeed, if resources are available for so many replicates, it might make sense to replicate different experimental points to check whether errors are even over the response surface.

1. We know the position of the axial points for a central composite design with all factorial points.

    (a) We can tell whether the design is rotatable but do not have enough information to tell whether it is orthogonal.
    (b) We can tell whether the design is orthogonal but do not have enough information to tell whether it is rotatable.
    (c) We do not have enough information to tell whether the design is orthogonal or rotatable.

2. Not all rotatable designs are orthogonal, but all orthogonal designs are rotatable.

    (a) True
    (b) False

2.4.4 Modelling

Once the design is performed, it is then possible to calculate the values of the coefficients in the model using regression and design matrices or almost any standard statistical procedure and assess the significance of each term using ANOVA, F-tests and t-tests if felt appropriate. We can then answer, for example, whether an interaction is significant, or whether to take into account quadratic terms: these may then allow us to further simplify our model or else to ask

Table 2.32 Three possible two-factor central composite designs.

Design A

−1	−1	Rotatability	✓
−1	1	Orthogonality	×
1	−1	N_c	6
1	1	a	1.414
−1.414	0		
1.414	0	Lack-of-fit (df)	3
0	−1.414	Replicates (df)	5
0	1.414		
0	0		
0	0		
0	0		
0	0		
0	0		
0	0		

Design B

−1	−1	Rotatability	×
−1	1	Orthogonality	✓
1	−1	N_c	6
1	1	a	1.320
−1.320	0		
1.320	0	Lack-of-fit (df)	3
0	−1.320	Replicates (df)	5
0	1.320		
0	0		
0	0		
0	0		
0	0		
0	0		
0	0		

Design 3

−1	−1	Rotatability	✓
−1	1	Orthogonality	✓
1	−1	N_c	8
1	1	a	1.414
−1.414	0		
1.414	0	Lack-of-fit (df)	3
0	−1.414	Replicates (df)	7
0	1.414		
0	0		
0	0		
0	0		
0	0		
0	0		
0	0		
0	0		
0	0		

Table 2.33 Position of the axial points for rotatability and orthogonality for central composite designs with varying number of replicates (one less than the number of central points).

	Rotatability	Orthogonality		
		R		
k		3	4	5
2	1.414	1.210	1.267	1.320
3	1.682	1.428	1.486	1.541
4	2.000	1.607	1.664	1.719
5	2.378	1.764	1.820	1.873
5 (half factorial)	2.000	1.719	1.771	1.820

questions that can be given a physical interpretation. Sometimes the significance of terms is indicated in an ANOVA table using *s, the more the greater the significance of each term (or lower p value). It is important to remember that ANOVA tables assume that the underlying errors are normally distributed.

It is also important to recognise that these designs are mainly employed in order to produce a detailed model and also to look at interactions and higher order (quadratic) terms. The number of experiments becomes excessive if the number of factors is large. If more than about five significant factors are to be studied, it is best to narrow down the problem first using exploratory designs, although the possibility of using fractional factorials on the corners helps. Remember also that it is conventional (but not always essential) to ignore interaction terms above second order.

After the experiments have been performed, it is then possible to produce a detailed mathematical model of the response surface. If the purpose is optimisation, it might then be useful, for example, by using contour or 3D plots, to determine the position of the optimum. For relatively straightforward cases, partial derivatives can be employed to solve the equations, as illustrated in Problems 7 and 16; however, if there are a large number of terms, an analytical solution can be difficult and there can also be more than one optimum. It is always recommended to try to look at the system graphically, even if there are too many factors to visualise the whole of the experimental space at once. It is also important to realise that there may be other issues that influence our definition of an optimum, such as expense of raw materials, availability of key components or even time. Sometimes a design can be used to model several responses, and each one can be analysed separately, perhaps one might be the yield of a reaction, another the cost of raw materials and another the level of impurities in a produce. Chemometricians should resist the temptation to insist on a single categorical 'correct' answer.

1. The significance of terms in a model obtained using a central composite design can be evaluated using ANOVA.
    (a) True
    (b) False

2.4.5 Statistical Factors

Another important use of central composite designs is to determine a good range of compounds for testing, as in QSAR. Consider the case in Figure 2.4. Rather than the axes being physical variables such as concentrations, they can be abstract mathematical or statistical variables such as principal components (see Chapter 4). These could come from molecular property descriptors, for example, bond lengths and angles, hydrophobicity, dipole moments and so on. Consider, for example, a database of several hundred compounds. Perhaps selection is interesting for biological tests. It may be very expensive to test all compounds; hence, a sensible strategy is to reduce the number of compounds to a selection. Taking the first two PCs as the factors, a selection of nine representative compounds can be obtained using a central composite design as follows:

- Determine the scores of the principal components of the original data set.
- Scale each PC, for example, so that the highest score equals +1 and the lowest score equals −1.

- Then, choose those compounds whose scores are closest to the desired values. For example, in the case in Figure 2.4, choose a compound whose score is closest to (−1,−1) for the bottom left-hand corner and closest to (0,0) for the centre point.
- Perform experimental tests on this subset of compounds and then use some form of modelling to relate the desired activity to structural data. Note that this modelling does not have to be multi-linear modelling, as discussed in this section, but could also be PLS, as introduced in Chapter 6.

1. It is possible for factors to be multivariate combinations of raw variables.
   (a) True
   (b) False

2.5 Mixture Designs

Chemists and statisticians use the term *mixture* in different ways. To a chemist, any combination of several substances is a mixture. In more formal statistical terms, however, a mixture involves a set of factors whose total is a constant sum, this property is often called *closure* and will be discussed in completely different contexts in the area of scaling data before principal components analysis (Sections 4.6 and 7.2.4). Hence, in statistics (and chemometrics), a solvent system in HPLC or a blend of components in products, such as paints, drugs or food, is considered a mixture, as each component can be expressed as a proportion and the total adds up to 1 or 100%. The response could be a chromatographic separation, or the taste of a foodstuff or physical properties of a manufactured material. Often the aim of modelling is to find an optimum blend of components that taste best, or provide the best chromatographic separation or the material that is most durable.

Compositional mixture experiments involve some quite specialist techniques and a whole range of considerations must be made before designing and analysing such experiments. The principal consideration is that the value of each factor is constrained. Take, for example, a three-component mixture of acetone, methanol and water, which may be solvents used as the mobile phase for a chromatographic separation. If we know that there is 80% water in the mixture, there can be no more than 20% acetone or methanol in the mixture. If there is also 15% acetone, the amount of methanol is fixed at 5%. In fact, although there are three components in the mixtures, these translate into two independent factors.

2.5.1 Mixture Space

Most chemists represent their experimental conditions in mixture space, which corresponds to all possible allowed proportions of components that add up to 100%. A three-component mixture can be represented by a triangle (Figure 2.30),

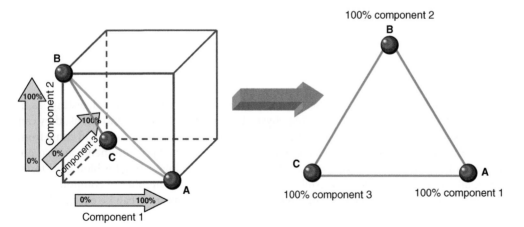

Figure 2.30 Three-component mixture space.

which is a two-dimensional cross-section of a three-dimensional space, represented by a cube, showing the allowed region in which the proportions of the three components add up to 100%. Points within this triangle or mixture space represent possible mixtures or blends.

- The three corners correspond to single components,
- points along the edges correspond to binary mixtures,
- points inside the triangle correspond to ternary mixtures,
- the centre of the triangle corresponds to an equal mixture of all the three components and
- all points within the triangle are physically allowable blends.

As the number of components increases, so does the dimensionality of the mixture space. Physically meaningful mixtures can be represented as points in this space.

- For two components, the mixture space is simply a straight line,
- for three components a triangle and
- for four components a tetrahedron.

Each object (pictured in Figure 2.31) is called a *simplex* – the simplest possible object in space of a given dimensionality: the dimensionality is one less than the number of components in a mixture, so a tetrahedron (three dimensions) represents a four-component mixture.

A number of common designs can be envisaged as ways of determining a sensible number and arrangement of points within the simplex.

1. A four-component mixture can be represented by a tetrahedron. Possible blends consisting of three of the components are represented

   (a) On the corners
   (b) On the faces
   (c) Within the tetrahedron

2.5.2 Simplex Centroid

2.5.2.1 Design

These designs are probably the most widespread. For k components, they involve performing $2^k - 1$ experiments; that is, for four components, 15 experiments are performed. It involves all possible combinations of the proportions 1, $\frac{1}{2}$ to $1/k$ and is best illustrated by an example. A three-component design consists of

- three single-component combinations,
- three binary combinations and
- one ternary combination.

These experiments are represented graphically in mixture space in Figure 2.32 and tabulated in Table 2.34.

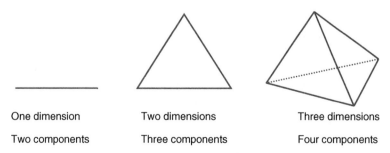

| One dimension | Two dimensions | Three dimensions |
| Two components | Three components | Four components |

Figure 2.31 Simplex in one, two and three dimensions.

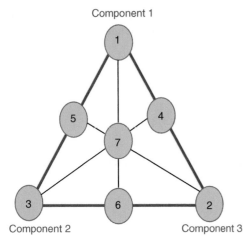

Component 1

Component 2   Component 3

Figure 2.32 Three-component simplex centroid design.

Table 2.34 Three-component simplex centroid mixture design.

Experiment	Component 1	Component 2	Component 3	
1	1	0	0	⎫
2	0	1	0	⎬ Single component
3	0	0	1	⎭
4	1/2	1/2	0	⎫
5	1/2	0	1/2	⎬ Binary
6	0	1/2	1/2	⎭
7	1/3	1/3	1/3	⎬ Ternary

1. How many experiments are there for a four-component simplex centroid design?

   (a) 7
   (b) 12
   (c) 15

2.5.2.2 Model

Just as previously, a model and design matrix can be obtained. However, the nature of the model requires some detailed thought. Consider trying to estimate the model for a three-component design of the form

$$y = c_0 + c_1 x_1 + c_2 x_2 + c_3 x_3 + c_{11} x_1^2 + c_{22} x_2^2 + c_{33} x_3^2 + c_{12} x_1 x_2 + c_{13} x_1 x_3 + c_{23} x_2 x_3$$

This model consists of 10 terms, impossible if only seven experiments are performed. How can the number of terms be reduced? Arbitrarily removing three terms such as the quadratic or interaction terms has little theoretical justification. A major problem with the equation above is that the value of x_3 depends on x_1 and x_2, as it equals $1 - x_1 - x_2$ so there are, in fact, only two independent factors. If a design matrix consisting of the first four terms of the equation mentioned above was set up, it would not have an inverse, and the calculation is impossible. The solution is to set up a reduced model. Consider, instead, a model consisting only of the first three terms:

$$y = a_0 + a_1 x_1 + a_2 x_2$$

This is, in effect, equivalent to a model containing just the three single-component terms without an intercept as

$$y = a_0(x_1 + x_2 + x_3) + a_1 x_1 + a_2 x_2 = (a_0 + a_1)x_1 + (a_0 + a_2)x_2 + a_0 x_3 = b_1 x_1 + b_2 x_2 + b_3 x_3$$

It is not possible to produce a model containing both the intercept and the three single-component terms. Closed data sets, as in mixtures, have a whole series of interesting mathematical properties, but it is primarily important simply to watch for these anomalies.

The two common types of models, one with an intercept and one without, are related. Models excluding the intercept are often referred to as *Sheffé models*, whereas those with the intercept are referred to as *Cox models*. Normally, a full Sheffé model includes all higher order interaction terms, and for this design is given by

$$y = b_1 x_1 + b_2 x_2 + b_3 x_3 + b_{12} x_1 x_2 + b_{13} x_1 x_3 + b_{23} x_2 x_3 + b_{123} x_1 x_2 x_3$$

As seven experiments have been performed, all the seven terms can be calculated, namely

- three one-factor terms,
- three two-factor interactions and
- one three-factor interaction.

The design matrix is given in Table 2.35, and being a square matrix, the coefficients can easily be determined using the inverse.

The full seven-term Cox model is given by

$$y = a_0 + a_1 x_1 + a_2 x_2 + a_{11} x_1^2 + a_{22} x_2^2 + a_{12} x_1 x_2 + a_{1122}(x_1^2 x_2 + x_2^2 x_1)$$

Note that only x_1 and x_2 are involved in the model because x_3 is dependent on the other two terms. We could relate the models as $x_1 + x_2 + x_3 = 1$. Note the rather ugly last term that is required if we want to fit a full model. Obviously, for both types of models, we can omit terms to provide some degrees of freedom for the lack-of-fit.

1. Intercept terms are used in
   (a) The full Sheffé model
   (b) The full Cox model
   (c) Both types of model

2.5.2.3 Multi-component Designs

A full simplex centroid design for k components consists of $2^k - 1$ experiments of which there are

- k single blends,
- $k \times (k-1)/2$ binary blends, each component being present in a proportion of ½,
- $k!/[(k-m)!m!]$ blends containing m components (these can be predicted by the binomial theorem or Pascal's triangle), each component being present in a proportion of $1/m$ and
- finally 1 blend consisting of all components, each component being present in a proportion of $1/k$.

Each type of blend yields an equivalent number of interaction terms in the Sheffé model. Hence, for a five-component mixture and three-component blends, there will be $5!/[(5-3)!3!] = 10$ mixtures such as (1/3 1/3 1/3 0 0) containing all possible combinations and 10 terms such as $b_1 b_2 b_3$.

It is normal to use all possible interaction terms in the mixture model, although this does not leave any degrees of freedom for determining lack-of-fit. Reducing the number of higher order interactions in the model but maintaining

Table 2.35 Design matrix for a three-factor simplex centroid design.

x_1	x_2	x_3	$x_1 x_2$	$x_1 x_3$	$x_2 x_3$	$x_1 x_2 x_3$
1.000	0.000	0.000	0.000	0.000	0.000	0.000
0.000	1.000	0.000	0.000	0.000	0.000	0.000
0.000	0.000	1.000	0.000	0.000	0.000	0.000
0.500	0.500	0.000	0.250	0.000	0.000	0.000
0.500	0.000	0.500	0.000	0.250	0.000	0.000
0.000	0.500	0.500	0.000	0.000	0.250	0.000
0.333	0.333	0.333	0.111	0.111	0.111	0.037

the full design is possible; however, this must be carefully thought because each term can also be re-expressed, in part, as lower order interactions using the Cox model. This will, although, allow the calculation of some measure of confidence in predictions. It is important to recognise that the columns of the mixture design matrix are not orthogonal and can never be because the proportion of each component depends on all others; hence, there will always be some correlation between the factors.

For multi-component mixtures, it is often impracticable to perform a full simplex centroid design; one approach is to simply remove higher order terms, not only from the model but also from the design. A five-component design containing up to second-order terms is presented in Table 2.36. Such designs can be denoted as {k,m} simplex centroid designs, where k is the number of components in the mixture and m the highest order interaction. Note that at least binary interactions are required for squared terms (in the Cox model) and so for optimisation.

1. How many possible ternary blends are there for a six-component mixture?
   (a) 20
   (b) 15
   (c) 10

2. How many experiments are required for a {4,2} simplex centroid design?
   (a) 6
   (b) 10
   (c) 11

2.5.3 Simplex Lattice

Another class of designs called *simplex lattice* has been developed and is often preferable to the reduced simplex centroid design when it is required to reduce the number of interaction terms: they span the mixture space more evenly.

A {k,m} simplex lattice design consists of all possible combinations of 0, 1/m, 2/m ... m/m or a total of

$$N = (k + m - 1)!/[(k - 1)!m!]$$

experiments where there are k factors. A {3,3} simplex lattice design can be set up analogous to the {3,3} simplex centroid design given in Table 2.34. There are

Table 2.36 A {5,2} simplex centroid design.

Comp 1	Comp 2	Comp 3	Comp 4	Comp 5
1	0	0	0	0
0	1	0	0	0
0	0	1	0	0
0	0	0	1	0
0	0	0	0	1
1/2	1/2	0	0	0
1/2	0	1/2	0	0
1/2	0	0	1/2	0
1/2	0	0	0	1/2
0	1/2	1/2	0	0
0	1/2	0	1/2	0
0	1/2	0	0	1/2
0	0	1/2	1/2	0
0	0	1/2	0	1/2
0	0	0	1/2	1/2

- three single-factor experiments,
- six experiments where one factor is at 2/3 and the other at 1/3 and
- one experiment where all factors are at 1/3,

resulting in $5!/(2!3!) = 10$ experiments in total, as illustrated in Table 2.37 and Figure 2.33. Note that there are now more experiments than are required for a full Sheffé model; hence, some information about the significance of each parameter could be obtained; however, no replicates are measured. Generally, chemists mainly use mixture models for the purpose of optimisation or graphical presentation of results. Table 2.38 lists how many experiments are required for a variety of $\{k,m\}$ simplex lattice designs.

Table 2.37 Two-component simplex lattice design.

Experiment	Component 1	Component 2	Component 3	
1	1	0	0	⎫ Single component
2	0	1	0	
3	0	0	1	⎭
4	1/3	2/3	0	⎫
5	2/3	1/3	0	
6	1/3	0	2/3	⎬ Binary
7	2/3	0	1/3	
8	0	2/3	1/3	
9	0	1/3	2/3	⎭
10	1/3	1/3	1/3	⎬ Ternary

Figure 2.33 Three-component simplex lattice design.

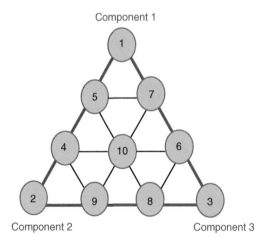

Table 2.38 Number of experiments required for various simplex lattice designs, with different numbers of components and interactions.

components (k)	interactions (m)				
	2	3	4	5	6
2	3				
3	6	10			
4	10	20	35		
5	15	35	70	126	
6	21	56	126	252	462

1. How many experiments are needed for a {5,3} simplex lattice design?

   (a) 21

   (b) 35

2.5.4 Constraints

In chemistry, there are frequent constraints on the proportions of each component. For example, it might be of interest to study the effect of changing the proportion of ingredients in a cake. Sugar will be one ingredient, but there is no point in baking a cake using 100% sugar and 0% of each other ingredients. A more sensible approach is to put a constraint on the amount of sugar, perhaps between 2% and 5%, and look for solutions in this reduced mixture space. A good design will only test blends within the specified regions.

Constrained mixture designs are often quite difficult to set up, but there are four fundamental situations, exemplified in Figure 2.34, each of which requires a different strategy.

- *Only a lower bound* for each component is specified in advance.
    - The first step is to determine whether the proposed lower bounds are feasible. The sum of the lower bounds must be less than 1. For three factors, lower bounds of 0.5, 0.1 and 0.2 are satisfactory, whereas lower bounds of 0.3, 0.4 and 0.5 are not as they add up to more than 1.
    - The next step is to determine new upper bounds. For each component, these are one minus the sum of the lower bounds for all other components. If the lower bounds for three components are 0.5, 0.1 and 0.2, then the upper bound for the first component is $1 - 0.1 - 0.2 = 0.7$; hence, the upper bound of one component plus the lower bounds of the other two must equal 1.

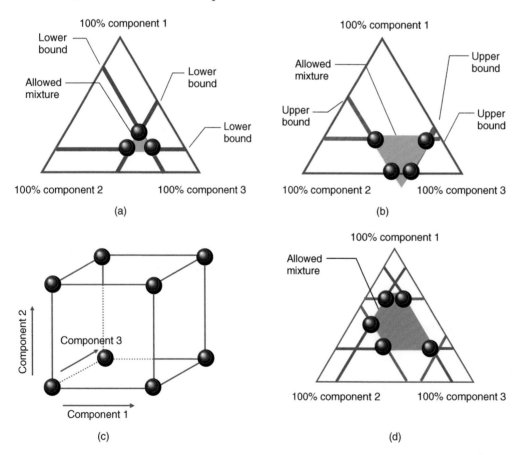

Figure 2.34 Four situations encountered in constrained mixture designs. (a) Lower bounds defined, (b) upper bounds defined, (c) upper and lower bounds defined, fourth factor as filler and (d) upper and lower bounds defined.

Table 2.39 Constrained mixture design with three lower bounds.

	Simple centroid design			Constrained design		
	Factor 1	Factor 2	Factor 3	Factor 1	Factor 2	Factor 3
	1.000	0.000	0.000	0.700	0.100	0.200
	0.000	1.000	0.000	0.500	0.300	0.200
	0.000	0.000	1.000	0.500	0.100	0.400
	0.500	0.500	0.000	0.600	0.200	0.200
	0.500	0.000	0.500	0.600	0.100	0.300
	0.000	0.500	0.500	0.500	0.200	0.300
	0.333	0.333	0.333	0.567	0.167	0.267
L	0.5	0.1	0.2			
U	0.7	0.3	0.4			

- The third step is to take a standard design and recalculate the conditions as follows:

$$x_{new,f} = x_{old,f}(U_f - L_f) + L_f$$

where L_f and U_f are the lower and upper bounds for component f. This is illustrated in Table 2.39.

The experiments fall in exactly the same pattern as the original mixture space. Some authors call the vertices of the mixture space 'pseudo-components'; hence, the first pseudo-component consists of 70% of pure component 1, 10% of pure component 2 and 20% of pure component 3. Any standard design can now be employed. It is also possible to perform all the modelling on the pseudo-components and convert back to the true proportions at the end.

- *An upper bound* is placed on each factor in advance. Note that not all possible combinations of upper bounds are possible. The constrained mixture space often becomes somewhat more complex dependent on the nature of the upper bounds. The trick is to find the extreme corners of a polygon in mixture space, perform experiments at these corners, midway along the edges and, if desired, in the centre of the design. There are no hard and fast rules as the theory behind these designs is quite complex. Recommended guidance is provided below for two situations. The methods are illustrated in Table 2.40 for a three-component design.
  - If the sum of all $(k-1)$ upper bounds is less than 1, and 1 minus the $(k-1)$ upper bounds is less than the kth upper bound, then do as follows:
    a. Set up k experiments where all but one factor is its upper bound (the first three in Table 2.40(a)). These are the extreme vertices of the constrained mixture space.
    b. Then, set up binary intermediate experiments, simply the average of two of the k extremes.
    c. If desired, set up ternary experiments and so on.
  - If this condition is not met, the constrained mixture space will resemble an irregular polygon as in Figure 2.34(b). An example is illustrated in Table 2.40(b).
    a. Find the extreme vertices for those combinations of $(k-1)$ components that are less than 1, of which there are two in this example.
    b. Each missing vertex (one in this case) increases the number of new vertices by 1. If, for example, it is impossible to simultaneously reach maxima for components 2 and 3, create one new vertex with component 2 at its highest level (U_2), component 1 at 0 and component 3 at $(1 - U_2)$, with another vertex for component 3 at U_3, component 1 at 0 and component 2 at $(1 - U_3)$.
    c. If there are v vertices, calculate extra experimental points between the vertices. As the figure formed by the vertices in (b) has four sides, there will be four extra experiments, making eight in total. This is equivalent to performing one experiment on each corner of the mixture space in Figure 2.34(b) and one experiment on each edge.
    d. Occasionally, one or more experiments are performed in the middle of the new mixture space, which is the average of the v vertices.

Table 2.40 Constrained mixture designs with upper bounds established in advance.

(a)

Upper bounds	0.3	0.4	0.5	
1	0.3	0.4	0.3	Components 1 and 2 high
2	0.3	0.2	0.5	Components 1 and 3 high
3	0.1	0.4	0.5	Components 2 and 3 high
4	0.3	0.3	0.4	Average of experiments 1 and 2
5	0.2	0.4	0.4	Average of experiments 1 and 3
6	0.2	0.3	0.5	Average of experiments 2 and 3
7	0.233	0.333	0.433	Average of experiments 1, 2 and 3

(b)

Upper bounds	0.7	0.5	0.2	
1	0.7	0.1	0.2	Components 1 and 3 high
2	0.3	0.5	0.2	Components 2 and 3 high
3	0.7	0.3	0.0	Component 1 high, Component 2 as high as possible
4	0.5	0.5	0.0	Component 2 high, Component 1 as high as possible
5	0.7	0.2	0.1	Average of experiments 1 and 3
6	0.4	0.5	0.1	Average of experiments 2 and 4
7	0.5	0.3	0.2	Average of experiments 1 and 2
8	0.6	0.4	0.0	Average of experiments 3 and 4

Note that in some circumstances, a three-component constrained mixture space may be described by a hexagon, resulting in 12 experiments on the edges. Provided there are no more than four components, the constrained mixture space is often best visualised graphically, and an even distribution of experimental points can be determined by geometric means.

- *Each component has an upper and lower bound and an additional (k + 1)th component is added* (the fourth in this example) so that the total comes to 100%; this additional component is called a *filler*. An example might be where the fourth component is water, the others being solvents, buffer solutions and so on. This is quite common in chromatography; for example, if the main solvent is aqueous. Standard designs such as factorial designs can be employed for the three components in Figure 2.34(c), with the proportion of the final component computed from the remainder, given by $(1 - x_1 - x_2 - x_3)$. Of course, such designs will only be available if the upper bounds are low enough that their sum is no more than (often much less than) 1. However, in some applications, it is quite common to have some background filler, for example flour in baking of a cake, and active ingredients that are present in quite small amounts.
- *Upper and lower bounds defined in advance.* In order to reach this condition, the sum of the upper bound for each component plus the lower bounds for the remaining components must not be greater than 1; that is, for three-component component,

$$U_1 + L_2 + L_3 \leq 1$$

and so on for components 2 and 3. Note that the sum of all the upper bounds together must be at least equal to 1. Another condition for three components is that

$$L_1 + U_2 + U_3 \geq 1$$

Otherwise the lower bound for component 1 can never be achieved, similar conditions applying to the other components. These equations can be extended to designs with more components. Two examples are illustrated in Table 2.41, one feasible and the other not feasible.

Table 2.41 Example of simultaneous constraints in mixture designs.

Impossible conditions

L	0.1	0.5	0.4
U	0.6	0.7	0.8

Possible conditions

L	0.1	0.0	0.2
U	0.4	0.6	0.7

In such cases, the rules for setting up the mixture design are, in fact, quite straightforward for three components, provided the conditions are met.

– Determine how many vertices, the maximum will be 6 for three components. If the sum of the upper bound for one component and the lower bounds for the remaining components equal 1, then the number of vertices is reduced by 1. The number of vertices also reduces if the sum of the lower bound of one component and the upper bounds of the remaining components equals 1. Call this number v. Normally, one will not obtain conditions for three components for which there are less than three vertices, if any less, the limits are too restrictive to show much variation.

– Each vertex corresponds to the upper bound for one component, the lower bound for another component and the final component is the remainder, after subtracting from 1.

– Order the vertices so that the level of one component remains constant between vertices.

– Double the number of experiments, by taking the average between each successive vertex (and also the average between the first and the last), to provide $2v$ experiments. These correspond to experiments on the edges of the mixture space.

– Finally, it is usual to perform an experiment in the centre, which is simply the average of all the vertices.

Table 2.42 illustrates two constrained mixture designs, one with six and the other with five vertices. The logic can be extended to several components but can be quite complicated. If you are using a very large number of components all with constraints as can sometimes be the case, for example, in fuel or food chemistry where a lot of ingredients may influence the quality of the product, it is probably best to look at the original literature, as designs for multi-factor constrained mixtures are very complex. There is insufficient space in this introductory text to describe all the possibilities in detail. Sometimes constraints might be placed on one or two components, or one component could have an upper limit, another a lower limit and so on. There are no hard and fast rules; however, when the number of components is sufficiently small, it is important to try to visualise the design. The trick is to try to obtain a fairly even distribution of experimental points over the mixture space. Some techniques, which will include feasible design points, do not have this property.

1. Consider the following proposed upper and lower limits for a constrained mixture design

Lower	0.1	0.3	0.1
Upper	0.6	0.6	0.3

   (a) The design is not feasible
   (b) The design has six vertices and is feasible
   (c) The design has five vertices and is feasible
   (d) The design has four vertices and is feasible

2. A constrained mixture design for three components has upper bounds of 0.6, 0.3 and 0.2 for components 1–3. What is the lower bound for component 2?

   (a) 0.2
   (b) 0.1
   (c) 0.0

Table 2.42 Constrained mixture design where both upper and lower limits are known in advance.

Six vertices			
Lower	0.1	0.2	0.3
Upper	0.4	0.5	0.6

Step 1

$0.4 + 0.2 + 0.3 = 0.9$

$0.1 + 0.5 + 0.3 = 0.9$

$0.1 + 0.2 + 0.6 = 0.9$

$0.4 + 0.5 + 0.3 = 1.2$

$0.4 + 0.2 + 0.6 = 1.2$

$0.1 + 0.5 + 0.6 = 1.2$

so $v = 6$

Steps 2 and 3 Vertices			
A	0.4	0.2	0.4
B	0.4	0.3	0.3
C	0.1	0.5	0.4
D	0.2	0.5	0.3
E	0.1	0.3	0.6
F	0.2	0.2	0.6

Steps 4 and 5 Design				
1	A	0.4	0.2	0.4
2	Average A&B	0.4	0.25	0.35
3	B	0.4	0.3	0.3
4	Average B&C	0.25	0.4	0.35
5	C	0.1	0.5	0.4
6	Average C&D	0.15	0.5	0.35
7	D	0.2	0.5	0.3
8	Average D&E	0.15	0.4	0.45
9	E	0.1	0.3	0.6
10	Average E&F	0.15	0.25	0.6
11	F	0.2	0.2	0.6
12	Average F&A	0.3	0.2	0.5
13	Centre	0.2333	0.3333	0.4333

(*Continued*)

Experimental Design | 81

Table 2.42 (Continued)

Five vertices			
Lower	0.1	0.3	0
Upper	0.7	0.6	0.4

Step 1

$0.7 + 0.3 + 0.0 = 1.0$

$0.1 + 0.6 + 0.0 = 0.7$

$0.1 + 0.3 + 0.4 = 0.8$

$0.7 + 0.6 + 0 = 1.3$

$0.7 + 0.3 + 0.4 = 1.4$

$0.1 + 0.6 + 0.4 = 1.1$

so $v = 5$

Steps 2 and 3 Vertices

A	0.7	0.3	0.0
B	0.4	0.6	0.0
C	0.1	0.6	0.3
D	0.1	0.5	0.4
E	0.3	0.3	0.4

Steps 4 and 5 Design

1	A	0.7	0.3	0.0
2	Average A&B	0.55	0.45	0.0
3	B	0.4	0.6	0.0
4	Average B&C	0.25	0.6	0.15
5	C	0.1	0.6	0.3
6	Average C&D	0.1	0.55	0.35
7	D	0.1	0.5	0.4
8	Average D&E	0.2	0.4	0.4
9	E	0.3	0.3	0.4
10	Average E&A	0.5	0.3	0.2
11	Centre	0.32	0.46	0.22

2.5.5 Process Variables

Finally, it is useful to briefly mention designs for which there are two types of variables: conventional (often called *process*) variables, such as pH and temperature, and mixture variables, such as solvent composition. A typical experimental design is represented in Figure 2.35, in the case of two process variables and three mixture variables consisting of 28 experiments. Such designs are relatively straightforward to set up, using the principles of this and earlier sections, but care should be taken when calculating a model, which can become very complex. The interested reader is strongly advised to check the detailed literature as it is easy to get very confused when analysing such types of design, although it is important not to get put off, many problems in chemistry involve both types of variables and as there are often interactions, between mixture and process variables (a simple example is that the pH dependence of a reaction depends on solvent composition), such situations can be quite common.

1. A design involving varying solvent composition consisting of methanol and acetone and changing pH and temperature.

    (a) There are one mixture and two process variables.

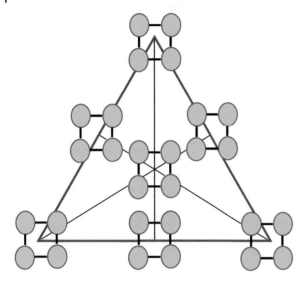

Figure 2.35 Mixture design with process variables.

(b) There are two mixture and two process variables.
(c) All variables are process variables.

2.6 Simplex Optimisation

Experimental designs can be employed for a large variety of purposes, one of the most successful being optimisation. Traditional statistical approaches normally involve forming a mathematical model of a process, and then, either computationally or algebraically, optimising this model to determine the best conditions. There are many applications, however, in which obtaining a mathematical relationship between the response and the factors that influence it are not of primary interest. Is it necessary to model precisely how pH and temperature influence the yield of a reaction? When shimming an NMR machine, is it really important to know the precise relationship between field homogeneity and resolution? In engineering, especially, methods for optimisation have been developed, which do not require a mathematical model of the system. The philosophy is to perform a series of experiments, changing the values of the control parameters, until a desired response is obtained. Statisticians may not like this approach, as it is not normally possible to calculate confidence in the model and the methods may fall down when experiments are highly irreproducible, but in practice, sequential optimisation has been very successfully applied throughout chemistry.

One of the most popular approaches is called *simplex optimisation*. A simplex is the simplest possible object in N dimensional space, for example, a line in one dimension and a triangle in two dimensions, as introduced previously (Figure 2.31). Simplex optimisation implies that a series of experiments are performed on the corners of such a figure. Most simple descriptions are of two-factor designs, where the simplex is a triangle, but, of course, there is no restriction on the number of factors.

2.6.1 Fixed Sized Simplex

The most common, and easiest to understand, method of simplex optimisation is called the *fixed sized simplex*. It is best described as a series of rules.

The main steps are as follows, exemplified by a two-factor experiment.

- Define how many factors are of interest, which we will call k.
- Perform $k + 1$ (=3 in our case) experiments on the vertices of a simplex (or triangle for two factors) in factor space. The conditions for these experiments depend on the step-size. This defines the final 'resolution' of the optimum. The smaller the step-size, the better the optimum can be defined, but the more the experiments necessary. A typical initial simplex might consist of the three experiments, for example
  - pH 3 temperature 30 °C
  - pH 3.01 temperature 31 °C
  - pH 3.02 temperature 30 °C

Figure 2.36 Initial experiments (a, b and c) on the edge of a simplex: two factors and the new conditions if experiment results in the worst response.

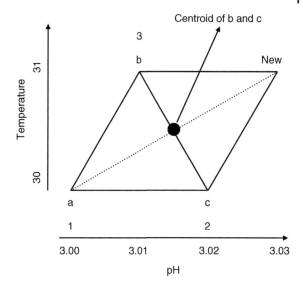

Such a triangle is illustrated in Figure 2.36. It is important to establish sensible initial conditions, especially the spacing between the experiments; in this example, one is searching very narrow pH and temperature ranges, and if the optimum is far from these conditions, the optimisation will take a long time.

- Rank the response (e.g. the yield or rate of a reaction) from 1 (worst) to $k + 1$ (best) over each of the initial conditions. Note that the response does not need to be quantitative, it could be qualitative, for example, which food tastes best. In vector form, the conditions for the nth response are given by $\boldsymbol{x}_n$, where the higher the value of n, the better the response, for example, $\boldsymbol{x}_3 = (3.01\ 31)$ implies that the best response was at pH 3.01 and 31 °C.
- Establish new conditions for the next experiment as follows:

$$\boldsymbol{x}_{new} = \boldsymbol{c} + \boldsymbol{c} - \boldsymbol{x}_1$$

where $\boldsymbol{c}$ is the centroid of the responses 2 to $k + 1$ (excluding the worst response), defined by the average of these responses represented in vector form, an alternative expression for the new conditions is $\boldsymbol{x}_{new} = \boldsymbol{x}_2 + \boldsymbol{x}_3 - \boldsymbol{x}_1$ when there are two factors. In the example above
  - if the worst response is at $\boldsymbol{x}_1 = (3.00\ 30)$
  - the centroid of the remaining responses is $\boldsymbol{c} = ((3.01 + 3.02)/2\ (30 + 31)/2) = (3.015\ 30.5)$
  - so the new response is $\boldsymbol{x}_{new} = (3.015\ 30.5) + (30.015\ 30.5) - (3.00\ 30) = (30.03\ 31)$

This is illustrated in Figure 2.36, with the centroid indicated. The new experimental conditions are often represented by reflection of the worst conditions in the centroid of the remaining conditions. Keep the points $\boldsymbol{x}_{new}$ and the kth (=2) best responses from the previous simplex, resulting in $k + 1$ new responses. The worst response from the previous simplex is rejected.

- Continue as in the two steps mentioned above, unless the new conditions result in a response that is worse than the remaining k (=2) conditions, that is, $y_{new} < y_2$ where y is the corresponding response and the aim is maximisation. In this case, return to the previous conditions and calculate

$$\boldsymbol{x}_{new} = \boldsymbol{c} + \boldsymbol{c} - \boldsymbol{x}_2$$

where $\boldsymbol{c}$ is the centroid of the responses 1 and 3 to $k + 1$ (excluding the second worst response) and can also be expressed by $\boldsymbol{x}_{new} = \boldsymbol{x}_1 + \boldsymbol{x}_3 - \boldsymbol{x}_2$, for two factors.
In the case illustrated in Figure 2.36, this would simply involve reflecting point 2 in the centroid of points 1 and 3. Keep these new conditions together with the worst and the $k - 1$ best responses from the previous simplex. The second worst response from the previous simplex is rejected; hence, in the case of three factors, we keep old responses 1, 3 and the new one, rather than old responses 2, 3 and the new one.

- Check for convergence. When the simplex is at an optimum, it normally oscillates around in a triangle or hexagon. If the same conditions reappear, stop. There are a variety of stopping rules, but it should generally be obvious when optimisation has been achieved. If you are writing a robust package, you will need to take a lot of rules into consideration, but if you are doing the experiments manually, it is simply normal to check what is happening.

The progress of a fixed sized simplex is illustrated in Figure 2.37.

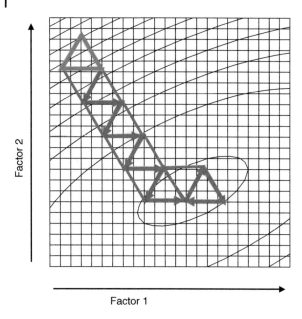

Figure 2.37 Progress of a fixed sized simplex.

Factor 2

Factor 1

1. Six factors are being optimised. How many points are in the simplex?
   (a) 3
   (b) 5
   (c) 6
   (d) 7

2. The step-size of a simplex is defined by the initial conditions.
   (a) True
   (b) False

2.6.2 Elaborations

Many elaborations have been developed over the years. One of the most important is the $k + 1$ rule. If a vertex has remained part of the simplex for $k + 1$ steps, perform the experiment again. The reason for this is that response surfaces may be noisy, so an unduly optimistic response could have been obtained because of experimental variability. This is especially important when the response surface is flat near the optimum. Another important issue relates to boundary conditions. Sometimes there are physical reasons why a condition cannot cross a boundary, an obvious case being a negative concentration. It is not always easy to deal with such situations, but it is possible to define $x_{new} = c + c - x_2$ rather than $x_{new} = c + c - x_1$ under such circumstances. If the simplex constantly tries to cross a boundary, either the constraints are a little unrealistic and so should be changed or the behaviour near the boundary needs further investigation. Starting a new simplex near the boundary with a small step-size may solve the problem.

1. Significant experimental irreproducibility may cause oscillation around an optimum.
   (a) True
   (b) False

2.6.3 Modified Simplex

A weakness with the standard method for simplex optimisation is a dependence on the initial step-size, which is defined by the initial conditions. For example, in Figure 2.36, we set a very small step-size for both variables; this may be fine if we are quite sure we are near the optimum, otherwise a bigger triangle would reach the optimum quicker. However, the problem is that the bigger step-size may miss the optimum altogether. An alternative method is called the *modified*

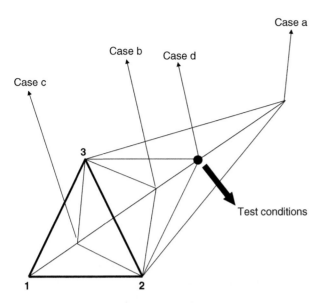

Figure 2.38 Modified simplex; the original simplex is indicated in bold, with the responses ordered from 1 (worse) to 3 (best). The test conditions are indicated.

simplex algorithm and allows the step-size to be altered, reduced as the optimum is reached or increased when far from the optimum.

For the modified simplex, we change the step $x_{new} = c + c - x_1$ of the fixed sized simplex as follows. The new response at point x_{test} is determined according to one of the four cases illustrated in Figure 2.38.

- If the response is better than all the other responses in the previous simplex, that is, $y_{test} > y_{k+1}$ then expand the simplex, so that

$$x_{new} = c + \alpha(c - x_1)$$

where α is a number greater than 1, typically equal to 2.

- If the response is better than the worst of the other responses in the previous simplex, but worse than the second worst, that is, $y_1 < y_{test} < y_2$, then *contract* the simplex but in the direction of this new response

$$x_{new} = c + \beta(c - x_1)$$

where β is a number less than 1, typically equal to 0.5.

- If the response is worse than the other responses, that is, $y_{test} < y_1$, then contract the simplex but in the opposite direction of this new response

$$x_{new} = c - \beta(c - x_1)$$

where β is a number less than 1, typically equal to 0.5.

- In all other cases, simply calculate

$$x_{new} = x_{test} = c + c - x_1$$

as in the normal (fixed-sized) simplex.

- Then, perform another experiment at x_{new} and keep this new experiment plus the k (=2 when there are three factors), best previous experiments from the previous simplex to give a new simplex.

- If the value of the response at the new vertex is less than that of the remaining k responses, we still return to the original simplex and reject the second best response, repeating the calculation as mentioned above.

There are yet further sophistications such as the super-modified simplex, which allows mathematical modelling of the shape of the response surface to provide guidelines as per the choice of the next simplex. Simplex optimisation is only one of several computational approaches to optimisation, including evolutionary optimisation, and steepest ascent methods. However, it has been much used in chemistry, largely due to the work of S. Deming and colleagues, being one of the first systematic approaches applied to the optimisation of real chemical data.

1. The modified simplex allows the step-size to both expand and contract.

    (a) False, it can only expand to reach an optimum faster.

    (b) False, it can only contract when close to an optimum to define it better.

    (c) True.

2.6.4 Limitations

In many well-behaved cases, simplex performs well and is quite an efficient approach for optimisation. There are, however, a number of limitations.

- If there is a large amount of experimental error, then the response is not very reproducible. This can cause problems, for example, when searching a fairly flat response surface.
- Sensible initial conditions and scaling (coding) of the factors are essential. This can only come from empirical chemical knowledge.
- If there are serious discontinuities in the response surface, this cannot always be taken into account.
- There is no modelling information. Simplex does not aim to predict unknown responses, produce a mathematical model or test the significance of the model using ANOVA. There is no indication of the size of interactions or related effects.

There is some controversy as to whether simplex methods should genuinely be considered as experimental designs, rather than algorithms for optimisation. Some statisticians often totally ignore this approach, and, indeed, many books and courses of experimental design in chemistry will omit simplex methods altogether, concentrating exclusively on approaches for mathematical modelling of the response surface. However, engineers and programmers have employed simplex and related approaches for optimisation for many years, and these methods have been much used, for example, in spectroscopy and chromatography, thus should be considered by the chemist. As a practical tool where the detailed mathematical relationship between response and underlying variables is not of primary concern, the methods described above are very valuable. They are also easy to implement computationally and to automate and simple to understand.

1. Simplex designs assume there are no interactions.

    (a) True

    (b) False

Problems

2.1 A Two-Factor, Two-Level Design

Section 2.2.3   Section 2.3.1

The following represents the yield of a reaction recorded at two catalyst concentrations and two reaction times

Concentration (mM)	Time (h)	Yield
0.1	2	29.8
0.1	4	22.6
0.2	2	32.6
0.2	4	26.2

1. Obtain the design matrix from the raw data, D, containing four coefficients of the form

$$y = b_0 + b_1 x_1 + b_2 x_2 + b_{12} x_1 x_2$$

2. By using this design matrix, calculate the relationship between the yield (y) and the two factors from the relationship $b = D^{-1}y$.
3. Repeat the calculations in question 2 mentioned above, but using the coded values of the design matrix.

2.2 Use of a Fractional Factorial Design to Study Factors that Influence NO Emissions in a Combustor
Section 2.2.3   Section 2.3.2
It is desired to reduce the level of NO in combustion processes for environmental reasons. Five possible factors are to be studied. The amount of NO is measured as mg/MJ fuel. A fractional factorial design was performed. The following data were obtained, using coded values for each factor.

Load	Air:fuel ratio	Primary air (%)	NH_3 (dm^3/h)	Lower secondary air (%)	NO
−1	−1	−1	−1	1	109
1	−1	−1	1	−1	26
−1	1	−1	1	−1	31
1	1	−1	−1	1	176
−1	−1	1	1	1	41
1	−1	1	−1	−1	75
−1	1	1	−1	−1	106
1	1	1	1	1	160

1. Calculate the coded values for the intercept, the linear and all two-factor interaction terms. You should obtain a matrix of 16 terms.
2. Demonstrate that there are only eight unique possible combinations in the 16 columns and indicate which terms are confounded.
3. Set up the design matrix inclusive of the intercept and five linear terms.
4. Determine the six terms arising from question 3 using the pseudo-inverse. Interpret the magnitude of the terms and comment on their significance.
5. Predict the eight responses using $\hat{y} = Db$ and calculate the percentage root mean square error, adjusted for degrees of freedom, relative to the average response.

2.3 Equivalence of Mixture Models
Section 2.5.2.2
The following data are obtained for a simple mixture design.

Factor 1	Factor 2	Factor 3	Response
1	0	0	41
0	1	0	12
0	0	1	18
0.5	0.5	0	29
0.5	0	0.5	24
0	0.5	0.5	17

1. The data are to be fitted to a model of the form

$$y = b_1 x_1 + b_2 x_2 + b_3 x_3 + b_{12} x_1 x_2 + b_{13} x_1 x_3 + b_{23} x_2 x_3.$$

Set up the design matrix, and by calculating $D^{-1}y$, determine the six coefficients.

2. An alternative model is of the form

$$y = a_0 + a_1x_1 + a_2x_2 + a_{11}x_1^2 + a_{22}x_2^2 + a_{12}x_1x_2.$$

Calculate the coefficients for this model.

3. Show, algebraically, the relationship between the two sets of coefficients, by substituting

$$x_3 = 1 - x_1 - x_2$$

into the equation for the model 1 mentioned above. Verify that the numerical terms do indeed obey this relationship and comment.

2.4 Construction of Mixture Designs
Section 2.5.3   Section 2.5.4

1. How many experiments are required for {5,1}, {5,2} and {5,3} simplex lattice designs?
2. Construct a {5,3} simplex lattice design.
3. How many combinations are required in a full five-factor simplex centroid design? Construct this design.
4. Construct a {3,3} simplex lattice design.
5. Repeat the above design using the following lower bound constraints.

$$x_1 \geq 0.0$$
$$x_2 \geq 0.3$$
$$x_3 \geq 0.4$$

2.5 Normal Probability Plots
Section 2.2.4.5

The following is a table of responses of eight experiments at coded levels of three variables, A, B, and C.

A	B	C	Response
−1	−1	−1	10
1	−1	−1	9.5
−1	1	−1	11
1	1	−1	10.7
−1	−1	1	9.3
1	−1	1	8.8
−1	1	1	11.9
1	1	1	11.7

1. It is desired to model the intercept, all single, two- and three-factor coefficients. Show that there are only eight coefficients and explain why squared terms cannot be taken into account.
2. Set up the design matrix and calculate the coefficients. Do this without using the pseudo-inverse.
3. Excluding the intercept term, there are seven coefficients. A normal probability plot can be obtained as follows. First, rank the seven coefficients in order. Then, for each coefficient of rank p, calculate a probability $(p - 0.5)/7$. Convert these probabilities into expected proportions of the normal distribution for a reading of appropriate rank using an appropriate function in Excel. Plot the values of each of the seven effects (horizontal axis) against the expected proportion of normal distribution for a reading of given rank.
4. From the normal probability plot, several terms are significant. Which are they?
5. Explain why normal probability plots work.

2.6 Use of a Saturated Factorial Design to Study Factors in the Stability of a Drug
Section 2.3.1   Section 2.2.3

The aim of the study is to determine factors that influence the stability of a drug, diethylpropion as measured by HPLC after 24 h. The higher the percentage, the better the stability.

Three factors are considered.

Factor	Level (−)	Level (+)
Moisture (%)	57	75
Dosage form	Powder	Capsule
Clorazepate (%)	0	0.7

A full factorial design is performed, with the following results, using coded values for each factor.

Factor 1	Factor 2	Factor 3	Response
−1	−1	−1	90.8
1	−1	−1	88.9
−1	1	−1	87.5
1	1	−1	83.5
−1	−1	1	91.0
1	−1	1	74.5
−1	1	1	91.4
1	1	1	67.9

1. Determine the design matrix corresponding to the model below, using coded values throughout.

$$y = b_0 + b_1 x_1 + b_2 x_2 + b_3 x_3 + b_{12} x_1 x_2 + b_{13} x_1 x_3 + b_{23} x_2 x_3 + b_{123} x_1 x_2 x_3$$

2. Using the inverse of the design matrix, determine the coefficients $b = D^{-1} y$.
3. Which of the coefficients do you feel are significant? Is there any specific interaction term that is significant?
4. The three main factors are all negative, which, without considering the interaction terms, would suggest that the best response is when all factors are at their lowest level. However, the response for the first experiment is not the highest, and this suggests that for best performance at least one factor must be at a high level. Interpret this in the light of the coefficients.
5. A fractional factorial design could have been performed using four experiments. Explain why, in this case, such a design would have missed key information.
6. Explain why the inverse of the design matrix can be used to calculate the terms in the model, rather than using the pseudo-inverse $b = (D'D)^{-1} D'y$. What changes in the design or model would require using the pseudo-inverse in the calculations?
7. Show that the coefficients in question 2 could have been calculated by multiplying the responses by the coded value of each term, summing all eight values, and dividing by 8. Demonstrate that the same answer is obtained for b_1 using both methods of calculation, and explain why.
8. From this exploratory design, it appears that two major factors and their interaction are most significant. Propose a two-factor central composite design that could be used to obtain more detailed information. How would you deal with the third original factor?

2.7 Optimisation of Assay Conditions for tRNAs Using a Central Composite Design
Section 2.4   Section 2.2.3   Section 2.2.2   Section 2.2.4.4   Section 2.2.4.3
The influence of three factors, namely pH, enzyme concentration and amino acid concentration, is to be studied on the esterification of tRNA arginyl-tRNA synthetase by counting the radioactivity of the final product, using ^{14}C-labelled arginine. The higher the count, the better the conditions.

The factors are coded at five levels as follows.

	Level	−1.7	−1	0	1	1.7
Factor 1	Enzyme (µg protein)	3.2	6.0	10.0	14.0	16.8
Factor 2	Arginine (pmoles)	860	1000	1200	1400	1540
Factor 3	pH	6.6	7.0	7.5	8.0	8.4

The results of the experiments are as follows.

Factor 1	Factor 2	Factor 3	Counts
1	1	1	4930
1	1	−1	4810
1	−1	1	5128
1	−1	−1	4983
−1	1	1	4599
−1	1	−1	4599
−1	−1	1	4573
−1	−1	−1	4422
1.7	0	0	4891
−1.7	0	0	4704
0	1.7	0	4566
0	−1.7	0	4695
0	0	1.7	4872
0	0	−1.7	4773
0	0	0	5063
0	0	0	4968
0	0	0	5035
0	0	0	5122
0	0	0	4970
0	0	0	4925

1. Using a model of the form

$$\hat{y} = b_0 + b_1 x_1 + b_2 x_2 + b_3 x_3 + b_{11} x_1^2 + b_{22} x_2^2 + b_{33} x_3^2 + b_{12} x_1 x_2 + b_{13} x_1 x_3 + b_{23} x_2 x_3$$

set up the design matrix D.
2. How many degrees of freedom are required for the model? How many are available for replication and how many are left to determine the significance of the lack-of-fit?
3. Determine the coefficients of the model using the pseudo-inverse $b = (D'D)^{-1} D'y$ where y is the vector of responses.
4. Determine the 20 predicted responses by $\hat{y} = Db$ and the overall sum of square residual error and the root mean square residual error (divide by the residual degrees of freedom). Express the latter error as a percentage of the standard deviation of the measurements. Why is it more appropriate to use a standard deviation rather than a mean in this case?
5. Determine the sum of square replicate error and, from question 4, the sum of square lack-of-fit error. Divide the sum of square residual, lack-of-fit and replicate errors by their appropriate degrees of freedom and construct a simple ANOVA table with these three errors, and compute the F-ratio.

6. Determine the variance of each of the 10 parameters in the model as follows. Compute the matrix $(\mathbf{D'D})^{-1}$ and take the diagonal elements for each parameter. Multiply these by the mean square residual error obtained in question 5 mentioned above.
7. Calculate the t-statistic for each of the 10 parameters in the model and determine which are most significant.
8. Select the intercept and five other most significant coefficients and determine a new model. Calculate the new sum of square residual error and comment.
9. Using partial derivatives, determine the optimum conditions for the enzyme assay using coded values of the three factors. Convert these to the raw experimental conditions.

2.8 Simplex Optimisation
Section 2.6
Two variables, a and b, influence a response y. These variables may, for example, correspond to pH and temperature, influencing synthetic yield. It is the aim of optimisation to find the values of a and b that give the minimum value of y.
The theoretical dependence of the response on the variables is

$$y = 2 + a^2 - 2a + 2b^2 - 3b + (a-2)(b-3)$$

Assume that this dependence is unknown in advance, but use it to generate the response for any value of the variables. Assume there is no noise in the system.
1. Using partial derivatives, show that the minimum value of y is obtained when $a = 15/7$ and compute the value of b and y at this minimum.
2. Perform simplex optimisation using as a starting point

a	b
0	0
1	0
0.5	0.866

This is done by generating the equation for y, and watching how y changes with each new set of conditions a and b. You should reach a point where the response oscillates; although the oscillation is not close to the minimum, the values of a and b giving the best overall response should be reasonable. Record each move of the simplex and the response obtained.
3. What are the estimated values of a, b and y at the minimum and why do they differ from those in question 1?
4. Perform a simplex using a smaller step-size, namely starting at

A	b
0	0
0.5	0
0.25	0.433

What are the values of a, b and y and why are they much closer to the true minimum?

2.9 Error Analysis for Simple Response Modelling
Section 2.2.2   Section 2.2.3
The following represents 12 experiments involving two factors x_1 and x_2, together with the response y.

x_1	x_2	y
0	0	5.4384
0	0	4.9845
0	0	4.3228
0	0	5.2538
−1	−1	8.7288
−1	1	0.7971
1	−1	10.8833
1	1	11.1540
1	0	12.4607
−1	0	6.3716
0	−1	6.1280
0	1	2.1698

1. By constructing the design matrix and then using the pseudo-inverse, calculate the coefficients for the best-fit model given by the equation

$$y = b_0 + b_1 x_1 + b_2 x_2 + b_{11} x_1^2 + b_{22} x_2^2 + b_{12} x_1 x_2$$

2. From these coefficients, calculate the 12 predicted responses and the residual (modelling) error as the sum of squares of the residuals.
3. Calculate the contribution to this error of the replicates simply by calculating the average response over the four replicates and then subtracting each replicate response and summing the squares of these residuals.
4. Calculate the sum of square lack-of-fit error by subtracting the value in question 3 from that in question 2.
5. Divide the lack-of-fit and replicate errors by their respective degrees of freedom and comment.

2.10 The Application of a Plackett–Burman Design to the Screening of Factors Influencing a Chemical Reaction
Section 2.3.3
7
The yield of a reaction of the form $A + B \rightarrow C$ is to be studied as influenced by 10 possible experimental conditions, listed below

Factor		Units	Low	High
x_1	% NaOH	%	40	50
x_2	Temperature	°C	80	110
x_3	Nature of catalyst		A	B
x_4	Stirring		Without	With
x_5	Reaction time	min	90	210
x_6	Volume of solvent	ml	100	200
x_7	Volume of NaOH	ml	30	60
x_8	Substrate/NaOH ratio	mol/ml	0.5×10^{-3}	1×10^{-3}
x_9	Catalyst/substrate ratio	mol/ml	4×10^{-3}	6×10^{-3}
x_{10}	Reagent/substrate ratio	mol/mol	1	1.25

The design, including an eleventh dummy factor, is given below, with the observed yields.

Expt	x_1	x_2	x_3	x_4	x_5	x_6	x_7	x_8	x_9	x_{10}	x_{11}	Yield
1	−	−	−	−	−	−	−	−	−	−	−	15
2	+	+	−	+	+	+	−	−	−	+	−	42
3	−	+	+	−	+	+	+	−	−	−	+	3
4	+	−	+	+	−	+	+	+	−	−	−	57
5	−	+	−	+	+	−	+	+	+	−	−	38
6	−	−	+	−	+	+	−	+	+	+	−	37
7	−	−	−	+	−	+	+	−	+	+	+	74
8	+	−	−	−	+	−	+	+	−	+	+	54
9	+	+	−	−	−	+	−	+	+	−	+	56
10	+	+	+	−	−	−	+	−	+	+	−	64
11	−	+	+	+	−	−	−	+	−	+	+	65
12	+	−	+	+	+	−	−	−	+	−	+	59

1. Why is a dummy factor employed? Why is a Plackett–Burman design more desirable than a two-level fractional factorial in this case?
2. Verify that all the columns are orthogonal to each other.
3. Set up a design matrix, D, and determine the coefficients b_0 to b_{11}.
4. An alternative method for calculating the coefficients for factorial designs such as the Plackett–Burman design is to multiply the yields of each experiment by the levels of the corresponding factor, summing these and dividing by 12. Verify that this provides the same answer as using the inverse matrix for factor 1.
5. A simple method for reducing the number of experimental conditions for further study is to look at the size of the factors and eliminate those that are less than the dummy factor. How many factors remain and what are they?

2.11 Use of a Constrained Mixture Design to Investigate the Conductivity of a Molten Salt System
Section 2.5.4 Section 2.5.2.2
A molten salt system consisting of three components is prepared, and the aim is to investigate the conductivity according to the relative proportion of each component. The three components are given below.

Component		Lower limit	Upper limit
x_1	$NdCl_3$	0.2	0.9
x_2	LiCl	0.1	0.8
x_3	KCl	0.0	0.7

The experiment is coded to give pseudo-components so that a value of 1 corresponds to the upper limit (see above) and a value of 0 to the lower limit of each component. The experimental results are as follows.

z_1	z_2	z_3	Conductivity (Ω^{-1} cm^{-1})
1	0	0	3.98
0	1	0	2.63
0	0	1	2.21
0.5	0.5	0	5.54
0.5	0	0.5	4.00
0	0.5	0.5	2.33
0.3333	0.3333	0.3333	3.23

1. Represent the constrained mixture space, diagrammatically, in the original mixture space. Explain why the constraints are possible and why the new reduced mixture space remains a triangle.
2. Produce a design matrix consisting of seven columns in the true mixture space as follows. The true composition of a component 1 is given by $Z_1 (U_1 - L_1) + L_1$ where U and L are the upper and lower bounds for the component. Convert all the three columns of the matrix given above using this equation and then set up a design matrix, containing three single-factor terms, and all possible two- and three-factor interaction terms (using Sheffé model).
3. Calculate the model linking the conductivity to the proportions of the three salts.
4. Predict the conductivity when the proportion of the salts is 0.209, 0.146 and 0.645.

2.12 Use of Experimental Design and Principal Components Analysis for Reduction of Number of Chromatographic Tests

Section 2.4.5 Section 4.3 Section 4.6.4 Section 4.8.1

The following table represents the result of a number of tests performed on eight chromatographic columns, involving performing chromatography on eight compounds at pH 3 in methanol mobile phase and measuring four peak-shaped parameters. Note that you may have to transpose the matrix in Excel for further work. The aim is to reduce the number of experimental tests necessary using experimental design. Each test is denoted by a mnemonic. The first letter (e.g. P) stands for a compound, the second part of the name, k, N, N(df), or As standing for four peak-shaped/retention time measurements.

	Inertsil ODS	Inertsil ODS-2	Inertsil ODS-3	Kromasil C-18	Kromasil C8	Symmetry C18	Supelco ABZ+	Purospher
Pk	0.25	0.19	0.26	0.3	0.28	0.54	0.03	0.04
PN	10 200	6 930	7 420	2 980	2 890	4 160	6 890	6 960
PN(df)	2 650	2 820	2 320	293	229	944	3 660	2 780
PAs	2.27	2.11	2.53	5.35	6.46	3.13	1.96	2.08
Nk	0.25	0.12	0.24	0.22	0.21	0.45	0	0
NN	12 000	8 370	9 460	13 900	16 800	4 170	13 800	8 260
NN(df)	6 160	4 600	4 880	5 330	6 500	490	6 020	3 450
NAs	1.73	1.82	1.91	2.12	1.78	5.61	2.03	2.05
Ak	2.6	1.69	2.82	2.76	2.57	2.38	0.67	0.29
AN	10 700	14 400	11 200	10 200	13 800	11 300	11 700	7 160
AN(df)	7 790	9 770	7 150	4 380	5 910	6 380	7 000	2 880
AAs	1.21	1.48	1.64	2.03	2.08	1.59	1.65	2.08
Ck	0.89	0.47	0.95	0.82	0.71	0.87	0.19	0.07
CN	10 200	10 100	8 500	9 540	12 600	9 690	10 700	5 300
CN(df)	7 830	7 280	6 990	6 840	8 340	6 790	7 250	3 070
CAs	1.18	1.42	1.28	1.37	1.58	1.38	1.49	1.66
Qk	12.3	5.22	10.57	8.08	8.43	6.6	1.83	2.17
QN	8 800	13 300	10 400	10 300	11 900	9 000	7 610	2 540
QN(df)	7 820	11 200	7 810	7 410	8 630	5 250	5 560	941
QAs	1.07	1.27	1.51	1.44	1.48	1.77	1.36	2.27
Bk	0.79	0.46	0.8	0.77	0.74	0.87	0.18	0
BN	15 900	12 000	10 200	11 200	14 300	10 300	11 300	4 570
BN(df)	7 370	6 550	5 930	4 560	6 000	3 690	5 320	2 060
Bas	1.54	1.79	1.74	2.06	2.03	2.13	1.97	1.67
Dk	2.64	1.72	2.73	2.75	2.27	2.54	0.55	0.35
DN	9 280	12 100	9 810	7 070	13 100	10 000	10 500	6 630

	Inertsil ODS	Inertsil ODS-2	Inertsil ODS-3	Kromasil C-18	Kromasil C8	Symmetry C18	Supelco ABZ+	Purospher
DN(df)	5 030	8 960	6 660	2 270	7 800	7 060	7 130	3 990
Das	1.71	1.39	1.6	2.64	1.79	1.39	1.49	1.57
Rk	8.62	5.02	9.1	9.25	6.67	7.9	1.8	1.45
RN	9 660	13 900	11 600	7 710	13 500	11 000	9 680	5 140
RN(df)	8 410	10 900	7 770	3 460	9 640	8 530	6 980	3 270
RAs	1.16	1.39	1.65	2.17	1.5	1.28	1.41	1.56

1. Transpose the data so that the 32 tests correspond to columns of a matrix (variables) and the eight chromatographic columns to the rows of a matrix (objects). Standardise each column by subtracting the mean and dividing by the population standard deviation (Section 4.6.4). Why is it important to standardise this data?
2. Perform PCA (principal components analysis) on this data and retain the first three loadings (methods for performing PCA are discussed in Section 4.3; see also Section A.2.1 and relevant Sections A.4 and A.5 if you are using Excel or Matlab).
3. Take the three loading vectors and transform to a common scale as follows. For each loading vector, select the most positive and most negative value and code these to +1 and −1, respectively. Scale all the intermediate values in a similar manner, leading to a new scaled loading matrix of 32 columns and three rows. Produce the new scaled loading vectors.
4. Select a factorial design as follows, with one extra point in the centre, to obtain a range of tests, which is a representative subset of the original tests.

Design point	PC1	PC2	PC3
1	−	−	−
2	+	−	−
3	−	+	−
4	+	+	−
5	−	−	+
6	+	−	+
7	−	+	+
8	+	+	+
9	0	0	0

Calculate the Euclidean distance of each of the 32 scaled loadings from each of the nine design points; for example, the first design point calculates the Euclidean distance of the loadings scaled as in question 3 from the point $(-1,-1,-1)$, by the equation $d_1 = \sqrt{(p_{11} + 1)^2 + (p_{12} + 1)^2 + (p_{13} + 1)^2}$ (Section 4.8.1).
5. Indicate the chromatographic parameters closest to the nine design points. Hence, recommend a reduced number of chromatographic tests and comment on the strategy.

2.13 A Mixture Design with Constraints
Section 2.5.4
It is desired to perform a three-factor mixture design with constraints on each factor as follows:

	x_1	x_2	x_3
Lower	0.0	0.2	0.3
Upper	0.4	0.6	0.7

1. The mixture design is normally represented as an irregular polygon, with, in this case, six vertices. Calculate the percentage of each factor at the six co-ordinates.
2. It is desired to perform 13 experiments, namely, on the six corners, in the middle of the six edges and in the centre. Produce a table of 13 mixtures.
3. Represent the experiment diagrammatically.

2.14 Construction of Five-Level Calibration Designs
Section 2.3.4

The aim is to construct a five-level partial factorial (or calibration) design involving 25 experiments and up to 14 factors, each at levels −2, −1, 0, 1 and 2. Note that this design is only one of many possible such designs.

1. Construct the experimental conditions for the first factor using the following rules:
   - The first experiment is at level −2.
   - This level is repeated for experiments 2, 8, 14 and 20.
   - The levels for experiments 3–7 are given as follows (0, 2, 0, 0, 1):
   - A cyclic permuter of the form $0 \to -1 \to 1 \to 2 \to 0$ is then used. Each block of experiments 9–13, 15–19 and 21–25 are related by this permuter, each block being one permutation away from the previous block; hence, experiments 9 and 10 are at levels −1 and 0, for example.
2. Construct the experimental conditions for the other 13 factors as follows:
   - Experiment 1 is always at level −2 for all factors.
   - The conditions for experiments 2–24 for the other factors are simply the cyclic permutation of the previous factor as explained in Section 2.3.4 and produce the matrix of experimental conditions.
3. What is the difference vector used in this design?
4. Calculate the correlation coefficients between all pairs of factors 1–14. Plot the two graphs of the levels of factor 1 versus factors 2 and 7. Comment.

2.15 A Four-Component Mixture Design Used for Blending of Olive Oils
Section 2.5.2.2

Fourteen blends of olive oils from four cultivars A to D are mixed together in the design below presented together with a taste panel score for each blend. The higher the score, the better the taste of the olive oil.

A	B	C	D	Score
1	0	0	0	6.86
0	1	0	0	6.50
0	0	1	0	7.29
0	0	0	1	5.88
0.5	0.5	0	0	7.31
0.5	0	0.5	0	6.94
0.5	0	0	0.5	7.38
0	0.5	0.5	0	7.00
0	0.5	0	0.5	7.13
0	0	0.5	0.5	7.31
0.33333	0.33333	0.33333	0	7.56
0.33333	0.33333	0	0.33333	7.25
0.33333	0	0.33333	0.33333	7.31
0	0.33333	0.33333	0.33333	7.38

1. It is desired to produce a model containing 14 terms, namely four linear, six two-component and four three-component terms. What is the equation for this model?

2. Set up the design matrix and calculate the coefficients.
3. A good way to visualise the data is via contours in a mixture triangle, allowing three components to vary and constraining the fourth to be constant. Using step-size of 0.05, calculate the estimated responses from the model in question 3 when D is absent and $A + B + C = 1$. A table of 231 numbers should be produced. Using a contour plot, visualise this data. If you use Excel, upper right-hand half of the plot may contain meaningless data; to remove this, simply cover up this part of the contour plot by a white triangle. In modern versions of Matlab and some other software packages, triangular contour plots can be obtained straightforwardly. Comment on the optimal blend using the contour plot when D is absent.
4. Repeat the contour plot in question 3 for the following: (i) $A + B + D = 1$, (ii) $B + C + D = 1$, (iii) $A + C + D = 1$ and comment.
5. Why, in this example, is a strategy of visualisation of the mixture contours probably more informative than calculating a single optimum?

2.16 Central Composite Design Used to Study the Extraction of Olive Seeds in a Soxhlet Extractor
Section 2.4 Section 2.2.2
Three factors, namely (1) irradiation power as a percentage; (2) irradiation time in s; and (3) number of cycles are used to study the focussed microwave-assisted Soxhlet extraction of olive oil seeds, the response measuring the percentage recovery, which is to be optimised. A central composite design is set up to perform the experiments. The results are given below, using coded values of the variables.

Factor 1	Factor 2	Factor 3	Response
−1	−1	−1	46.64
−1	−1	1	47.23
−1	1	−1	45.51
−1	1	1	48.58
1	−1	−1	42.55
1	−1	1	44.68
1	1	−1	42.01
1	1	1	43.03
−1	0	0	49.18
1	0	0	44.59
0	−1	0	49.22
0	1	0	47.89
0	0	−1	48.93
0	0	1	49.93
0	0	0	50.51
0	0	0	49.33
0	0	0	49.01
0	0	0	49.93
0	0	0	49.63
0	0	0	50.54

1. A 10-parameter model is to be fitted to the data, consisting of the intercept, all single-factor linear and quadratic terms and all two-factor interaction terms. Set up the design matrix, and by using the pseudo-inverse, calculate the coefficients of the model using coded values.
2. The true values of the factors are given in the table below.

Variable	−1	+1
Power (%)	30	60
Time (s)	20	30
Cycles	5	7

Re-express the model in question 1 in terms of the true values of each variable, rather than the coded values.

3. Using the model in question 1 and the coded design matrix, calculate the 20 predicted responses, and the total sum of square error for the 20 experiments.

4. Determine the sum of squares replicate error as follows: (i) Calculate the mean response for the six replicates. (ii) Calculate the difference between the true and average response, square these and sum the six numbers.

5. Determine the sum of square lack-of-fit error as follows. (i) Replace the six replicate responses by the average response for the replicates. (ii) Using the 20 responses (with the replicates averaged) and the corresponding predicted responses, calculate the differences, square them and sum them.

6. Verify that the sums of squares in questions 4 and 5 add up to the total error obtained in question 3.

7. How many degrees of freedom are available for assessment of the replicate and lack-of-fit errors? Using this information, comment on whether the lack-of-fit is significant, and hence whether the model is adequate.

8. The significance each term can be determined by omitting the term from the overall model. Assess the significance of the linear term due to the first factor and the interaction term between the first and third factors in this way. Calculate a new design matrix with nine rather than 10 columns, removing the relevant column, and also remove the corresponding coefficients from the equation. Determine the new predicted responses using nine factors and calculate the increase in sum of square error over that obtained in question 3. Comment on the significance of these two terms.

9. Using coded values, determine the optimum conditions as follows. Discard the two interaction terms that are least significant, resulting in eight remaining terms in the equation. Obtain the partial derivatives with respect to each of the three variables and set up three equations equal to 0. Show that the optimum value of the third factor is given by $-b_3/(2b_{33})$ where the coefficients correspond to the linear and quadratic terms in the equations. Hence, calculate the optimum coded values for each of the three factors.

10. Determine the optimum true values corresponding to the conditions obtained in question 9. What is the percentage recovery at this optimum? Comment.

2.17 A Three-Component Mixture Design
Section 2.5.2
A three-factor mixture simplex centroid mixture design is performed, with the results as given below.

x_1	x_2	x_3	Response
1	0	0	9
0	1	0	12
0	0	1	17
0.5	0.5	0	3
0.5	0	0.5	18
0	0.5	0.5	14
0.3333	0.3333	0.3333	11

1. A seven-term model consisting of three linear terms, three two-factor interaction terms and one three-factor interaction term is fit to the data. Give the equation for this model, compute the design matrix and calculate the coefficients.

2. Instead of seven terms, it is decided to fit the model only to the three linear terms. Calculate these coefficients using only three terms in the model employing the pseudo-inverse. Determine the root mean square error for the predicted responses, comment on the difference in the linear terms in question 1 and the significance of the interaction terms.

3. It is possible to convert the model of question 1 to a seven-term model in two independent factors, consisting of two linear terms, two quadratic terms, two linear interaction terms and a quadratic term of the form $x_1 x_2 (x_1 + x_2)$. Show how the models relate algebraically.

4. For the model in question 3, set up the design matrix, calculate the new coefficients and show how these relate to the coefficients calculated in question 1 using the relationship obtained in question 3.

5. The matrices in questions 1, 2 and 4 all have inverses. However, a model that consisted of an intercept term and three linear terms would not, and it is impossible to use regression analysis to fit the data under such circumstances. Explain these observations.

3

Signal Processing

3.1 Introduction

Sequential signals are surprisingly widespread in the laboratory and require a large number of methods for analysis. Most data are obtained via computerised instruments such as near infrared (NIR), high performance liquid chromatography (HPLC) or nuclear magnetic resonance (NMR), and the nature of information such as peak integrals, shifts and positions is often dependent on how the information from the computer is first processed. An appreciation of this step is often essential before applying further multivariate methods such as pattern recognition or classification. Spectra and chromatograms are examples of series that are sequential in time or frequency. However, time series also occur very widely, for example, in industrial process control and natural processes.

3.1.1 Environmental and Geological Processes

An important source of data involves recording of samples regularly with time. Classically, such time series occur in environmental chemistry and geochemistry. A river might be sampled for the presence of pollutants such as polyaromatic hydrocarbons or heavy metals at different times of the year. Is there a trend and can this be related to seasonal factors? Different and fascinating processes occur in rocks, where depth in the sediment relates to burial time. For example, isotope ratios are a function of climate, as relative evaporation rates of different isotopes are dependent on temperature: certain specific cyclical changes in the Earth's rotation have resulted in the ice ages and so climate changes, leaving a systematic chemical record. A whole series of methods for time series analysis primarily based on the idea of correlograms (Section 3.4) can be applied to explore such types of cyclicity, which are often quite hard to elucidate. Many of these approaches were first used by economists and geologists who also encountered related problems.

One of the difficulties is that long-term and interesting trends are often buried within short-term random fluctuations. Statisticians distinguish among various types of noise that interfere with the signal, as discussed in Section 3.2.3.

In addition to obtaining correlograms, a large battery of methods are available to smooth time series, many of which are based on the so-called 'windows', whereby data are smoothed over a number of points in time. A simple method is to take the average reading over five points in time, but sometimes, this could miss out important information about cyclicity, especially for a process that is sampled slowly compared with the rate of oscillation. A number of linear filters have been developed, which are applicable to this type of data (Section 3.3); this procedure is often being described as convolution.

3.1.2 Industrial Process Control

In industry, sequential series may occur in the manufacturing process of a product. It could be crucial that a drug has a certain well-defined composition; otherwise an entire batch is unmarketable. Sampling the product regularly in time is essential for two reasons. The first is monitoring, simply to determine whether the quality is within the acceptable limits. The second is for control, to predict the future and check whether the process is getting out of control. It is costly to destroy a batch and not economically satisfactory to obtain information about acceptability several days after the event. As soon as the process begins to go wrong, it is often advisable to stop the plant and investigate. However, too many false alarms can be equally inefficient. A whole series of methods have been developed for the control of manufacturing processes, an area where chemometrics can often play a key and crucial role. In this chapter, we will not be discussing statistical control charts in detail, the whole topic being worthy of a book in its own rights. However, a number of methods outlined in this chapter are useful for handling such sequential processes,

Chemometrics: Data Driven Extraction for Science, Second Edition. Richard G. Brereton.
© 2018 John Wiley & Sons Ltd. Published 2018 by John Wiley & Sons Ltd.
Companion website: http://booksupport.wiley.com

especially to determine whether there are long-term trends that are gradually influencing the composition or nature of a manufactured product. Several linear filters together with modifications such as running median smoothing (RMS) (Section 3.3) can be employed under such circumstances. Chemometricians are especially interested in the extension to multivariate methods, for example, monitoring a spectrum as recorded regularly in time, which will be outlined in detail in later chapters.

3.1.3 Chromatograms and Spectra

The most common applications of methods for handling sequential series in the laboratory arise in chromatography and spectroscopy and will be emphasised in this chapter. An important aim is to smooth a chromatogram. A number of methods such as the Savitzky–Golay filter have been developed here (Section 3.3.1.2). A problem is that if a chromatogram is smoothed too much, the peaks are blurred and lose resolution, negating the benefits; thus, optimal filters have been developed that remove noise without broadening peaks excessively.

Another common need is to increase resolution, and sometimes spectra are routinely displayed in the derivative mode (e.g. electron spin resonance spectroscopy): there are a number of rapid computational methods for such calculations that do not emphasise noise too much (Section 3.3.2). Other approaches based on curve fitting and Fourier filters are also very common.

3.1.4 Fourier Transforms

The Fourier transform (FT) has revolutionised spectroscopy such as NMR and IR since the 1950s. The raw data are not obtained as a comprehensible spectrum but as a time series, where all spectroscopic information is mixed up and a mathematical transformation is required to obtain a comprehensible spectrum. One reason for performing FT spectroscopy is that a spectrum of acceptable signal-to-noise ratio can be recorded much more rapidly than via conventional spectrometers, often a hundred times more rapidly. This has allowed the development, for example, of ^{13}C NMR, as a routine analytical tool, because the low abundance of ^{13}C is compensated by faster data acquisition. However, special methods are required to convert this 'time domain' information (called a *free induction decay* (FID) in NMR parlance) into a 'frequency domain' spectrum, which can be interpreted directly (see Section 3.5.1).

Parallel with Fourier transform, spectroscopy has arisen a large number of approaches for enhancement of the quality of such data, often called *Fourier deconvolution*, involving manipulating the time series before Fourier transformation (Section 3.5.2). Many of these filters have their origins in engineering and are often described as digital filters. These are quite different to the classical methods for time series analysis used in economics or geology. Sometimes, it is even possible to take non-Fourier data, such as a normal spectrum, and Fourier transform it back to a time series, then use Fourier deconvolution and Fourier transform back again, often called *Fourier self-deconvolution*.

Fourier filters can be related to linear methods discussed in Section 3.3 by the convolution theorem (as discussed in Section 3.5.3).

3.1.5 Advanced Methods

In data analysis, there will always be new computational approaches that promote great interest among statisticians and computer scientists. Much frontline research in chemometrics is involved refining such methods, but it takes several years before the practical worth or otherwise novel data analytical approaches is demonstrated.

Methods for so-called *non-linear deconvolution* have been developed over the past few years, one of the most well-known methods being maximum entropy (Section 3.6.4). This latter approach was first used in infrared astronomy to deblur weak images of the sky and been successfully applied, for example, to police photography to determine car number plates from poor photographic images of a moving car in a crime. Enhancing the quality of a spectrum can also be regarded as a form of image enhancement and thus use similar computational approaches. A very successful application is in NMR imaging for medical tomography. The methods are called *non-linear* because they do not insist that the improved image is a linear function of the original data. A number of other approaches are also available in the literature, but maximum entropy has received much publicity largely because of the readily available software.

Wavelet transforms (Section 3.6.2) involve fitting a spectrum or chromatogram to a series of functions based on a basic shape called a *wavelet*, of which there are several in the literature. These transforms have the advantage in that, instead of storing, for example, 1024 spectral data points, it may be possible to retain only a few most significant wavelets and still not lose much information. This can result in both data compression and denoising of data.

Rapid algorithms for real-time filtering have attracted much interest among engineers and can be used to follow a process by smoothing the data as it occur. The Kalman filter is one such method (Section 3.6.1) that has been reported extensively in the analytical chemistry literature. Early textbooks of chemometrics often featured Kalman filters; although they still remain a tool, they are less prominent in the chemometrics literature nowadays but are widely used by engineers.

3.2 Basics

3.2.1 Peak shapes

Chromatograms and spectra are normally considered to consist of a series of peaks, superimposed upon noise. Each peak usually arises either from a characteristic absorption (such as a chromophore) or a characteristic compound. In most cases, the underlying peaks are distorted for a variety of reasons such as noise, blurring or overlap with neighbouring peaks. A major aim of chemometric methods is to obtain the underlying, undistorted information.

Peaks can be characterised in a number of ways, but a common approach, for symmetrical peaks as illustrated in Figure 3.1, is to characterise each peak by

1) a position at the centre (e.g. the elution time or spectral frequency),
2) a width, normally at half height and
3) an area.

For symmetrical peaks, the position in the centre is also normally the position of maximum height (we will define peaks as being unimodal) and the mean. Peak shapes can usually be related to statistical distributions, and in quantum mechanical terms are in fact such.

The relationship between area and peak height is dependent on peak shape, as discussed below, although heights are often easier to measure. If a series of peaks have the same shape, then the ratios of heights are proportional to ratios of areas. However, area is usually a better measure of chemical properties such as concentration and it is important to obtain precise information about peak shapes before relying on heights, for example, as raw data for pattern recognition programs.

Sometimes, the width at a different percentage of the peak height is cited rather than the half width. A further common measure is when the peak has decayed to a small percentage of the overall height (e.g. 1%), which is often taken as the total width of the peak, or alternatively has decayed to a size that relates to the noise level, for example, the root mean square noise.

In many cases of spectroscopy, peak shapes can be very precisely predicted, for example, from quantum mechanics, such as in NMR or visible spectroscopy. In other situations, the peak shape is dependent on complex physical processes, for example, in chromatography, and can only be modelled empirically. In the latter situation, it is not always practicable to obtain an exact model, and a number of closely similar empirical estimates will give equally useful information.

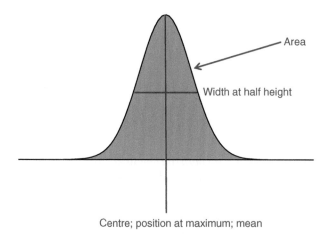

Figure 3.1 Main parameters that characterise a symmetric peak.

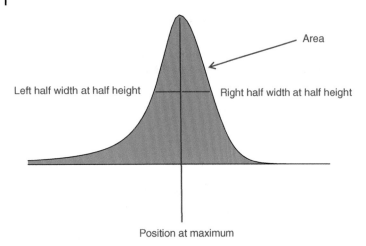

Figure 3.2 Main parameters that characterise an asymmetric peak.

For asymmetric peak shapes, it is normal to use the position maximum rather than the mean and to quote the left and right half widths at half height, as illustrated in Figure 3.2.

A few common peak shapes cover most situations. If these general peak shapes are not suitable for a particular purpose, it is probably best to consult specialised literature on the particular measurement technique. Note that unless there is a good reason, for example, from knowledge of quantum mechanics or a specific type of spectroscopy, it is usual to empirically model peak shapes with basic functions.

1. The relative peak heights of resolved compounds in a chromatogram are always proportional to relative concentrations.
   (a) True
   (b) False

3.2.1.1 Gaussians

These peak shapes are common in most types of chromatography and spectroscopy. A simplified formula for a Gaussian is given by

$$x_i = A \exp(-(x_i - x_0)^2/s^2)$$

where

- A is the height at the centre,
- x_0 is the position of the centre and
- s relates to the peak width.

Gaussians are based on a normal distribution where

- x_0 corresponds to the mean of a series of measurements and
- $s/\sqrt{2}$ to the standard deviation.

It can be shown that

- the width at half height of a Gaussian peak is given by $\Delta_{1/2} = 2s\sqrt{\ln 2}$ and
- the area by $\sqrt{\pi}A\,s$ using the equation presented above: note that this depends on *both* the height and the width. Note that the height and area are in the original units; hence, if the width is measured in seconds and height in absorbance unit (AU), it will be in units of AU $*$ s.

Notice that Gaussians are also the statistical basis of the normal distribution, see Section A.3.2, but the equation for the standard normal distribution is normally scaled so that the area under the curve equals 1. For signal analysis, we will use the simplified expression above.

1. A Gaussian peak has a height of 5 units and a width at half height of 2 units. Its area in the original units^2 is

    (a) 10
    (b) 10.64
    (c) 16.15

2. A peak has an area of 25 units^2; the intensity is measured in AUs and the horizontal axis in minutes. What is its area in AU * s?

    (a) 0.416
    (b) 25
    (c) 1500

3.2.1.2 Lorentzians

The Lorentzian peak shape corresponds to a statistical function called the *Cauchy distribution*. It is less common but often arises in certain types of spectroscopy such as NMR. A simplified formula for a Lorentzian is given by

$$x_i = A/(1 + (x_i - x_0)^2/s^2)$$

where

- A is the height at the centre,
- x_0 is the position of the centre and
- s relates to the peak width.

It can be shown that

- the width at half height of a Lorentzian peak is given by $\Delta_{1/2} = 2s$ and
- the area by $\pi A s$: note that this depends on *both* the height and the width as per Gaussian.

The main difference between Gaussian and Lorentzian peak shapes is that the latter has a bigger tail, as illustrated in Figure 3.3, for two peaks with identical half widths and heights.

1. A Lorentzian peak has a height of 5 units and a width at half height of 2 units. Its area in the original units^2 is

    (a) 8.86
    (b) 10
    (c) 15.71

2. Lorentzian peaks have more pronounced tails than Gaussian peaks.

    (a) True
    (b) False

3.2.1.3 Asymmetric Peak Shapes

In many forms of chromatography, it is hard to obtain symmetric peak shapes. Although a number of quite sophisticated models are available, a very simple first approximation is that of a Gaussian/Lorentzian peak shape. Figure 3.4(a)

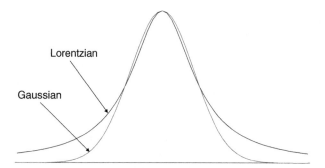

Figure 3.3 Gaussian and Lorentzian peak shapes of equal half heights.

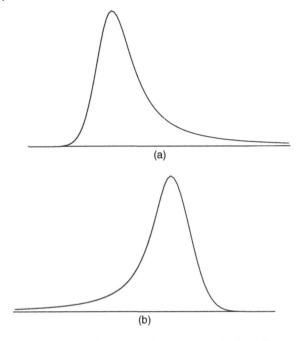

(a)

(b)

Figure 3.4 Asymmetric peak shapes often described by a Gaussian/Lorentzian model. (a) Tailing: left is Gaussian and right is Lorentzian (b) Fronting: left is Lorentzian and right is Gaussian.

represents a tailing peak shape, in which the left-hand side can be modelled by a Gaussian and the right-hand side by a Lorentzian. A fronting peak is illustrated in Figure 3.4(b): such peaks are much rarer.

1. The left-hand side of asymmetric peak is well modelled by a Lorentzian but the right-hand side by a Gaussian. The peak is
   (a) Tailing
   (b) Fronting

3.2.1.4 Use of Peak Shape Information
Peak shape information can be employed in two principal ways.

- *Curve fitting* is quite common. There are a variety of computational algorithms, most involving some type of least squares minimisation. If there are suspected (or known) to be three peaks in a cluster, of Gaussian shape, then nine parameters need to be found, namely the three peak positions, peak widths and peak heights. In any curve fitting, it is important to determine whether there is certain knowledge of the peak shapes, how many peaks are there and are of certain features, for example, the positions of each component, in advance. It is also important to appreciate that much chemical data are not of sufficient quality for very detailed models. In chromatography, an empirical approach is normally adequate: over-modelling can be dangerous. The result of the curve fitting can be a better description of the system; for example, by knowing peak areas, it may be possible to determine relative concentrations of components in a mixture. Note that if there is no prior knowledge of a system that could consist of a complex cluster of unresolved peaks, many curve fitting algorithms will fail; hence, it is usual to constrain the methods, for example, using known information to introduction restrictions to the model.
- Simulations also have an important role in chemometrics. Such simulations are a way of trying to understand a system. If the result of a chemometric method (such as multivariate curve resolution – see Chapter 7) results in reconstructions of peaks that are close to the real data, then the underlying peak shapes provide a good description. Simulations are also used to explore how well different techniques work and under what circumstances they break down.

A typical chromatogram or spectrum consists of several peaks, at different positions, of different intensities and sometimes of different shapes. Figure 3.5 represents a cluster of three peaks, together with their total intensity. Although the right-hand side peak pair is quite easy to resolve visually, this is not true for the left-hand side peak pair, and it would be especially hard to identify the position and intensity of the first peak of the cluster without using some form of data analysis.

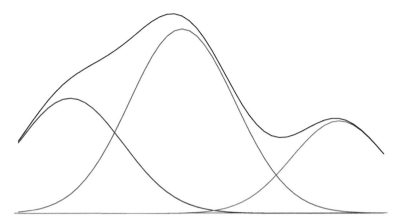

Figure 3.5 Three peaks forming a cluster.

1. Curve fitting is most appropriate when there is little or no information known about a cluster of partially overlapping peaks.

   (a) True
   (b) False

3.2.2 Digitisation

Almost all modern laboratory-based data are now obtained via computers and are acquired in a digitised rather than an analogue form. It is always important to understand how digital resolution influences the ability to resolve out peaks.

Many techniques for recording information result in only a small number of data points per peak. A typical NMR peak may be only a couple of hertz at half width, especially using well-resolved instrumentation. Yet, a spectrum recorded at 500 MHz, where 16 K (=16 384) data points are used to represent 10 ppm (or 5000 Hz), involves each data point representing $3.28 = 16\,384/5000$ Hz. A 2 Hz peak width is represented by only 6.56 data points. In coupled chromatography, a typical sampling rate may be 2 s; yet, peak half widths may be 20 s or less and interesting compounds separated by 30 s. Poor digital resolution can influence the ability to obtain information. It is useful to be able to determine how serious these errors are.

Consider a Gaussian peak, with a true width at half height of 16.65 units and a height of 1 unit. The theoretical area can be calculated using the equations of Section 3.2.1.1.

- The width at half height is given by $2s\sqrt{\ln 2}$, so that $s = 16.65/(2\sqrt{\ln 2}) = 10$.
- The area is given by $\sqrt{\pi}A\,s$, but $A = 1$, so that the area is $\sqrt{\pi}10 = 17.725\,\text{units}^2$.

Typical units might be AU $*$ s if the sampling time is in s and the intensity in absorption units.

Consider the effect of digitising this peak at different rates, as indicated in Table 3.1 and illustrated in Figure 3.6. An easy way of determining integrated intensities is simply to sum the product of the intensity at each data point (x_i) by the sampling interval (δ) over a sufficiently wide range, that is, to calculate $\delta\sum x_i$. The estimates are given in Table 3.1, and it can be seen that for the worst digitised peak (at 20 units, or once per half height), the estimated integral is 20.157, an error of 13.7%.

A feature of Table 3.1 is that acquisition of data starts at exactly two data points in each case. In practice, the precise start of acquisition cannot easily be controlled and is often irreproducible, and it is easy to show that when poorly digitised, estimated integrals and apparent peak shapes will depend on this offset. In practice, the instrumental operator will notice a bigger variation in estimated integrals if digital resolution is low. Although peak widths must approach digital resolution for significant errors in integration, in some techniques such as gas chromatography mass spectrometry (GC-MS) or nuclear magnetic resonance (NMR), this condition is often obtained. In many situations, instrumental software is used to smooth or interpolate the data, and many users are unaware that this step has automatically taken place. These simple algorithms can result in considerable further distortions in quantitative parameters.

A second factor that can influence quantitation is digital resolution in the intensity and direction (or vertical scale in the graph). This is due to the analogue to digital converter (ADC) and sometimes can be experimentally corrected by

Table 3.1 Reducing digital resolution.

1 point in 8		1 point in 12		1 point in 20	
Time	Intensity	Time	Intensity	Time	Intensity
2	0.0000	2	0.0000	2	0.0000
10	0.0000	14	0.0000	22	0.0000
18	0.0000	26	0.0000	42	0.0392
26	0.0000	38	0.0079	62	0.9608
34	0.0012	50	0.3679	82	0.0079
42	0.0392	62	0.9608	102	0.0000
50	0.3679	74	0.1409		
58	0.9608	86	0.0012		
66	0.6977	98	0.0000		
74	0.1409	110	0.0000		
82	0.0079				
90	0.0001				
98	0.0000				
106	0.0000				
114	0.0000				
Integral	17.725		17.743		20.157

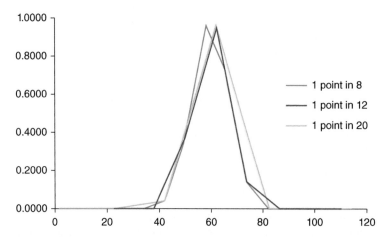

Figure 3.6 Influence on the appearance of a peak as digital resolution is reduced corresponding to Table 3.1.

changing the receiver gain. However, for most modern instrumentation, this limitation is not so serious and, therefore, will not be discussed in detail below, but is illustrated in Problem 3.7.

1. In NMR, a peak has a width at half height of 4 Hz. A spectrum is represented by 2^{13} data points recorded over 6000 Hz. Its width at half height in data points is
   (a) 10.92
   (b) 5.46
   (c) 2.73

3.2.3 Noise

Imposed on signals is noise. In basic statistics, the nature and origin of noise are often unknown but assumed to obey a normal distribution. Indeed, many statistical tests such as the *t*-test and *F*-test (see Sections A.3.4 and A.3.5) assume this and are only approximations in the absence of experimental study of such noise distributions. However, a theorem called the *central limit theorem* suggests that most symmetrical distributions do tend towards a normal distribution in the centre, which is, therefore, a good approximation in the absence of other information.

In laboratory-based chemistry, there are two fundamental sources of noise in instrumental measurements.

- The first involves sample preparation, for example, dilution, weighing and extraction efficiency. We will not discuss these types of errors in this chapter, but can be minimised by good analytical procedures.
- The second is inherent to the measurement technique. No instrument is perfect; thus, the signal is imposed upon noise. The observed signal is given by

$$x = \tilde{x} + e$$

where $\tilde{x}$ is the 'perfect' or true signal and e represents noise. The aim of most signal-processing techniques is to obtain estimates of the true underlying signal in the absence of noise, that is, to separate the signal from the noise. The 'tilde' on top of the 'x' is to be distinguished from the 'hat' that refers to the estimated signal, often obtained from regression techniques including methods described in this chapter. Note that in this chapter, x will be used to denote the analytical signal or instrumental response, not y as in Chapter 2. This is to introduce a notation that is consistent with most of the open literature. Different investigators working in different areas of science often independently developed incompatible notation, and in an overview such as this text, it is preferable to stick reasonably closely to the generally accepted conventions to avoid confusion.

There are two main types of measurement noise.

3.2.3.1 Stationary Noise

The noise at each successive point (normally in time) does not depend on the noise at the previous point. In turn, there are two major types of stationary noise.

- *Homoscedastic noise.* This is the simplest to envisage. The features of the noise, normally the mean and standard deviation, remain constant over the entire data series. The most common type of noise is given by a normal distribution, usually with mean 0 unless there is a baseline problem, and standard deviation dependent on the instrument used. In most real-world situations, there are several sources of instrumental noise, but a combination of different symmetric noise distributions often tends towards a normal distribution.
- *Heteroscedastic noise.* This type of noise is dependent on signal intensity and is often proportional. The noise may still be represented by a normal distribution, but the standard deviation of that distribution is proportional to intensity of the underlying signal. A form of heteroscedastic noise often appears to arise if the data are transformed before processing, a common method being a logarithmic transform used in many types of spectroscopy such as UV/vis or IR spectroscopy, from transmittance to absorbance. The true noise distribution is imposed upon the raw data, but the transformed information distorts this.

Figure 3.7 illustrates the effect of both types of noise on a typical signal. It is important to recognise that several detailed models of noise are possible; however, in practice, it is not easy or interesting to perform sufficient experiments to determine such distributions. Indeed, it may be necessary to acquire several hundred or thousand spectra to obtain an adequate model, which represents overkill in most real-world situations. It is not possible to rely too heavily on published studies of instrumental noise distributions because each instrument is different and the experimental distribution is a balance between several sources, which differ in relative importance in each machine. In fact, as manufacture of instruments improves, certain types of noise are reduced in size and new effects come into play; hence, a thorough study of noise distributions performed a few decades ago is unlikely to be correct in detail on a more modern instrument. Many years ago, people studied instrumental noise distributions in more detail than is now necessary because the signal-to-noise ratios were lower, instruments changed less rapidly and they were often interested in improving the manufacture and performance of instruments.

In the absence of certain experimental knowledge, it is best to stick to a fairly straightforward distribution such as a normal distribution.

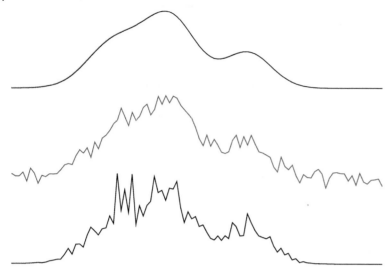

Figure 3.7 Examples of noise. From top to bottom: underlying signal, homoscedastic and heteroscedastic.

1. Heteroscedastic noise can be modelled by a normal distribution.
   (a) Never
   (b) Sometimes
   (c) Always

3.2.3.2 Correlated Noise

Sometimes, as a series is sampled, the level of noise in each sample depends on that of the previous sample. This is quite common in process control. For example, there may be problems in one aspect of the manufacturing procedure, an example being the proportion of an ingredient. If the proportion is in error by 0.5% at 2 in the afternoon, does this provide an indication of the error at 2.30?

Many such sources cannot be understood in great detail, but a generalised approach is that of *autoregressive moving average* (ARMA) noise.

- The *moving average* (MA) component relates the noise at time i to the values of the noise at previous times. A model of order p is given by $e_i = \sum_{t=0}^{t=p} c_{i-t} e_{i-t}$, where e_{i-t} is the noise at time $i-t$ and c_{i-t} a corresponding coefficient. A simple approach for simulating this type of noise is to put $p = 1$ and set the coefficient to 1. Under such circumstances, $e_i = g_i + e_{i-1}$ where g_i is generated using a normal distribution. Table 3.2 illustrates a stationary noise distribution and a MA distribution generated by simply adding successive values of the noise, so that, for example, the noise at time $= 4$ is given by $-0.00927 = 0.05075 - 0.06001$.
- The *autoregressive* component relates the noise to the observed response at one or more previous times. A model of order p is given by $x_i = \sum_{t=0}^{t=p} c_{i-t} x_{i-t} + e_i$. Note that in a full ARMA model, e_i itself is dependent on past values of noise.

There is a huge literature on ARMA processes, which are particularly important in the analysis of long-term trends such as in economics: it is quite likely that an underlying factor causing errors in estimates changes with time rather than fluctuating completely randomly. There have been developed a battery of specialised techniques to cope with such situations. The chemist must be aware of these noise models, especially when studying natural phenomena such as in environmental chemistry but also to a lesser extent in instrumental analysis. However, there is rarely sufficient experimental evidence to establish highly sophisticated noise models. It is although well advised, when studying a process, to determine whether a stationary noise distribution is adequate, especially if the results of simulations are to be relied upon; thus, an appreciation of basic methods for modelling noise is important. Very elaborate models are unlikely to be easy to verify experimentally. In areas such as geology or economics, the nature of the underlying noise can be important for modelling, but it is not so essential in chemometrics, although it is useful to be aware of the existence of such models if studying natural phenomenon.

Table 3.2 Stationary and moving average noise.

Time	Stationary	Moving average
1	−0.12775	
2	0.14249	0.01474
3	−0.06001	−0.04527
4	0.05075	0.00548
5	0.06168	0.06716
6	−0.14433	−0.07717
7	−0.10591	−0.18308
8	0.06473	−0.11835
9	0.05499	−0.06336
10	−0.00058	−0.06394
11	0.04383	−0.02011
12	−0.08401	−0.10412
13	0.21477	0.11065
14	−0.01069	0.09996
15	−0.08397	0.01599
16	−0.14516	−0.12917
17	0.11493	−0.01424
18	0.00830	−0.00595
19	0.13089	0.12495
20	0.03747	0.16241

1. Heteroscedastic noise can be calculated by considering only the current measurement, whereas calculation of ARMA noise requires knowledge of past measurements.

 (a) True
 (b) False

3.2.3.3 Signal-to-Noise Ratio

The signal-to-noise ratio is a useful parameter to measure. The higher this number, the more intense the signal is relative to the background. This measurement is essentially empirical, and the most common definition involves dividing the height of a relevant signal (normally the most intense if there are several in a data set) by the root mean square of the noise, measured in a region of the data where there is known to be no signal. Most common measures of signal-to-noise ratio assume that the noise is homoscedastic. There usually needs to be an area of the signal where there are no peaks present and the baseline has to be subtracted.

If it is assumed that noise is normally distributed, it is possible to determine confidence that a 'blip' in a spectrum or chromatogram represents a peak or is just due to noise, according to the number of standard deviations it is above the baseline. If, for example, a 'blip' is 1.5 standard deviations above a baseline, 7.7% of measurements (see Table A.1) will exceed this level; hence, there is a 7.7% chance that the data point is due to random noise. Usually, either a 5% (1.65 standard deviation), a 1% (2.33 standard deviation) or a 0.1% (3 standard deviations) cut-off is used and the latter is sometimes called the *limit of detection*. Measurements more intense than these limits are considered to be real peaks rather than just artefacts of the noise.

1. The root mean square noise level is 2 and a data point has an intensity of 7 above the baseline. Noise is assumed to be homoscedastic and normally distributed.

 (a) There is more than 99% confidence that this represents a true peak rather than noise.
 (b) There is more than 95% but less than 99% confidence that this represents a true peak rather than noise.
 (c) There is insufficient confidence that this represents a true peak.

3.2.4 Cyclicity

Not all chemometric data arise from spectroscopy or chromatography; some are from studying processes evolving over time, ranging from a few hours (e.g. a manufacturing process) to thousands of years (e.g. a geological process). Many techniques for studying such processes are common to those developed in analytical instrumentation.

In some cases, cyclic events occur, dependent, for example, on time of day, season of the year or cyclical temperature fluctuations (such as diurnal fluctuations). These can be modelled using sine functions and are the basis of time series analysis (Section 3.4). In addition, cyclicity is also observed in Fourier spectroscopy, and Fourier transform techniques (Section 3.5) may occasionally be combined with methods for time series analysis.

  1. Cyclical processes are often found in environmental chemistry.
      (a) True
      (b) False

3.3 Linear Filters

3.3.1 Smoothing Functions

A key need is to obtain a signal as informative as possible after removing the noise from a data set. When data are sequentially obtained, such as in time or frequency, the underlying signals often arise from a sum of smooth, monotonic functions, such as those described in Section 3.2.1, whereas the underlying noise is often an uncorrelated function. An important method for revealing the signals involves smoothing the data; the principle is that the noise will be smoothed away using quite mild methods, whilst the signal, being broader, will remain. This approach depends on the peaks having a half width of several data points: if digital resolution is very poor, signals may appear as spikes and be confused with noise.

It is important to determine the optimum filter for any particular application. Too much smoothing and the signal itself is reduced in intensity and resolution. Too little smoothing and noise remains. The optimum smoothing function depends on peak widths (in data points) as well as noise characteristics.

3.3.1.1 Moving Averages

Conceptually, the simplest methods are linear filters where the resultant smoothed data are given as a linear function of the original data. Normally, this involves using the surrounding data points, for example, using a function of the three points in Figure 3.8, to recalculate a value for the central point i. Algebraically, such functions are expressed by

$$x_{i,new} = \sum_{j=-p}^{p} c_j x_{i+j}$$

One of the simplest is a three-point MA. Each point is replaced by the average of itself and the points before and after, so in the above equation,

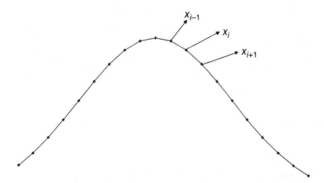

Figure 3.8 Selection of points to be used in a three-point moving average filter.

- $p = 1$
- $c_j = 1/3$ for all three points.

The filter can be extended to a five-point MA ($p = 2$, $c = 1/5$), seven-point MA and so on.

- The more the points in the filter, the greater the reduction in noise, but the higher the chance of blurring the signal.
- The number of points p in the filter is often called a *window* or a *filter width*.

The filter is moved along the time series or spectrum, each data point being replaced successively by the corresponding filtered data point. The optimal filter depends on the noise distribution and signal width. It is best to experiment with a number of different filter widths to find the optimum. The $(p-1)/2$ points at the beginning and end of a series are usually removed; for example, if using a five-point window, points 1–2 are removed, and the first new filtered point is number 3.

1. A time series consists of 80 data points. It is filtered using a seven-point moving average. The resultant time series consists of

   (a) 77 points
   (b) 74 points
   (c) 68 points

3.3.1.2 Savitzky–Golay Filters, Hanning and Hamming Windows

MA filters have the disadvantage that they use a linear approximation for the data. However, peaks are often best approximated by polynomial curves. This is particularly true at the centre of a peak, where a linear model will always underestimate the intensity. Quadratic, cubic or even quartic models provide better approximations. The principle of MAs can be extended to polynomials. A seven-point cubic filter, for example, is used to fit a model

$$x_i = b_0 + b_1 i + b_2 i^2 + b_3 i^3$$

using a seven-point window, replacing the centre point by its best-fit estimate. The window is moved along the data, point by point, the calculation being repeated each time.

However, regression is computationally intense, and it would be computationally slow to perform this calculation in full to simply improve the appearance of a spectrum or chromatogram, which may consist of thousands of data points. The user wants to be able to select a menu item or icon on a screen and almost instantaneously visualise an improved picture. Savitzky and Golay in 1964 presented an alternative and simplified method of determining the new value of x_i simply by re-expressing the calculation as a sum of coefficients. These Savitzky–Golay filters are normally represented in a tabular form (see Table 3.3). To determine a coefficient c_j,

Table 3.3 Savitzky–Golay coefficients c_{i+j} for smoothing.

Window size	5	7	9	7	9
j		Quadratic/cubic		Quartic/quintic	
−4			−21		15
−3		−2	14	5	−55
−2	−3	3	39	−30	30
−1	12	6	54	75	135
0	17	7	59	131	179
1	12	6	54	75	135
2	−3	3	39	−30	30
3		−2	14	5	−55
4			−21		15
Normalisation constant	35	21	231	231	429

- decide on the order of the model (quadratic and cubic models give identical results as do quartic and quintic models),
- decide on the window size and
- determine c_j by selecting the appropriate number from Table 3.3 and dividing by the corresponding normalisation constant.

Several other MA methods have been proposed in the literature, two of the best known being the Hanning window (named after Julius Von Hann) (for which three points has weights 0.25, 0.5 and 0.25) and the Hamming window (named after R.W. Hamming) (for which five points has weights 0.0357, 0.2411, 0.4464, 0.2411 and 0.0357) – not to be confused but very similar in effects. These windows can be calculated for any size, but we recommend these two basic filter sizes.

Notice that although quadratic, cubic or higher approximations of the data are employed, the filters are still called linear because each filtered point is a linear combination of the original data.

1. A linear filter is defined as
   (a) A filter that uses linear combinations of original data to obtain a local model of a sequential series even though the model may be polynomial.
   (b) A filter that always results in a linear model of a sequential series.

3.3.1.3 Calculation of Linear Filters

The calculation of MA and Savitzky–Golay filters is illustrated in Table 3.4.

- The first point of the three-point MA (see column 2) is simply given by

$$-0.049 = (0.079 - 0.060 - 0.166)/3$$

- The first point of the seven-point Savitzky–Golay quadratic/cubic filtered data can be calculated as follows. From Table 3.3, obtain the seven coefficients, namely $c_{-3} = c_3 = -2/21 = -0.095$, $c_{-2} = c_2 = 3/21 = 0.143$ and $c_{-1} = c_1 = 6/21 = 0.286$, $c_0 = 7/21 = 0.333$.

  Multiply these coefficients by the raw data and sum to get the smoothed value of the data

$$x_{i,new} = -0.095 \times 0.079 + 0.143 \times -0.060 + 0.286 \times -0.166 + 0.333 \times -0.113$$
$$+ 0.286 \times 0.111 + 0.143 \times 0.145 - 0.095 \times 0.212$$
$$= -0.069$$

Figure 3.9(a) is a representation of the raw data. The result of using MA filters is shown in Figure 3.9(b). A three-point MA preserves the resolution (just), but a five-point MA loses this and the cluster appears to be composed of only one peak. In contrast, the five- and seven-point quadratic/cubic Savitzky–Golay filters (Figure 3.9(c)) preserve resolution whilst reducing noise and only starts to lose resolution when using a nine-point function.

1. Seven successive data points are as follows:

   | −0.008 | 0.299 | 0.410 | 0.361 | 0.175 | 0.101 | 0.184 |

   Using a seven-point cubic Savitzky–Golay filter, the central point after smoothing is
   (a) 0.120
   (b) 0.307
   (c) 0.328
   (d) 6.891

3.3.1.4 Running Median Smoothing

Most conventional filters involve computing local multi-linear models; however, in certain areas such as process analysis, there can be spikes (or outliers) in the data, which are unlikely to be part of a continuous process. An alternative method involves using RMS functions, which calculate the median rather than the mean over a window. An example of a process is given in Table 3.5. A five-point MA and five-point RMS smoothing function is compared. A check on

Table 3.4 Results of various filters on a data set.

Raw data	Moving average			Quadratic/cubic Savitzky–Golay		
	3-point	5-point	7-point	5-point	7-point	9-point
0.079						
−0.060	−0.049					
−0.166	−0.113	−0.030		−0.156		
−0.113	−0.056	−0.017	0.030	−0.081	−0.069	
0.111	0.048	0.038	0.067	0.061	0.026	−0.005
0.145	0.156	0.140	0.168	0.161	0.128	0.093
0.212	0.233	0.291	0.338	0.206	0.231	0.288
0.343	0.400	0.474	0.477	0.360	0.433	0.504
0.644	0.670	0.617	0.541	0.689	0.692	0.649
1.024	0.844	0.686	0.597	0.937	0.829	0.754
0.863	0.814	0.724	0.635	0.859	0.829	0.765
0.555	0.651	0.692	0.672	0.620	0.682	0.722
0.536	0.524	0.607	0.650	0.491	0.539	0.628
0.482	0.538	0.533	0.553	0.533	0.520	0.540
0.597	0.525	0.490	0.438	0.550	0.545	0.474
0.495	0.478	0.395	0.381	0.516	0.445	0.421
0.342	0.299	0.330	0.318	0.292	0.326	0.335
0.061	0.186	0.229	0.242	0.150	0.194	0.219
0.156	0.102	0.120	0.157	0.103	0.089	0.081
0.090	0.065	0.053	0.118	0.074	0.016	0.041
−0.050	0.016	0.085	0.081	−0.023	0.051	0.046
0.007	0.059	0.070	0.080	0.047	0.055	0.070
0.220	0.103	0.063	0.071	0.136	0.083	0.072
0.081	0.120	0.091	0.063	0.126	0.122	0.102
0.058	0.076	0.096	0.054	0.065	0.114	0.097
0.089	0.060	0.031	0.051	0.077	0.033	0.054
0.033	0.005	0.011	0.015	0.006	0.007	
−0.107	−0.030	−0.007		−0.051		
−0.016	−0.052					
−0.032						

the calculation of the two different filters is as follows:

- The five-point MA filter at time 4 is −0.010, calculated by taking the mean values for times 2–6, that is,

$$-0.010 = (0.010 - 0.087 - 0.028 + 0.021 + 0.035)/5$$

- The five-point RMS filter at time 4 is 0.010. This is calculating by arranging the readings for time 2–6 in order as follows: −0.087, −0.028, 0.010, 0.021, 0.035, and selecting the middle value.

The results are presented in Figure 3.10. Underlying trends are not obvious from inspection of the raw data. Of course, further mathematical analysis might reveal a systematic trend, but in most situations, the first inspection is graphical. The five-point MA does suggest a systematic process, but it is not at all clear whether the underlying process increases monotonically with time, or increases and then decreases. The five-point RMS suggests a process increasing with time and is much smoother than the result of a MA filter.

Each type of smoothing function removes different features in the data, and often a combination of several approaches is recommended especially for real-world problems. Dealing with outliers is an important issue: sometimes, these points are due to measurement errors. Many processes take time to deviate from the expected value, a sudden glitch in the system unlikely to be a real effect. Often, a combination of filters is recommended, for example, a five-point median smoothing followed by a three-point Hanning window. These methods are very easy to implement computationally, and it is possible to view the results of different filters simultaneously.

1. Running median smoothing can result in flat peak shapes.
    (a) True
    (b) False

3.3.2   Derivatives

The methods described in Section 3.3.1 are primarily concerned with removing noise. Most such methods either leave peak widths unchanged or increased, equivalent to blurring. In signal analysis, an important separate need is to increase resolution. In Section 3.5.2, we will discuss the use of filters combined with Fourier transformation. In Chapter 7, we

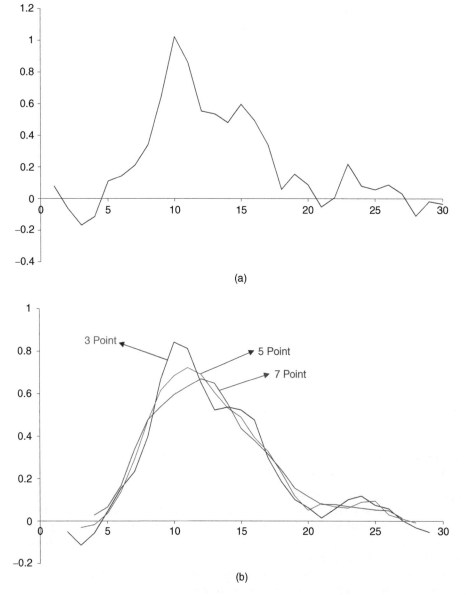

(a)

(b)

Figure 3.9 Filtering of data. (a) Raw data, (b) moving average filters, (c) quadratic/cubic Savitzky–Golay filters.

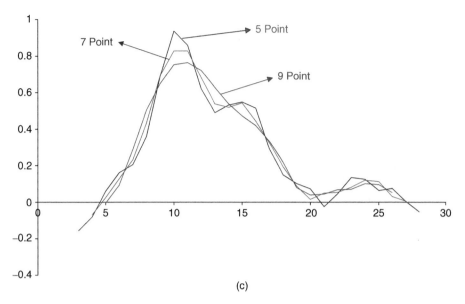

(c)

Figure 3.9 (*Continued*)

Table 3.5 A sequential process: illustration of moving average and median smoothing.

Time	Data	5-point MA	5-point RMS
1	0.133		
2	0.010		
3	−0.087	0.010	0.010
4	−0.028	−0.010	0.010
5	0.021	0.048	0.021
6	0.035	0.047	0.021
7	0.298	0.073	0.035
8	−0.092	0.067	0.035
9	0.104	0.109	0.104
10	−0.008	0.094	0.104
11	0.245	0.207	0.223
12	0.223	0.225	0.223
13	0.473	0.251	0.223
14	0.193	0.246	0.223
15	0.120	0.351	0.223
16	0.223	0.275	0.193
17	0.745	0.274	0.190
18	0.092	0.330	0.223
19	0.190	0.266	0.190
20	0.398	0.167	0.190
21	−0.095	0.190	0.207
22	0.250	0.200	0.239
23	0.207	0.152	0.207
24	0.239		
25	0.160		

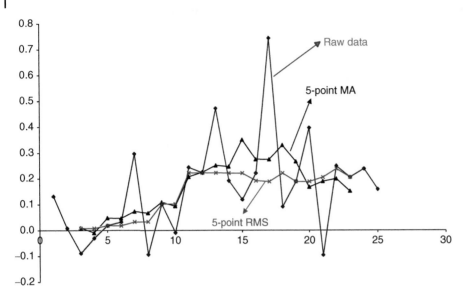

Figure 3.10 Comparison of moving average and running median smoothing.

will discuss how to improve resolution when there is an extra dimension to the data (e.g. multivariate curve resolution). However, a simple and frequently used approach is to calculate derivatives. The principle is that inflection points in partially resolved peaks become turning points in the derivatives. The first and second derivatives of a pure Gaussian are presented in Figure 3.11.

- The first derivative equals zero at the centre of the peak and is a good way of accurately pinpointing the position of a broad peak. It exhibits two turning points.
- The second derivative is a minimum at the centre of the peak, crosses zero at the positions of the turning points for the first derivative and exhibits two further turning points farther apart than in the first derivative.
- The apparent peak width is reduced using derivatives.

The properties are most useful when there are several closely overlapping peaks, and higher order derivatives are often employed, for example in electron spin resonance and electronic absorption spectroscopy to improve resolution. Figure 3.12 illustrates the first and second derivatives of two closely overlapping peaks. The second derivative clearly indicates two peaks and allows one to pinpoint their positions. The first derivative would suggest that the peak is not pure but, in this case, probably does not provide definitive evidence. It is, of course, possible to continue and calculate the third derivative, fourth derivative and so on, but the patterns can become quite complicated.

There are, however, two disadvantages of using derivatives. First, they can be computationally intense, as a fresh calculation is required for each data point in a spectrum or chromatogram. Second, and most importantly, they amplify noise substantially, and, therefore, require low signal-to-noise ratios. These limitations can be overcome by using Savitzky–Golay coefficients similar to those described in Section 3.3.1.3, which involve rapid calculation of smoothed higher derivatives. The coefficients for a number of window sizes and approximations are presented in Table 3.6. This is a common method for the determination of derivatives and implemented in many software packages.

1. Savitzky–Golay derivative functions are computationally rapid methods for calculating derivatives but they amplify noise.
   (a) True
   (b) False

3.3.3 Convolution

Common principles occur in different areas of science, often under different names and are introduced in conceptually different guises. In many cases, the driving force is the expectations of readers, who may be potential users of techniques, customers on courses or programmers, often with a variety of backgrounds such as engineering, analytical

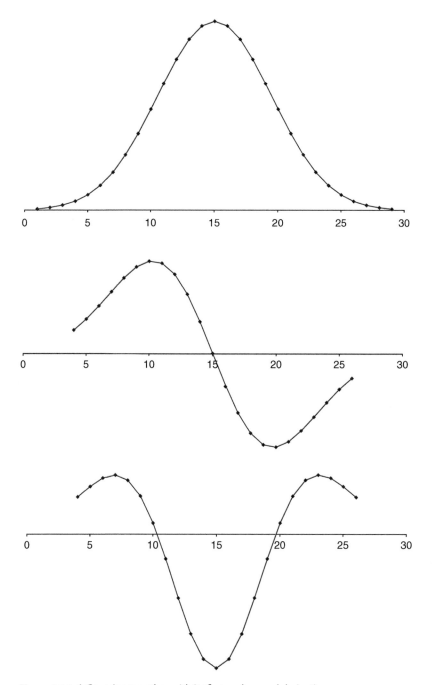

Figure 3.11 A Gaussian together with its first and second derivative.

chemistry, physics and so on. Sometimes, even the marketplace forces different approaches: students attend courses with varying levels of background knowledge and will not necessarily opt (or pay) for courses that are based on certain requirements. This is especially important in the interface between mathematical and experimental science.

Smoothing functions can be introduced in various ways, for example, as sums of coefficients or as a method for fitting local polynomials. In the signal analysis literature, primarily dominated by engineers, linear filters are often considered a form of convolution. The principles of convolution are straightforward. Two functions, f and g, are convoluted to give h if

$$h_i = \sum_{j=-p}^{j=p} f_j g_{i+j}$$

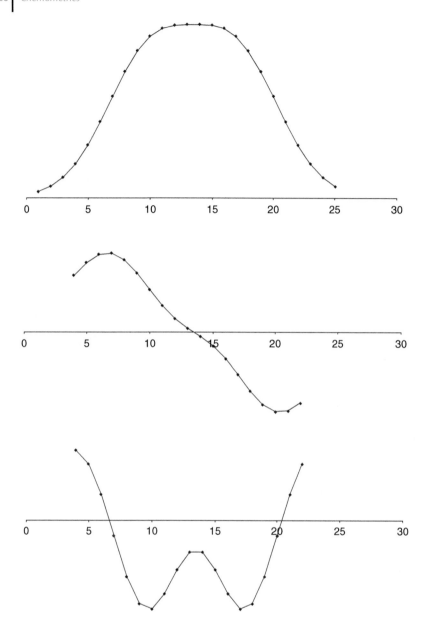

Figure 3.12 Two closely overlapping peaks together with their first and second derivatives.

Sometimes, this operation is written using a convolution operator denoted by a '*', so that

$$h(i) = f(i) * g(i)$$

This process of convolution is exactly equivalent to digital filtering, in the example given above

$$x_{new}(i) = x(i) * g(i)$$

where $g(i)$ is a filter function. It is, of course, possible to convolute any two functions with each other, providing each is of the same size. It is possible to visualise these filter functions graphically. Figure 3.13 illustrates the convolution function (or window) for a three-point MA, a Hanning window and a five-point Savitzky–Golay second-derivative quadratic/cubic filter. The resultant spectrum is the convolution of such functions with the raw data.

Convolution is a convenient general mathematical way of dealing with a number of methods for signal enhancement. We mention this for completion as some readers will come across this in the literature.

Table 3.6 Savitzky–Golay coefficients for derivatives.

Window size

First derivatives

J	5	7	9	5	7	9
	Quadratic			Cubic/quartic		
−4			−4			86
−3		−3	−3		22	−142
−2	−2	−2	−2	1	−67	−193
−1	−1	−1	−1	−8	−58	−126
0	0	0	0	0	0	0
1	1	1	1	8	58	126
2	2	2	2	1	67	193
3		3	3		−22	142
4			4			−86
Normalisation	10	28	60	12	252	1 188

Second derivatives

J	5	7	9	5	7	9
	Quadratic/cubic			Quartic/quintic		
−4			28			−4 158
−3		5	7		−117	12 243
−2	2	0	−8	−3	603	4 983
−1	−1	−3	−17	48	−171	−6 963
0	−2	−4	−20	−90	−630	−12 210
1	−1	−3	−17	48	−171	−6 963
2	2	0	−8	−3	603	4 983
3		5	7		−117	12 243
4			28			−4 158
Normalisation	7	42	462	36	1 188	56 628

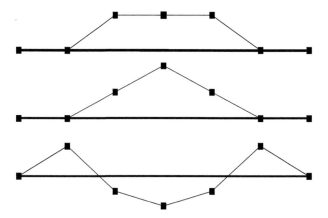

Figure 3.13 From top to bottom, a three-point moving average, a Hanning window and a five-point Savitzky–Golay quadratic second-derivative window convolution functions.

1. A Savitzky–Golay filter can be considered a convolution function.
   (a) True
   (b) False

3.4 Correlograms and Time Series Analysis

Time series analysis has a long statistical vintage, with major early applications in economics and engineering. The aim is to study cyclical trends in processes. In the methods of Section 3.3, we were mainly concerned with peaks arising from chromatography or spectroscopy or else processes that occur in manufacturing. There were no underlying cyclical features. However, in certain circumstances, features can re-occur at regular intervals. These could arise from a geological process, a manufacturing plant or environmental monitoring, the cyclic changes being due to season of the year, time of day or even hourly events.

The aim of time series analysis is to mainly reveal the cyclical trends in a data set. These will be buried within non-cyclical phenomena and also various sources of noise. In spectroscopy, where the noise distributions are well understood and primarily stationary, Fourier transforms are the method of choice. However, when studying natural processes, there are likely to be a much larger number of factors influencing the response, including often correlated (or ARMA) noise, as discussed in Section 3.2.3.2. Under such circumstances, time series analysis is preferable and can reveal quite weak cyclicity. The disadvantage is that original intensities are lost, the resultant information being primarily about how strong the evidence is that a particular process exhibits cyclicity. Most methods for time series analysis involve the calculation of a correlogram at some stage.

3.4.1 Auto-correlograms

Consider the information depicted in Figure 3.14, which represents a process changing with time. It appears that there is some cyclicity, but this is buried within the noise. The numerical data are presented in Table 3.7.

An auto-correlogram involves calculating the correlation coefficient between a time series and itself, shifted by a given number of data points called a 'lag'. If there are I data points in the original time series, then a correlation coefficient for a lag of l points will involve $I - l$ data points. Hence, in the case of Table 3.7, there are 30 points in the original data set but only 25 points in the data set for which $l = 5$. Point number 1 in the shifted data set corresponds to point number 6 in the original data set. A common equation for the correlation coefficient for a time series lagged by l is given by

$$r_l = \frac{\sum_{i-1}^{I-l}(x_i - \overline{x})(x_{i+l} - \overline{x})}{\sum_{i=1}^{I}(x_i - \overline{x})^2}$$

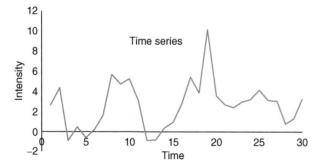

Figure 3.14 A time series.

Table 3.7 Data in Figure 3.14 together with the data lagged by five points in time.

i	Data, $l = 0$	Data, $l = 5$
1	2.768	0.262
2	4.431	1.744
3	−0.811	5.740
4	0.538	4.832
5	−0.577	5.308
6	0.262	3.166
7	1.744	−0.812
8	5.740	−0.776
9	4.832	0.379
10	5.308	0.987
11	3.166	2.747
12	−0.812	5.480
13	−0.776	3.911
14	0.379	10.200
15	0.987	3.601
16	2.747	2.718
17	5.480	2.413
18	3.911	3.008
19	10.200	3.231
20	3.601	4.190
21	2.718	3.167
22	2.413	3.066
23	3.008	0.825
24	3.231	1.338
25	4.190	3.276
26	3.167	
27	3.066	
28	0.825	
29	1.338	
30	3.276	

where $\bar{x}$ is the mean of the I data points. Note that this simplified equation divides by the sum of squares of all the I data points and assumes that the mean is invariant. It can also be described as the covariance between the lagged data points divided by the variance of all the full time series. The equation can be rewritten as

$$r_l = \frac{c_{xx,l}}{s^2}$$

where $c_{xx,l}$ is the covariance of x with itself lagged by l points (see Section A.3.1.3 for definition) divided by its variance.

A more complicated equation is sometimes used:

$$r_l = \frac{\sum\limits_{i-1}^{I-l}(x_i - \bar{x}_{1 \text{ to } I-l})(x_{i+l} - \bar{x}_{l+1 \text{ to } I})}{\sqrt{\sum\limits_{i=1}^{I-l}(x_i - \bar{x}_{1 \text{ to } I-l})^2 \sum\limits_{i=l+1}^{Il}(x_i - \bar{x}_{l+1 \text{ to } I})^2}}$$

in which it is assumed that the mean and variance (or standard deviation) vary according to the amount of lag. In practice, unless the lag is large relative to the time series, there is very little difference between these two equations, and below we will use the simpler equation.

In addition, there are a number of other computational alternatives to these two formulae.

There are a number of properties of the correlogram.

- For a lag of 0, the correlation coefficient is 1.
- It is possible to have negative lags as well as positive lags, but for an auto-correlogram, $r_l = r_{-l}$ and sometimes only one half of the correlogram is displayed.
- The closer the correlation coefficient is to 1, the more similar the two series. If a high correlation is observed for a large lag, this indicates cyclicity.
- As the lag increases, the number of data points used to calculate the correlation coefficient decreases; hence, r_l becomes less informative and more dependent on noise. Large values of l are not advisable, a good compromise is to calculate the correlogram for values of l between half or two-thirds of I.

The resultant auto-correlogram for the data in Table 3.7 is presented in Figure 3.15. The cyclic pattern is now much clearer than in the original data. Note that the graph is symmetric about the origin as expected, and the maximum lag used in this example equals 20 points, being 30 points in the original data.

An auto-correlogram emphasises only cyclical features. Sometimes, there are non-cyclical trends superimposed over the time series. Such situations regularly occur in economics. Consider trying to determine the factors relating to expenditure in a seaside resort. A cyclical factor will undoubtedly be seasonal, being more business in the summer. However, other factors such as interest rates, exchange rates and long-term economic activity will also come into play and the information will be mixed up in the resultant statistics. Expenditure could also be divided into food, accommodation, clothes and so on. Each will be influenced to different extent by seasonality. Correlograms specifically emphasise the cyclical causes of expenditure. In chemistry, they are most valuable when time-dependent noise interferes with stationary noise, for example in the environmental chemistry of a river where there may be specific types of pollutants or changes in chemicals that occur spasmodically but once discharged take time to dissipate as well, for example, in diurnal cyclical factors.

The correlogram can be further processed by Fourier transformation, smoothing functions or a combination of both, whose techniques are discussed in Sections 3.3 and 3.5. Sometimes, the results can be represented in the form of probabilities, for example, the chance that there is really a genuine underlying cyclical trend of a given frequency. Such calculations, although, make certain definitive assumptions about the underlying noise distributions and experimental error and cannot always be generalised and would be rare in chemistry, although there is a large field of time series analysis that is well developed in geology, economics and so on.

1. Auto-correlograms are always symmetrical around zero lag.
   - (a) True
   - (b) False

2. A time series consists of 40 data points. An auto-correlogram is calculated for lag up to 25, including negative lags. The total number of points in the correlogram is
   - (a) 40
   - (b) 50
   - (c) 51
   - (d) 81

3.4.2 Cross-correlograms

It is possible to extend these principles to the comparison of two independent time series. Consider measuring the levels of Ag and Ni in a river with time. Although each may show a cyclical trend, are there trends common to both metals? The cross-correlation function between two series represented by x and y can be calculated for a lag of l,

$$r_l = \frac{c_{xy,l}}{s_x s_y}$$

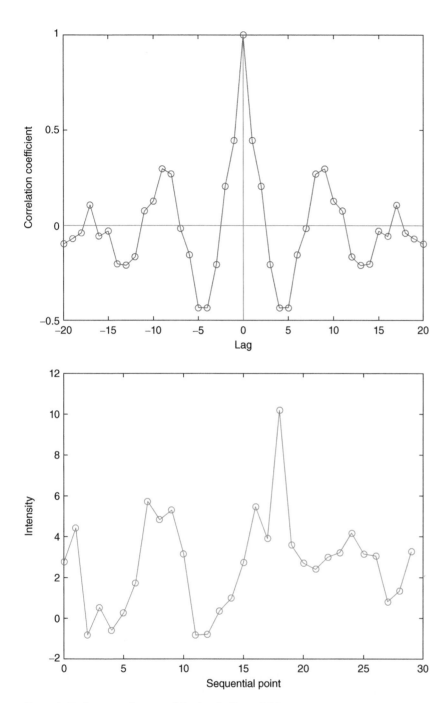

Figure 3.15 Auto-correlogram of the data in Figure 3.14.

where $c_{xy,l}$ is the covariance between the functions at lag l, which can be given by

$$c_{xy,l} = \sum_{i=1}^{I-l}(x_i - \overline{x})(y_{i+l} - \overline{y})/(I - l) \quad \text{for } l \geq 0$$

$$c_{xy,l} = \sum_{i=1}^{I-l}(x_{i+l} - \overline{x})(y_i - \overline{y})/(I - l) \quad \text{for } l \leq 0$$

and s corresponds to the appropriate standard deviations (see Section A.3.1.3 for more details about the covariance). There are a number of equivalent computational formulae available in the literature. Note that the average of x and y

Table 3.8 Two time series, for which the cross-correlogram is presented in Figure 3.16.

Series 1	Series 2
2.768	1.061
2.583	1.876
0.116	0.824
−0.110	1.598
0.278	1.985
2.089	2.796
1.306	0.599
2.743	1.036
4.197	2.490
5.154	4.447
3.015	3.722
1.747	3.454
0.254	1.961
1.196	1.903
3.298	2.591
3.739	2.032
4.192	2.485
1.256	0.549
2.656	3.363
1.564	3.271
3.698	5.405
2.922	3.629
4.136	3.429
4.488	2.780
5.731	4.024
4.559	3.852
4.103	4.810
2.488	4.195
2.588	4.295
3.625	4.332

should strictly be recalculated according to the number of data points in the window, but in practice, providing the window is not too small, the overall averages are usually acceptable, as also used for the simplified formula for the auto-correlogram.

The cross-correlogram is no longer symmetric about 0; hence, a negative lag does not give the same result as a positive lag. Table 3.8 is of two time series, 1 and 2. The raw time series and the corresponding cross-correlogram are presented in Figure 3.16. The raw time series appear to exhibit a long-term increase but it is not entirely obvious that there are common cyclical features. The correlogram suggests that both contain a cyclical trend of around eight data points, as the correlograms exhibit a strong minimum at $l = \pm 8$. The correlograms may, for example, represent two different elements, which happen to show a common cyclical trend: the cross-correlogram asks whether there is some common cyclicity between two different factors or variables.

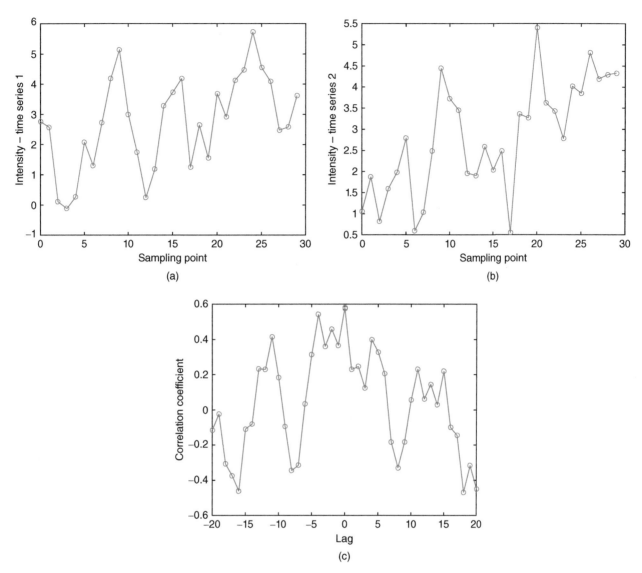

Figure 3.16 Two time series (a,b) and their corresponding cross-correlogram (c).

1. Cross-correlograms are always symmetrical around zero lag.
    (a) True
    (b) False

3.4.3 Multivariate Correlograms

In the real-world, there may be a large number of variables that change with time, for example, the composition of a manufactured product. In a chemical plant, the resultant material could depend on a huge number of factors such as the quality of the raw material, the performance of the apparatus, even the time of day which could relate to who is on shift, or small changes in power supplies. Instead of monitoring each factor individually, it is common to obtain an overall statistical indicator, typically the scores of a principal component (see Chapter 4), but sometimes other statistical indicators such as the Mahalanobis distance from the centroid of a batch of typical in control or acceptable samples are also obtained (see Chapter 5). The correlogram is computed from this mathematical summary of the raw data rather than the concentration of an individual constituent.

1. Principal component scores can show cyclicity.
   (a) True
   (b) False

3.5 Fourier Transform Techniques

The mathematics of Fourier transformation has been well established for two centuries, but early computational algorithms were first described in the 1960s, a prime method being the Cooley–Tukey algorithm. Originally employed in physics and engineering, Fourier transform (FT) techniques are now essential tools of the chemists. Modern NMR, IR and X-ray spectroscopy, among others, depend on Fourier transform methods. FTs have been extended to two-dimensional time series, plus a wide variety of modifications, for example, phasing, resolution enhancement and applications to image analysis have been developed over the past decades. For certain types of instrumentation, most notably NMR, there is a vast literature on enhancements to FT methods, but below we primarily focus on the main approaches, allowing the reader to delve into the specialist literature where appropriate.

3.5.1 Fourier Transforms

3.5.1.1 General Principles

The original literature on Fourier series and transform techniques involved primarily applications to continuous data sets. However, in chemical instrumentation, data are not sampled continuously but at regular intervals of time, so all data are digitised. The discrete Fourier transform (DFT) is used to process such data and will be described below. It is important to recognise that DFTs have specific properties that distinguish them from continuous FTs.

DFTs involve transformation between two types of data. In FTNMR, the raw data are acquired at regular intervals of time, often called the *time domain* or more specifically described by a Free Induction Decay (FID). FTNMR has been developed over the years because it is much quicker to obtain data than using conventional (continuous wave) methods. An entire spectrum can be sampled in a few seconds, rather than minutes, speeding up the procedure of data acquisition by one or two orders of magnitude. This has meant that it is possible to record spectra of small quantities of compounds or of natural abundance of isotopes such as ^{13}C, now routine in modern laboratories.

The trouble with this is that the time domain is not easy to interpret, and here arises the need for DFTs. Each peak in a spectrum can be described by three parameters, namely a height, width and position, as discussed in Section 3.2.1. In addition, each peak has a shape; in NMR, this is Lorentzian. A spectrum consists of a sum of peaks and is often referred to as the *frequency domain*. However, raw data, for example, in NMR, are recorded in the *time domain* and each frequency domain peak corresponds to a time series characterised by

- an initial intensity,
- an oscillation rate and
- a decay rate.

The time domain consists of a sum of time series, each corresponding to a peak in the spectrum. Noise is superimposed on this time series. Fourier transforms convert the time series into a recognisable spectrum as indicated in Figure 3.17. Each parameter in the time domain corresponds to a parameter in the frequency domain as indicated in Table 3.9.

- The faster the rate of oscillation in the time series, the farther away the peak is from the origin in the spectrum.
- The faster the rate of decay in the time series, the broader the peak in the spectrum.
- The higher the initial intensity in the time series, the greater the area of the transformed peak.

The peak shape in the frequency domain relates to the decay curve (or mechanism) in the time domain. The time domain equivalent to a Lorentzian peak is

$$f(t) = A\cos(\omega t)e^{-t/s}$$

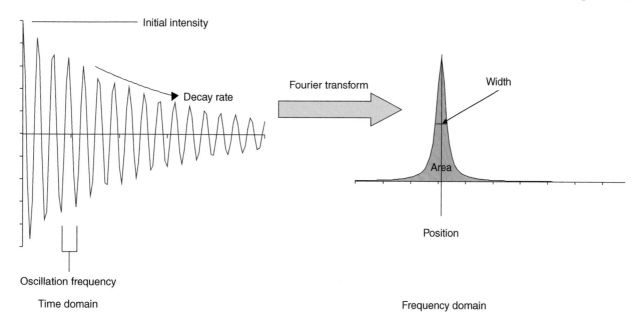

Figure 3.17 Fourier transformation from a time domain to a frequency domain.

Table 3.9 Equivalence between parameters in the time domain and frequency domain.

Time domain	Frequency domain
Initial intensity	Peak area
Oscillation frequency	Peak position
Decay rate	Peak width

where A is the initial height (corresponding to the area in the transform), ω to the oscillation frequency (corresponding to the position in the transform) and s to the decay rate (corresponding to the peak width in the transform). The key to the line shape is the exponential decay mechanism, and it can be shown that a decaying exponential transforms into a Lorentzian. Each type of time series has an equivalent to peak shape in the frequency domain, and together, these are called a *Fourier pair*. It can be shown that a Gaussian in the frequency domain corresponds to a Gaussian in the time domain and an infinitely sharp spike in the frequency domain to a non-decaying signal in the time domain.

In the vast majority of spectra, there will be several peaks, and the time series appear much more complex than in Figure 3.17, consisting of several superimposed curves, as exemplified in Figure 3.18. The beauty of Fourier transform

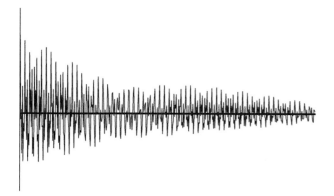

Figure 3.18 Typical time series consisting of several components.

spectroscopy is that all the peaks can be observed simultaneously, thus allowing rapid acquisition of data, but a mathematical transform is required to make the data comprehensible.

1. The broader a peak is in a spectrum, the faster its corresponding time domain profile will decay.
   (a) True
   (b) False

3.5.1.2 Fourier Transform Methods

The process of Fourier transformation converts the raw data (e.g. a time series) into two frequency domain spectra, one of which is called a *real spectrum* and the other called *imaginary* (this terminology comes from complex numbers). The true spectrum is represented only by half the transformed data, as indicated in Figure 3.19. Hence, if there are 1024 data points in the original time series, 512 will correspond to the real transform and 512 to the imaginary transform.

The mathematics of Fourier transformation is not too difficult to understand, but it is important to realise that authors use slightly different terminology, and definitions, especially with regard to constants in the transform. When reading a paper or text, consider these factors very carefully and always check that the result is realistic. We will adopt a number of definitions as follows.

The *forward* transform converts a purely *real* series into both a real and an imaginary transform, whose spectrum may be defined by

$$F(\omega) = RL(\omega) - i\,IM(\omega)$$

where F is the Fourier transform, ω is the frequency in the spectrum, i the square root of -1 and RL and IM the two halves of the transform.

The *real* part is obtained by performing a *cosine* transform on the original data, given by (in its simplest form)

$$RL(n) = \sum_{m=0}^{M-1} f(m)\cos(nm/M)$$

and the *imaginary* part by performing a *sine* transform

$$IM(n) = \sum_{m=0}^{M-1} f(m)\sin(nm/M)$$

These are sometimes expressed as $F(n) = \sum_{n=0}^{M-1} \exp(-i\pi nm/M)$, where i is the square root of -1 and $F(n) = RL(n) - i\,IM(n)$ (there is a relationship for complex exponentials called *Euler's formula*).

These terms need some definition.

- There are M data points in the original (time series) data,
- m refers to each point in the time series,
- n is a particular point in the transform, usually, in our case, a spectrum and
- the angles are in cycles per second.

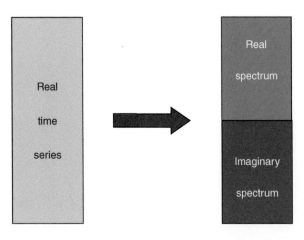

Figure 3.19 Transformation of a real time series to real and imaginary pairs.

If you use radians, you must multiply the angles by 2π and if degrees by 360°, but the equations above are presented in a simple way. There are quite a number of methods for determining the units of the transformed data, provided we are transforming a purely real time series to a real spectrum of half the size ($M/2$), then if the sampling interval in the time domain is δt s, the interval of each data point in the frequency domain is $\delta\omega = 1/(M\,\delta t)$ Hz (= cycles per second). To give an example, if we record 8000 data points in total in the time domain at intervals of 0.001 s (so the total acquisition time is 8 s), then the real spectrum will consist of 4000 data points at intervals of $1/(8000 \times 0.001) = 0.125$ Hz, thus will cover a total range of detectable frequencies over $4000 \times 0.125 = 500$ Hz. The rationale behind these numbers will be described in Section 3.5.1.4. Some books contain equations that appear more complicated to those presented here because they transform from time to frequency units rather than that from data points.

An *inverse* transform converts the real and imaginary pairs into a real series and is of the form

$$f(t) = \mathrm{rl}(t) + i\,\mathrm{im}(t)$$

Notice the '+' sign. Otherwise, the transform is similar to the forward transform, the real part involving the multiplication of a cosine wave with the spectrum. Sometimes, a factor of $1/N$, where there are N data points in the transformed data, is applied to the inverse transform, so that a combination of forward and inverse transforms gives the original time series.

FTs are best understood by a simple numerical example. For simplicity, we will give an example where there is a purely real spectrum and both real and imaginary time series – the opposite to normal but perfectly reasonable: in the case of Fourier self-convolution (Section 3.5.2.3), this indeed is the procedure. We will show only the real half of the transformed time series. Consider a spike as shown in Figure 3.20. The spectrum is of zero intensity except at one point, $m = 2$. We assume that there are M (=20) points numbered from 0 to 19 in the spectrum.

What happens to the first 10 points of the transform? The values are given by

$$RL(n) = \sum_{m=0}^{19} f(m)\cos(nm/M)$$

As $f(m)$ equals 0 except where $m = 2$, when it equals 10, the equation simplifies furthermore so that

$$RL(n) = 10\;\cos(2n/20)$$

The angular units of the cosine are cycles per unit time; hence, this angle must be multiplied by 2π to convert into radians. (While employing computer packages for trigonometry, always check whether units are in degrees, radians or cycles: this is simple to do; the cosine of 360° equals the cosine of 2π radians, which equals the cosine of 1 cycle and equals 1.) As shown in Figure 3.20, there is half a cycle every 10 data points, as $2 \times 10/20 = 1$, and the initial intensity equals 10 because this is the area of the spike (obtaining by summing the intensity in the spectrum over all data points). It should be evident that the farther the spike from the origin, the more the cycles in the transform. Similar calculations can be employed to demonstrate other properties of Fourier transforms as discussed above.

1. A time series consists of 5000 data points, acquired at regular intervals of 0.002 s. The range of detectable spectral frequencies is

   (a) 200 Hz
   (b) 250 Hz
   (c) 500 Hz

2. A spectrum consists of a spike at data point 5. If the spectrum is real and recorded over 50 data points, the resultant real transform consists of a cosine function with the following number of maxima.

   (a) 2
   (b) 3
   (c) 5
   (d) 6

3.5.1.3 Real and Imaginary Pairs

In the Fourier transform of a real time series, the peak shapes in the real and imaginary halves of the spectrum differ. Ideally, the real spectrum corresponds to an *absorption* line shape and the imaginary spectrum to a *dispersion* line

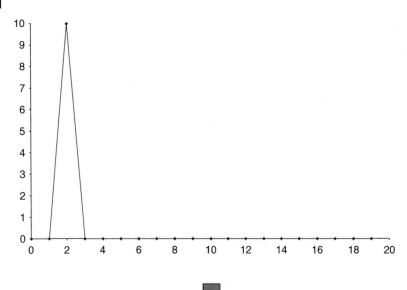

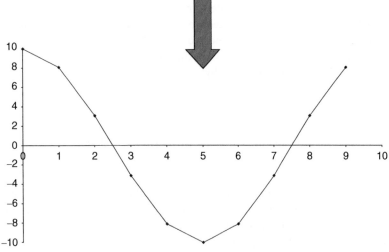

Figure 3.20 Fourier transform of a spike.

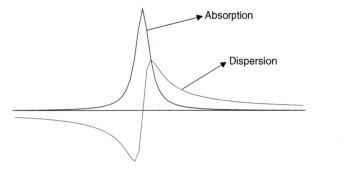

Figure 3.21 Absorption and dispersion line shapes.

shape, as illustrated in Figure 3.21. The absorption line shape is equivalent to a fundamental peak shape such as a Lorentzian or Gaussian, whereas the dispersion line shape resembles a little a first derivative.

However, often these two peak shapes are mixed together in the real spectrum, due to small imperfections in acquiring the data, called *phase errors*. The reason for this is that data acquisition does not always start exactly at the top of the cosine wave, and in practice, the term $\cos(\omega t)$ should be substituted by $\cos(\omega t + \phi)$. The angle ϕ is called the *phase*

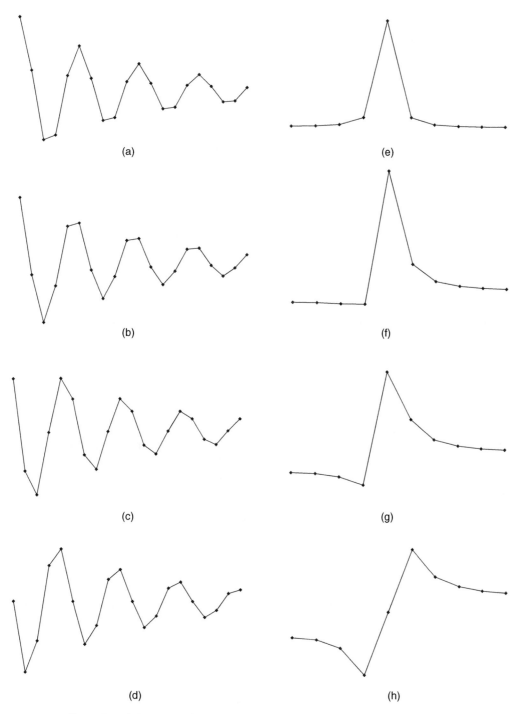

Figure 3.22 Illustration of phase errors (time series (a–d) and real transform (e–h)).

angle. As a phase angle in a time series of −90° converts a cosine wave into a sine wave, the consequence of phase errors is to mix the sine and cosine components of the real and imaginary transforms for a perfect peak shape. As this angle changes, the shape of the real spectrum gradually distorts, as illustrated in Figure 3.22. There are various different types of phase errors. A zero-order phase error is one that is constant through a spectrum, whereas a first-order phase error varies linearly from one end of a spectrum to the other, so that $\phi = \phi_0 + \phi_1\omega$ and is dependent on ω. Higher order phase errors are possible, for example, when looking at images of the body or of food.

There are a variety of solutions to this problem, a common one being to correct this by adding together proportions of the real and imaginary data until an absorption peak shape is achieved using an angle ϕ, so that

$$\text{ABS} = \cos(\phi)\,\text{RL} + \sin(\phi)\,\text{IM}$$

Ideally, this angle should equal the phase angle, which is usually experimentally unknown. Sometimes, phasing is fairly tedious experimentally and can change across a spectrum, and there are, in addition to manual methods, a number of algorithms for phasing. For complex problems such as two-dimensional Fourier transforms, phasing can be quite difficult.

An alternative is to take the absolute value, or magnitude, spectrum defined by

$$\text{MAG} = \sqrt{\text{RL}^2 + \text{IM}^2}$$

The power spectrum is sometimes also used, which is the square of the magnitude spectrum. Although the magnitude spectrum is easy to calculate and always positive, it is important to realise that it is not quantitative: the peak area of a two-component mixture is not equal to the sum of peak areas of each individual component, the reason being that the sum of squares of two numbers is not equal to the square of their sum. As, sometimes, spectroscopic peak areas (or heights) are used for chemometric pattern recognition studies, this limitation is important to appreciate, although it is not important in, for example, imaging.

1. For a phase angle of −90°, the real spectrum corresponds to the absorption spectrum and the imaginary to the dispersion spectrum.
   (a) True
   (b) False

3.5.1.4 Sampling Rates and Nyquist Frequency

An important property of DFTs relates to the rate which data are sampled. Consider the time series in Figure 3.23, each square indicating a sampling point. If it is sampled at half the rate, it will appear that there is no oscillation, as every alternative data point will be eliminated. Therefore, there is no way of distinguishing such a series from a zero frequency series. The oscillation frequency in Figure 3.23 is called the *Nyquist frequency*. Anything that oscillates faster than this frequency will be indistinguishable from a sine wave at a lower frequency. The rate of sampling establishes the range of observable frequencies. The higher the sampling rate in time, the greater the range of observable frequencies. In order to increase the spectral width, a higher sampling rate is required; hence, more data points must be collected per unit time. The equation

$$M = 2\,S\,T$$

links the number of data points acquired (e.g. $M = 4000$) in the time domain, the range of observable frequencies (e.g. $S = 500\,\text{Hz}$) and the acquisition time (e.g. $T = 4\,\text{s}$). Higher frequencies are said to be 'folded over' or 'aliasing'

Figure 3.23 A sparsely sampled time series sampled at the Nyquist frequency. Blue: underlying time series, red observed time series if sparsely sampled.

and appear to be at lower frequencies, as they are undistinguishable. If $S = 500$ Hz, a peak oscillating at 600 Hz will appear indistinguishable from a peak at 400 Hz in the transform. Notice that this relationship determines how a sampling rate in the time domain results in a digital resolution in the frequency or spectral domain (see Section 3.5.1.2). In the time domain, if samples are taken every $\delta t = T/M$ s, in the frequency domain, we obtain a data point every $\delta \omega = 2S/M = 1/T = 1/(M\, \delta t)$ Hz in the spectral domain. Notice that in certain spectroscopies (such as quadrature detection FTNMR), it is possible to record both two time domain signals (treated mathematically as real and imaginary time series) and transform these into real and imaginary spectra. In such cases, only $M/2$ points are recorded in time, so the sampling frequency in the time domain is halved.

The Nyquist frequency is not only important in instrumental analysis. Consider sampling a geological core, where depth relates to time, to determine whether the change in concentrations of a compound, or isotopic ratios, exhibits cyclicity. A finite amount of core is needed to obtain adequate quality samples, which means there is a limitation in samples per length of core. This, in turn, limits the maximum frequency that can be observed. More intense sampling may require a more sensitive analytical technique; thus, for a given method, there is a limitation to the range of frequencies that can be observed.

Note that slightly different terminologies are used for different techniques, but the description above is based on NMR. The units for the range of spectral frequencies are expressed as hertz (or cycles per second) for the equations used above, but of course must be changed as appropriate.

1. A spectrum consists of 2048 real points in the frequency domain and a 1000 Hz range of observable frequencies. What was the acquisition time?

   (a) 2.048 s
   (b) 1.024 s

3.5.1.5 Fourier Algorithms

A final consideration relates to algorithms used for Fourier transforms. DFT methods became widespread in the 1960s partly because Cooley and Tukey developed a rapid computational method, the fast Fourier transform (FFT). This method required the number of sampling points to be a power of 2, for example, 1024, 2048 and so on, and many chemists still associate powers of 2 with Fourier transformation. However, there is no special restriction on the number of data points in a time series, the only consideration is relating to the speed of computation. The method for Fourier transformation introduced above is slow for large data sets, and early computers were much more limited in capabilities, but it is not always necessary to use rapid algorithms in modern day applications unless the amount of data is really large. There is a huge technical literature on Fourier transform algorithms, but it is important to recognise that an algorithm is simply a means to an end, and not an end in itself.

1. It is necessary for the number of data points in a time series to be a power of 2 when using the Cooley–Tukey algorithm for Fourier transformation.

   (a) True
   (b) False

3.5.2 Fourier Filters

In Section 3.3, we discussed a number of linear filter functions that can be used to enhance the quality of spectra and chromatograms. When performing Fourier transforms, it is possible to apply filters to the raw (time domain) data before Fourier transformation, and this is a common method in spectroscopy to enhance resolution or signal-to-noise ratio, as an alternative to applying filters directly to the spectral domain. There is a huge literature on this; thus, we only describe some of the most common methods below.

3.5.2.1 Exponential Filters

The width of a peak in a spectrum primarily depends on the decay rate in the time domain. The faster the decay, the broader the peak. Figure 3.24 illustrates a broad peak together with its corresponding time domain. If it is desired to increase resolution, a simple approach could be to change the shape of the time domain function so that the decay is

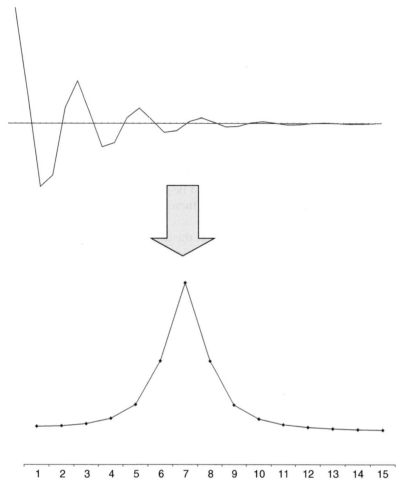

Figure 3.24 Fourier transformation of a rapidly decaying time series.

slower. In some forms of spectroscopy (such as NMR), the time series contains a term due to exponential decay and can be characterised by

$$f(t) = A\cos(\omega t)e^{-t/s} = A\cos(\omega t)e^{-\lambda t}$$

as described in Section 3.5.1.1. The larger the magnitude of λ, the more rapid the decay, hence the broader the peak. Multiplying the time series by a positive exponential of the form

$$g(t) = e^{+\kappa t}$$

changes the decay rate to give a new time series

$$h(t) = f(t)g(t) = A\cos(\omega t)e^{-\lambda t}e^{+\kappa t}$$

The exponential decay constant is now equal to $-\lambda + \kappa$. Provided $\kappa < \lambda$, the rate of decay is reduced, and, as indicated in Figure 3.25, results in a narrower line width in the transform, thus would be reflected in improved resolution if there is a cluster of several overlapping peaks.

1. Multiplying a time series by a positive exponential function increases peak widths in the corresponding frequency domain.

   (a) True
   (b) False

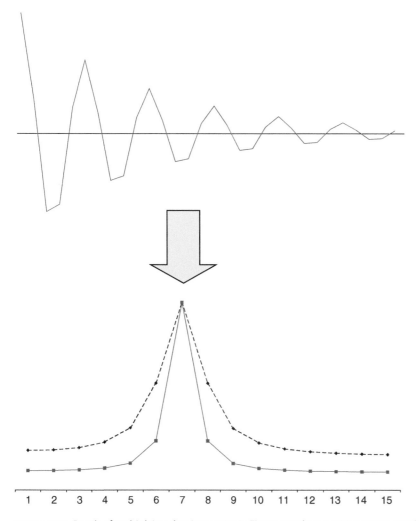

Figure 3.25 Result of multiplying the time series in Figure 3.24 by a positive exponential, the transform of the original time series being represented by a dotted blue line.

3.5.2.2 Influence of Noise

Theoretically, it is possible to conceive multiplying the original time series by increasingly positive exponentials until peaks are one data point wide. Clearly, there is a flaw in our argument; otherwise, it would be possible to obtain indefinitely narrow peaks and achieve any desired resolution.

The difficulty is that real spectra always contain noise. Figure 3.26 represents a noisy time series, together with the exponentially filtered data. The filtered time series amplifies noise substantially, which can interfere with signals. Although the peak width of the new transform has indeed decreased, the noise has increased. In addition to making peaks hard to identify, noise also reduces the ability to determine integrals and concentrations and sometimes to accurately pinpoint peak positions.

How can this be solved? Clearly, there are limits to the amount of peak sharpening that is practicable, but the filter function can be improved so that noise reduction and resolution enhancement are applied simultaneously. One common method is to multiply the time series by a double exponential filter of the form

$$g(t) = e^{+\kappa t - \nu t^2}$$

where the first (linear) term of the exponential increases with time and enhances resolution, and the second (quadratic) term decreases noise. Provided the values of κ and ν are chosen correctly, the result will be an increase in resolution without an increase in noise. The main aim is to emphasise the middle of the time series whilst reducing the end. These two terms could be optimised theoretically if peak widths and noise levels are known in advance, but, in most practical cases, they are chosen empirically. The effect on the noisy data in Figure 3.26 is illustrated in Figure 3.27, for a typical

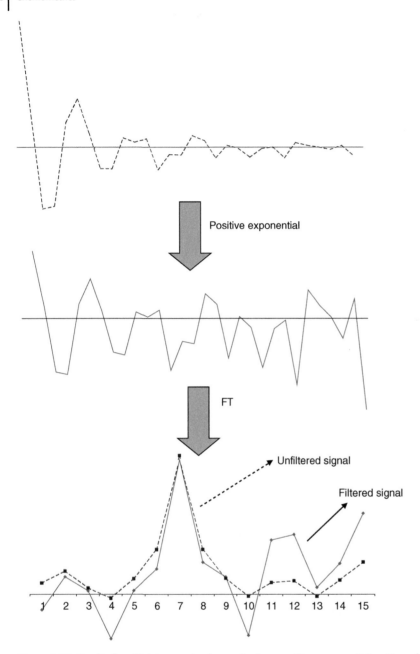

Figure 3.26 Result of multiplying a noisy time series by a positive exponential and transforming the new signal.

double exponential filter, the dotted line representing the result of the single exponential filter. The time series decays more slowly than the original, but there is not much increase in noise. The peak shape in the transform is almost as narrow as that obtained using a single exponential, but noise is dramatically reduced.

A large number of so-called *matched* or *optimal filters* have been proposed in the literature, many specific to a particular kind of data, but the general principles are to obtain increased resolution without introducing too much noise. It is important to recognise that these filters can distort peak shapes. Although there is a substantial literature on this subject, the best approach is to tackle the problem empirically rather than relying on elaborate rules. Figure 3.28 shows the result of applying a simple double exponential function to a typical time series. Note the bell-shaped function, which is usual. The original spectrum suggests that there is a cluster of peaks, but only two clear peaks are visible. Applying a filter function suggests that there are at least four underlying peaks in the spectrum, although there is some distortion of the data in the middle, probably a result of a function that is slightly too severe. For a comprehensive discussion of Fourier filters, it is best to refer to the literature for a specific type of instrumental data.

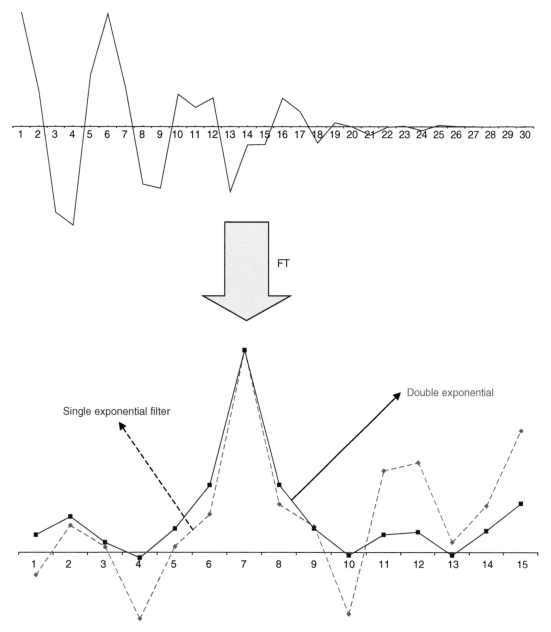

Figure 3.27 Multiplying the data in Figure 3.25 by a double exponential.

1. A typical double exponential filter consists of two exponent terms, the linear one to reduce noise and the quadratic to increase resolution.
   (a) True
   (b) False

3.5.2.3 Fourier Self-Deconvolution

In many forms of spectroscopy such as NMR and IR, data are often acquired directly as a time series and must be Fourier transformed to obtain an interpretable spectrum. However, any spectrum or chromatogram can be processed using Fourier filters, even if not acquired as a time series. The secret is to inverse transform (see Section 3.5.1.2) back to a time series.

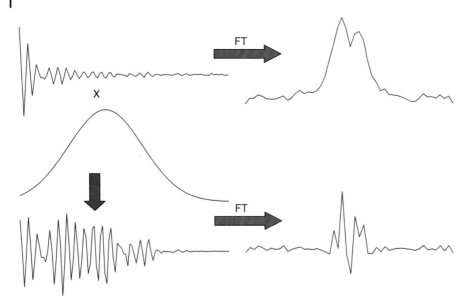

Figure 3.28 Use of a double exponential filter.

Normally, three steps are employed, as illustrated in Figure 3.29.

- Transform the spectrum into a time series. This time series does not physically exist but can be handled by a computer.
- Then apply a Fourier filter to the time series.
- Finally transform the spectrum back, resulting in improved quality.

This procedure is called *Fourier self-deconvolution* and is an alternative to digital filters of Section 3.3.

1. Any frequency domain signal can be improved in shape using Fourier filters.

   (a) True
   (b) False

3.5.3 Convolution Theorem

Some people get confused by the difference between Fourier filters, linear smoothing and resolution functions. In fact, both methods are equivalent and are related by the convolution theorem, and both have similar aims to improve the quality of spectroscopic or chromatographic or time series data.

The principles of convolution have been discussion in Section 3.3.3. Two functions, f and g, are said to be convoluted to give h, if

$$h_i = \sum_{j=-p}^{j=p} f_j \, g_{i+j}$$

Convolution involves moving a window or digital filter function (such as a Savitzky–Golay or MA) along a series of data such as a spectrum, multiplying the data by that function at each successive data point. In the applications discussed in this section, window functions will always be applied in the spectral domain rather than time domain. A three-point MA involves multiplying each set of three points in a spectrum by a function containing the values (1/3, 1/3, 1/3), and the spectrum is said to be convoluted by the MA filter function.

Filtering a time series, using Fourier time domain filters, however, involves multiplying the entire time series by a single function, so that

$$H_i = F_i G_i$$

Figure 3.29 Fourier self-deconvolution of a peak cluster.

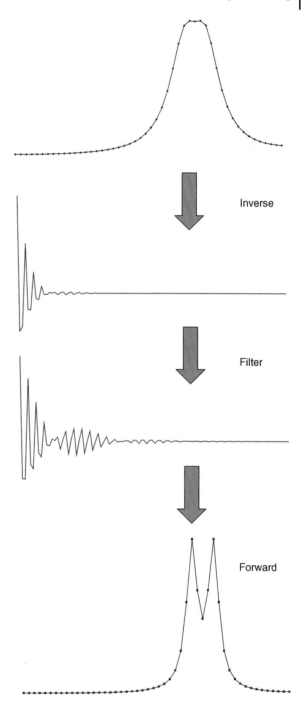

The convolution theorem states that f, g and h are Fourier transforms of F, G and H. Hence, linear filters as applied directly to spectroscopic data have their equivalence as Fourier filters in the time domain; in other words, convolution in one domain is equivalent to multiplication in the other domain. This approach is best largely dependent on computational complexity and convenience. For example, both MAs and exponential Fourier filters are easy to apply, so are simple approaches, one applied direct to the frequency spectrum and the other to the raw time series. Convoluting a spectrum with the Fourier transform of an exponential decay is a difficult procedure; thus, the choice of domain is made according to how easy the calculations are.

1. In the equation $h_i = \sum_{j=-p}^{j=p} f_j g_{i+j}$ as discussed above.

   (a) f represents the spectral domain and g a window function
   (b) g represents the spectral domain and f a window function
   (c) f represents the time domain and g a window function
   (d) g represents the time domain and f a window function

3.6 Additional Methods

A number of more sophisticated methods have been implemented over the past few decades, although some of the ideas such as Bayes' theorem are very much older. In certain instances, a more specialised approach is appropriate and also generates much interest in the literature. There are particular situations, for example, where data are very noisy or incomplete, or where rapid calculations are required, which require particular solutions. The methods listed below are topical and implemented within a number of common software packages. They do not represent a comprehensive review but are added for completion, as they are regularly reported in the chemometrics literature and are often available in common software packages. The discussion on Bayesian methods is introduced within this section, although can be more widely associated with most statistics: however, unlike, for example, the machine learning or applied statistics community, they are rarely used within chemometrics; thus, we restrict to its application to signal analysis to avoid repetition throughout the text.

3.6.1 Kalman Filters

The Kalman filter has its origin in the need for rapid online curve fitting. In some situations such as chemical kinetics, it is desirable to calculate a model whilst the reaction is taking place rather than wait until the end. In online applications such as process control, it may be useful to see a smoothed curve as the process is taking place, in real time, rather than later. The general philosophy is that, as something evolves with time, more information becomes available so that the model can be refined. As each successive sample is recorded, the model improves. It is possible to predict the response from information provided at previous sample times and see how this differs from the observed response, thus changing the model.

Kalman filters are quite complex to implement computationally, but the principles are as follows and will be illustrated by the case where a single response (y) depends on a single factor (x). There are three main steps.

- Model the current data point (i), for example, calculate $\hat{y}_{i|i-1} = x_i b_{i-1}$ using a polynomial in x, and methods introduced in Chapter 2. The parameters b_{i-1} are initially guesses, which are refined with time. The '|' symbol means that the model of y_i is based on the first $i-1$ data points and x_i is a row vector consisting of the terms in a model (usually but not exclusively polynomial) and b is a column vector. For example, if $x_i = 1.5$ (e.g. the time a sample is measured), then a three-parameter quadratic model of the form $y_i = b_0 + b_1 x + b_2 x_2^2$ gives the vector $x_i = (1, 1.5, 2.25)$.
- This next step is to see how well this model predicts the current data point and calculate $d_i = y_i - \hat{y}_{i|i-1}$, which is called the *innovation*. The closer these values, the better the model.
- Finally, refine the model by recalculating the coefficients

$$b_i = b_{i-1} + k_i d_i$$

If the estimated and observed values of y are identical, the value of b will be unchanged. If the observed value is more than the estimated value, it makes sense to increase the size of the coefficients to compensate. The column vector k_i is called the *gain* vector. There are a number of ways of calculating this, but the larger it is, the greater the uncertainty in the data.

A common (but complicated way) of calculating the gain vector is as follows.

- Start with a matrix V_{i-1}, which represents the variance (or error) of the coefficients. This is a square matrix, with the number or rows and columns equal to the number of coefficients in the model. Hence, if there are five coefficients, there will be 25 elements in the matrix. The higher these numbers, the less certain the prediction of the coefficients. Start with a diagonal matrix containing some high numbers.

- Guess a number r that represents the approximate error at each point. This could be the root mean square replicate error. This number is not too crucial, and it can be set as a constant throughout the calculation.
- The vector k_i is given by

$$k_i = \frac{V_{i-1}x_i'}{x_1 V_{i-1}x_i' - r} = \frac{V_{i-1}x_i'}{q - r}$$

- The new matrix for the current data point V_i is given by

$$V_i = V_{i-1} - k_i x_i V_{i-1}$$

The magnitude of the elements of this matrix should reduce with time, as the measurements become more certain, which means a consequential reduction in k and convergence of the coefficients in b.

Although it is not always necessary to understand the computational details, it is important to appreciate the application of the method. Table 3.10 represents the progress of such a calculation.

- A model of the form $y_i = b_0 + b_1 x + b_2 x^2$ is to be set up, there being three coefficients.
- The initial guess of the three coefficients is 0.000. Therefore, the guess of the response when $x = 0$ is 0, and the innovation is $0.840 - 0.000$ (or the observed minus the predicted using the initial model).
- Start with a matrix $V_i = \begin{bmatrix} 100 & 0 & 0 \\ 0 & 100 & 0 \\ 0 & 0 & 100 \end{bmatrix}$ the diagonal numbers representing high uncertainty in measurements of the parameters, given the experimental numbers.

Table 3.10 Kalman filter calculation.

x_i	y_i	b_0	b_1	b_2	$\hat{y}_i$	k_i'			d_i
0	0.840	0.841	0.000	0.000	0.000	1.001	0.000	0.000	
1	0.737	0.841	−0.052	−0.052	0.841	−0.001	0.501	0.501	−0.104
2	0.498	0.841	−0.036	−0.068	0.530	0.001	−0.505	0.502	−0.032
3	0.296	0.849	−0.114	−0.025	0.124	0.051	−0.451	0.250	0.172
4	0.393	0.883	−0.259	0.031	0.003	0.086	−0.372	0.143	0.390
5	0.620	0.910	−0.334	0.053	0.371	0.107	−0.304	0.089	0.249
6	0.260	0.842	−0.192	0.020	0.829	0.119	−0.250	0.060	−0.569
7	0.910	0.898	−0.286	0.038	0.458	0.125	−0.208	0.042	0.452
8	0.124	0.778	−0.120	0.010	1.068	0.127	−0.176	0.030	−0.944
9	0.795	0.817	−0.166	0.017	0.490	0.127	−0.150	0.023	0.305
10	0.436	0.767	−0.115	0.010	0.831	0.126	−0.129	0.017	−0.395
11	0.246	0.712	−0.064	0.004	0.693	0.124	−0.113	0.014	−0.447
12	0.058	0.662	−0.024	−0.001	0.469	0.121	−0.099	0.011	−0.411
13	−0.412	0.589	0.031	−0.006	0.211	0.118	−0.088	0.009	−0.623
14	0.067	0.623	0.007	−0.004	−0.236	0.115	−0.078	0.007	0.303
15	−0.580	0.582	0.033	−0.006	−0.210	0.112	−0.070	0.006	−0.370
16	−0.324	0.605	0.020	−0.005	−0.541	0.108	−0.063	0.005	0.217
17	−0.896	0.575	0.036	−0.007	−0.606	0.105	−0.057	0.004	−0.290
18	−1.549	0.510	0.069	−0.009	−0.919	0.102	−0.052	0.004	−0.630
19	−1.353	0.518	0.065	−0.009	−1.426	0.099	−0.047	0.003	0.073
20	−1.642	0.521	0.064	−0.009	−1.675	0.097	−0.043	0.003	0.033
21	−2.190	0.499	0.073	−0.009	−1.954	0.094	−0.040	0.002	−0.236
22	−2.206	0.513	0.068	−0.009	−2.359	0.091	−0.037	0.002	0.153

- Use a value of r of 0.1. Again this is a guess, but given the scatter of the experimental points, it looks as if this is a reasonable number. In fact, values greater or smaller than 10-fold do not make a major impact on the resultant model, although they do influence the first few estimates.

As more samples are obtained, the following can be observed:

- the size of k decreases,
- the values of the coefficients converge and
- there is a better fit to the experimental data.

Figure 3.30 shows the progress of the filter. The earlier points are very noisy and deviate considerably from the experimental data, whereas the later points represent quite a smooth curve. In Figure 3.31, the progress of the three coefficients is presented, the graphs normalised to a common scale for clarity. Convergence takes about 20 iterations. The final answer of $y_i = 0.513 + 0.068x - 0.009x^2$ is obtained in this case.

It is important to recognise that Kalman filters are computationally elaborate and are not really useful unless there is a special reason for performing online calculations. It is possible to take the entire X and y data in Table 3.10 and

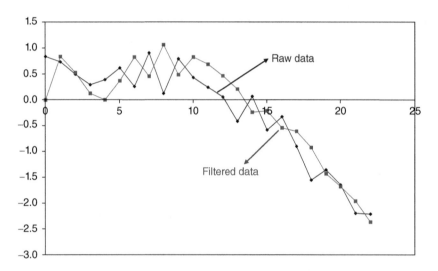

Figure 3.30 Progress of the Kalman filter, showing the fitted and raw data.

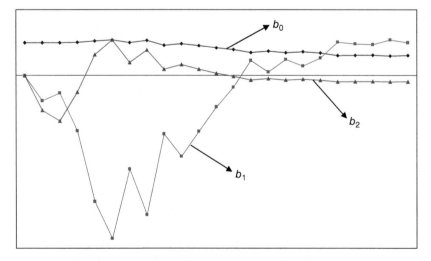

Figure 3.31 Change in the three coefficients predicted by the Kalman filter with time.

perform multiple linear regression as discussed in Chapter 2, so that

$$y = Xb$$

or

$$b = (X'X)^{-1}X'y$$

using the standard equation for the pseudo-inverse, giving an equation of $\hat{y}_i = 0.512 + 0.068x - 0.009x_2^2$ only very slightly different to the prediction by Kalman filters when $i = 22$. If all the data are available, there is little point using Kalman filter, the method is mainly useful for online predictions; however, they can be useful for smoothing 'on the fly' and with modern computers can be implemented very fast, so that, for example, a smoothed profile is presented on a screen in real time during a process.

Kalman filters can be extended to more complex situations with many variables and many responses. The model does not need to be multi-linear but, for example, may be exponential (e.g. in kinetics). Although the equations increase considerably in complexity, the basic ideas are the same. Although Kalman filters are in fact widely used in engineering, they are no longer very common in the chemometrics literature; thus, we will not discuss the more elaborate enhancements.

1. A process is modelled at point $i - 1$ by $\hat{y} = 0.5 + 0.2x - 0.05x^2$. The estimated value at point $i - 1$ is 0.8 and the observed value at point i is 0.95. If $k' = (0.03\ 0.06\ {-}0.04)$, what is the new model at point i?

   (a) $\hat{y} = 0.4955 + 0.191x + 0.01x^2$
   (b) $\hat{y} = 0.5045 + 0.209x - 0.11x^2$
   (c) $\hat{y} = 0.53 + 0.26x - 0.45x^2$
   (d) $\hat{y} = 0.47 + 0.14x + 0.35x^2$

2. A process is modelled by a full third-order polynomial. V has dimensions

   (a) 3×1
   (b) 3×3
   (c) 4×4

3.6.2 Wavelet Transforms

Another topical method in chemical signal processing is the wavelet transform. The general principles are discussed below, without providing detailed information about algorithms, and provide a general understanding of the approach: wavelets are implemented in several chemometric packages; hence, many people have come across them. We will restrict the discussion to the DWT (discrete wavelet transform).

The first transform now classed as a wavelet transform was described in the early twentieth century by Haar and is known as a Haar transform. During the mid-1980s, a collection of related methods were described, together called *wavelet transforms*, of which the Haar was the simplest. As is usual in the machine learning literature, grants, PhDs and conference presentations are obtained by describing increasingly complex algorithms. For complex data such as image processing, for example, in medicine or facial recognition, some of the more elaborate methods provide real benefit, but most chemometrics data are in fact relatively simple. In chemometrics, wavelet transforms are used for two primary purposes, to denoise data and to compress data.

The principle of the Haar transform is in fact very simple. For DWTs, the number of data points, which we will denote as x, is a power of 2 and we will restrict discussion to this situation. Consider the data of the left-hand column in Table 3.11. A simple approach of reducing the 16 numbers to 8 would be to average each two successive data points, or in our example we add these (we will discuss scaling below). This provides eight *approximation coefficients*, as in the centre column. By the process, the data have been compressed from 16 to 8 data points, and noise has been reduced by a two-point MA. This represents a level 1 wavelet transform. The data can be smoothed and compressed furthermore to four data points (in our example, this is probably too far) via a level 2 wavelet transform. The results are presented graphically in Figure 3.32. For the level 1 transform, peaks can be clearly distinguished. Obviously, in most practical situations, there will be far greater digital resolution; thus, it may be easier to distinguish peak shapes. However, if we used the level 1 transform, we would be able to distinguish signals better and reduce the amount of data we store.

Normally, to keep the scale of the data constant, the transformed data are multiplied, in this case by $(1/\sqrt{2})^l$ for level l. This makes the sum of squares at each level approximately constant: in our example, it is 5.911 for the raw data, 5.644 for level 1 and 5.264 for level 2. The data transform can be expressed in matrix terms, where the original data x has

Table 3.11 Numerical example for wavelet transform: left raw data, centre transformed data after level 1 wavelet and right after level 2 wavelet, without scaling.

0.271	0.409	1.419
0.138	1.011	1.227
0.588	0.916	3.035
0.422	0.311	2.886
0.545	1.235	
0.370	1.800	
0.192	1.786	
0.120	1.100	
0.487		
0.748		
0.956		
0.844		
1.142		
0.644		
0.727		
0.373		

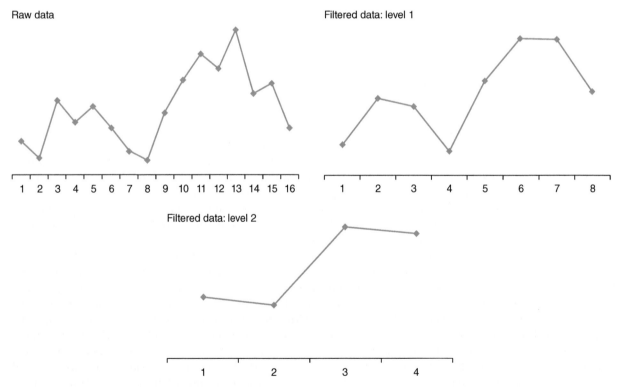

Figure 3.32 Raw data and wavelet filtered data in Table 3.11.

dimensions 16×1 and is pre-multiplied by a matrix w of dimensions 8×16 in this case for a level 1 transform to give a new smoothed and compressed d_1 data set of dimensions 8×1, hence $d_1 = w\,x$. The matrix w is of the form

$$\frac{1}{\sqrt{2}}\begin{bmatrix} 1 & 1 & 0 & 0 & 0 & 0 & 0 & 0 & 0 & 0 & 0 & 0 & 0 & 0 & 0 & 0 \\ 0 & 0 & 1 & 1 & 0 & 0 & 0 & 0 & 0 & 0 & 0 & 0 & 0 & 0 & 0 & 0 \\ 0 & 0 & 0 & 0 & 1 & 1 & 0 & 0 & 0 & 0 & 0 & 0 & 0 & 0 & 0 & 0 \\ 0 & 0 & 0 & 0 & 0 & 0 & 1 & 1 & 0 & 0 & 0 & 0 & 0 & 0 & 0 & 0 \\ 0 & 0 & 0 & 0 & 0 & 0 & 0 & 0 & 1 & 1 & 0 & 0 & 0 & 0 & 0 & 0 \\ 0 & 0 & 0 & 0 & 0 & 0 & 0 & 0 & 0 & 0 & 1 & 1 & 0 & 0 & 0 & 0 \\ 0 & 0 & 0 & 0 & 0 & 0 & 0 & 0 & 0 & 0 & 0 & 0 & 1 & 1 & 0 & 0 \\ 0 & 0 & 0 & 0 & 0 & 0 & 0 & 0 & 0 & 0 & 0 & 0 & 0 & 0 & 1 & 1 \end{bmatrix}$$

as can be verified. The coefficient pairs in each row are often called the *father wavelet*.

Now comes the tricky part. If we wanted to construct the original data back from this, we have some missing information. The missing information could be obtained by subtracting each sequential value to give what is often called the *mother wavelet*. Instead of the matrix above, we use

$$\frac{1}{\sqrt{2}}\begin{bmatrix} 1 & -1 & 0 & 0 & 0 & 0 & 0 & 0 & 0 & 0 & 0 & 0 & 0 & 0 & 0 & 0 \\ 0 & 0 & 1 & -1 & 0 & 0 & 0 & 0 & 0 & 0 & 0 & 0 & 0 & 0 & 0 & 0 \\ 0 & 0 & 0 & 0 & 1 & -1 & 0 & 0 & 0 & 0 & 0 & 0 & 0 & 0 & 0 & 0 \\ 0 & 0 & 0 & 0 & 0 & 0 & 1 & -1 & 0 & 0 & 0 & 0 & 0 & 0 & 0 & 0 \\ 0 & 0 & 0 & 0 & 0 & 0 & 0 & 0 & 1 & -1 & 0 & 0 & 0 & 0 & 0 & 0 \\ 0 & 0 & 0 & 0 & 0 & 0 & 0 & 0 & 0 & 0 & 1 & -1 & 0 & 0 & 0 & 0 \\ 0 & 0 & 0 & 0 & 0 & 0 & 0 & 0 & 0 & 0 & 0 & 0 & 1 & -1 & 0 & 0 \\ 0 & 0 & 0 & 0 & 0 & 0 & 0 & 0 & 0 & 0 & 0 & 0 & 0 & 0 & 1 & -1 \end{bmatrix}$$

to give the *detail coefficients*. Similar transforms can be visualised for level 2 wavelet transform, level 3 wavelet transform and so on, either in several steps of as a single transform.

Often, these functions are visualised. For the level 1 wavelet, we can, in our case, visualise eight father and eight mother wavelets, or 2^{n-l} where l is the level and the original data consist of 2^n data points, as shown in Figure 3.33. For level 2, these are expanded; each level has corresponding father and mother wavelets of increasing width but of identical shape. The scaling factor of a level l wavelet is $(1/\sqrt{2})^l$. These can be used to calculate the smoothed data at any level using just one stage or matrix transform. The mother wavelets are most commonly visualised but are not usually valuable and discarded in chemometrics.

There are, of course, many other types of wavelet functions apart from the Haar wavelet. The mother wavelet does not need to be symmetrical but always integrates to zero. The father wavelet is not always flat, but the mother and father wavelets are always represented in pairs, and each successive level involves contraction or expansion of the wavelet functions as appropriate. However, in this section, we will restrict the discussion on Haar wavelet, which is the most widespread and common.

1. Raw data consists of 64 points. How many level 2 wavelets are there?

   (a) 32
   (b) 16
   (c) 8
   (d) 4

2. Mother wavelets are always symmetrical.

   (a) True
   (b) False

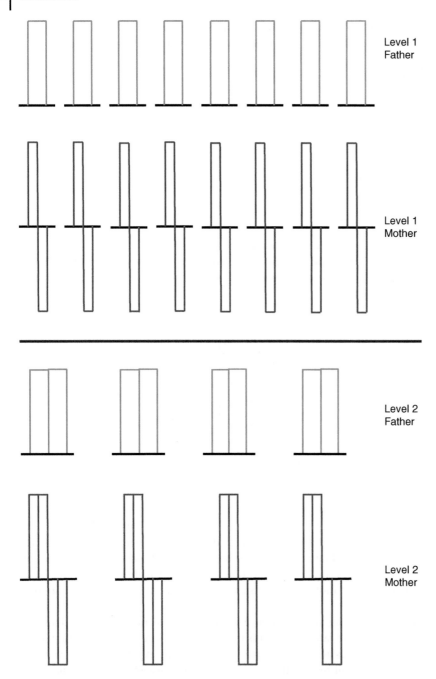

Figure 3.33 Haar wavelets of levels 1 and 2 corresponding to data in Table 3.11.

3.6.3 Bayes' Theorem

Bayesian approaches are now commonplace in most areas of modern statistics and machine learning, but less well used in chemometrics, partly because of the difficulty of incorporating probabilities into algorithms such as partial least squares (Chapter 6), hence the relatively limited discussion in this text. The principle is that before we perform an experiment, we should know something about a system; this prior knowledge can be converted into a probability. The default spectrum might be a flat distribution. As more information is gathered, we build up a more detailed picture of a system, which allows us to predict information usually to greater degrees of certainty. Each observation alters the probability. Many statisticians consider this as a common sense. For example, if we want to predict the temperature at the North Pole in winter, we have prior information that we can use before meteorological modelling, which then improves our predictions and so on. In statistics, these predictions are usually expressed in probabilities; for example, we may want to give an estimate that we are 99% confidence that the temperature on a given day will be within a defined range. This range can be obtained just from historic data, but the closer we get to our day, the more we can gather information on local weather patterns and thus the range may narrow.

Bayes' theorem can be presented in the following form.

Probability (answer given new information) $\propto$

Probability (answer given old information) $\times$ Probability (new information given answer)

There are various ways of mathematically defining this dependent on what our prior and new information is, but one way is as follows:

$$p(A|B) = p(B|A)p(A)/(p(A)p(B|A) + p(B)p(B| \sim A))$$

where '|' means given, for example, the probability that a person has a disease (A), a given diagnostic test shows positive (B) (not all tests or diagnoses are 100% perfect) and $\sim$ means not. $p(A)$ is the *prior* probability before the test has been taken; this might be the proportion of patients coming to a clinic that are diseased, or even the probability obtained from previous diagnoses; for example, we might know that 70% of people exhibiting a given set of symptoms are diseased but need to send the patient to a clinic for more confirmatory tests. $p(A|B)$ is the *posterior* probability; $p(B|A)$ relates to how good the test is, called the likelihood, and $p(B|\sim A)$ is the chance that a person actually tests positive for the disease even if they do not have it.

Let us consider a numerical example.

- We consider that if a patient has been referred to a surgery, his chances of having a disease is 0.80, this is the prior probability of 0.8.
- He or she tests positive.
- But from the previous experience, we know that 0.9 people have the disease test positive or $p(B|A)$.
- We also know that 0.3 of patients that do not have the disease also test positive or $p(B| \sim A)$. Note that 0.9 and 0.3 do not add up to 1.
- In our case, as there are only two choices, $p(B) = 1 - p(A)$

Using our equation, we find that the *posterior* probability has risen to $0.8 \times 0.9/(0.8 \times 0.9 + 0.2 \times 0.3) = 0.923$ from the *prior* probability of 0.8. If a patient comes to the surgery and the tests post, this means that their chances of having the disease is now 0.923.

This can be easily checked; for example, a totally non-specific test that both 70% of people that have the disease and 70% that do not test positive will result in a posterior probability of $0.8 \times 0.7/(0.8 \times 0.7 + 0.2 \times 0.7) = 0.8$ or is unchanged as the test does not provide any further useful evidence.

The equation can be expressed in several alternative forms.

When applied to signal processing, which is one of the commonest applications in chemometrics, various definitions are necessary.

- *Data* are experimental observations, for example the measurement of a time series or FID before Fourier transformation. *Data space* contains a data set for each experiment.
- A *map* is the desired result, for example a clean and noise free spectrum, or the concentration of several compounds in a mixture. *Map space* exists in a similar manner to data space.
- An operation or *transformation*, such as Fourier transformation or factor analysis, links these two spaces.

The aim of the experimenter is to obtain an estimate of map space as good as possible, consistent with his or her knowledge of the system. Normally, there are two types of knowledge.

- *Prior* knowledge is available before the experiment. There is almost always some information available about chemical data. An example is that a true spectrum will always be positive: we can reject statistical solutions that result in negative intensities. Sometimes, much more detailed information such as line shapes or compound concentrations is known. The previous map can be considered a probability distribution.
- *Experimental information*. This refines the prior knowledge to give a *posterior* model of the system.

Bayes' can be presented using the terminology discussed above:

p(map | experiment) $\propto$ p(map | prior info) $\times$ p(experiment | map)

Many scientists ignore the prior information, and for cases where data are quite good, this can be perfectly acceptable. However, chemical data analysis is most useful, where the answer is not so obvious, and the data are difficult to analyse.

The Bayesian method allows prior information or measurements or hypotheses to be taken into account. It also allows continuing experimentation, improving a model all the time, as the Bayesian model can be refined each time more data are obtained. A spectrum can be considered a probability distribution; thus, the aim of experimentation is to refine a prior spectral model.

It is important to emphasise that Bayesian methods are applicable throughout science and are introduced in this section primarily because of their relationship with maximum entropy in signal processing, as discussed below, but should not be viewed as an exclusive tool for signal analysis. This text is primarily about modern chemometric trends, and Bayesian methods are rarely encountered in the chemometrics literature, whilst very widespread in mainstream statistics.

1. 70% of all samples sent to a laboratory are suspected to be adulterated. A spectroscopic test suggests that if a sample is indeed adulterated, this test correctly identifies 80% of adulterated samples, but incorrectly 30% of non-adulterated samples. What is the posterior probability that this sample is actually adulterated?

   (a) 0.727
   (b) 0.903
   (c) 0.861
   (d) 0.823

3.6.4 Maximum Entropy

Over the past decades, there has been substantial scientific interest in the application of maximum entropy techniques with notable successes, for the chemist, in areas such as NMR spectroscopy and crystallography. Maxent has had a long statistical vintage, one of the modern pioneers being Jaynes, but the first significant scientific applications were in the area of deblurring of infrared images of the sky, involving development of the first modern computational algorithm, in the early 1980s. Since then, there has been an explosion of interest and several implementations available within commercial instrumentation. The most spectacular successes have been in the area of image analysis, for example NMR tomography, as well as forensic applications such as obtaining clear car number plates from hazy police photos. In addition, there has been a very solid and large literature in the area of analytical chemistry.

3.6.4.1 Definition

Maxent is one method for determining the probability of a model. A simple example involves the toss of a six-sided unbiased die. What is the most likely underlying frequency distribution, and how can each possible distribution be measured? Figure 3.34 illustrates a flat distribution and Figure 3.35 illustrates a skew distribution (expressed as proportions). The concept of entropy can be introduced in a simple form and is defined by

$$S = -\sum_{i=1}^{I} p_i \log(p_i)$$

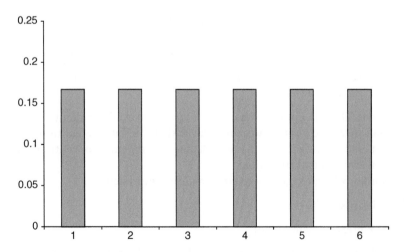

Figure 3.34 Frequency distribution for the toss of a die.

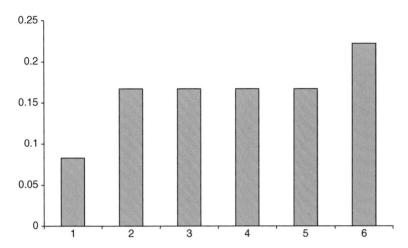

Figure 3.35 Another, but less likely, frequency distribution for toss of a die.

Table 3.12 Maximum entropy calculation for unbiased die, logarithms to the base 10.

	P		$p \log(p)$	
	Figure 3.34	Figure 3.35	Figure 3.34	Figure 3.35
1	0.167	0.083	0.130	0.090
2	0.167	0.167	0.130	0.130
3	0.167	0.167	0.130	0.130
4	0.167	0.167	0.130	0.130
5	0.167	0.167	0.130	0.130
6	0.167	0.222	0.130	0.145
Entropy			0.778	0.754

where p_i is the probability of outcome i. In the case of our die, there are six outcomes, and each outcome has a probability of 1/6 (Figure 3.34). The distribution with maximum entropy is the most likely underlying distribution. Table 3.12 consists of the entropy calculation for the two distributions and demonstrates that the even distribution results in the highest entropy and is best, given evidence available. In the absence of experimental information, a flat distribution is indeed the most likely. There is no reason why any one number on the die should be favoured above other numbers, unless it is biased. These distributions can be likened to spectra sampled at six data points – if there is no other information, the spectrum with maximum entropy is a flat distribution.

However, constraints can be added. For example, it might be known that the die is actually a biased die with a mean of 4.5 instead of 3.5, lots of experiments suggest this. What distribution is expected now? Consider distributions A and B in Table 3.13. Which are more likely? Maximum entropy will select distribution B. It is rather unlikely (unless we know something) that the numbers 1 and 2 will never appear. Note that the value of $0 \log(0)$ is 0 and that in this example, logarithms are calculated to the base 10, although using natural logarithms is equally acceptable. Of course, in this simple example, we do not include any knowledge about the distribution of the faces of the die, and we suspect that uneven weight carried by the die causes this deviation. We could then include more information, perhaps that the farther a face is from the weight, the less more likely it is to land upwards, which could help refine the distributions further.

A spectrum or chromatogram can be considered as a probability distribution. If the data are sampled at 1000 different points, then the intensity at each data point is a probability. For a flat spectrum, the intensity at each point in the spectrum equals 0.001; hence, the entropy is given by

$$S = -\sum_{i=1}^{1000} 0.001 \log(0.001) = -1000 \times 0.001 \times (-3) = 3$$

Table 3.13 Maximum entropy calculation for biased die.

	p		p log(p)	
	A	**B**	**A**	**B**
1	0.00	0.0238	0.000	0.039
2	0.00	0.0809	0.000	0.088
3	0.25	0.1380	0.151	0.119
4	0.25	0.1951	0.151	0.138
5	0.25	0.2522	0.151	0.151
6	0.25	0.3093	0.151	0.158
Entropy			0.602	0.693

This, in fact, is the maximum entropy solution but does not yet take account of experimental data but is the most likely distribution in the absence of more information or is prior probability distribution.

It is important to realise that there are a number of other definitions of entropy in the literature, only the most common being described in this chapter.

1. A six-sided die is tossed 10 times. It comes up 1, 3, 4 and 5 two times, and 2 and 6 one time. What is the entropy of the distribution, using logs to the base 10.
   - (a) −0.759
   - (b) 0.759
   - (c) −2.408
   - (d) 2.408

2. Using the maximum entropy criterion, and the hypothesis that the die is unbiased, compare the result in question 1 to a result where the die came up 1, 3, 4, 5 and 6 two times and never 2. Which is the most likely distribution?
   - (a) The distribution of question 1
   - (b) The distribution of question 2

3.6.4.2 Modelling

In practice, there are an infinite, or at least very large, number of statistically identical models that can be obtained from a system. If I know a chromatographic peak consists of two components, I can come up any number of ways of fitting the chromatogram, all with identical least squares fit to the data. In the absence of further information, a smoother solution is preferable and most definitions of entropy will pick such an answer.

Although in the absence of any information at all, a flat spectrum or chromatogram is the best answer, experimentation will change the solutions considerably, and should pick two underlying peaks that fit the data well consistent with maximum entropy. Into the entropic model, information can be built relating to knowledge of the system. Normally, a parameter is calculated as

Entropy function − statistical fit function

High entropy is good, but not at the cost of a numerically poor fit to the data; however, a model that fits the original (and possibly very noisy) experiment well (e.g. a polynomial with low least square errors) is not a good model if the entropy is too high. The statistical fit can involve a least squares function such as χ^2, which is hoped to minimise. In practice, for a number of models with identical fit to the data, the one with maximum entropy is the most likely. Maximum entropy is used to calculate a *prior* probability (see discussion on Bayes' theorem) and experimentation refines this to give a *posterior* probability. Of course, it is possible to refine the model still further by performing yet more experiments, using the posterior probabilities of the first set of experiments as prior probabilities for the next experiments. In reality, this is what many scientists do, continuing experimentation until they reach a desired level of confidence, the Bayesian method simply refining the solutions.

For relatively sophisticated applications, it is necessary to implement the method as a computational algorithm, being a number of packages available in instrumental software. The algorithms are quite complex and out of the scope of this introductory text; however, it is worth understanding the general principles. One of the biggest successes has been the application to FTNMR, the implementation being as follows.

- Guess the solution, for example, a spectrum using whatever knowledge is available. In NMR, it is possible to start with the prior guess of a flat spectrum in the absence of other information.
- Then, take this guess and transform it into data space, for example, Fourier transforming the spectrum to a time series.
- Using a statistic such as the χ^2 statistic, see how well this guess compares with the experimental data.
- Refine this guess of data space and try to reduce the statistic by a set amount. There will, of course, be a large number of possible solutions, select the solution with maximum entropy.
- Then, repeat the cycle but using the new solution until a good fit to the data is available.

It is important to realise that least squares and maximum entropy solutions often provide different best answers and move the solution in opposite directions; therefore, a balance is required. Maximum entropy algorithms are often regarded as a form of *non-linear deconvolution*. For linear methods, the new (improved) data set can be expressed as linear functions of the original data, as discussed in Section 3.3, whereas non-linear solutions cannot. Chemical knowledge often favours non-linear answers: for example, we know that most underlying spectra are all positive; yet, solutions involving linear sums of terms may often produce negative answers, which have little meaning.

1. On the whole, the higher the entropy of a spectrum, the better the statistical goodness of fit to the experimental data.
   (a) True
   (b) False

Problems

3.1 Savitzky–Golay and Moving Average Smoothing Functions
   Section 3.3.1.2   Section 3.3.1.1
   A data set is recorded over 26 sequential points to give the following data.

   0.0168
   0.0591
   −0.0009
   0.0106
   0.0425
   0.0236
   0.0807
   0.1164
   0.7459
   0.7938
   1.0467
   0.9737
   0.7517
   0.7801
   0.5595
   0.6675
   0.7158
   0.5168
   0.1234
   0.1256
   0.0720

−0.1366

−0.1765

  0.0333

  0.0286

−0.0582

1. Produce a graph of the raw data. Verify that there appear two peaks, but quite substantial noise. An aim is to smooth away the noise but preserving resolution.
2. Smooth the data in the following five ways: (a) five-point moving average, (b) seven-point moving average, (c) five-point quadratic Savitzky–Golay filter, (d) seven-point quadratic Savitzky–Golay filter and (e) nine-point Savitzky–Golay filter. Present the results numerically and in the form of two graphs, the first involving superimposing (a) and (b) and the second involving superimposing (c), (d) and (e).
3. Comment on the differences between the five smoothed data sets in question 2. Which filter would you choose as the optimum?

3.2 Fourier Functions
Section 3.5

The following represent four real functions, sampled over 32 data points, numbered from 0 to 31.

Sample	A	B	C	D
0	0	0	0	0
1	0	0	0	0
2	0	0	0.25	0
3	0	0	0.5	0
4	0	0	0.25	0.111
5	0	0.25	0	0.222
6	1	0.5	0	0.333
7	0	0.25	0	0.222
8	0	0	0	0.111
9	0	0	0	0
10	0	0	0	0
11	0	0	0	0
12	0	0	0	0
13	0	0	0	0
14	0	0	0	0
15	0	0	0	0
16	0	0	0	0
17	0	0	0	0
18	0	0	0	0
19	0	0	0	0
20	0	0	0	0
21	0	0	0	0
22	0	0	0	0
23	0	0	0	0
24	0	0	0	0
25	0	0	0	0
26	0	0	0	0
27	0	0	0	0

Sample	A	B	C	D
28	0	0	0	0
29	0	0	0	0
30	0	0	0	0
31	0	0	0	0

1. Plot graphs of these functions and comment on the main differences.
2. Calculate the real transform over the points 0–15 of each of the four functions by using the following equation

$$RL(n) = \sum_{m=0}^{M-1} f(m) \cos(nm/M)$$

where $M = 32$, n runs from 0 to 15 and m from 0 to 31.

(If you use angles in radians, you should include the factor of 2π in the equation.)

3. Plot the graphs of the four real transforms.
4. How many oscillations are in the transform for A? Why is this? Comment on the reason why the graph does not decay.
5. What is the main difference between the transforms of A, B and D, and why is this so?
6. What is the difference between the transforms of B and C and why?
7. Calculate the imaginary transform of A, replacing cosine by sine in the equation above and plot a graph of the result. Comment on the difference in appearance between the real and imaginary transforms.

3.3 Cross-correlograms
Section 3.4.2
Two time series, A and B, are recorded as follows:

A	B
6.851	3.721
2.382	0.024
2.629	5.189
3.047	−1.022
−2.598	−0.975
−0.449	−0.194
−0.031	−4.755
−7.578	1.733
−4.253	−1.964
2.598	0.434
1.655	2.505
4.980	−1.926
5.227	3.973
−2.149	−0.588
0.000	0.782
2.149	1.563
−5.227	−4.321
−4.980	0.517
−1.655	−3.914
−2.598	−0.782

A	B
4.253	2.939
7.578	−0.169
0.031	5.730
0.449	−0.154
2.598	−0.434
−3.047	−0.387
−2.629	−5.537
−2.382	0.951
−6.851	−2.157

1. Plot superimposed graphs of each time series.
2. Calculate the cross-correlogram of these time series, by lagging the second time series between −20 and 20 points relative to the first time series. To perform this calculation for a lag of +5 points, shift the second time series so that the first point in time (3.721) is aligned with the sixth point (−0.449) of the first series, shifting all other points as appropriate, and calculate the correlation coefficient of the 25 lagged points of series B with the first 25 points of series A.
3. Plot a graph of the cross-correlogram. Are there any frequencies common to both time series?

3.4 An Introduction to Maximum Entropy

Section 3.6.4

The value of entropy can be defined, for a discrete distribution, by where there are i states and p_i is the probability of each state. In this problem, use probabilities to the base 10 for comparison.

The following are three possible models of a spectrum, recorded at 20 wavelengths.

A	B	C
0.105	0.000	0.118
0.210	0.000	0.207
0.368	0.000	0.332
0.570	0.000	0.487
0.779	0.002	0.659
0.939	0.011	0.831
1.000	0.044	0.987
0.939	0.135	1.115
0.779	0.325	1.211
0.570	0.607	1.265
0.368	0.882	1.266
0.210	1.000	1.201
0.105	0.882	1.067
0.047	0.607	0.879
0.018	0.325	0.666
0.006	0.135	0.462
0.002	0.044	0.291
0.001	0.011	0.167
0.000	0.002	0.087
0.000	0.000	0.041

1. The spectral models may be regarded as a series of 20 probabilities of absorbance at each wavelength. Hence, if the total absorbance over 20 wavelengths summed to x, then the probability at each wavelength is simply the absorbance divided by x. Convert the three models into three probability vectors.
2. Plot a graph of the three models.
3. Explain why only positive values of absorbance are expected for ideal models.
4. Calculate the entropy for each of the three models.
5. The most likely model, in the absence of other information, is the one with the most positive entropy. Discuss the relative entropies of the three models.
6. What other information is normally used when maximum entropy is applied to chromatography or spectroscopy?

3.5 Some Simple Smoothing Methods for Time Series
Section 3.3.1.4 Section 3.3.1.2
The following represents a time series recorded at 40 points in time. The aim of this problem is to look at a few smoothing functions, both in the text and extended.

16.148
17.770
16.507
16.760
16.668
16.433
16.721
16.865
15.456
17.589
16.628
16.922
17.655
16.479
16.578
16.240
17.478
17.281
16.625
17.111
16.454
16.253
17.140
16.691
16.307
17.487
17.429
16.704
16.124
17.312
17.176
17.229
17.243

17.176

16.682

16.557

17.463

17.341

17.334

16.095

1. Plot a graph of the raw data.
2. Calculate three- and five-point median smoothing functions (denoted by '3' and '5') on the data (to do this, replace each point by the median of a span of N points) and plot the resultant graphs.
3. Re-smooth the three-point median smoothed data by a further three-point median smoothing function (denoted by '33') and then further by a Hanning window of the form $\hat{x}_i = 0.25x_{i-j} + 0.5x_i + 0.25x_{i+j}$ (denoted by '33H'), plotting both graphs as appropriate.
4. For the four smoothed data sets in (2) and (3), calculate the 'rough', which is calculated by subtracting the smooth from the original data, and plot appropriate graphs.
5. Smooth the rough obtained from the '33' data set in question 2 by a Hanning window and plot a graph.
6. If necessary, superimpose selected graphs computed above on top of the graph of the original data, comment on the results and state where you think there may be problems with the process, and whether these are single discontinuities or deviations over a period of time.

3.6 Multivariate Correlograms
Section 3.4.1   Section 3.4.3   Section 4.3

The following data represent six measurements on a sample taken at 30 points in time.

0.151	0.070	1.111	−3.179	−3.209	−0.830
−8.764	−0.662	−10.746	−0.920	−8.387	−10.730
8.478	−1.145	−3.412	1.517	−3.730	11.387
−10.455	−8.662	−15.665	−8.423	−8.677	−5.209
−14.561	−12.673	−24.221	−7.229	−15.360	−13.078
0.144	−8.365	−7.637	1.851	−6.773	−5.180
11.169	−7.514	4.265	1.166	−2.531	5.570
−0.169	−10.222	−6.537	0.643	−7.554	−7.441
−5.384	10.033	−3.394	−2.838	−1.695	1.787
−16.154	−1.021	−19.710	−7.255	−10.494	−8.910
−12.160	−11.327	−20.675	−13.737	−14.167	−4.166
−13.621	−7.623	−14.346	−4.428	−8.877	−15.555
4.985	14.100	3.218	14.014	1.588	−0.403
−2.004	0.032	−7.789	1.958	−8.476	−5.601
−0.631	14.561	5.529	1.573	4.462	4.209
0.120	4.931	−2.821	2.159	0.503	4.237
−6.289	−10.162	−14.459	−9.184	−9.207	−0.314
12.109	9.828	4.683	5.089	2.167	17.125
13.028	0.420	14.478	9.405	8.417	4.700
−0.927	−9.735	−0.106	−3.990	−3.830	−6.613
3.493	−3.541	−0.747	6.717	−1.275	−3.854
−4.282	−3.337	2.726	−4.215	4.459	−2.810
−16.353	0.135	−14.026	−7.458	−5.406	−9.251
−12.018	−0.437	−7.208	−5.956	−2.120	−8.024
10.809	3.737	8.370	6.779	3.963	7.699

−8.223	6.303	2.492	−5.042	−0.044	−7.220
−10.299	8.805	−8.334	−3.614	−7.137	−6.348
−17.484	6.710	0.535	−9.090	4.366	−11.242
5.400	−4.558	10.991	−7.394	9.058	11.433
−10.446	−0.690	1.412	−11.214	4.081	−2.988

1. Perform principal components analysis (PCA) (uncentred) on the data, and plot a graph of the scores of the first four PCs against time (this technique is described in Chapter 4 in more detail).
2. Do you think there is any cyclicity in the data? Why?
3. Calculate the correlogram of the first PC, using lags of 0–20 points. In order to determine the correlation coefficient of a lag of two points, calculate the correlation between points 3–30 and 1–28. The correlogram is simply a graph of the correlation coefficient against lag number.
4. From the correlogram, if there is cyclicity, determine the approximate frequency of this cyclicity and explain.

3.7 Simple Integration Errors When Digitisation is Poor
Section 3.2.1.1 Section 3.2.2
A Gaussian peak, whose shape is given by

$$A = 2e^{-(7-x)^2}$$

where A is the intensity as a function of position (x) is recorded.
1. What is the expected integral of this peak?
2. What is the exact width of the peak at half height?
3. The peak is recorded at every x value between 1 and 20. The integral is computed simply by adding the intensity at each of these values. What is the estimated integral?
4. The detector is slightly misaligned, so that the data are not recorded at integral values of x. What are the integrals if the detector records the data at (a) 1.2, 2.2 … 20.2, (b) 1.5, 2.5 …. 20.5 and (c) 1.7, 2.7 … 20.7?
5. There is a poor ADC resolution in the intensity and direction: 5 bits represent a true reading of 2, so that a true value of 2 is represented by 11111 in the digital recorder. The true reading is always rounded down to the nearest integer. This means that possible levels are 0/31 (=binary 00000), 2/31 (=00001), 4/31 and so on. Hence, a true reading of 1.2 would be rounded down to 18/31 or 1.1613. Explain the principle of ADC resolution and show why this is so.
6. Calculate the estimated integrals for the case in (3) and the three cases in (4).
   (Hint: if using Excel, you can use the INT function).

3.8 First and Second Derivatives of UV/vis Spectra Using the Savitzky–Golay Method
Section 3.3.1.2
Three spectra have been obtained, A consisting of pure compound 1, B of a mixture and C of pure compound 2. The data, together with wavelengths, normalised to a constant intensity of 1 are presented below.

Wavelength	A	B	C
220	0.891	1.000	1.000
224	1.000	0.973	0.865
228	0.893	0.838	0.727
232	0.592	0.575	0.534
236	0.225	0.288	0.347
240	0.108	0.217	0.322
244	0.100	0.244	0.370
248	0.113	0.267	0.398
252	0.132	0.262	0.376
256	0.158	0.244	0.324

Wavelength	A	B	C
260	0.204	0.251	0.306
264	0.258	0.311	0.357
268	0.334	0.414	0.466
272	0.422	0.536	0.595
276	0.520	0.659	0.721
280	0.621	0.762	0.814
284	0.711	0.831	0.854
288	0.786	0.852	0.834
292	0.830	0.829	0.763
296	0.838	0.777	0.674
300	0.808	0.710	0.589
304	0.725	0.636	0.529
308	0.606	0.551	0.480
312	0.477	0.461	0.433
316	0.342	0.359	0.372
320	0.207	0.248	0.295
324	0.113	0.161	0.226
328	0.072	0.107	0.170
332	0.058	0.070	0.122
336	0.053	0.044	0.082
340	0.051	0.026	0.056
344	0.051	0.016	0.041
348	0.051	0.010	0.033

1. Produce and superimpose the graphs of the raw spectra. Comment.
2. Calculate the five-point Savitzky–Golay quadratic first and second derivative of A. Plot the graphs and interpret them; compare both first and second derivatives and discuss the appearance in terms of the number and positions of the peaks.
3. Repeat this for spectrum C. Why is the pattern more complex? Interpret the graphs.
4. Calculate the five-point Savitzky–Golay quadratic second derivatives of all the three spectra and superimpose the resultant graphs. Repeat for the seven-point derivatives. Which graph is clearer, five- or seven-point derivatives? Interpret the results for spectrum B. Do the derivatives show that it is clearly a mixture? Comment on the appearance of the region between 270 and 310 nm and compare to the original spectra.

3.9 Fourier Analysis of NMR Signals
Section 3.5.1.4   Section 3.5.1.2   Section 3.5.1.3
The data below consists of 72 sequential readings in time (organised in columns for clarity), which represent a raw time series (or FID) acquired over a region of an NMR spectrum. The first column represents the first 20 points in time, the second points 21–40 and so on.

−2 732.61	−35.90	−1 546.37	267.40
−14 083.58	845.21	−213.23	121.18
−7 571.03	−1 171.34	1 203.41	11.60
5 041.98	−148.79	267.88	230.14
5 042.45	2 326.34	−521.55	−171.80
2 189.62	611.59	45.08	−648.30
1 318.62	−2 884.74	−249.54	−258.94

−96.36	−2 828.83	−1 027.97	264.47
−2 120.29	−598.94	−39.75	92.67
−409.82	1 010.06	1 068.85	199.36
3 007.13	2 165.89	160.62	−330.19
5 042.53	1 827.65	−872.29	991.12
3 438.08	−786.26	−382.11	
−2 854.03	−2 026.73	−150.49	
−9 292.98	−132.10	−460.37	
−6 550.05	932.92	256.68	
3 218.65	−305.54	989.48	
7 492.84	−394.40	−159.55	
1 839.61	616.13	−1 373.90	
−2 210.89	−306.17	−725.96	

1. The data were acquired at intervals of 0.008124 s. What is the spectral width of the Fourier transform, taking into account that only half the points are represented in the transform? What is the digital resolution of the transform?

2. Plot a graph of the original data, converting the horizontal axis into seconds.

3. In a simple form, the real transform can be expressed by

$$RL(n) = \sum_{m=0}^{M-1} f(m) \cos (nm/M)$$

Define the parameters in the equation in terms of the data set discussed in this problem. What is the equivalent equation for the imaginary transform?

4. Perform the real and imaginary transforms on this data (note that you may have to write a small program to do this, but it can be laid out in a spreadsheet without a program). Notice that n and m should start at 0 rather than 1, and if angles are calculated in radians, it is necessary to include a factor of 2π. Plot the real and imaginary transforms using a scale of hertz for the horizontal axis.

5. Comment on the phasing of the transform and produce a graph of the absolute value spectrum.

6. Phasing involves finding an angle ψ such that

$$ABS = \cos(\psi)\, RL + \sin(\psi)\, IM$$

A first approximation is that this angle is constant throughout a spectrum. By looking at the phase of the imaginary transform, obtaining in question 4 above, can you produce a first guess of this angle? Produce the result of phasing using this angle and comment.

7. How might you overcome the remaining problem of phasing?

4

Principal Component Analysis and Unsupervised Pattern Recognition

4.1 Introduction

Multivariate analysis is the core of chemometrics, and most of the first papers in the literature in the 1970s used simple multivariate methods to explore complex chemical data sets. The key concept is that many studies in chemistry and associate laboratory measurements (e.g. in biology or medicine) involve recording many variables per sample. For example, a spectrum may be recorded at hundreds of wavelengths, or a chromatogram at many elution times. Hence, for each sample, we have a lot of information. Traditionally, one or two variables (e.g. the intensity at characteristic wavelengths) were used to describe a sample. However, this poses the problem of choosing the best variables, and often, there is a lot of information mixed up in the signal. A mixture may consist of half a dozen or even more compounds: most materials of interest, whether the extract from a plant, or a fruit juice, or a piece of wood or a bodily fluid, consist, in fact, of many compounds. They may not all absorb at a single spectroscopic wavelength; hence, the composition cannot necessarily be obtained just by monitoring single wavelengths. Often, we do not want to know the full chemical composition of materials: this may be very complicated and time consuming and the aim of an investigation might be primarily to identify, for example, the origin of the sample: samples of similar provenance may group together; hence, we may just want to visualise how our sample fits in to an existing set.

There are numerous ways we might want to deal with this multivariate data. The methods described in this chapter are unsupervised; that is, they do not assume any pre-defined pattern or grouping and usually result in visualisation in the form of graphs. Common to all these approaches, data are represented in matrix form, usually with the rows corresponding to single samples (or observations, e.g., a spectrum or a chromatogram) and columns to variables (e.g. wavelengths or elemental composition) and the cells usually equalling intensity.

We will discuss two main groups of techniques in this chapter.

4.1.1 Exploratory Data Analysis

Exploratory data analysis (EDA) mainly consists of principal components analysis (PCA) and factor analysis (FA). Classical statistical examples come from biology and psychology. Psychometricians have, for many years, had the need to translate numbers, such as answers to questions in tests, into relationships between individuals. How verbal ability, numeracy and the ability to think in three dimensions can be predicted from a test? Can different people be grouped by these abilities? And does this grouping reflect the backgrounds of the people taking the test? Are there differences according to educational background, age, sex or even linguistic group?

In chemistry, we too need to ask similar questions, but the raw data are often chromatographic or spectroscopic. An example is animal pheromones: animals recognise each other more by smell than by sight, and different animals often lay scent trails, sometimes in their urine. The chromatogram of a urine sample may contain several hundred compounds, and it is often not obvious to the untrained observer to decide which samples are most significant. Sometimes, the most potent compounds are only present in small quantities. Yet, animals can often detect through scent marking that whether there is one in-heat member of the opposite sex looking for a mate, or whether there is a dangerous intruder entering their territory. EDA of chromatograms of urine samples can highlight differences in chromatograms of different social groups or different sexes and gives a simple visual idea to the main relationships between these samples. Sections 4.2–4.6 cover these approaches.

Chemometrics: Data Driven Extraction for Science, Second Edition. Richard G. Brereton.
© 2018 John Wiley & Sons Ltd. Published 2018 by John Wiley & Sons Ltd.
Companion website: http://booksupport.wiley.com

4.1.2 Cluster Analysis

A more formal method of treating samples is cluster analysis, the most widespread method for unsupervised pattern recognition in chemometrics. Many methods have their origins in numerical taxonomy. Biologists measure features in different organisms, for example, various body length parameters. Using a couple of dozen features, it is possible to see which species are most similar and draw a picture of these similarities, called a *dendrogram*, in which more closely related species are closer to each other. The main branches of the dendrogram can represent bigger divisions, such as subspecies, species, genera and families.

These principles can be directly applied to chemistry. It is possible to determine similarities in amino acid sequences of myoglobin in a variety of species. The more similar the species, the closer the relationship: chemical similarity mirrors biological similarity. Sometimes, the amount of information is huge; for example, in large genomic or crystallographic databases, cluster analysis is the only practicable way of searching for similarities.

Unsupervised pattern recognition differs from EDA in that the aim of the methods is to detect similarities between species, whereas using EDA, there is no particular prejudice whether or how many groups will be found. Cluster analysis is described in more detail in Section 4.8.

4.2 The Concept and Need for Principal Components Analysis

PCA is probably the most widespread multivariate chemometric technique, and because of the importance of multivariate measurements in chemistry, it is regarded, by many, as the technique that most significantly changed the chemist's view of data analysis.

4.2.1 History

There are numerous claims to the first use of PCA in the literature. Probably, the most famous early paper was by Pearson in 1901. However, the fundamental ideas are based on approaches well known to physicists and mathematicians for much longer, namely those of eigenanalysis. In fact, some school mathematics syllabuses teach ideas about matrices that are relevant to modern chemistry. An early description of the method in physics was provided by Cauchy in 1829. It has been claimed that the earliest non-specific reference to PCA in the literature was in 1878, although the author of the paper almost certainly did not realise the potential and was dealing mainly with a problem of linear calibration.

It is generally accepted that the revolution in the use of multivariate methods took place in the 1920s and 1930s. Psychometrics is one of the earliest applications of applied multivariate analysis that developed during this period and one important area involves relating answers in tests to underlying factors, for example, verbal and numerical ability as illustrated in Figure 4.1. PCA relates a data matrix consisting of these answers to a number of psychological 'factors'. In certain areas of statistics, concepts of FA and PCA are intertwined, but in chemometrics, both approaches

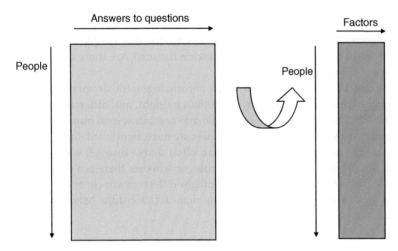

Figure 4.1 Factor analysis in psychology.

have different implications, PCA involves using abstract functions of the data to look at patterns, whereas FA involves obtaining information such as spectra that can be directly related to the chemistry.

Natural scientists of all disciplines, from biologists, geologists and chemists, have caught onto these approaches over the past few decades. Within the chemical community, the first major applications of PCA were reported in the 1970s and form the foundation of many modern chemometric methods described in this and subsequent chapters.

1. Principal components analysis and factor analysis have the same meaning in chemistry.
   (a) True
   (b) False

4.2.2 Multivariate Data Matrices

A key idea is that most chemical measurements are inherently *multivariate*. This means that more than one measurement can be made on a single sample. An obvious example is spectroscopy: we can record a spectrum at hundreds of wavelength on a single sample. Many traditional chemical approaches are *univariate*, in which only one wavelength (or measurement) is used per sample, but this misses much information. Another important application is quantitative structure–property relationships, in which many physical measurements are available on a number of candidate compounds (bond lengths, dipole moments, bond angles, etc.). Can we predict, statistically, the biological activity of a compound? Can this assist in pharmaceutical drug development? Several pieces of information are available. PCA is one of the several multivariate methods that allows us to explore patterns in these data, similar to exploring patterns in psychometric data. Which compounds behave similarly? Which people belong to a similar group? How can this behaviour be predicted from available information?

As an example, consider a metabolomics experiment in which mice of identical genetic makeup (haplotypes) were fed different diets at different periods of time. Extracts of their urine were analysed by NMR. The results can be organised as a matrix, whose rows represent the individual mice (or samples) and the columns the NMR peaks (or variables), as illustrated in Figure 4.2. The individual numbers in the matrix are called *elements* – not to be confused with chemical elements. The data are multivariate because several pieces of information were recorded per sample. If we analysed 200 mouse samples and 1000 NMR peaks, we say that the data are organised into a 200×1000 matrix, with the number of rows cited first.

What might we want to ask about the data? Which mice are similar as judged from their NMR profile of urine extracts? Is diet or feeding time important? Can we see trends according to diet and accordingly how long they were fed? Which NMR signals were markers for diet and can we relate these to specific chemicals? We may have sampled a mouse several times, is the identity of the mouse more important than the diet it was fed? These questions are not about grouping the mice into different classes, but about answering questions whether there are actually any interesting trends, after which we may then want to go on and perform more detailed models or additional experiments. The information about the mice (or their treatment) is mixed up, and it is unlikely that any single NMR peak on its own would be able to answer all our questions. We want a visual or exploratory method to try to convert the huge amount of NMR data into something we can visualise.

Of course, sometimes, the trends can be related to a specific chemical species, for example, in coupled chromatography or equilibrium studies. PCA can still be used as an important exploratory technique, but the principal components, rather than primarily being used for visualisation, can be related to directly interpretable chemical trends. In this chapter, we illustrate the techniques using both types of case studies. In general terms for the former type of data,

Figure 4.2 Matrix representation of results from a metabolomics experiment.

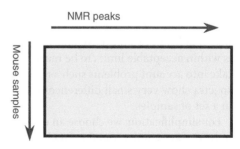

we may then go on to use the more statistical methods discussed in Chapter 5, whereas for the latter type of data, we may be able to get physically interpretable information as discussed in Chapter 7.

---

1. An experiment is used to obtain 300 samples of serum extracts from patients both with a disease of varying degrees and controls and measure 30 GCMS peaks. The data can be organised as a matrix of dimensions as follows.

    (a) 300×30
    (b) 30×300

4.2.3   Case Studies

In order to illustrate the main concepts of PCA, we will introduce three case studies.

4.2.3.1   Case Study 1: Resolution of Overlapping Chromatographic Peaks

The first case study is from coupled chromatography. It is not necessary to understand the detailed chemical motivations behind the chromatographic technique. This case study represents the intensity of spectra recorded at a number of wavelengths, which are also sequentially related to time. In fact, many of the earliest applications of PCA in chemistry in the 1960s and 1970s were to overlapping analytical peaks such as in chromatography.

The case study involves a high-performance liquid chromatography–diode array detector (HPLC–DAD) chromatogram sampled at $i = 30$ points in time (each at 1 s intervals) and $j = 28$ wavelengths of approximately 4.8 nm intervals, as presented in Table 4.1 (note that the wavelengths are rounded to the nearest nanometer for simplicity, but the original data were not collected at exact nanometer intervals). Absorbances are presented in absorbance units (AU). For readers not familiar with this application, the data set can be considered to consist of a series of 30 spectra recorded sequentially in time, arising from a mixture of compounds each of which has its own characteristic underlying unimodal time profile (often called an *elution profile*).

The data can be represented by a 30×28 matrix, the rows corresponding to elution times and the columns corresponding to wavelengths. Calling this matrix X, and each element x_{ij}, the profile chromatogram at time i

$$X_i = \sum_{j=1}^{28} x_{ij}$$

is given in Figure 4.3 and consists of at least two co-eluting peaks.

---

1. A diode array chromatogram is recorded at 5 s intervals over 5 min, starting at 5 s, and at 5 nm intervals between 200 and 300 nm. It can be represented by a multivariate matrix of dimensions

    (a) 60×20
    (b) 20×60
    (c) 60×21
    (d) 21×60

4.2.3.2   Case Study 2: Near Infrared Spectroscopy of Food

Near infrared (NIR) spectroscopy is one of the first techniques to use chemometrics, its use dating back from the 1970s. It has an important role in food, pharmaceuticals and petrochemicals. In NIR, it is often not possible to examine the contribution of individual compounds, but it is primarily used to study complex mixtures. Many substances such as foods contain a large number of chemicals: the interest is usually not so much to determine the relative amounts of each in a sample but to study ensemble properties, for example, the provenance of a food stuff or whether a pharmaceutical is within acceptable limits to be marketed or the quality of a fuel. Usually, NIR spectra have first to be transformed to take into account problems such as baseline effects, and in the example we use, this has already been done. Often, NIR spectra show very small differences that are not easily detectable to the eye, which are diagnostic of different origins in a set of samples.

For simplification, we choose an example in which we have reduced the digital resolution to 19 nm, so that there are only 32 wavelengths, and the data have already been pre-processed so that baseline effects have been removed. There

Table 4.1 Case study 1: a chromatogram recorded at 30 points in time and 28 wavelengths.

nm	220	225	230	234	239	244	249	253	258	263	268	272	277	282	287	291	296	301	306	310	315	320	325	329	334	339	344	349
1	0.006	0.004	0.003	0.003	0.003	0.004	0.004	0.003	0.002	0.003	0.004	0.005	0.006	0.006	0.005	0.004	0.003	0.002	0.002	0.003	0.003	0.002	0.002	0.001	0.001	0.000	0.000	0.000
2	0.040	0.029	0.023	0.021	0.026	0.030	0.029	0.023	0.018	0.021	0.029	0.038	0.045	0.046	0.040	0.030	0.021	0.017	0.017	0.019	0.019	0.017	0.013	0.009	0.005	0.002	0.001	0.000
3	0.159	0.115	0.091	0.085	0.101	0.120	0.117	0.090	0.071	0.084	0.116	0.153	0.178	0.182	0.158	0.120	0.083	0.066	0.069	0.075	0.075	0.067	0.053	0.035	0.019	0.008	0.003	0.001
4	0.367	0.267	0.212	0.198	0.236	0.280	0.271	0.209	0.165	0.194	0.270	0.354	0.413	0.422	0.368	0.279	0.194	0.155	0.160	0.173	0.174	0.157	0.123	0.081	0.043	0.019	0.008	0.003
5	0.552	0.405	0.321	0.296	0.352	0.419	0.405	0.312	0.247	0.291	0.405	0.532	0.621	0.635	0.555	0.422	0.296	0.237	0.244	0.262	0.262	0.234	0.183	0.120	0.065	0.029	0.011	0.004
6	0.634	0.477	0.372	0.323	0.377	0.449	0.435	0.338	0.271	0.321	0.447	0.588	0.689	0.710	0.629	0.490	0.356	0.289	0.289	0.301	0.292	0.255	0.196	0.129	0.069	0.031	0.012	0.005
7	0.687	0.553	0.412	0.299	0.329	0.391	0.382	0.305	0.255	0.310	0.433	0.571	0.676	0.713	0.657	0.546	0.430	0.362	0.340	0.325	0.290	0.234	0.172	0.111	0.060	0.026	0.010	0.004
8	0.795	0.699	0.494	0.270	0.263	0.311	0.311	0.262	0.240	0.304	0.426	0.565	0.682	0.744	0.731	0.662	0.571	0.497	0.438	0.378	0.299	0.209	0.139	0.087	0.046	0.021	0.008	0.003
9	0.914	0.854	0.586	0.251	0.207	0.243	0.252	0.230	0.233	0.308	0.432	0.576	0.708	0.798	0.824	0.793	0.724	0.643	0.547	0.441	0.317	0.191	0.111	0.066	0.035	0.016	0.006	0.003
10	0.960	0.928	0.628	0.232	0.165	0.193	0.206	0.202	0.222	0.301	0.424	0.568	0.705	0.809	0.860	0.855	0.802	0.719	0.602	0.469	0.320	0.174	0.091	0.051	0.026	0.012	0.005	0.002
11	0.924	0.902	0.606	0.206	0.133	0.155	0.170	0.174	0.201	0.277	0.391	0.524	0.655	0.760	0.819	0.826	0.786	0.707	0.587	0.450	0.299	0.154	0.074	0.040	0.021	0.009	0.004	0.002
12	0.834	0.815	0.544	0.178	0.108	0.126	0.140	0.147	0.174	0.243	0.342	0.460	0.576	0.672	0.729	0.741	0.709	0.638	0.528	0.402	0.264	0.132	0.061	0.032	0.016	0.007	0.003	0.002
13	0.725	0.704	0.468	0.150	0.089	0.103	0.115	0.123	0.148	0.206	0.291	0.391	0.490	0.573	0.624	0.636	0.610	0.550	0.455	0.345	0.225	0.111	0.050	0.026	0.013	0.006	0.003	0.001
14	0.615	0.596	0.395	0.125	0.073	0.085	0.095	0.102	0.123	0.173	0.244	0.327	0.411	0.481	0.525	0.535	0.514	0.463	0.383	0.290	0.189	0.093	0.042	0.021	0.011	0.005	0.002	0.001
15	0.519	0.500	0.331	0.105	0.061	0.071	0.080	0.086	0.103	0.144	0.203	0.273	0.343	0.402	0.438	0.447	0.429	0.387	0.320	0.242	0.158	0.077	0.035	0.018	0.009	0.004	0.002	0.001
16	0.437	0.419	0.277	0.088	0.052	0.060	0.067	0.072	0.086	0.121	0.171	0.229	0.288	0.337	0.367	0.374	0.359	0.324	0.268	0.203	0.132	0.065	0.029	0.015	0.008	0.004	0.002	0.001
17	0.369	0.354	0.234	0.075	0.044	0.051	0.057	0.061	0.073	0.102	0.144	0.193	0.243	0.284	0.309	0.315	0.302	0.272	0.225	0.171	0.112	0.055	0.025	0.013	0.007	0.003	0.002	0.001
18	0.314	0.300	0.198	0.064	0.038	0.044	0.049	0.052	0.062	0.087	0.122	0.165	0.206	0.241	0.263	0.267	0.256	0.231	0.191	0.145	0.095	0.047	0.021	0.011	0.006	0.003	0.001	0.001
19	0.269	0.257	0.170	0.055	0.033	0.038	0.042	0.045	0.053	0.074	0.105	0.141	0.177	0.207	0.225	0.229	0.219	0.197	0.163	0.124	0.081	0.041	0.019	0.010	0.005	0.002	0.001	0.001
20	0.233	0.222	0.147	0.048	0.029	0.033	0.037	0.039	0.046	0.064	0.091	0.122	0.153	0.179	0.194	0.197	0.189	0.170	0.141	0.107	0.070	0.035	0.016	0.009	0.004	0.002	0.001	0.001
21	0.203	0.193	0.127	0.042	0.025	0.029	0.033	0.034	0.040	0.056	0.079	0.107	0.133	0.156	0.169	0.171	0.164	0.147	0.122	0.093	0.061	0.031	0.014	0.008	0.004	0.002	0.001	0.001
22	0.178	0.169	0.112	0.037	0.022	0.026	0.029	0.030	0.036	0.049	0.070	0.094	0.117	0.137	0.148	0.150	0.143	0.129	0.107	0.082	0.054	0.027	0.013	0.007	0.004	0.002	0.001	0.000
23	0.157	0.149	0.099	0.032	0.020	0.023	0.026	0.027	0.031	0.044	0.062	0.083	0.104	0.121	0.131	0.132	0.126	0.114	0.094	0.072	0.048	0.024	0.011	0.006	0.003	0.002	0.001	0.000
24	0.140	0.132	0.088	0.029	0.018	0.021	0.023	0.024	0.028	0.039	0.055	0.074	0.092	0.107	0.116	0.118	0.112	0.101	0.084	0.064	0.042	0.021	0.010	0.005	0.003	0.001	0.001	0.000
25	0.125	0.118	0.078	0.026	0.016	0.019	0.021	0.021	0.025	0.035	0.049	0.066	0.082	0.096	0.104	0.105	0.100	0.090	0.075	0.057	0.038	0.019	0.009	0.005	0.003	0.001	0.001	0.000
26	0.112	0.106	0.070	0.023	0.015	0.017	0.019	0.019	0.023	0.031	0.044	0.059	0.074	0.086	0.093	0.094	0.090	0.081	0.067	0.051	0.034	0.017	0.008	0.004	0.002	0.001	0.001	0.000
27	0.101	0.096	0.063	0.021	0.013	0.015	0.017	0.018	0.020	0.028	0.040	0.054	0.067	0.078	0.084	0.085	0.081	0.073	0.061	0.046	0.031	0.016	0.008	0.004	0.002	0.001	0.001	0.000
28	0.092	0.087	0.057	0.019	0.012	0.014	0.015	0.016	0.019	0.026	0.036	0.049	0.061	0.071	0.076	0.077	0.073	0.066	0.055	0.042	0.028	0.014	0.007	0.004	0.002	0.001	0.000	0.000
29	0.084	0.079	0.052	0.017	0.011	0.013	0.014	0.015	0.017	0.023	0.033	0.044	0.055	0.064	0.070	0.070	0.067	0.060	0.050	0.038	0.025	0.013	0.006	0.003	0.002	0.001	0.000	0.000
30	0.076	0.072	0.048	0.016	0.010	0.012	0.013	0.013	0.015	0.021	0.030	0.041	0.051	0.059	0.064	0.064	0.061	0.055	0.046	0.035	0.023	0.012	0.006	0.003	0.002	0.001	0.000	0.000

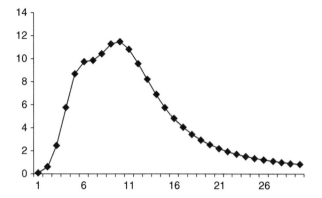

Figure 4.3 Case study 1: chromatographic peak profiles, involving summing intensities of the data from Table 4.1 over all wavelengths.

are numerous quite specialist approaches for preparing NIR data, which are out of the scope of this book, to which the reader is referred to the literature. For illustrative purposes, we are primarily interested in multivariate analysis of the data. The data consist of 72 spectra of edible oils, from four origins, 18 of corn oils, 30 of olive oils, 16 of safflower oils and 8 of corn margarines. The numerical data are presented in Table 4.2, presented as a matrix of 72 rows and 32 columns: each row represents a spectrum of an extract from one of the four groups and each column a wavelength. The elements of the matrix are intensities in AU.

The spectra are presented in Figure 4.4. From this, it can be seen that there are only very small differences between the 72 spectra. The aim of chemometrics is to determine the origins of a sample from its spectrum, for example, which type of oil it belongs to, in this case. By viewing a single spectrum, it is not obvious, but due to the relative stability of NIR spectra (after instrumental artefacts are removed), we will see that using chemometrics, the different groups can be distinguished easily, despite a tiny spectral difference.

1. A series of 50 NIR spectra are recorded, between 800 and 1600 nm inclusive at 10 nm intervals. They can be represented as a data matrix of dimensions

    (a) 50×81
    (b) 50×80
    (c) 81×50
    (d) 80×50

4.2.3.3 Case Study 3: Properties of the Elements

In case studies 1 and 2, the columns (or variables) are sequentially related, as they represent successive wavelengths, and in case study 1, the rows are also sequentially related, representing successive elution times. There is, of course, no requirement of any sequential relationship between successive columns and variables.

Case study 3 is presented in Table 4.3, which consists of the five properties of 27 elements. The second column (group) is not a property but a descriptor. Can we use these five properties to show how elements are related in their chemical reactivity? This 27×5 data matrix is quite different from the ones above. The variables represent quite different things on different scales and thus, as discussed below, have to be handled a bit differently to the previous examples where all measurements are in the same units (AUs). However, in common with the other examples, it is our desire to end up with a graphical simplification of the data to see what are the main trends.

In order to handle this, the data have to be converted to a similar scale, and this is done by standardisation, a technique described later in Section 4.6.4.

1. For a data matrix to be handled by PCA

    (a) All measurements in a row must be in the same physical unit, but this is not a requirement for a column.
    (b) All measurements in a column must be in the same physical unit, but this is not a requirement for a row.
    (c) All measurements in the data matrix must be in the same physical unit.

Table 4.2 Case study 2: NIR spectra of 72 oils in AU recorded at 32 wavelengths, consisting of four groups A: corn oil, B: olive oil, C: safflower oil, D: corn margarine, after baseline correction and suitable pre-processing.

nm	1499	1470	1441	1412	1383	1354	1325	1296	1267	1238	1209	1180	1151	1122	1093	1064	1035	1006	977	948	919	890	861	832	803	774	745	716	687	658	629	600
A	0.051	0.384	0.700	0.398	0.385	0.493	0.398	0.486	0.680	0.944	0.829	1.370	1.792	0.829	0.919	0.419	0.337	0.276	0.251	0.225	0.170	0.145	0.134	0.117	0.070	0.107	0.221	0.667	0.409	0.289	0.312	0.502
A	0.048	0.388	0.697	0.394	0.383	0.488	0.393	0.482	0.679	0.944	0.830	1.366	1.798	0.832	0.916	0.419	0.335	0.273	0.251	0.224	0.169	0.144	0.133	0.116	0.066	0.108	0.231	0.669	0.413	0.293	0.319	0.512
A	0.017	0.357	0.698	0.383	0.380	0.492	0.401	0.490	0.683	0.948	0.833	1.372	1.796	0.833	0.920	0.421	0.339	0.277	0.255	0.229	0.173	0.147	0.136	0.120	0.069	0.110	0.222	0.669	0.409	0.289	0.312	0.502
A	0.047	0.381	0.696	0.395	0.381	0.490	0.395	0.485	0.681	0.943	0.831	1.367	1.799	0.830	0.917	0.417	0.334	0.274	0.253	0.226	0.170	0.145	0.134	0.116	0.067	0.107	0.237	0.667	0.409	0.291	0.318	0.507
A	0.053	0.394	0.696	0.396	0.384	0.488	0.393	0.482	0.678	0.944	0.830	1.362	1.793	0.833	0.917	0.419	0.335	0.274	0.254	0.226	0.170	0.145	0.133	0.116	0.067	0.107	0.225	0.666	0.410	0.291	0.317	0.509
A	0.041	0.375	0.700	0.396	0.384	0.495	0.400	0.489	0.681	0.946	0.833	1.374	1.795	0.831	0.918	0.420	0.338	0.276	0.256	0.228	0.172	0.147	0.136	0.117	0.068	0.108	0.221	0.666	0.405	0.283	0.310	0.498
A	0.051	0.383	0.701	0.398	0.383	0.492	0.399	0.486	0.678	0.944	0.829	1.364	1.801	0.829	0.915	0.418	0.334	0.274	0.249	0.224	0.170	0.145	0.136	0.117	0.069	0.107	0.222	0.668	0.409	0.292	0.315	0.510
A	0.053	0.395	0.698	0.397	0.385	0.491	0.394	0.483	0.677	0.943	0.830	1.368	1.798	0.830	0.915	0.419	0.334	0.274	0.251	0.225	0.170	0.145	0.135	0.117	0.069	0.108	0.224	0.666	0.410	0.291	0.317	0.508
A	0.053	0.390	0.702	0.401	0.388	0.494	0.399	0.487	0.679	0.943	0.829	1.366	1.788	0.827	0.915	0.419	0.335	0.274	0.252	0.226	0.171	0.147	0.136	0.118	0.068	0.107	0.219	0.667	0.408	0.286	0.307	0.495
A	0.035	0.373	0.698	0.390	0.382	0.491	0.398	0.486	0.681	0.946	0.831	1.368	1.794	0.832	0.918	0.420	0.337	0.275	0.252	0.227	0.172	0.146	0.135	0.118	0.068	0.108	0.225	0.669	0.410	0.290	0.314	0.506
A	0.067	0.400	0.701	0.405	0.387	0.491	0.395	0.482	0.679	0.943	0.828	1.366	1.794	0.828	0.918	0.418	0.336	0.274	0.248	0.223	0.169	0.143	0.133	0.116	0.069	0.106	0.223	0.667	0.410	0.289	0.313	0.504
A	0.003	0.340	0.701	0.379	0.380	0.496	0.407	0.496	0.686	0.951	0.834	1.374	1.789	0.831	0.922	0.422	0.341	0.280	0.257	0.233	0.176	0.149	0.138	0.122	0.072	0.110	0.213	0.668	0.406	0.285	0.304	0.493
A	0.046	0.388	0.700	0.397	0.387	0.493	0.398	0.487	0.677	0.946	0.833	1.368	1.791	0.834	0.918	0.422	0.338	0.277	0.257	0.228	0.172	0.147	0.135	0.118	0.067	0.109	0.210	0.664	0.406	0.284	0.309	0.501
A	0.044	0.376	0.698	0.395	0.381	0.491	0.397	0.487	0.682	0.944	0.831	1.370	1.795	0.830	0.917	0.418	0.335	0.274	0.253	0.226	0.171	0.146	0.134	0.116	0.068	0.107	0.235	0.667	0.408	0.288	0.316	0.504
A	0.033	0.359	0.702	0.394	0.382	0.497	0.404	0.493	0.685	0.947	0.835	1.374	1.791	0.829	0.918	0.420	0.337	0.277	0.256	0.228	0.173	0.148	0.136	0.118	0.069	0.108	0.228	0.667	0.404	0.279	0.308	0.494
A	0.054	0.399	0.701	0.403	0.391	0.495	0.397	0.487	0.680	0.943	0.831	1.369	1.791	0.829	0.915	0.420	0.336	0.275	0.253	0.228	0.172	0.148	0.135	0.118	0.067	0.107	0.218	0.667	0.407	0.282	0.302	0.488
A	0.052	0.389	0.700	0.399	0.385	0.492	0.397	0.486	0.678	0.944	0.830	1.367	1.795	0.829	0.915	0.419	0.334	0.274	0.251	0.225	0.170	0.146	0.135	0.117	0.068	0.107	0.222	0.667	0.409	0.289	0.313	0.505
A	0.051	0.378	0.704	0.402	0.386	0.496	0.403	0.490	0.680	0.945	0.828	1.368	1.797	0.826	0.914	0.418	0.335	0.274	0.250	0.226	0.171	0.147	0.136	0.118	0.069	0.107	0.216	0.669	0.407	0.287	0.305	0.498
B	0.068	0.419	0.697	0.398	0.388	0.516	0.414	0.511	0.661	0.952	0.834	1.364	1.788	0.819	0.882	0.434	0.335	0.295	0.254	0.221	0.156	0.155	0.154	0.128	0.098	0.117	0.208	0.606	0.337	0.267	0.317	0.496
B	0.077	0.434	0.699	0.403	0.393	0.516	0.412	0.509	0.659	0.950	0.836	1.359	1.793	0.817	0.879	0.432	0.334	0.292	0.252	0.219	0.154	0.153	0.152	0.126	0.096	0.116	0.212	0.606	0.339	0.268	0.319	0.497
B	0.073	0.425	0.702	0.402	0.393	0.518	0.416	0.513	0.660	0.951	0.837	1.359	1.794	0.819	0.881	0.433	0.334	0.293	0.254	0.217	0.151	0.154	0.153	0.127	0.096	0.117	0.207	0.602	0.336	0.264	0.320	0.490
B	0.070	0.423	0.708	0.397	0.387	0.515	0.412	0.509	0.654	0.952	0.831	1.363	1.785	0.822	0.880	0.430	0.328	0.291	0.249	0.217	0.150	0.152	0.154	0.130	0.101	0.117	0.211	0.612	0.343	0.271	0.322	0.504
B	0.075	0.427	0.709	0.401	0.392	0.516	0.412	0.508	0.654	0.952	0.830	1.361	1.790	0.819	0.878	0.430	0.328	0.291	0.248	0.221	0.151	0.153	0.153	0.128	0.099	0.116	0.210	0.609	0.339	0.272	0.320	0.501
B	0.076	0.434	0.711	0.404	0.394	0.520	0.416	0.513	0.656	0.954	0.831	1.358	1.786	0.818	0.878	0.430	0.329	0.292	0.251	0.218	0.157	0.155	0.154	0.129	0.100	0.117	0.215	0.608	0.339	0.263	0.311	0.489
B	0.076	0.433	0.700	0.396	0.387	0.511	0.412	0.511	0.663	0.952	0.840	1.366	1.788	0.818	0.883	0.432	0.333	0.293	0.254	0.220	0.153	0.155	0.154	0.128	0.099	0.117	0.209	0.608	0.339	0.268	0.321	0.504
B	0.078	0.429	0.702	0.403	0.395	0.520	0.412	0.511	0.663	0.953	0.840	1.367	1.784	0.813	0.879	0.429	0.334	0.295	0.253	0.219	0.155	0.156	0.156	0.129	0.099	0.116	0.206	0.600	0.332	0.260	0.308	0.484
B	0.075	0.415	0.705	0.406	0.396	0.511	0.412	0.516	0.665	0.952	0.839	1.363	1.782	0.813	0.879	0.429	0.330	0.294	0.253	0.219	0.153	0.157	0.154	0.127	0.097	0.117	0.209	0.605	0.338	0.265	0.310	0.483
B	0.067	0.436	0.702	0.403	0.390	0.514	0.417	0.510	0.656	0.950	0.830	1.364	1.796	0.814	0.878	0.430	0.331	0.293	0.252	0.223	0.153	0.154	0.154	0.126	0.096	0.115	0.211	0.603	0.341	0.269	0.319	0.498
B	0.077	0.423	0.701	0.406	0.389	0.517	0.414	0.509	0.655	0.949	0.838	1.360	1.794	0.821	0.879	0.434	0.330	0.292	0.251	0.219	0.157	0.155	0.153	0.128	0.097	0.115	0.210	0.606	0.340	0.268	0.319	0.494
B	0.067	0.431	0.705	0.399	0.392	0.516	0.413	0.514	0.658	0.952	0.836	1.363	1.786	0.817	0.879	0.433	0.334	0.295	0.254	0.222	0.156	0.155	0.154	0.128	0.099	0.116	0.205	0.604	0.336	0.264	0.312	0.491
B	0.072	0.415	0.702	0.402	0.394	0.517	0.415	0.511	0.659	0.954	0.834	1.359	1.799	0.816	0.883	0.434	0.333	0.294	0.253	0.220	0.157	0.154	0.153	0.127	0.097	0.116	0.207	0.604	0.336	0.265	0.313	0.492
B	0.070	0.424	0.697	0.400	0.392	0.521	0.421	0.517	0.660	0.951	0.837	1.355	1.794	0.823	0.883	0.436	0.335	0.295	0.255	0.220	0.152	0.154	0.154	0.128	0.097	0.114	0.201	0.599	0.334	0.260	0.303	0.482
B	0.073	0.417	0.703	0.405	0.390	0.518	0.415	0.512	0.662	0.952	0.839	1.359	1.794	0.821	0.883	0.435	0.329	0.293	0.250	0.217	0.151	0.153	0.154	0.130	0.100	0.117	0.209	0.610	0.341	0.265	0.314	0.494
B	0.074	0.434	0.708	0.399	0.390	0.515	0.411	0.507	0.655	0.953	0.831	1.361	1.792	0.820	0.881	0.430	0.328	0.291	0.247	0.217	0.151	0.154	0.153	0.129	0.100	0.116	0.210	0.610	0.339	0.273	0.321	0.504
B	0.073	0.428	0.709	0.401	0.391	0.517	0.413	0.510	0.654	0.953	0.831	1.361	1.788	0.820	0.882	0.430	0.328	0.292	0.249	0.217	0.153	0.153	0.154	0.129	0.100	0.117	0.210	0.610	0.341	0.268	0.318	0.498

(Continued)

Table 4.2 (Continued)

nm	1499	1470	1441	1412	1383	1354	1325	1296	1267	1238	1209	1180	1151	1122	1093	1064	1035	1006	977	948	919	890	861	832	803	774	745	716	687	658	629	600
B	0.069	0.423	0.703	0.403	0.394	0.515	0.415	0.514	0.664	0.952	0.840	1.366	1.785	0.814	0.880	0.430	0.331	0.291	0.251	0.220	0.155	0.154	0.152	0.128	0.097	0.117	0.210	0.608	0.338	0.266	0.315	0.495
B	0.069	0.425	0.702	0.402	0.393	0.514	0.413	0.513	0.663	0.952	0.840	1.366	1.785	0.814	0.880	0.430	0.331	0.291	0.251	0.220	0.155	0.154	0.152	0.127	0.097	0.117	0.211	0.608	0.339	0.267	0.318	0.498
B	0.071	0.428	0.703	0.404	0.395	0.515	0.415	0.514	0.664	0.952	0.839	1.366	1.785	0.813	0.879	0.429	0.330	0.290	0.250	0.220	0.154	0.153	0.152	0.127	0.097	0.117	0.209	0.608	0.339	0.266	0.316	0.495
B	0.077	0.432	0.702	0.403	0.392	0.518	0.415	0.512	0.657	0.951	0.833	1.362	1.791	0.818	0.879	0.433	0.332	0.294	0.252	0.220	0.154	0.155	0.155	0.128	0.099	0.117	0.208	0.603	0.335	0.264	0.314	0.491
B	0.076	0.430	0.702	0.403	0.391	0.517	0.414	0.511	0.656	0.951	0.832	1.362	1.792	0.817	0.879	0.432	0.331	0.293	0.252	0.219	0.153	0.155	0.154	0.128	0.099	0.117	0.211	0.605	0.336	0.267	0.316	0.494
B	0.077	0.431	0.702	0.404	0.392	0.518	0.415	0.512	0.657	0.951	0.833	1.363	1.791	0.817	0.879	0.432	0.332	0.294	0.252	0.219	0.154	0.155	0.155	0.128	0.099	0.117	0.209	0.604	0.335	0.265	0.314	0.491
B	0.071	0.421	0.698	0.400	0.388	0.516	0.414	0.511	0.659	0.950	0.837	1.360	1.797	0.819	0.883	0.435	0.334	0.293	0.253	0.221	0.156	0.156	0.154	0.126	0.096	0.116	0.210	0.604	0.340	0.266	0.315	0.491
B	0.074	0.425	0.698	0.402	0.389	0.516	0.414	0.511	0.659	0.949	0.837	1.360	1.797	0.819	0.883	0.435	0.334	0.293	0.253	0.221	0.156	0.156	0.154	0.126	0.096	0.116	0.210	0.605	0.339	0.266	0.315	0.490
B	0.071	0.420	0.698	0.400	0.387	0.516	0.414	0.511	0.659	0.950	0.838	1.359	1.797	0.821	0.883	0.436	0.334	0.293	0.252	0.220	0.155	0.155	0.153	0.126	0.096	0.116	0.210	0.604	0.340	0.268	0.317	0.494
C	0.037	0.365	0.694	0.393	0.377	0.479	0.386	0.473	0.682	0.941	0.824	1.371	1.799	0.841	0.937	0.413	0.341	0.271	0.263	0.234	0.177	0.145	0.131	0.117	0.060	0.108	0.232	0.690	0.433	0.294	0.309	0.500
C	0.035	0.358	0.691	0.391	0.375	0.478	0.388	0.474	0.683	0.943	0.827	1.363	1.807	0.844	0.936	0.413	0.340	0.267	0.260	0.233	0.177	0.143	0.130	0.117	0.060	0.109	0.228	0.688	0.436	0.299	0.315	0.510
C	0.044	0.374	0.698	0.400	0.384	0.483	0.391	0.478	0.686	0.944	0.829	1.372	1.797	0.840	0.934	0.412	0.339	0.267	0.260	0.234	0.177	0.143	0.129	0.115	0.058	0.107	0.228	0.686	0.429	0.288	0.300	0.489
C	0.041	0.366	0.698	0.400	0.382	0.496	0.391	0.476	0.685	0.943	0.837	1.369	1.799	0.841	0.934	0.411	0.338	0.267	0.258	0.233	0.175	0.143	0.129	0.114	0.056	0.106	0.229	0.685	0.432	0.291	0.303	0.497
C	0.034	0.355	0.696	0.394	0.378	0.482	0.390	0.476	0.685	0.943	0.827	1.370	1.802	0.841	0.936	0.412	0.339	0.268	0.260	0.233	0.177	0.144	0.130	0.118	0.059	0.108	0.232	0.690	0.434	0.293	0.308	0.499
C	0.041	0.365	0.697	0.397	0.377	0.480	0.386	0.471	0.681	0.940	0.826	1.369	1.798	0.847	0.936	0.412	0.335	0.265	0.252	0.226	0.175	0.143	0.129	0.114	0.056	0.106	0.232	0.694	0.442	0.300	0.312	0.504
C	0.039	0.366	0.697	0.394	0.377	0.481	0.388	0.473	0.682	0.939	0.828	1.370	1.791	0.847	0.939	0.414	0.338	0.266	0.255	0.228	0.176	0.142	0.128	0.113	0.057	0.107	0.232	0.697	0.438	0.298	0.311	0.494
C	0.038	0.354	0.700	0.397	0.379	0.486	0.394	0.478	0.685	0.939	0.829	1.365	1.799	0.848	0.938	0.415	0.340	0.269	0.257	0.229	0.177	0.144	0.130	0.114	0.058	0.106	0.227	0.686	0.436	0.290	0.299	0.486
C	0.038	0.364	0.695	0.395	0.379	0.481	0.390	0.476	0.684	0.943	0.827	1.367	1.802	0.841	0.936	0.412	0.340	0.268	0.260	0.233	0.177	0.143	0.130	0.117	0.059	0.107	0.230	0.688	0.432	0.293	0.307	0.498
C	0.036	0.360	0.694	0.393	0.377	0.480	0.388	0.474	0.683	0.942	0.827	1.367	1.804	0.842	0.936	0.413	0.340	0.269	0.261	0.233	0.177	0.143	0.130	0.117	0.060	0.108	0.231	0.689	0.434	0.295	0.310	0.502
C	0.038	0.363	0.695	0.395	0.379	0.483	0.389	0.475	0.684	0.943	0.829	1.368	1.802	0.841	0.936	0.412	0.340	0.268	0.260	0.233	0.177	0.143	0.130	0.117	0.059	0.107	0.230	0.688	0.433	0.293	0.307	0.499
C	0.041	0.366	0.697	0.398	0.382	0.490	0.391	0.477	0.685	0.944	0.834	1.368	1.799	0.840	0.935	0.411	0.339	0.267	0.258	0.233	0.176	0.143	0.129	0.115	0.057	0.106	0.228	0.686	0.432	0.292	0.304	0.497
C	0.038	0.366	0.696	0.396	0.379	0.484	0.388	0.474	0.683	0.941	0.828	1.369	1.802	0.841	0.936	0.412	0.340	0.270	0.262	0.234	0.177	0.144	0.130	0.116	0.059	0.107	0.233	0.689	0.432	0.292	0.305	0.496
C	0.039	0.360	0.698	0.396	0.378	0.483	0.390	0.475	0.683	0.943	0.828	1.367	1.797	0.847	0.938	0.414	0.339	0.268	0.255	0.228	0.176	0.143	0.129	0.114	0.057	0.106	0.230	0.692	0.438	0.295	0.306	0.492
C	0.047	0.376	0.696	0.398	0.374	0.475	0.378	0.463	0.676	0.931	0.822	1.361	1.816	0.846	0.931	0.407	0.328	0.259	0.245	0.221	0.172	0.144	0.131	0.115	0.055	0.105	0.235	0.696	0.451	0.310	0.323	0.529
C	0.039	0.362	0.698	0.395	0.378	0.482	0.389	0.474	0.683	0.943	0.827	1.366	1.796	0.847	0.938	0.414	0.338	0.267	0.255	0.228	0.176	0.142	0.129	0.114	0.057	0.107	0.231	0.694	0.438	0.296	0.308	0.493
D	0.051	0.392	0.711	0.389	0.380	0.510	0.408	0.505	0.669	0.934	0.853	1.347	1.810	0.831	0.917	0.447	0.340	0.286	0.291	0.250	0.161	0.148	0.139	0.115	0.075	0.106	0.194	0.573	0.318	0.265	0.316	0.504
D	0.054	0.397	0.711	0.396	0.387	0.514	0.410	0.508	0.671	0.931	0.850	1.355	1.810	0.820	0.916	0.443	0.338	0.284	0.292	0.251	0.163	0.149	0.140	0.117	0.077	0.108	0.196	0.578	0.315	0.258	0.306	0.485
D	0.053	0.395	0.710	0.394	0.385	0.512	0.409	0.506	0.672	0.931	0.850	1.350	1.814	0.822	0.918	0.444	0.339	0.286	0.292	0.251	0.162	0.149	0.141	0.118	0.077	0.108	0.202	0.577	0.314	0.258	0.307	0.486
D	0.054	0.398	0.711	0.395	0.386	0.512	0.409	0.507	0.671	0.931	0.850	1.350	1.814	0.823	0.917	0.444	0.338	0.285	0.292	0.251	0.164	0.149	0.141	0.117	0.077	0.108	0.199	0.580	0.315	0.257	0.306	0.486
D	0.053	0.397	0.711	0.395	0.385	0.512	0.409	0.507	0.671	0.931	0.850	1.350	1.810	0.823	0.917	0.444	0.339	0.285	0.292	0.251	0.163	0.149	0.140	0.117	0.077	0.108	0.199	0.578	0.315	0.258	0.307	0.487
D	0.052	0.395	0.711	0.393	0.384	0.511	0.409	0.506	0.671	0.932	0.851	1.348	1.811	0.825	0.918	0.445	0.339	0.286	0.291	0.251	0.162	0.149	0.140	0.117	0.076	0.108	0.199	0.576	0.315	0.260	0.309	0.491
D	0.052	0.395	0.711	0.393	0.384	0.512	0.409	0.506	0.671	0.932	0.851	1.348	1.811	0.825	0.917	0.445	0.339	0.286	0.292	0.251	0.162	0.149	0.140	0.117	0.076	0.108	0.198	0.576	0.316	0.260	0.309	0.492
D	0.052	0.395	0.711	0.393	0.384	0.512	0.409	0.506	0.671	0.932	0.851	1.349	1.810	0.824	0.917	0.445	0.339	0.285	0.292	0.251	0.162	0.149	0.140	0.117	0.076	0.108	0.198	0.576	0.316	0.260	0.309	0.491

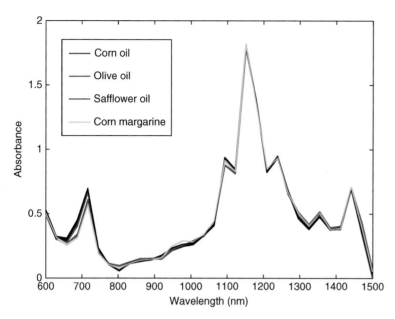

Figure 4.4 Case study 2: superimposed NIR spectra corresponding to the data in Table 4.2.

4.2.4 Aims of PCA

There are two principal needs in chemistry. The first is when PCs can be related to physically interpretable, underlying processes or compounds; thus, the number or quantitative values of the PCs can be directly interpreted in terms of the underlying chemistry. The second is when PCA is primarily used in a qualitative way for visualisation and simplification of data.

In the case of the example of case study 1, we would like to extract information from the two-way chromatogram, which relates to a discrete number of co-eluting compounds in the chromatographic mixture.

- The number of significant PCs is ideally equal to the number of significant compounds. If there are three compounds in the mixture, then we expect that there are only three significant PCs.
- Each PC is characterised by two pieces of information, the *scores*, which, in the case of chromatography, relate to the elution profiles, and the *loadings*, which relate to the spectra.

Below we will discuss in more detail how this information is obtained. However, the ultimate information is physically interpretable.

Case studies 2 and 3 primarily use PCA in a descriptive way. We are, for example, interested in whether the NIR spectra can be used to distinguish four groups of samples and, if so, what wavelengths are characteristic of each group. We are also interested in whether the elements can be distinguished from their physical properties. PCA is used to convert a table of numbers into something pictorial. Of course, where a sample fits into a graph obtained by PCA can also tell us a lot about its provenance.

1. Principal components analysis is applied to the identified GCMS peaks of extracts from 100 patients, 50 with a disease and 50 controls without the disease.
   (a) PCA is primarily used to determine a quantitative model as to whether patients have a disease or not
   (b) PCA is primarily used to determine a qualitative picture of whether the two groups can be separated, and if so which peaks are likely to be responsible

4.3 Principal Components Analysis: The Method

4.3.1 Scores and Loadings

PCA involves an abstract mathematical transformation of the original data matrix, usually obtained experimentally, but sometimes transformed mathematically first, as discussed in Section 4.6, which takes the form

$$X = TP + E$$

Table 4.3 Case study 3: properties of some elements.

Element	Group	Melting point (K)	Boiling point (K)	Density	Oxidation number	Electronegativity
Li	1	453.69	1615	534	1	0.98
Na	1	371	1156	970	1	0.93
K	1	336.5	1032	860	1	0.82
Rb	1	312.5	961	1530	1	0.82
Cs	1	301.6	944	1870	1	0.79
Be	2	1550	3243	1800	2	1.57
Mg	2	924	1380	1741	2	1.31
Ca	2	1120	1760	1540	2	1
Sr	2	1042	1657	2600	2	0.95
F	3	53.5	85	1.7	−1	3.98
Cl	3	172.1	238.5	3.2	−1	3.16
Br	3	265.9	331.9	3100	−1	2.96
I	3	386.6	457.4	4940	−1	2.66
He	4	0.9	4.2	0.2	0	0
Ne	4	24.5	27.2	0.8	0	0
Ar	4	83.7	87.4	1.7	0	0
Kr	4	116.5	120.8	3.5	0	0
Xe	4	161.2	166	5.5	0	0
Zn	5	692.6	1180	7140	2	1.6
Co	5	1765	3170	8900	3	1.8
Cu	5	1356	2868	8930	2	1.9
Fe	5	1808	3300	7870	2	1.8
Mn	5	1517	2370	7440	2	1.5
Ni	5	1726	3005	8900	2	1.8
Bi	6	544.4	1837	9780	3	2.02
Pb	6	600.61	2022	11340	2	1.8
Tl	6	577	1746	11850	3	1.62

where

- T are called the *scores* and have as many rows as the original data matrix,
- P are called the *loadings* and have as many columns as the original data matrix,
- the number of columns in the matrix T equals the number of rows in the matrix P and
- E is called an *error matrix*, of the same dimensions as X.

Some people transpose P; that is, swap the rows and columns around, but we will stick to the equation above in this book for consistency. It is possible to calculate scores and loadings matrices as large as desired, providing the 'common' dimension is no larger than the smallest dimension of the original data matrix X and corresponds to the number of PCs that are calculated.

Hence, if the original data matrix has dimensions 30×28 (or $I \times J$), no more than 28 (non-zero) PCs can be calculated. If the number of PCs is denoted by A, then this number can be no larger than 28.

- The dimensions of T will be $30 \times A$.
- The dimensions of P will be $A \times 28$.

Each scores matrix consists of a series of column vectors, and each loadings matrix, as defined in this section, consists of a series of row vectors. These vectors can also be called *eigenvectors*. Many authors denote these vectors by t_a and p_a, where a is the number of the principal component (1, 2, 3 up to A). The scores matrices T and P are composed of several such vectors, one for each principal component. The first scores vector and the first loadings vector are often

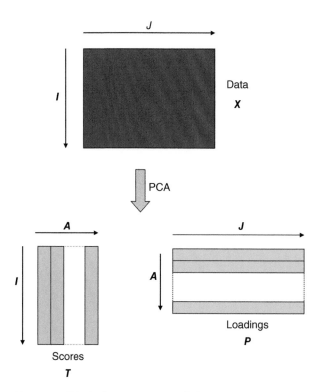

Figure 4.5 Principal components analysis.

called the *eigenvectors* of the first principal component. This is illustrated in Figure 4.5. Each successive component is characterised by a pair of eigenvectors.

The first three scores and loadings vectors for the data in Table 4.1 (case study 1) are presented in Table 4.4 for the first three PCs ($A = 3$).

Principal components have certain mathematical properties, and the aim of PCA algorithms is to find new variables (the PCs) that satisfy these properties. Different algorithms result in slightly different properties for the resultant eigenvectors, also called *scores* and *loadings*. Below we will discuss the properties if the nonlinear iterative partial least squares (NIPALS) algorithm is employed, which is probably the commonest algorithm used in chemometrics.

Scores and loadings possess a number of features. It is important to recognise that PCs are simply abstract mathematical entities.

- All scores and loadings vectors have the property.
  - the sums

$$\sum_{i=1}^{I} t_{ia} \cdot t_{ib} = 0 \quad \text{and} \quad \sum_{j=1}^{J} p_{aj} \cdot p_{bj} = 0$$

  where $a \neq b$, and t and p correspond to the elements of the corresponding eigenvectors. Some authors state that the scores and loadings vectors are mutually *orthogonal*, as some of the terminology of chemometrics arises from multivariate statistics where people like to think of PCs as vectors in multi-dimensional space, each variable representing an axis; hence, some of the geometric analogies have been incorporated into the mainstream literature. If the columns are mean-centred (i.e. the average of each column has been subtracted from itself), then
  - the correlation coefficient between any two scores vectors is equal to 0.
- Each loadings vector is also *normalised*. There are various different definitions of a normalised vector, but we use $\sum_{j=1}^{J} p_{aj}^2 = 1$. Note that there are several algorithms for PCA; using the SVD (singular value decomposition) method, the scores are also normalised. However, in this section, we will restrict the calculations to the NIPALS methods. It is sometimes stated that the loadings vectors are *orthonormal* combining both properties.

Table 4.4 Scores and loadings for case study 1.

Scores			Loadings		
0.017	0.006	−0.001	0.348	−0.103	−0.847
0.128	0.046	0.000	0.318	−0.254	−0.214
0.507	0.182	−0.002	0.220	−0.110	−0.011
1.177	0.422	−0.001	0.101	0.186	−0.022
1.773	0.626	0.001	0.088	0.312	−0.028
2.011	0.639	0.000	0.104	0.374	−0.031
2.102	0.459	−0.004	0.106	0.345	−0.018
2.334	0.180	−0.003	0.094	0.232	0.008
2.624	−0.080	0.007	0.093	0.132	0.041
2.733	−0.244	0.018	0.121	0.123	0.048
2.602	−0.309	0.016	0.170	0.166	0.060
2.320	−0.310	0.006	0.226	0.210	0.080
1.991	−0.280	−0.004	0.276	0.210	0.114
1.676	−0.241	−0.009	0.308	0.142	0.117
1.402	−0.202	−0.012	0.314	−0.002	0.156
1.176	−0.169	−0.012	0.297	−0.166	0.212
0.991	−0.141	−0.012	0.267	−0.284	0.213
0.842	−0.118	−0.011	0.236	−0.290	0.207
0.721	−0.100	−0.009	0.203	−0.185	0.149
0.623	−0.086	−0.009	0.166	−0.052	0.107
0.542	−0.073	−0.008	0.123	0.070	0.042
0.476	−0.063	−0.007	0.078	0.155	0.000
0.420	−0.055	−0.006	0.047	0.158	−0.018
0.373	−0.049	−0.006	0.029	0.111	−0.018
0.333	−0.043	−0.005	0.015	0.061	−0.021
0.299	−0.039	−0.005	0.007	0.027	−0.013
0.271	−0.034	−0.004	0.003	0.010	−0.017
0.245	−0.031	−0.004	0.001	0.003	−0.003
0.223	−0.028	−0.004			
0.204	−0.026	−0.003			

- Some people also use the square matrix $T'T$, which has the properties that all elements are zero except those along the diagonals, the value of the diagonal elements relating to the size (or importance) of each successive PC. The square matrix PP' has a special property that it is an identity matrix, the dimensions are equal to the number of PCs in the model.

After PCA, the original variables (e.g. absorbances recorded at 28 wavelengths) are reduced to a number of significant principal components (e.g. 3). PCA can be used as a form of variable reduction, reducing the large original data set (recorded at 28 wavelengths for case study 1) to a much smaller more manageable data set (e.g. consisting of three principal components), which can be interpreted more easily, as illustrated in Figure 4.6. The loadings can be visualised as representing the means to this end.

The original data are said to be mathematically modelled by the PCs. Using A PCs, it is possible to establish a model for each element of X of the form

$$x_{ij} = \sum_{a=1}^{A} t_{ia}p_{aj} + e_{ij} = \hat{x}_{ij} + e_{ij}$$

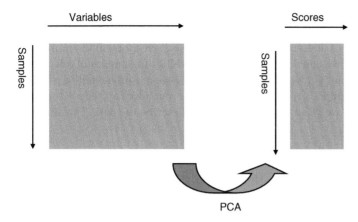

Figure 4.6 PCA as a form of variable reduction.

which is the non-matrix version of the fundamental PCA equation above. Hence, the estimated value of x for the data in Table 4.1 at the 10th wavelength (263 nm) and 8th point in time (true value of 0.304) is given by

- 2.334*0.121 = 0.282 for a one-component model and
- 2.334*0.121 + 0.180*0.123 = 0.304 for a two-component model,

suggesting that two PCs do provide a good estimate of the intensity at this wavelength and point in time.

1. Fifty samples of plasma extracts are recorded spectroscopically using 150 Raman wavelengths. Four principal components are calculated. Using the definitions in this section
    (a) The scores matrix has dimensions 50×4 and the loadings matrix 150×4.
    (b) The scores matrix has dimensions 50×4 and the loadings matrix 4×150.
    (c) The scores matrix has dimensions 4×50 and the loadings matrix 4×150.
    (d) The scores matrix has dimensions 4×50 and the loadings matrix 150×4.

2. Using the NIPALS algorithm
    (a) Scores are normalised but loadings are not.
    (b) Scores are not normalised but loadings are.
    (c) Both scores and loadings are normalised.

3. If data are uncentred
    (a) Scores of different PCs are orthogonal and uncorrelated.
    (b) Scores of different PCs are neither orthogonal nor uncorrelated.
    (c) Scores of different PCs are orthogonal but usually have some correlation.

4.3.2 Rank and Eigenvalues

A fundamental next step is to determine the number of significant PCs that can best characterise a matrix. In a series of mixture spectra or portion of a chromatogram, this should, ideally, correspond to the number of compounds under observation, and there has been a great deal of chemometric literature on methods for determining how many PCs are best for modelling a data set.

The *rank* of a matrix is a mathematical concept that equals to the number of components in a data set and is the number of PCs that exactly model a matrix, so that E is zero, and is usually equal to the smaller of the number of samples or variables, unless there are correlations.

However, in chemistry, we often talk of the approximate or chemical rank of a matrix. This represents the number of components deemed significant, for example, so that the error matrix is relatively very small. There is a huge literature here, and it is important to find how many components are significant, they may, for example, represent the number of significant compounds in a mixture. We may have recorded 50 spectra over 100 wavelengths, but there are only six compounds. It can also be used to simplify data and remove noise. The chemical rank is usually regarded as the number of significant PCs.

When using PCA for EDA, the number of significant PCs does not usually even have to correspond to the number of compounds in a mixture (if appropriate). For example, if we consider case study 2 (the NIR of a samples of oils), there may be hundreds of detectable compounds, but we are not primarily interested in this: we are interested in whether we can distinguish these groups. The number of significant PCs is an indication of how much we can reduce the data, for example, from 32 wavelengths to 2 or 3 new 'factors' or PCs that are sufficient to describe the data. The aim is to find A so that $X \approx TP$. The matrix E is considered noise.

4.3.2.1 Eigenvalues

Normally after PCA, the *size* of each component can be measured. This is often called an *eigenvalue*: the earlier (and more significant) the components, the larger their size. There are a number of definitions in the literature, but a simple one defines the eigenvalue of a PC as the sum of squares of the scores, so that

$$g_a = \sum_{i=1}^{I} t_{ia}^2$$

where g_a is the ath eigenvalue.

The *sum* of all non-zero eigenvalues for a data matrix equals the sum of squares of the entire data matrix so that

$$\sum_{a=1}^{K} g_a = \sum_{i=1}^{I} \sum_{j=1}^{J} x_{ij}^2$$

where K is the smaller of I or J. Note that if the data are pre-processed before PCA, as discussed in Section 4.6, x must likewise be pre-processed for this property to hold.

Frequently, eigenvalues are presented as percentages, for example, of the sum of squares of the entire (pre-processed) data set, or

$$V_a = 100 \frac{g_a}{\sum_{i=1}^{I} \sum_{j=1}^{J} x_{ij}^2}$$

Successive eigenvalues correspond to smaller percentages.

The *cumulative* percentage eigenvalue is often used to determine (approximately) the proportion of the data modelled using PCA and is given by $\sum_{a=1}^{A} g_a$. The closer this the total sum of squares of the (pre-processed) data, the more faithful the model. The percentage can be plotted against the number of eigenvalues in the PC model.

It is an interesting feature that the *residual sum of squares* (RSSs)

$$\text{RSS}_A = \sum_{i=1}^{I} \sum_{j=1}^{J} x_{ij}^2 - \sum_{a=1}^{A} g_a$$

after A eigenvalues also equals the sum of squares for the error matrix, between the PC model and the raw data, whose elements are defined by

$$e_{ij} = x_{ij} - \hat{x}_{ij} = x_{ij} - \sum_{a=1}^{A} t_{ia} p_{aj}$$

or

$$\sum_{i=1}^{I} \sum_{j=1}^{J} x_{ij}^2 - \sum_{a=1}^{A} g_a = \sum_{i=1}^{I} \sum_{j=1}^{J} e_{ij}^2$$

because the product of the elements of any two different eigenvectors is 0, as discussed in Section 4.3.1.

The first three eigenvalues for the data in Table 4.1 are presented in Table 4.5. The total sum of squares of the entire data set is 61.00, allowing the various percentages to be calculated. It can be seen that two eigenvalues represent more than 99.99% of the data in this case. In fact, the interpretation of the size of an eigenvalue depends, in part, on the

Table 4.5 Eigenvalues for case study 1 (raw data).

g_a	V_a	Cumulative (%)
59.21	97.058	97.058
1.79	2.939	99.997
0.0018	0.003	100.000

nature of the pre-processing. However, as the chromatogram probably consists of only two compounds in the cluster, this conclusion is physically reasonable.

For case study 3 (the properties of the elements), we see quite a different pattern. First of all, the data are standardised before PCA. This procedure will be discussed in Section 4.6.4. The first four eigenvalues are presented in Table 4.6. We now see that there is no obvious cut-off as the cumulative percentage of the data increases gradually. There are only five possible PCs because there are only five variables, and we find that four PCs model 99.15% of the data: in contrast to case study 1 where two PCs are sufficient to model 99.99% of the data out of a possible 32 non-zero PCs.

Using the size of eigenvalues, we can try to estimate the number of significant components in the data set.

A simple rule might be to reject PCs whose cumulative eigenvalues account for less than a certain percentage (e.g. 5%) of the data; in the case of Table 4.6, this would suggest that the first three components are significant. For Table 4.5, this would suggest that only one PC should be retained. However, in the latter case, we would be incorrect, as the original information was not centred before PCA and the first component is mainly influenced by the overall size of the data set. Centring the columns for case study 1 (as discussed in Section 4.6.3) reduces the total sum of squares of the data set from 61.00 to 24.39. The eigenvalues from the mean-centred principal components are presented in Table 4.7 and now the first eigenvalue contributes less, hence suggests that two components are required to model at least 95% of the data. There is no general guidance whether to use centred or raw data when determining the number of significant components, the most appropriate method being dependent on the nature of the data and one's experience.

More elaborate information can be obtained by looking at the size of the error matrix as defined above. The sum of squares of the matrix E can be expressed as the difference between the sum of squares of the matrices X and $\hat{X}$. Consider Table 4.6, after three components are calculated,

- the sum of squares of $\hat{X}$ equals 129.94 (or the sum of the first three eigenvalues $= 87.51 + 29.57 + 12.86$).

Table 4.6 Eigenvalues for case study 3 (standardised data).

g_a	V_a	Cumulative (%)
87.509	64.821	64.821
29.574	21.907	86.728
12.857	9.524	96.252
3.916	2.901	99.153

Table 4.7 Size of eigenvalues for case study 1 after column centring.

g_a	V_a	Cumulative (%)
22.6	92.657	92.657
1.79	7.338	99.995
0.0103	0.004	100.000

However,

- the sum of squares of the original data X equals 135, as the data have been standardised and there are 27×5 measurements (for standardised data, the total sum of squares equals the number of elements in the data, as discussed in Section 4.6.4).

Therefore,

- the sum of squares of the error matrix E equals $135 - 129.94$ or 5.06: this number is also equal to the sum of eigenvalues 4 and 5, as there are only two non-zero eigenvalues in the data.

Sometimes, the eigenvalues can be interpreted in physical terms. For example,

- the data set in Table 4.1 consists of 30 spectra recorded in time at 28 wavelengths,
- the error matrix is of size 30×28, consisting of 840 elements,
- but the error sum of squares after $a = 1$ PC equals $60.99 - 59.21 = 1.78 \text{ AU}^2$,
- so the root mean square error is equal to $\sqrt{1.78/840} = 0.046$ (in fact, some chemometricians adjust this for the loss of degrees of freedom due to the calculation of one PC but because 840 is a large number, this adjustment is small and we will stick to the convention in this book of dividing x errors simply by the number of elements in the data matrix).

Is this a physically sensible number? This depends on the original units of measurement and what the instrumental noise characteristics are. If it is known that the root mean square noise is about 0.05 units, then it seems sensible. If the noise level, however, is substantially lower, then not enough PCs have been calculated. In fact, most modern chromatographic instruments can determine peak intensities much more accurately than 0.05 AU; hence, this would suggest that a second PC is required.

The principle of examining the size of successive eigenvalues can be extended, and in spectroscopy, a large number of so-called *indicator* functions have been proposed, many by Malinowski, whose text on FA is a classic. Most functions involve producing graphs of functions of eigenvalues and predicting the number of significant components using various criteria. Over the past decade, several new functions have been proposed, some based on distributions such as the F-test. For more statistical applications such as quantitative structure–activity relationships (QSAR), these indicator functions are not so applicable, but in spectroscopy and certain forms of chromatography where there are normally a physically defined number of factors and well-understood error (or noise) distributions, such approaches are valuable. From a simple rule of thumb, knowing (or estimating) the noise distribution, for example, from a portion of a chromatogram or spectrum where there is known to be no compounds (or a blank), and then determining how many eigenvalues are required to reduce the level of error, allows an estimate of significant components.

1. The total sum of squares of a data matrix is equal to 735.21, and its first eigenvalue is equal to 512.60 and its second to 48.91. Therefore, the sum of squares of the remaining error matrix is
    - (a) 173.70
    - (b) 48.91
    - (c) 22.61

2. A series of spectra are recorded for 50 samples at 32 wavelengths. The data are standardised and the sum of the first three eigenvalues is 1257. The per cent of the data represented by the first three eigenvalues is
    - (a) 88.6%
    - (b) 78.6%

4.3.2.2 Cross-validation

A complementary series of methods for determining the number of significant components is based on cross-validation. It is assumed that significant components model 'data', whereas later (and redundant) components model 'noise'. Auto-predictive models involve fitting PCs to the entire data set and always provide a closer fit to the data the more the components are calculated. Hence, the residual error will be smaller if 10 rather than nine PCs are calculated. This does not necessarily indicate that it is correct to retain all 10 PCs; the later PCs may model noise that we do not want.

The significance of each PC can be tested by observing how well an 'unknown' sample is predicted. The commonest form of cross-validation is called *leave one out* (LOO) cross-validation, in which a single sample is removed from the data set at a time. There are several other approaches, but for brevity, we will describe only one in detail in this section. For example, if there are 10 objects, perform PC on nine samples, and see how well the property/ies of the remaining sample is predicted.

The following steps are normally employed.

- Initially, leave out sample 1 ($=i$).
- Perform PCA on the remaining $I-1$ samples, for example, samples 2–10. For efficiency, it is possible to calculate several PCs ($=A$) simultaneously. Obtain the scores T and loadings P. Notice that there will be different scores and loadings matrices according to which sample is removed.
- Next, determine what would be the scores for the left out sample i simply by

$$\hat{t}_i = x_i P'$$

Notice that this equation is quite simple and is obtained from standard multiple linear regression $\hat{t}_i = x_i P'(PP')^{-1}$, but as the loadings are orthonormal, $(PP')^{-1}$ is a unit matrix.

- Then, calculate the model for sample i for a PCs by

$$^{a,cv}\hat{x}_i = {}^{a}\hat{t}_i{}^{a}P$$

where the superscript a refers to the model using the first a PCs; hence, $^{a}\hat{x}_i$ has the dimension $1 \times J$, $^{a}\hat{t}_i$ has dimensions $1 \times a$ (i.e. is a scalar if only one PC is retained) and consists of the first a scores obtained in step 3, and ^{a}P has dimensions $a \times J$ and consists of the first a rows of the loadings matrix.

- Next, repeat this by leaving another sample out and going to step 2 until all samples have been removed once.
- The error, often called the *predicted residual error sum of squares* or PRESS, is then calculated as

$$\text{PRESS}_a = \sum_{i=1}^{I} \sum_{j=1}^{J} (^{a,cv}\hat{x}_{ij} - x_{ij})^2$$

This is simply the sum of squares difference between the observed and true values for each object using an a PC model.

The PRESS errors can then be compared with the *residual sum of squares* (RSS) errors each object for straight PCA (sometimes called the *auto-prediction* or *training set error*), given by

$$\text{RSS}_a = \sum_{i=1}^{I} \sum_{j}^{J} x_{ij}^2 - \sum_{k=1}^{a} g_k$$

or

$$\text{RSS}_a = \sum_{i=1}^{I} \sum_{j=1}^{J} (^{a,auto}\hat{x}_{ij} - x_{ij})^2$$

All equations presented above assume no mean centring or data pre-processing, further steps are required involving subtracting the mean of $I-1$ samples each time a sample is left out if mean centred. If the data are pre-processed before cross-validation, it is essential that both PRESS and RSS are presented on the same scale. A problem is that if one takes a subset of the original data, the mean and standard deviation will differ for each group; hence, it is safest to convert all the data to the original units for calculation of errors. The computational method can be quite complex and there are no generally accepted conventions but we recommend the following.

- Pre-process the entire data set.
- Perform PCA on the entire data set to give predicted $\hat{X}$ in pre-processed units (e.g. mean-centred or standardised).
- Convert this matrix back to the original units.
- Determine the RSS in the original units.

One sample is then taken out at a time.

- Take one sample out and determine statistics such as means or standard deviations for the remaining $I-1$ samples.

- Then, pre-process these remaining samples, according to the statistics obtained above for $I-1$ samples, and perform PCA on this data.
- Then, scale the remaining Ith sample using the mean and standard deviation (as appropriate) obtained above for $I-1$ samples.
- Then, obtain the predicted scores $\hat{t}_i$ for this sample, using the loadings and the scaled vector x_i obtained in the two previous steps.
- Then, predict the vector $\hat{x}_i$ by multiplying $\hat{t}_i p$, where the loadings have been determined from the $I-1$ pre-processed samples.
- Now rescale the predicted vector to the original units.
- Next, remove another sample and repeat the steps until each sample is removed once.
- Finally, calculate PRESS values in the original units.

There are quite a number of variations in cross-validation, especially methods for calculating errors, each group or programmer has their own favourite. For brevity, we recommend one single approach. Note that although some steps are common, it is normal to use different criteria when using cross-validation in multivariate calibration, as described in Section 6.6.2. Do not get surprised if different packages provide what appear to be different numerical answers for the estimation of similar parameters and always try to understand what the developer of the software has intended; normally, quite extensive documentation is available.

There are various ways of interpreting these two errors numerically, but a common approach is to compare the PRESS error using $a+1$ PCs to the RSS using a PCs. If the latter error is significantly larger, then the extra PC is modelling only noise, so is not significant. Sometimes, this is mathematically defined by computing the ratio $\text{PRESS}_a/\text{PRESS}_{a-1}$, and if this exceeds 1, use $a-1$ PCs in the model. If the errors are quite close in size, it is safest to continue checking further components; normally, there will be a sharp difference when sufficient components have been computed. Often, PRESS will start increasing after the optimum number of components has been calculated.

It is easiest to understand the principle using a small numerical example. As the data sets of case studies 1 to 2 are rather large, a simulation will be introduced. Table 4.8 is of data set consisting of 10 objects and eight variables. In Table 4.9(a), the scores and loadings for eight PCs (the number is limited by the variables) using only samples 2–10 are presented – note that the data are non-centred. Table 4.9(b) illustrates the calculation of the sum of square cross-validated error for sample 1 as increasing number of PCs are calculated. In Table 4.10(a), these errors are summarised for all samples. In Table 4.10(b), eigenvalues are calculated, together with the RSSs as increasing number of PCs are computed for both auto-predictive and cross-validated models. The latter can be obtained by summing the rows in Table 4.10(a). RSS decreases continuously, whereas PRESS levels off. This information can be illustrated graphically (see Figure 4.7): the vertical sale is usually presented logarithmically, which takes into account the very high first eigenvalues, quite usual in cases where the data are uncentred, and so the first eigenvalue is mainly one of size and can appear (falsely) to dominate the data. Using the above criteria, the PRESS value of the fourth PC is greater than the RSS of the third PC, so an optimal model would appear to consist of three PCs. A simple graphical approach, taking the optimum number of PCs to be where the graph of PRESS levels off or increases would likewise suggests that there

Table 4.8 Cross-validation example.

	A	B	C	D	E	F	G	H
1	89.821	59.760	68.502	48.099	56.296	95.478	71.116	95.701
2	97.599	88.842	95.203	71.796	97.880	113.122	72.172	92.310
3	91.043	79.551	104.336	55.900	107.807	91.229	60.906	97.735
4	30.015	22.517	60.330	21.886	53.049	23.127	12.067	37.204
5	37.438	38.294	50.967	29.938	60.807	31.974	17.472	35.718
6	83.442	48.037	59.176	47.027	43.554	84.609	67.567	81.807
7	71.200	47.990	86.850	35.600	86.857	57.643	38.631	67.779
8	37.969	15.468	33.195	12.294	32.042	25.887	27.050	37.399
9	34.604	68.132	63.888	48.687	86.538	63.560	35.904	40.778
10	74.856	36.043	61.235	37.381	53.980	64.714	48.673	73.166

Table 4.9 Calculation of cross-validated error for sample 1.

(a) Scores and loadings for first eight PCs on nine samples, excluding sample 1

Scores

259.25	9.63	20.36	2.29	−3.80	0.04	−2.13	0.03
248.37	−8.48	−5.07	−3.38	1.92	−5.81	0.53	−0.46
96.43	−24.99	−20.08	8.34	2.97	0.12	0.33	0.29
109.79	−23.52	−3.19	−0.38	−5.57	0.38	3.54	1.41
181.87	46.76	4.34	2.51	2.44	0.30	0.65	1.63
180.04	−16.41	−20.74	−2.09	−1.57	1.91	−3.55	0.16
80.31	8.27	−13.88	−5.92	2.75	2.54	0.60	1.17
157.45	−34.71	27.41	−1.10	4.03	2.69	0.80	−0.46
161.67	23.85	−12.29	0.32	−1.12	2.19	2.14	−2.63

Loadings

0.379	0.384	−0.338	−0.198	−0.703	0.123	−0.136	0.167
0.309	−0.213	0.523	−0.201	−0.147	−0.604	−0.050	0.396
0.407	−0.322	−0.406	0.516	0.233	−0.037	−0.404	0.286
0.247	−0.021	0.339	0.569	−0.228	0.323	0.574	0.118
0.412	−0.633	−0.068	−0.457	−0.007	0.326	0.166	−0.289
0.388	0.274	0.431	0.157	0.064	0.054	−0.450	−0.595
0.263	0.378	0.152	−0.313	0.541	0.405	0.012	0.459
0.381	0.291	−0.346	−0.011	0.286	−0.491	0.506	−0.267

(b) Predictions for sample 1

	PC1 $\longrightarrow$ PC7								
Predicted scores	207.655	43.985	4.453	−1.055	4.665	−6.632	0.329		
Predictions	A	B	C	D	E	F	G	H	Sum of square error
PC1	78.702	64.124	84.419	51.361	85.480	80.624	54.634	79.109	2025.93
	95.607	54.750	70.255	50.454	57.648	92.684	71.238	91.909	91.24
	94.102	57.078	68.449	51.964	57.346	94.602	71.916	90.369	71.41
	94.310	57.290	67.905	51.364	57.828	94.436	72.245	90.380	70.29
	91.032	56.605	68.992	50.301	57.796	94.734	74.767	91.716	48.53
	90.216	60.610	69.237	48.160	55.634	94.372	72.078	94.972	4.54
	90.171	60.593	69.104	48.349	55.688	94.224	72.082	95.138	4.43

Table 4.10 Calculation of RSS and PRESS.

(a) Summary of cross-validated sum of square errors

Object	1	2	3	4	5	6	7	8	9	10
PC1	2025.9	681.1	494.5	1344.6	842	2185.2	1184.2	297.1	2704	653.5
	91.2	673	118.1	651.5	66.5	67.4	675.4	269.8	1655.4	283.1
	71.4	91.6	72.7	160.1	56.7	49.5	52.5	64.6	171.6	40.3
	70.3	89.1	69.7	159	56.5	36.2	51.4	62.1	168.5	39.3
	48.5	59.4	55.5	157.4	46.7	36.1	39.4	49.9	160.8	29.9
	4.5	40.8	8.8	154.5	39.5	19.5	38.2	18.9	148.4	26.5
	4.4	0.1	2.1	115.2	30.5	18.5	27.6	10	105.1	22.6

(b) RSS and PRESS calculations

Eigenvalues	RSS	PRESS	$PRESS_a/RSS_{a-1}$
316 522.1	10 110.9	12 412.1	
7324.6	2786.3	4551.5	0.450
2408.7	377.7	830.9	0.298
136.0	241.7	802.2	2.124
117.7	123.9	683.7	2.829
72.9	51.1	499.7	4.031
36.1	15.0	336.2	6.586
15.0	0.0	n/a	

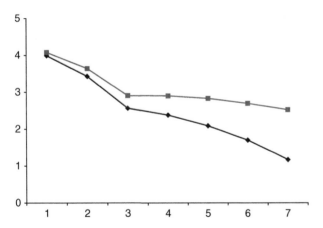

Figure 4.7 Graph of log of PRESS (top) and RSS (bottom) for data set in Table 4.8.

are three PCs in the model. Sometimes, PRESS values increase after the optimum number of components has been calculated, but this is not so in this example.

There are, of course, many other modifications of cross-validation, two of the most common are listed below.

- Instead of removing one object at a time, remove a block of objects, for example, four objects, and then cycle round so that each object is part of a group. This can speed up the cross-validation algorithm. However, with modern fast computers, this enhancement is less needed.
- Remove portions of the data rather than individual samples. Statisticians have developed a number of approaches, and some traditional chemometric software uses this method. This involves removing a certain number of measurements and replacing them by guesses, for example, the standard deviation of the column, performing PCA and then determining how well these measurements are predicted. If too many PCs have been employed, the measurements are not well predicted.

As in the case of most chemometric methods, there are innumerable variations on the theme, and it is important to be careful to check every author in detail. However, the LOO method described above is popular and relatively easy to implement and understand.

1. Comparing PRESS and RSS which is true.
   (a) For the same number of PCs, PRESS is likely to be larger than RSS.
   (b) For the same number of PCs, PRESS is likely to be less than RSS.
   (c) There is not much relationship between the two.

4.4 Factor Analysis

Principal components are primarily abstract mathematical entities with certain well-defined properties such as orthogonality. Chemists, however, often like to think in terms of physically interpretable components, such as spectra or concentration profiles, and these are often called *factors*. Note that there is a difference between the chemist's and statistician's terminology: many statisticians would call principal components factors instead.

As an illustration, we will use the case of coupled chromatography, such as HPLC–DAD, as in case study 1. For a simple chromatogram, the underlying data set can be described as a sum of responses for each significant compound in the data, which are characterised by (a) an elution profile and (b) a spectrum, plus noise or instrumental error. In matrix terms, this can be written as

$$X = CS + E$$

where

- X is the original data matrix or coupled chromatogram,
- C is a matrix consisting of the elution profiles of each compound,
- S is a matrix consisting of the spectra of each compound and
- E is an error matrix.

Consider the matrix of case study 1, a portion of a chromatogram recorded over 30 points in time and 28 wavelengths, consisting of two partially overlapping compounds.

- X is a matrix of 30 rows and 28 columns,
- C is a matrix of 30 rows and two columns, each column corresponding to the elution profile of a single compound,
- S is a matrix of two rows and 28 columns, each row corresponding to the spectrum of a single compound,
- E is a matrix of the same size as X.

If we observe X, can we then predict C and S? In previous chapters, we have used a 'hat' notation to indicate a *prediction*, so it is also possible to write the above equation as

$$X \approx \widehat{C}\widehat{S}$$

Ideally, the predicted spectra and chromatographic elution profiles are close to the true ones, but it is important to realise that we can *never directly or perfectly* observe the underlying data. There will always be measurement error even in practical spectroscopy. Chromatographic peaks may be partially overlapping or even embedded, which means that chemometric methods will help resolve the chromatogram into individual components.

One aim of chemometrics is to obtain these predictions after first treating the chromatogram as a multivariate data matrix and then performing PCA. Each compound in the mixture is a 'chemical' factor with its associated spectra and elution profile, which can be related to principal components, or 'abstract' factors, by a mathematical transformation.

It is possible to relate PCs to chemical information, such as, for example, elution profiles and spectra in diode array HPLC by

$$\widehat{X} = TP = \widehat{C}\widehat{S}$$

The conversion from 'abstract' to 'chemical' factors is sometimes called a *rotation* or *transformation* and examples will be discussed in more detail in Chapter 7, briefly it involves finding a square matrix R so that

$$TP = TRR^{-1}P = \widehat{C}\widehat{S} \quad \text{or} \quad \widehat{C} = TR \quad \text{and} \quad \widehat{S} = R^{-1}P$$

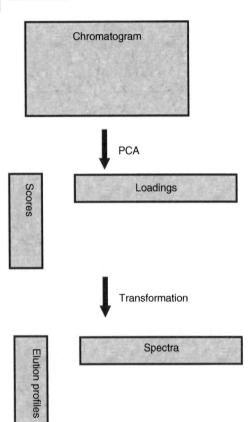

and is illustrated in Figure 4.8. Note that FA in chemometrics is by no means restricted to chromatography. An example is the pH titration profile of a number of species containing a different number of protons together with their spectra. Each equilibrium species has a pH titration profile and a characteristic spectrum. We will discuss other applications in Chapter 7.

FA is often called by a number of alternative names such as 'rotation' or 'transformation' but is a procedure used to relate the abstract PCs to meaningful chemical factors; the influence of Malinowski in the 1980s has introduced this terminology into chemometrics.

It is important to understand that the historic development of multivariate methods in chemometrics came from two very different perspectives. The first arose from physical chemistry and quantitative analytical chemistry. Multivariate methods were used to estimate physically interpretable things such as reaction rates of spectral profiles. The other came from statistics and PCs were primarily viewed as abstract entities helped to simplify and interpret data. The terminology of FA in chemistry mainly came from the former group.

1. *R* is used in the transformation from *T* (an abstract factor) to *Ĉ*, where there are 30 samples and three principal components

    (a) *R* is a 3×3 matrix
    (b) *R* is a 30×3 matrix
    (c) *R* is a 30×30 matrix

4.5 Graphical Representation of Scores and Loadings

Many revolutions in chemistry relate to the graphical presentation of information. For example, fundamental to the modern chemist's way of thinking is the ability to draw structures on paper in a convenient and meaningful manner. Years of debate preceded the general acceptance of the Kekulé structure for benzene: today's organic chemist can write down and understand complex structures of natural products without the need to plough through pages of numbers of orbital densities and bond lengths. Yet, underlying these representations are quantum mechanical probabilities; hence, the ability to convert from numbers to a simple diagram has allowed a large community to clearly think about chemical reactions.

Hence, with statistical data, and modern computers, it is easy to convert from numbers to graphs. Many modern multivariate chemometricians think geometrically as much as numerically, and concepts such as principal components are often treated as much as objects in an imaginary multivariate space than mathematical entities. The algebra of multi-dimensional space is the same as that of multivariate statistics. Older texts, of course, were written before the days of modern computing; hence, the ability to produce graphs was more limited. However, it is now possible to obtain a large number of graphs rapidly using simple software, much is even possible using Excel. There are many ways of visualising PCs. Below we will primarily look at graphs of the first two PCs, for simplicity.

4.5.1 Scores Plots

One of the simplest plots is that of the scores of one PC against the other. Figure 4.9 illustrates the PC plot of the first two PCs against each other obtained from case study 2, corresponding to plotting a graph of the first two columns of the matrix T. The horizontal axis is the scores for the first PC and the vertical axis for the second PC. No pre-processing has been performed. Note that the NIPALS algorithm has been used: this affects the scale of the axes; if SVD were used, the axes would be normalised; thus, when employing different packages, always check what algorithm has been employed. In this text, we restrict to NIPALS for simplicity.

From this 'picture', we can clearly see that the four groups are distinguished by the NIR spectra. Note that in the PCA algorithm, the sign of the PC cannot be controlled; hence, some packages may invert the axes because the sign of a square root can be either negative or positive.

It is possible to present the data in three dimensions, in this case adding the scores of PC3 as the third axis, as shown in Figure 4.10. Although these graphical representations look nice, in fact, the third PC adds little to the separation. The scores can also be presented in one dimension, as shown in Figure 4.11. We can see that all groups are in fact separated using PC1, but the distinction between safflower oil and corn oil is more evident using PC2; however, corn margarine and olive oil are not separable in PC2. If we had not coloured the groups, we may not have easily been able to distinguish all four using a single component, but once both are plotted, we can see all groups are easily distinguishable.

Case study 1 has different characteristics to case study 2. In this case, the observations are relating sequentially in time, whereas in case study 1, there is no specific sequence to the data.

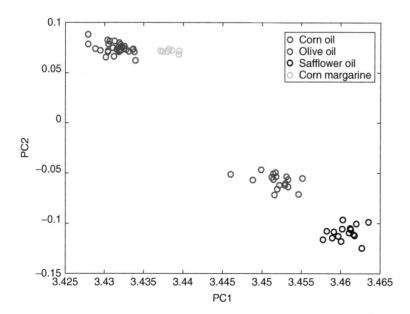

Figure 4.9 Plot of scores of PC2 versus PC1 for case study 2.

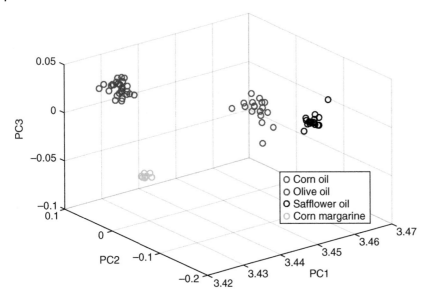

Figure 4.10 3D plot for the scores of case study 2.

The scores plot of the first two components against each other is presented in Figure 4.12 and can be interpreted in some detail as follows.

- The linear regions of the graph represent regions of the chromatogram where there are pure compounds.
- The curved portion represents a region of co-elution.
- The closer to the origin, the lower the intensity.

Hence, the PC plot suggests that the region between elution times 6 and 10 s (approximately) is one of co-elution. The reason why this method works is that the spectrum over the chromatogram changes with elution time. During co-elution, the spectral appearance changes most, and PCA uses this information.

How can these graphs help?

- The pure regions can inform us about the spectra of the pure compounds.
- The shape of the PC plot informs us of the amount of overlap and quality of chromatography.
- The number of bends in a PC plot can provide information about the number of different compounds in a complex multi-peak cluster.

In cases where there is a meaningful sequential order to a data set, as in spectroscopy or chromatography, but also, for example, where objects are related in time or pH such as in titrations or kinetics, it is also possible to plot the scores against sample number, see Figure 4.13. From this, it appears that the first PC primarily relates to the magnitude of the measurements, whereas the second discriminates between the two components in the mixture, being positive for the fastest eluting component and negative for the slowest component. Note that the appearance and interpretation of such plots depend crucially on data scaling, as will be discussed in Section 4.6. This will also be described in more detail in Section 7.2 in the context of evolutionary signals.

Case study 3 does not involve any sequential relationship between either the objects (chemical elements) or the variables (properties of these elements). In order to visualise the data correctly, it is first necessary to standardise the columns, which we will discuss in more detail in Section 4.6.4. The resultant plot of PC1 versus PC2 is presented in Figure 4.14.

Scores plots can be used to answer many different questions about the relationship between objects (or samples) and more examples are given in the problems at the end of this chapter. For case study 2, we are mainly interested in grouping of samples, whereas for case study 1, our prime interest is to find out what is happening during the chromatographic elution process. In many cases, scores plots are used for EDA and can answer numerous questions, for example, are there outliers, or are groups homogeneous or are there unsuspected subgroups. The largest components are not always the most discriminating, and it is sometimes useful to look at later components.

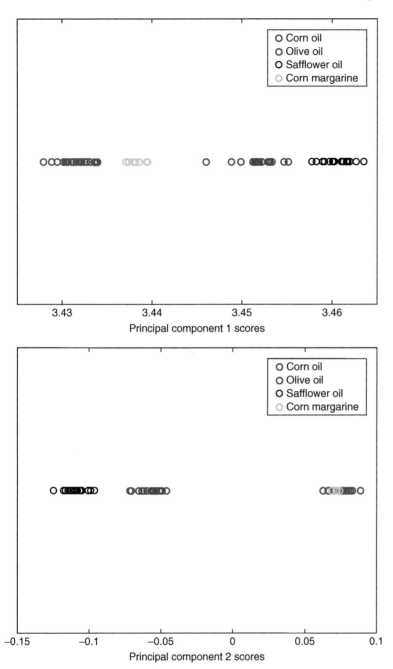

Figure 4.11 1D plot of the scores of PCs 1 and 2 for case study 1.

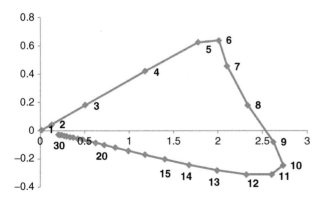

Figure 4.12 Scores of PC2 (vertical axis) versus PC1 (horizontal axis) for case study 1.

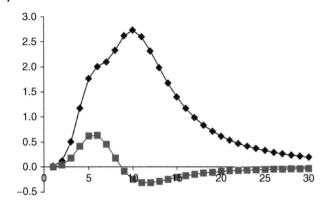

Figure 4.13 Scores of the first two PCs of case study 1 versus sample number.

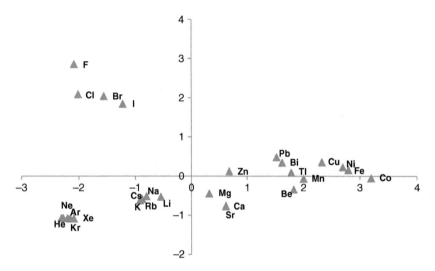

Figure 4.14 Scores of principal component 2 (vertical axis) versus principal component 1 (horizontal axis) for the standardised data of case study 3.

1. Four PCs are calculated. How many 2D dimensional graphs of the scores of one PC against another can be obtained?

   (a) 3
   (b) 4
   (c) 6

2. A chemical process is studying by spectroscopy as a function of pH. A graph of the scores of the first few components as a function of pH is likely to provide us with insights into the change in chemical species with pH.

   (a) True
   (b) False

4.5.2 Loadings Plots

It is not, however, only the scores that are of interest but also the loadings. Exactly, the same principles apply; for example, the value of the loadings at one PC can be plotted against that at the other PC.

The result for the first two PCs for case study 1 is shown in Figure 4.15. This figure looks quite complicated because both spectra overlap and absorb at similar wavelengths. The pure spectra are presented in Figure 4.16. Now we can understand a little more about these graphs. From this, we can see that the top right-hand corner of the scores plot corresponds to a direction for the fastest eluting compound (=**A**), whereas the bottom right-hand corner to the slowest eluting compound (=**B**). Similar interpretation can be obtained from the loadings plots. Wavelengths in the bottom

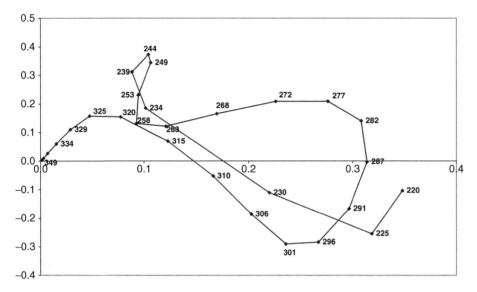

Figure 4.15 Loadings plot of PC2 (vertical axis) against PC1 (horizontal axis) for case study 1, with wavelengths indicated in nanometre.

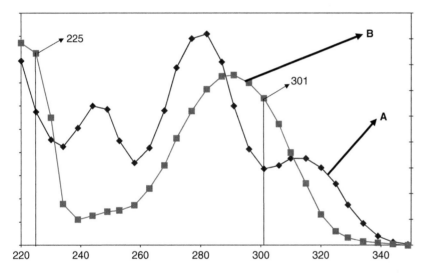

Figure 4.16 Pure spectra of compounds in case study 1.

half of the graph mainly correspond to **B**, for example, 301 and 225 nm. In Figure 4.16, these wavelengths are indicated and represent the maximum ratio of the spectral intensities for the **B** to **A**. In contrast, high wavelengths, above 325 nm, belong to **A** and are displayed in the top half of the graph. The characteristic peak for **A** at 244 nm is also obvious in the loadings plot.

Further interpretation is possible, but it can easily be seen that the loadings plots provide detailed information about which wavelengths are most associated with which compound. For complex multi-component clusters or spectral of mixtures, this information can be very valuable, especially if the pure components are not available.

The plot of the loadings of PC2 versus PC1 for case study 2 is given in Figure 4.17. The graph is not all that clear, but the parts of the spectrum represented in the top left-hand corner, for example, are negative for PC1 but positive for PC2. If we compare to the scores plot of Figure 4.9, we see that corn margarine and olive oil occupy this corner of the graph and hence these are likely to be diagnostic wavelengths in the loadings plots also. The trends can be better presented by a graph of loadings against wavelength, as shown in Figure 4.18. Loadings quite near 0 have little diagnostic influence, whereas loadings that are intense (either negative or positive) suggest wavelengths that are quite diagnostic. A lengthy interpretation is left to the reader, but, for example, we see that features around 700 nm appear quite important in PC2. Looking back at Figure 4.4, we see this region of the spectrum has a positive variation for corn oil and safflower oil,

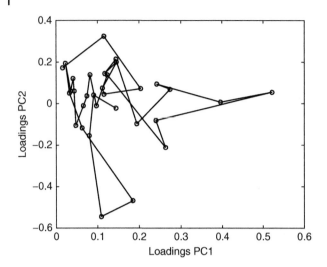

Figure 4.17 Loadings of PC2 versus PC1 for case study 2.

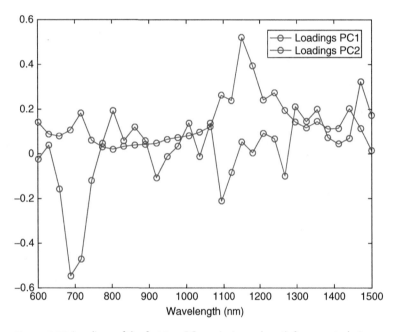

Figure 4.18 Loadings of the first two PCs against wavelength for case study 2.

which in turn have negative scores in PC2. Of course, by very close inspection of the original NIR spectra, we might be able to detect these trends, but PCA magnifies these to make them much more obvious and easier to visualise.

For case study 3, we can see the loadings plot in Figure 4.19 for the first two PCs. Note that later PCs could be visualised, but they provide us with limited additional information. We can see that, for example, melting point and boiling point provide very similar information. Electronegativity, however, provides quite different insights and is almost at right angles in the loadings plot to the other four properties. We can see that electronegativity has a high loading in the vertical axis (PC2), whereas the other properties mainly influence PC1 (the horizontal axis).

Further insight can be obtained by comparing the scores (Figure 4.14) to the loadings. The halides, for example, have a high score for PC2, which corresponds to electronegativity in the loadings plot, suggesting that the halides are characterised by high electronegativity. Elements with high PC1 scores (to the right of the plot) are more dense, as melting points and density have a corresponding high loading in PC1.

Loadings plots can be used to answer a lot of questions about the data and are a very flexible facility available in almost all chemometric software. They can be used for all sorts of problems, for example, to suggest biomarkers

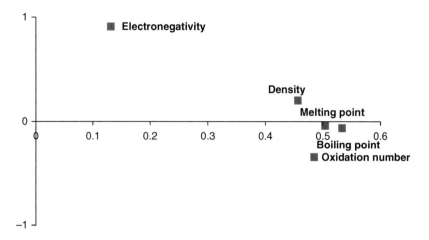

Figure 4.19 Loadings of principal component 2 versus principal component 1 for the standardised data of case study 3.

(which compounds have loadings corresponding to a specific group of samples), to look for diagnostic wavelengths or chromatographic peaks.

1. Loadings plots can be compared with scores plots to suggest diagnostic variables.
    (a) True
    (b) False

2. Three loadings are calculated for a data set. It is possible to obtain
    (a) One 3D loadings plot and two 2D loadings plots of different PCs against each other.
    (b) One 3D loadings plot and three 2D loadings plots of different PCs against each other.
    (c) One 3D loadings plot and one 2D loadings plot of different PCs against each other.
    (d) Three 3D loadings plots and two 2D loadings plots of different PCs against each other.

4.6 Pre-processing

All chemometric methods are influenced by the method for data pre-processing, or preparing information before application of mathematical algorithms. An understanding is essential for correct interpretation from multivariate data packages, but will be illustrated with reference to PCA, and is one of the first steps in data preparation. It is often called *scaling* or *data transformation* and the most appropriate choice can relate to the chemical or physical aim of the analysis. Pre-processing is normally performed before PCA, but in this chapter, it is introduced afterwards as it is hard to understand how it influences the resultant models without first appreciating the main concepts of PCA.

4.6.1 Transforming Individual Elements of a Matrix

One of the first considerations is whether individual elements of a matrix need transformation. The measurement units may mask the trends we are looking for. As an example, we often measure acidity or alkalinity in units of pH rather than $[H^+]$; hence, it is sometimes useful to transform raw measurements For example, we may be interested in the distribution of biomarkers using chromatographic peak intensities. Sometimes, minor metabolites are of low concentrations, but their relative variation may be as interesting as the major ones.

Logarithmic scaling is a common means of performing this transformation, where x_{ij} is replaced by $\log(x_{ij})$ either to the base 10 or as natural logs. The difficulty here is where some readings are recorded by zero: this is quite common if measurements are below the limit of detection, a logarithm of zero is undefined. Under such circumstances, these zero values are usually replaced by a small positive number, for example, half the lowest detected value for the variable in question, or a small baseline is added to every number. If there are too many zeroes in a data set, logarithmic scaling

Table 4.11 Example for logarithmic scaling; the first five samples belong to one group and the last five to a separate group.

(a) Raw data

1.321	4.203	2.150	28.687
3.880	4.825	3.593	28.595
2.314	5.607	3.134	33.927
4.553	5.516	4.136	24.576
3.871	4.762	3.593	38.639
5.553	4.495	4.885	35.625
6.109	4.727	5.289	30.656
5.764	4.717	5.163	25.710
5.839	7.142	5.896	35.666
5.804	6.010	5.285	31.648

(b) Logarithmically scaled data

0.121	0.624	0.332	1.458
0.589	0.684	0.555	1.456
0.364	0.749	0.496	1.531
0.658	0.742	0.617	1.391
0.588	0.678	0.556	1.587
0.744	0.653	0.689	1.552
0.786	0.675	0.723	1.487
0.761	0.674	0.713	1.410
0.766	0.854	0.771	1.552
0.764	0.779	0.723	1.500

is impracticable, but an alternative is to use power scaling, for example, replacing x_{ij} by its square root (other powers can be used), providing none of the numbers are negative.

As an example, see the data in Table 4.11, consisting of 10 samples characterised by four variables. In Figure 4.20, the scores of PC2 versus PC1 are presented both for the raw and the log scaled data. The first five samples belong to a different group to the last five and are coloured blue and red, respectively. We can see that for the raw data, there is no real discrete difference; in fact, three samples from the first (blue) group seem quite distinct from the other seven samples; hence, without hindsight, we may try to divide the data into a group of seven and another of three. However, when the data are log scaled, there appears a much more obvious distinction, with the five red samples forming a compact group in the top right and five blue samples an elongated group. Both groups can, in fact, be completely distinguished by their PC2 scores. In fact, a problem with the raw data is that variable 4 is of much higher intensity to the other three variables but does not provide much discriminatory power. Once logarithms are calculated, each variable has a similar influence and the discrimination between two groups is easier to see. An analogy is in metabolomics profiling, where variable 4 may represent an intense background peak such as a metabolite that dominates the analysis but one with no difference between two groups.

1. If a few readings are below detection limits, is it possible to perform logarithmic scaling on a data set.

    (a) No, because it is not possible to calculate logarithms of zero.
    (b) Yes, if not too many are zero, by replacing by a relatively small number.

2. Square root scaling is a useful alternative to log scaling.

    (a) Where there are many missing readings.
    (b) Where there are a significant number of zero readings.
    (c) Where there are negative readings.

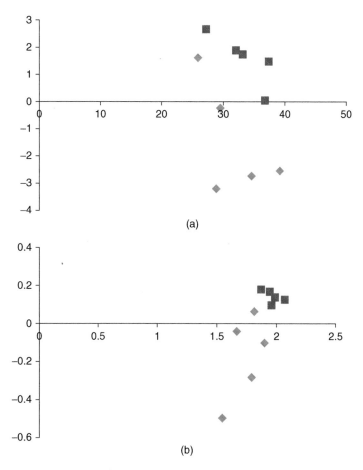

Figure 4.20 Scores of the first two PCs of the data in Table 4.11, (a) raw data (b) log scaled data.

4.6.2 Row Scaling

Another common problem is when the amount of a sample is hard to control. For example, it may be hard to obtain an identical amount of an extract. Under such circumstances, we are primarily interested in proportions rather than absolute amounts. We may be interested in a rock sample and the proportion of different elements found, but each sample may be of very different size. This problem is sometimes called one of closure and in fact can introduce some quite specific problems into the data.

However, it is usual under these circumstances to scale the rows (i.e. all the measurements on each sample) to a specific total. The simplest is to scale each row to a constant total so that x_{ij} is replaced by $x_{ij}\big/\sum_{j=1}^{J}x_{ij}$. The sum over all measured variables then becomes 1. There are other variants; for example, if there is a standard (internal or external), all peaks or concentrations are referenced to this standard, which may, for example, be an added compound of known concentration. Row scaling usually assumes that all values are positive. If the data have been first transformed using logarithms, great care should be taken, and if such a transformation is felt necessary, alternatives such as replacing x_{ij} by $\sqrt{x_{ij}}$ is normally preferred.

Row scaling to a constant total is sometimes also called *normalisation*, but this term can have many different meanings, so we avoid it. Row scaling can make a substantial difference to the resultant analysis. It is usually only sensible if all variables are of approximately the same intensity. If, for example, one variable was to correspond to the concentration of a compound that is found in high amounts (as variable 4 of the data in Table 4.11), scaling is dominated by this variable. There are various ways of overcoming this, for example, to log scale the data, or ignore some variables or even divide the data into groups. In spectroscopy where wavelengths are all measured on the same scale, row scaling to a constant total usually is quite robust.

An example is presented in Table 4.12 involving a 10×8 matrix (10 samples and eight variables). The rows are scaled to a constant total of 1. The graphs of the scores of PC2 versus PC1 are presented in Figure 4.21. Although the raw

Table 4.12 Example for row scaling.

(a) Raw data

8.47	11.20	14.67	18.57	20.51	17.98	12.03	6.18
0.34	0.43	0.53	0.65	0.73	0.66	0.48	0.27
10.41	12.35	14.59	17.43	19.52	18.68	14.48	8.84
2.60	3.00	3.44	4.04	4.54	4.47	3.60	2.28
3.79	4.25	4.74	5.45	6.16	6.23	5.22	3.42
5.15	5.53	5.84	6.45	7.37	7.86	7.04	4.87
8.88	9.26	9.37	10.04	11.56	12.87	12.09	8.65
14.86	14.79	14.00	14.16	16.56	19.87	20.07	15.08
13.02	12.72	11.70	11.51	13.56	16.83	17.53	13.42
7.00	6.71	5.97	5.69	6.77	8.74	9.40	7.33

(b) Row scaled data

0.077	0.102	0.134	0.169	0.187	0.164	0.110	0.056
0.084	0.104	0.129	0.159	0.177	0.162	0.117	0.067
0.090	0.106	0.125	0.150	0.168	0.161	0.125	0.076
0.093	0.107	0.123	0.144	0.162	0.160	0.129	0.082
0.096	0.108	0.121	0.139	0.157	0.159	0.133	0.087
0.103	0.110	0.116	0.129	0.147	0.157	0.140	0.097
0.107	0.112	0.113	0.121	0.140	0.156	0.146	0.105
0.115	0.114	0.108	0.109	0.128	0.154	0.155	0.117
0.118	0.115	0.106	0.104	0.123	0.153	0.159	0.122
0.121	0.116	0.104	0.099	0.117	0.152	0.163	0.127

data does suggest, when the 10 samples are connected in sequence, that there is some sort of trend, once the data are normalised, this is very clear. In fact, the data were simulated by multiplying a 10×2 *C* matrix by a 2×8 *S* matrix, each row of which was multiplied by another randomly generated scalar v_i, as illustrated in Table 4.13. Row scaling in this example dramatically reveals the underlying trends. Of course, this does not always happen, there may not necessarily be a sequential meaning to the data, for example; however, it is appropriate in certain circumstances.

Row scaling is normally performed after transformation, as discussed in the previous section, but before any methods for column scaling. Note that it is rarely meaningful to row scale where there are negative values.

1. Row scaling is the same as standardisation.
   (a) True
   (b) False

4.6.3 Mean Centring

It is, also, possible to mean centre the columns by subtracting the mean of each column (or variable) so that to transform x_{ij} to $x_{ij} - \bar{x}_j$.

In most traditional statistics, this is always done before PCA because the majority of statistical methods involve looking at variation around a mean. However, in some areas such as signal analysis, the key is to look at variation above a baseline (e.g. in chromatography and spectroscopy), so this default is not always necessary. We will illustrate this by the data of Table 4.14.

The scores plot of PC2 versus PC1 for the data in Table 4.14 before column centring is shown in Figure 4.22 and after on Figure 4.23. Note that the latter plot is not a simple transformation of Figure 4.22 about the origin. In fact, the process of centring is quite complicated, but in this case, the first PC approximates to the average intensity and by removing this we are looking at the variation around this mean. Hence, PC1 of the mean centred data is quite similar

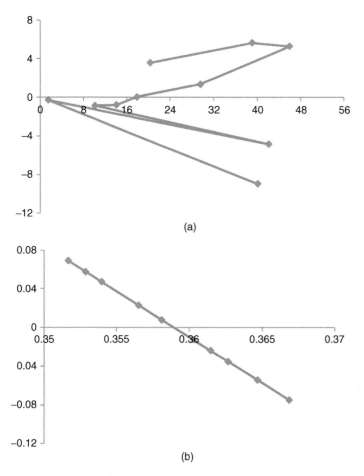

(a)

(b)

Figure 4.21 Scores of the first two PCs of the data in Table 4.12 (a) raw data (b) row scaled data to constant total.

PC scores plot of PC2 versus PC1 for raw data of Table 4.12

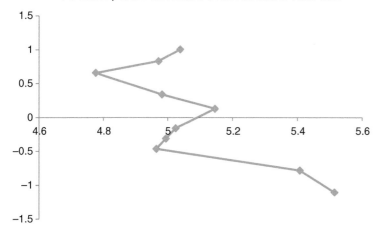

Figure 4.22 PC scores plot of PC2 versus PC1 for raw data of Table 4.12.

to PC2 of the uncentred data, as illustrated in Figure 4.23. This relationship is not always exact. It is important to understand that although for the example in this section it may well be desirable to remove the mean and so to centre the data matrix, in other cases, especially in chromatography and spectroscopy where variation above the baseline is of interest, this is not always appropriate. In Figure 4.24, we illustrate the plot of the scores of the first two PCs for case study 2, after centring, and compare to Figure 4.11 and see that there is very little difference, the scores of PC1 for both the centred and the uncentred data are related, so centring in this case primarily has the effect of shifting the origin.

Table 4.13 How the data in Table 4.12 were simulated as discussed in the text.

C matrix

0.22	1.63
0.5	1.39
0.7	1.1
0.85	1
1	0.9
1.3	0.72
1.46	0.55
1.71	0.3
1.92	0.21
2.1	0.1

S matrix

0.801	0.755	0.654	0.604	0.725	0.970	1.073	0.849
0.604	0.840	1.145	1.480	1.627	1.381	0.867	0.405

v

7.297
0.276
8.500
2.025
2.816
3.486
5.918
9.586
7.822
4.018

PC scores plot of PC2 versus PC1 for data after centring of Table 4.12

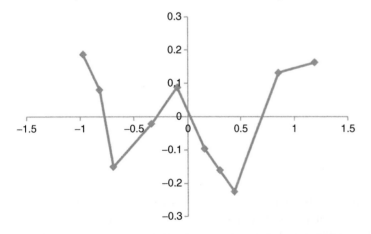

Figure 4.23 PC scores plot of PC2 versus PC1 for data after centring of Table 4.12.

Table 4.14 Example for Mean Centring.

1.161	1.535	2.010	2.545	2.811	2.464	1.648	0.847
1.240	1.545	1.918	2.359	2.624	2.405	1.741	0.987
1.225	1.453	1.717	2.051	2.297	2.198	1.704	1.040
1.285	1.482	1.700	1.993	2.243	2.205	1.778	1.127
1.344	1.511	1.684	1.936	2.189	2.213	1.852	1.214
1.476	1.587	1.674	1.851	2.113	2.255	2.018	1.396
1.501	1.565	1.584	1.696	1.953	2.176	2.043	1.463
1.551	1.543	1.461	1.477	1.727	2.073	2.094	1.574
1.664	1.626	1.495	1.471	1.733	2.152	2.241	1.716
1.742	1.670	1.487	1.417	1.684	2.175	2.339	1.824

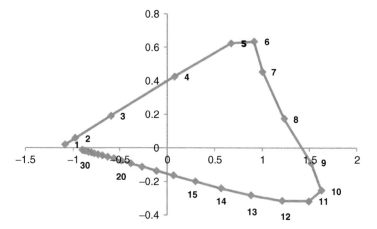

Figure 4.24 Scores plot of PC2 versus PC1 for case study 1 after centring.

Mean centring often has a significant influence on the relative size of the first eigenvalue, which may be reduced dramatically in relative size, and can influence the apparent number of significant components in a data set.

1. Is the prime effect of column mean centring to shift the origin of a PC scores plot?
    (a) Yes always.
    (b) It depends on the case study.
    (c) No.

4.6.4 Standardisation

Standardisation is another common method for data scaling, sometimes called *auto-scaling*, and occurs after mean centring: each variable is also divided by its standard deviation to transform x_{ij} to

$$\frac{x_{ij} - \overline{x}_j}{\sqrt{\sum_{i=1}^{I} (x_{ij} - \overline{x}_j)^2 / I}}$$

The population standard deviation is used rather than the sample standard deviation (see Section A.3.1.2), as this is used as a method of mathematical transformation rather than statistical estimation. Often, variables are on quite different scales, consider case study 3. If they are not standardised, numerically large variables will have a very large influence on the resultant PCs.

Table 4.15 Standardising the data of Table 4.11.

Mean	4.501	5.201	4.313	31.373
Standard deviation	1.564	0.835	1.122	4.355
	Standardised data			
	−2.033	−1.195	−1.927	−0.617
	−0.397	−0.449	−0.642	−0.638
	−1.399	0.487	−1.050	0.587
	0.034	0.378	−0.157	−1.561
	−0.403	−0.525	−0.641	1.669
	0.673	−0.845	0.510	0.977
	1.029	−0.567	0.870	−0.165
	0.808	−0.579	0.758	−1.300
	0.856	2.325	1.411	0.986
	0.833	0.970	0.867	0.063

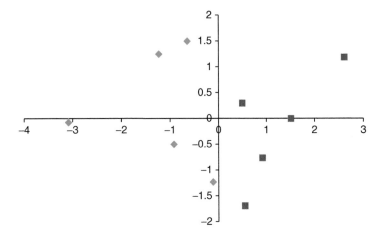

Figure 4.25 Plot of the scores of the first two PCs of the standardised data in Table 4.15.

As an example, we standardise the data in Table 4.11 as presented numerically in Table 4.15. Note an interesting numerical consequence that the sum of squares of each column equals I or the number of objects or 10 in this case and the total sum of squares of all variables equal IJ or 10×4 or 40 in this case. In the example, variable 4 dominates the analysis of the raw data and has very limited discriminatory power; it probably adds some information but is not so useful. Once PCA is performed, each variable has equal influence. We can see the result in Figure 4.25. Although for 10 objects it is not so obvious if they are already distinguished in hindsight, we can nevertheless see that there is now perfect discrimination using the scores of PC1.

Standardisation can be important even if variables are in the same units. Consider, for example, a case where the concentrations of 30 metabolites are monitored in a set of organisms. Some metabolites might be abundant in all samples, but their variation is not very significant. The change in concentration of the minor compounds might have a significant relationship with the underlying biology. If standardisation is not performed, PCA will be dominated by the most intense compounds.

In some cases, standardisation (or closely related scaling) is an essential first step in data analysis. In case study 3, the parameters are on very different scales and it is not meaningful to perform PCA unless they are all of comparable significance. Sometimes, the difference in numerical values for each variable is simply a consequence of how they are recorded, for example, do we record acidity as pH or $[H^+]$ or do we record a weight in kg or g?

Standardisation can also influence the appearance of loadings plots.

Standardisation should usually be done after all other common methods for data transformation, especially row scaling. Standardisation introduces negative numbers into a data set, whereas row scaling usually assumes that numbers are positive.

1. Standardised data are always mean centred.
    (a) True
    (b) False

2. A data set consists of 30 samples and 12 variables. The sum of squares of the standardised data set is
    (a) 30
    (b) 12
    (c) 360
    (d) 330

4.6.5 Further Methods

There is a very large battery of methods for data pre-processing, although the ones described above are the most common.

- It is possible to combine approaches, for example, first to scale the rows to a constant total and then standardise a data set.
- Weighting of each variable according to any external criterion of importance is sometimes employed. When we standardise a data set, we usually assume that each variable has equal importance.
- Block scaling can be used down the columns analogous to standardisation. If, for example, we record 20 HPLC peaks but 100 NIR wavelengths, we could scale each block so that the total sum of squares is 1 or a defined number, rather than IJ. This means that the NIR information does not overwhelm the HPLC information. This also, of course, means that each individual NIR measurement is 1/5 (=20/100) the significance of each HPLC measurement.
- The Box Cox transformation is sometimes used as an alternative to power and logarithmic transformations of individual elements. The aim is to transform each variable into a distribution resembling the normal distribution. For certain statistical tests, the assumption of normality is important.

Undoubtedly, however, the appearance and interpretation not only of PC plots but of almost all chemometric techniques depend on data pre-processing. The influence of pre-processing can be quite dramatic; hence, it is essential for the user of chemometric software to understand and question how and why the data have been scaled or transformed before interpreting the result from a package. More consequences are described in Chapter 7.

1. A data set consists of 50 samples, recorded at 80 UV wavelengths and 120 NIR wavelengths. It is desired to make the NIR and UV information blocks of equal significance.
    (a) Each UV measurement is weighted as being of 1.5 the importance of each NIR measurement.
    (b) Each UV measurement is weighted as being 2/3 the importance of each NIR measurement.
    (c) The individual measurements are standardised for each of the variables.

4.7 Comparing Multivariate Patterns

PC plots are often introduced only by reference to the independent loadings or scores plot of a single data set. Yet, there are common patterns within these different graphs. Consider taking measurements of the concentration of a mineral in a geochemical deposit. This information could be presented as a table of sampling sites and observed concentrations. However, a much more informative approach would be to produce a picture in which physical location and mineral concentration are superimposed, such as a coloured map, each different colour corresponding to a concentration range of the mineral. Two pieces of information are connected, namely geography and concentration. Hence, in many applications of multivariate analysis, one aim may be to connect the samples (e.g. geographical location/sampling site) represented by scores to the variables (e.g. chemical measurements) represented by loadings. Graphically, this requires the superimposition of two types of information.

Another common need is to compare two independent types of measurements. Consider recording the result of a taste panel for a type of food. Their scores relate to the underlying chemical or manufacturing process. A separate measurement could be chemical, such as a chromatographic or spectroscopic profile. Ideally, the chemical measurements will relate to the taste, can each type of measurement give similar information, and so, can we predict the taste by using analytical chemical techniques?

4.7.1 Biplots

A biplot involves superimposition of a scores and a loadings plot. In order to superimpose each plot on a sensible scale, one approach is to divide the scores to provide scaled scores, which are then superimposed on the loadings as follows:

$$\frac{t_{ia}}{\sum_{i=1}^{I} t_{ia}^2 / I}$$

Notice that if the scores are mean centred, the denominator equals the variance. Some authors use the expression in the denominator of this equation to represent an eigenvalue; hence, in certain articles, it is stated that the scores of each PC are divided by their eigenvalue. As is usual in chemometrics, it is important to recognise that there are many different schools of thought and incompatible definitions.

We will illustrate this by case study 3, as shown in Figure 4.26. At first, the results do not seem clear-cut. For example, electronegativity is at the top but overlapping with the most electronegative elements (the halides). However, this is because the property relates to all the elements. Those at the bottom of the scores plot have low electronegativity and those at the top of the scores plot have high electronegativity. Likewise, other properties such as MP and BP are associated with elements in the right-hand side of the figure, suggesting that these properties are most associated with PC1, a positive value implying high MP and so on, and a negative value low MP. Note that the sign of a PC can differ according to calculation method, but if inverted, both loadings and scores will change in sign, resulting in reflection around the axes.

It is not necessary to restrict biplots to two PCs, but, of course, when more than three are used, graphical representation becomes difficult, and numerical measures of fit between scores and loadings are often employed, using statistical software.

1. A biplot is a plot of one PC against another.
   (a) True
   (b) False

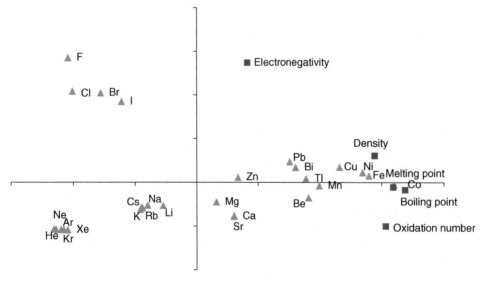

Figure 4.26 Biplot of scores of the first two PCs of case study 3.

4.7.2 Procrustes Analysis

Another important facility is to be able to compare different types of measurements. For example, we may want to determine whether MS and NMR measurements provide similar information. A statistical method called *procrustes analysis* will help us here.

Procrustes was a Greek god who kept a house by the side of the road where he offered hospitality to passing strangers, who were invited for a meal and a night's rest in his very special bed which Procrustes described as having the unique property that its length exactly matched whosoever lay down upon it. What he did not say was the method by which this 'one-size-fits-all' was achieved: as soon as the guest lay down, Procrustes went to work upon them, stretching them if they were too short for the bed or chopping off their legs if they were too long.

Similarly, procrustes analysis in statistics involves comparing two spatial representations of data, such as two PC scores plots. One such plot is the reference and a second plot is manipulated to resemble the reference plot as closely as possible. This manipulation is done mathematically, involving up to four main transformations.

- *Reflection.* This transformation is a consequence of the inability to control the sign of a principal component.
- *Rotation.*
- *Scaling (or stretching).* This transformation is because the scales of the two types of measurements may be very different.
- *Translation.*

If two data sets are already standardised, transformation 3 may not be necessary, and the fourth transformation is not often used.

The aim is to reduce the root mean square difference between the scores of the reference data set and the transformed data set

$$r = \sqrt{\sum_{i=1}^{I} \sum_{a=1}^{A} ({}^{ref}t_{ia} - {}^{trans}t_{ia})^2 / I}$$

The resultant view gives a consensus of the techniques, but it can also answer how similar two different techniques are.

It is not necessary to restrict each measurement technique to two PCs; indeed, in many practical cases, four or five PCs are employed. Computer software is available to compare scores plots and provide a numeric indicator of the closeness of the fit, but it is not easy to visualise. As PCs do not often have a physical meaning, it is important to recognise that, in some cases, it is necessary to include several PCs for a meaningful result. For example, if two data sets are characterised by four PCs, and each one is of approximately equal size, the first PC for the reference data set may correlate most closely with the third for the comparison data set, thus including only the first two components in the model could result in very misleading conclusions. It is usually a mistake to compare PCs of equivalent significance with each other, especially when their size is fairly similar.

Procrustes analysis can be used to answer quite sophisticated questions. For example, in sensory research, are the results of a taste panel comparable with chemical measurements? If so, can the rather expensive and time-consuming taste panel be replaced by chromatography? A second use of procrustes analysis is to reduce the number of tests: an example being of clinical tests. Sometimes, 50 or more bacteriological tests are performed but can these be reduced to 10 or less? A way to check this is by performing PCA on the results of all 50 tests and compare the scores plot when using a subset of 10 tests. If the two scores plots provide comparable information, the 10 selected tests are just as good as the full set of tests. This can be of significant economic benefit.

1. It is desired to reduce the number of tests used to determine the quality of a fuel. Procrustes analysis can be used for this purpose.
   - (a) True
   - (b) False

4.8 Unsupervised Pattern Recognition: Cluster Analysis

EDA such as PCA is primarily used to determine general relationships between data. Sometimes, more complex questions need to be answered, such as do the samples fall into groups? Cluster analysis is a well-established approach that

Table 4.16 Example for cluster analysis.

	A	B	C	D	E
1	0.9	0.5	0.2	1.6	1.5
2	0.3	0.2	0.6	0.7	0.1
3	0.7	0.2	0.1	0.9	0.1
4	0.1	0.4	1.1	1.3	0.2
5	1.0	0.7	2.0	2.2	0.4
6	0.3	0.1	0.3	0.5	0.1

was primarily developed by biologists to determine similarities between organisms. Numerical taxonomy emerged from a desire to determine relationships between different species, for example, genera, families and phyla. Many textbooks in biology show how organisms are related using family trees.

The chemist also wishes to relate samples in a similar manner. Can protein sequences from different animals be related and does this tell us about the molecular basis of evolution? Can the chemical fingerprint of wines be related and does this tell us about the origins and taste of a particular wine? Unsupervised pattern recognition employs a number of methods, primarily cluster analysis, to group different samples (or objects) using chemical measurements.

4.8.1 Similarity

The first step is to determine the similarity between objects. Table 4.16 consists of six objects (1–6) and five measurements (A–E). What are the similarities between the objects? Each object has a relationship with the remaining five objects. How can a numerical value of similarity be defined? A similarity matrix can be obtained, in which the similarity between each pair of objects is calculated using a numerical indicator. Notice that it is possible to pre-process data before calculation of a number of these measures (see Section 4.6).

Four of the most popular ways of determining how similar objects are to each other are as follows.

- *Correlation coefficient between samples.* A correlation coefficient of 1 implies that samples have identical characteristics, which all objects have with themselves. Some workers use the square or absolute value of a correlation coefficient, and it depends on the precise physical interpretation as to whether negative correlation coefficients imply similarity or dissimilarity. In this text, we assume that the more negative the correlation coefficient, the less similar the objects; sometimes the square of the correlation coefficient is used instead. The correlation matrix is presented in Table 4.17. Notice that the top right is not presented, as it is the same as the bottom left. The higher the correlation coefficient, the more similar the objects.
- *Euclidean distance.* The distance between the two samples k and l is defined by

$$d_{kl} = \sqrt{\sum_{j=1}^{J} \left(x_{kj} - x_{lj}\right)^2}$$

where there are j measurements, and x_{ij} is the jth measurement on sample i; for example, x_{23} is the third measurement on the second sample, equalling 0.6 in Table 4.16. The smaller this value, the more similar the samples; thus, this

Table 4.17 Correlation matrix.

	1	2	3	4	5	6
1	1.000					
2	−0.041	1.000				
3	0.503	0.490	1.000			
4	−0.018	0.925	0.257	1.000		
5	−0.078	0.999	0.452	0.927	1.000	
6	0.264	0.900	0.799	0.724	0.883	1.000

Table 4.18 Euclidean distance matrix.

	1	2	3	4	5	6
1	0.000					
2	1.838	0.000				
3	1.609	0.671	0.000			
4	1.800	0.837	1.253	0.000		
5	2.205	2.245	2.394	1.600	0.000	
6	1.924	0.374	0.608	1.192	2.592	0.000

distance measure works in an opposite manner to the correlation coefficient and, strictly speaking, is a dissimilarity measure. The results are presented in Table 4.18. Although correlation coefficients vary between -1 and $+1$, this is not true for the Euclidean distance, which has no limit, although it is always a positive number. Sometimes, the equation is presented in the matrix format

$$d_{kl}^2 = (\boldsymbol{x}_k - \boldsymbol{x}_l)(\boldsymbol{x}_k - \boldsymbol{x}_l)'$$

where the objects are row vectors, as given in Table 4.16; this method is easy to implement in Excel or Matlab.

- *Manhattan distance.* This is defined slightly differently to the Euclidean distance and is given by

$$d_{kl} = \sum_{j=1}^{J} |x_{kj} - x_{lj}|$$

The difference between the Euclidean and the Manhattan distance is illustrated in Figure 4.27. The values are given in Table 4.19; notice that the Manhattan distance will always be greater than (or in exceptional cases equal to) the Euclidean distance.

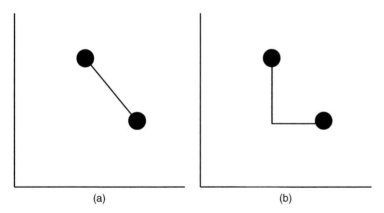

| (a) | (b) |

Figure 4.27 (a) Euclidean and (b) Manhattan distances.

Table 4.19 Manhattan distance matrix.

	1	2	3	4	5	6
1	0					
2	3.6	0				
3	2.7	1.1	0			
4	3.4	1.6	2.3	0		
5	3.8	4.4	4.3	3.2	0	
6	3.6	0.6	1.1	2.2	5.0	0

● *Mahalanobis distance*. This method is popular with many chemometricians, and, while superficially similar to the Euclidean distance, it takes into account that some variables may be correlated and so measure more or less the same properties. The distance between objects k and l is best defined in matrix terms by

$$d_{kl}^2 = (\boldsymbol{x}_k - \boldsymbol{x}_l)\boldsymbol{C}^{-1}(\boldsymbol{x}_k - \boldsymbol{x}_l)'$$

where $\boldsymbol{C}$ is the variance–covariance matrix of the variables, a matrix symmetric about the diagonal, whose elements represent the covariance between any two variables, of dimensions $J \times J$. See Section A.3.1 for definitions of these parameters; note that one should use the population rather than sample statistics. This measure has a similar computational formula to the Euclidean distance except that the inverse of the variance–covariance matrix is inserted as a scaling factor. However, the computational method discussed above cannot be applied where the number of variables exceeds the number of objects because the variance–covariance matrix would not have an inverse. There are some ways around this (e.g. when calculating spectral similarities where the number of wavelengths far exceeds the number of spectra), such as first performing PCA and then retaining the first few (or all non-zero) PCs or the most significant variables for subsequent analysis. In the case of Table 4.16, the Mahalanobis distance would not be a very useful measure unless either the number of samples is increased or the number of variables are decreased. This distance metric does have very important uses in chemometrics, but more commonly in the areas of supervised pattern recognition, as discussed in Chapter 5, where its properties will be described in more detail. Note in contrast that if the number of variables is very small, although the Mahalanobis distance is an appropriate measure, correlation coefficients are less useful.

There are several other related distance measures in the literature, but normally, quite good reasons are required if a very specialist measure is to be employed.

1. Two samples that are close to each other should have a relatively low distance but high correlation coefficient compared with others in a data set.
    (a) True
    (b) False

2. The Manhattan distance is always $\geq$ the Euclidean distance $\geq$ Mahalanobis distance.
    (a) True
    (b) False

4.8.2 Linkage

The next step is to link the objects. The most common approach is called *agglomerative clustering*, in which single objects are gradually connected to each other in groups. Any similarity measure can be used in the first step, but for simplicity, we will illustrate this using only the correlation coefficients presented in Table 4.17. Similar considerations apply to all the similarity measures introduced in Section 4.8.1, except that in the other cases the lower the distance, the more similar the objects, whereas a high correlation represents high similarity in our example.

● From the raw data, find the two objects that are most similar (closest together). According to Table 4.17, these are objects 2 and 5, as they have the highest correlation coefficient (=0.999) (remember that because only five measurements have been recorded, there are only 4 degrees of freedom for the calculation of correlation coefficients, which means quite high values can be obtained fairly easily).
● Next, form a 'group' consisting of these two most similar objects. Four of the original objects (1, 3 and 6) and a group consisting of objects 2 and 5 together remain, leaving a total of five new groups, four consisting of a single original object and one consisting of two 'clustered' objects.
● The tricky bit is to decide how to represent this new grouping. As in the case of distance measures, there are quite a few approaches. The main task is to recalculate the numerical similarity values between the new group and the remaining objects. There are three principal ways of doing this.
    − *Nearest neighbour*. The similarity of the new group from all the other groups is given by the *highest* similarity of either of the original objects to each other object. For example, object 6 has a correlation coefficient of 0.900 with object 2 and 0.883 with object 5. Hence, the correlation coefficient with the new combined group consisting of objects 2 and 5 is 0.900.

– *Farthest neighbour.* This is the opposite to nearest neighbour, and the *lowest* similarity is used, 0.883 in our case. Note that the farthest neighbour method of linkage refers only to the calculation of similarity measures after new groups are formed, and the two groups (or objects) with highest similarity are still always joined first.

– *Average linkage.* The average similarity is used, 0.892 in our case. There are, in fact, two different ways of doing this, according to the size of each group being joined together. Where they are of equal size (e.g. each consists of one object), both methods are equivalent. The two different ways are as follows.

Unweighted. If group A consists of N_A objects and group B of N_B objects, the new similarity measure is given by

$$s_{AB} = (N_A s_A + N_B s_B)/(N_A + N_B)$$

Weighted. The new similarity measure is given by

$$s_{AB} = (s_A + s_B)/2$$

The terminology indicates that for the unweighted method, the new similarity measure takes into consideration the number of objects in a group, the conventional terminology possibly being the opposite to what is expected. For the first link, each method provides identical results.

There are numerous other linkage methods, but it would be rare that a chemist needs to use too many combinations of similarity and linkage methods, but a good rule of thumb is to check the result of using a combination of approaches. However, when calculating a dendrogram using a package, always check what are the default steps and whether these are appropriate.

The new data matrix using nearest neighbour clustering is presented in Table 4.20, with the new values shaded. Remember that there are many similarity measures and methods for linking; hence, this table is only one possible way for handling the information.

Table 4.20 Nearest neighbour cluster analysis, using correlation coefficients for similarity measures, and data in Table 4.16.

	1	*2*	*3*	*4*	*5*	*6*
1	1.000					
2	−0.041	1.000				
3	0.503	0.490	1.000			
4	−0.018	0.925	0.257	1.000		
5	−0.078	**0.999**	0.452	0.927	1.000	
6	0.264	0.900	0.799	0.724	0.883	1.000
	1	*2&5*	*3*	*4*	*6*	
1	1.000					
2&5	−0.041	1.000				
3	0.503	0.490	1.000			
4	−0.018	**0.927**	0.257	1.000		
6	0.264	0.900	0.799	0.724	1.000	
	1	*2&5&4*	*3*	*6*		
1	1.000					
2&5&4	−0.018	1.000				
3	0.503	0.490	1.000			
6	0.264	**0.900**	0.799	1.000		
	1	*2&5&4&6*	*3*			
1	1.000					
2&5&4&6	0.264	1.000				
3	0.503	**0.799**	1.000			
	1	*2&5&4&6&3*				
1	1.000					
2&5&4&6&3	**0.503**	1.000				

1. Group A represents three objects and group B one object. The Euclidean distance of group A to group C is 4.53 and group B to group C is 7.81. What is the unweighted distance of the combined groups A and B to group C?

    (a) 6.17

    (b) 6.99

    (c) 5.35

    (d) Cannot say unless we know the number of objects in group C.

4.8.3 Next Steps

The next steps consist of continuing to group the data just as discussed above, until all objects have joined one large group. As there are six original objects, there will be five steps before achieving this. At each step, the most similar pair of objects or clusters is identified, and then they are combined into one new cluster, until all objects have been joined. The calculation is illustrated in Table 4.20, using nearest neighbour linkage, with the most similar objects at each step indicated in bold, and the new similarity measures shaded. In this particular example, all objects ultimately belong to the same cluster, although arguably object 1 (and possibly 3) does not have a very high similarity to the main group. In some cases, several clusters can be formed, although ultimately one large group is usually formed.

It is normal to then determine at what similarity measure each object joined a larger group, and so which objects resemble each other most.

1. How many linkage steps are there if clustering 20 objects?

    (a) 20

    (b) 19

    (c) Cannot say

4.8.4 Dendrograms

Often, the result of hierarchical clustering is presented in the form of a dendrogram (sometimes called a *tree diagram*). The objects are organised in a row, according to their similarities: the vertical axis represents the similarity measure at which each successive object joins a group. Using nearest neighbour linkage and correlation coefficients for similarities, the dendrogram in Table 4.20 is presented in Figure 4.28. It can be seen that object 1 is very different from the others. In this case, all the other objects appear to form a single group, but other clustering methods may give slightly different results. A good approach is to perform several different methods of cluster analysis and compare the results. If similar clusters are obtained, no matter which method is employed, we can rely on the results. Note that clusters can sometimes be reflected; for example, a cluster that is on the right side using one type of software may be on the left side using another but represent the same result, as there is no rule as to whether new branches form to the right or left.

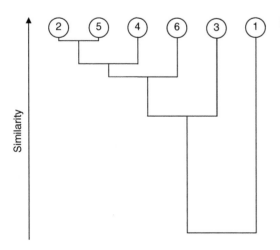

Figure 4.28 Dendrogram for cluster analysis example.

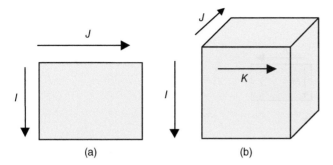

Figure 4.29 Two-way and three-way data.

1. A dendrogram will always have the same order of objects if the original data, linkage method and distance metric are the same.

    (a) True
    (b) False

4.9 Multi-way Pattern Recognition

Most traditional chemometrics is concerned with two-way data, often represented by matrices. Yet, over the past few years, a large interest in three-way chemical data has grown. Instead of organising the information as a two-dimensional array (Figure 4.29(a)), it falls into a three-dimensional 'tensor' or box (Figure 4.29(b)). Such data sets are surprisingly common. In Chapter 6, we will discuss multi-way partial least squares (PLS) (Section 6.5.3), the discussion in this section being restricted to EDA.

Consider, for example, an environmental chemical experiment in which the concentrations of six elements are measured at 20 sampling sites on 24 days in a year. There will be $20 \times 24 \times 6$ or 2880 measurements; however, these can be organised as a 'box' with 20 planes each corresponding to a sampling site and of dimensions 24×6 (Figure 4.30). Such data sets have been available for many years to psychologists and in sensory research. A typical example might involve a taste panel assessing 20 food products. Each food could involve the use of 10 judges who score eight attributes, resulting in a $20 \times 10 \times 8$ box. In psychology, we might be following the reactions of 15 individuals to five different tests on 10 different days, possibly each day under slightly different conditions, hence have a $15 \times 5 \times 10$ box. These problems involve finding the main factors that influence the taste of a food or the source of pollutant or the reactions of an individual and are a form of pattern recognition.

Three-dimensional analogies to principal components are required. There are no direct analogies to scores and loadings as in PCA; hence, the components in each of the three dimensions are often called *weights*. A number of methods are available to tackle this problem.

4.9.1 Tucker3 Models

These models involve calculating weight matrices corresponding to each of the three dimensions (e.g. sampling site, date and metal), together with a 'core' box or array, which provides a measure of magnitude. The three weight matrices

Figure 4.30 Possible method of arranging environmental sampling data.

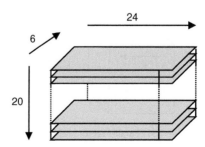

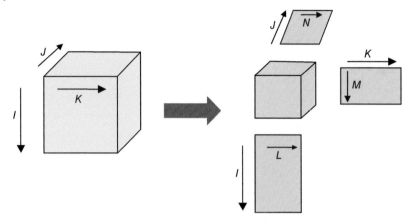

Figure 4.31 Tucker3 decomposition.

do not necessarily have the same dimensions; hence, for example, the number of significant components for the sampling sites may be different to those for the dates, unlike normal PCA where one of the dimensions of both the scores and loadings matrices must be identical. This model (or decomposition) is represented in Figure 4.31. The easiest mathematical approach is by expressing the model as a summation

$$x_{ijk} \approx \sum_{l=1}^{L} \sum_{m=1}^{M} \sum_{n=1}^{N} a_{il} b_{jm} c_{kn} z_{lmn}$$

where z represents what is often called a *core array* and a, b and c are functions relating to each of the three types of variables. Some authors use the concept of 'tensor multiplication' being a 3D analogy to 'matrix multiplication' in two dimensions; however, the details are confusing, and it is conceptually probably best to stick to summations, which computer programs do.

1. For a Tucker3 model, all dimensions of the core array must be equal.
    (a) True
    (b) False

4.9.2 Parallel Factor Analysis (PARAFAC)

Parallel factor analysis (PARAFAC) differs from the Tucker3 models in which each of the three dimensions contains the same number of components. Hence, the model can be represented as the sum of contributions due to g components, just as in normal PCA, as illustrated in Figure 4.32 and represented algebraically by

$$x_{ijk} \approx \sum_{g=1}^{G} a_{ig} b_{jg} c_{kg}$$

Each component can be characterised by one vector that is analogous to a scores vector and two vectors that are analogous to loadings, but some keep to the notation of 'weights' in three dimensions. Components can, in favourable

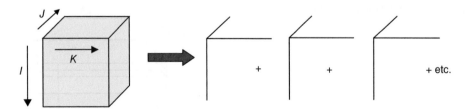

Figure 4.32 Parallel factor analysis (PARAFAC).

circumstances, be assigned a physical meaning. A simple example might involve following a reaction by recording a diode array HPLC chromatogram at different reaction times. A box whose dimensions are *reaction time × elution time × wavelength* can be used to represent the data. If there are four factors in the data, this would imply four significant compounds in a cluster in the HPLC (or four significant reactants), and the weights should correspond to the reaction profile, the chromatogram and the spectrum of each compound. Three-way methods in chemometrics had a special origin in fluorescence spectroscopy where two of the dimensions are emission and excitation wavelengths, which are quite stable spectroscopically. In chromatography, there may be difficulties aligning the chromatographic dimension, making PARAFAC somewhat less useful.

PARAFAC, however, is quite difficult to use and, although the results are easy to interpret physically, it is conceptually more complex than PCA. It can, however, lead to results that are directly interpretable, whereas the factors in PCA have a purely abstract meaning.

1.   The dimensions of the weights in a PARAFAC model must be the same.

    (a) True

    (b) False

4.9.3   Unfolding

Another approach is to simply 'unfold' the 'box' to give a long matrix. In the environmental chemistry example, instead of each sample being represented by a 24×6 matrix, it could be represented by a vector of length 144, each element consisting of the measurement of one element on one date, for example, the measurement of Cd concentration on July 15. Then, a matrix of dimensions 20 (sampling sites) $\times$ 144 (variables) is produced (Figure 4.33) and subjected to normal PCA. Note that a box can be subdivided into planes in three different ways (compare Figure 4.30 with Figure 4.33), according to which dimension is regarded as the 'major' dimension. While unfolding, it is also quite important to consider details of scaling and centring, which become far more complex in three dimensions as opposed to two. After unfolding, normal PCA can be performed. Components can be averaged over related variables; for example, we could take an average loading for Cd over all dates to give an overall picture of its influence on the observed data.

This comparatively simple approach is sometimes sufficient, but the PCA calculation neglects to take into account some relationships between the variables. For example, the relationship between concentration of Cd on July 15 and that on August 1, in an environmental analysis, is considered to be no stronger than the relationship between Cd concentration on July 15 and Hg on November 1 during the calculation of the components. However, after the calculations are performed, it is still possible to regroup the loadings and sometimes an easily understood method such as unfolded PCA can be of value.

1.   Forty mice are assayed daily over 2 weeks for the concentrations of 20 compounds. The data can be unfolded into a matrix of dimensions

    (a) 40×280

    (b) 40×20

    (c) $40 \times 14 \times 20$

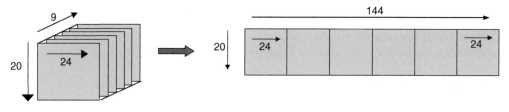

Figure 4.33 Unfolding.

Problems

4.1 Introductory PCA
Section 4.3.1   Section 4.3.2.1
The following is a data matrix consisting of seven samples and six variables:

2.7	4.3	5.7	2.3	4.6	1.4
2.6	3.7	7.6	9.1	7.4	1.8
4.3	8.1	4.2	5.7	8.4	2.4
2.5	3.5	6.5	5.4	5.6	1.5
4.0	6.2	5.4	3.7	7.4	3.2
3.1	5.3	6.3	8.4	8.9	2.4
3.2	5.0	6.3	5.3	7.8	1.7

The scores of the first two principal components on the centred data matrix are given as follows:

−4.0863	−1.6700
3.5206	−2.0486
−0.0119	3.7487
−0.7174	−2.3799
−1.8423	1.7281
3.1757	0.6012
−0.0384	0.0206

1. As $X \approx TP$, calculate the loadings for the first two PCs using the pseudo-inverse, remembering to centre the original data matrix first.
2. Demonstrate that the two scores vectors are orthogonal and the two loadings vectors are orthonormal. Remember that the answer will only to be within a certain degree of numerical accuracy.
3. Determine the eigenvalues and percentage variance of the first two principal components.

4.2 Introduction to Cluster Analysis
Section 4.8
The following data set consists of seven measurements (rows) on six objects A–F (columns):

A	B	C	D	E	F
0.9	0.3	0.7	0.5	1.0	0.3
0.5	0.2	0.2	0.4	0.7	0.1
0.2	0.6	0.1	1.1	2	0.3
1.6	0.7	0.9	1.3	2.2	0.5
1.5	0.1	0.1	0.2	0.4	0.1
0.4	0.9	0.7	1.8	3.7	0.4
1.5	0.3	0.3	0.6	1.1	0.2

1. Calculate the correlation matrix between the six objects.
2. Using the correlation matrix, perform cluster analysis using the furthest neighbour method. Illustrate each stage of linkage.
3. From the results in 2 draw a dendrogram and deduce which objects cluster closely into groups.

4.3 Certification of NIR Filters Using PC Scores Plots.
Section 4.3.1 Section 4.5.1 Section 4.3.2.1 Section 4.6.4

These data were obtained by the National Institute of Standards and Technology (US) while developing a transfer standard for verification and calibration of the x-axis of NIR spectrometers. Optical filters were prepared from two separate melts, 2035 and 2035a, of a rare earth glass. Filters from both melts provide seven well-suited adsorption bands of very similar but not quite identical location. One filter, Y, from one of the two melts was discovered to be unlabelled. Four 2035 filters and one 2035a filter were available at the time of this discovery. Six replicate spectra were taken from each filter. Band location data from these spectra are provided below, in cm^{-1}. The expected location uncertainties range from 0.03 to 0.3 cm^{-1}.

Type	#	P1	P2	P3	P4	P5	P6	P7
2035	18	5138.58	6804.70	7313.49	8178.65	8681.82	9293.94	10245.45
2035	18	5138.50	6804.81	7313.49	8178.71	8681.73	9293.93	10245.49
2035	18	5138.47	6804.87	7313.43	8178.82	8681.62	9293.82	10245.52
2035	18	5138.46	6804.88	7313.67	8178.80	8681.52	9293.89	10245.54
2035	18	5138.46	6804.96	7313.54	8178.82	8681.63	9293.79	10245.51
2035	18	5138.45	6804.95	7313.59	8178.82	8681.70	9293.89	10245.53
2035	101	5138.57	6804.77	7313.54	8178.69	8681.70	9293.90	10245.48
2035	101	5138.51	6804.82	7313.57	8178.75	8681.73	9293.88	10245.53
2035	101	5138.49	6804.91	7313.57	8178.82	8681.63	9293.80	10245.55
2035	101	5138.47	6804.88	7313.50	8178.84	8681.63	9293.78	10245.55
2035	101	5138.48	6804.97	7313.57	8178.80	8681.70	9293.79	10245.50
2035	101	5138.47	6804.99	7313.59	8178.84	8681.67	9293.82	10245.52
2035	102	5138.54	6804.77	7313.49	8178.69	8681.62	9293.88	10245.49
2035	102	5138.50	6804.89	7313.45	8178.78	8681.66	9293.82	10245.54
2035	102	5138.45	6804.95	7313.49	8178.77	8681.65	9293.69	10245.53
2035	102	5138.48	6804.96	7313.55	8178.81	8681.65	9293.80	10245.52
2035	102	5138.47	6805.00	7313.53	8178.83	8681.62	9293.80	10245.52
2035	102	5138.46	6804.97	7313.54	8178.83	8681.70	9293.81	10245.52
2035	103	5138.52	6804.73	7313.42	8178.75	8681.73	9293.93	10245.48
2035	103	5138.48	6804.90	7313.53	8178.78	8681.63	9293.84	10245.48
2035	103	5138.45	6804.93	7313.52	8178.73	8681.72	9293.83	10245.56
2035	103	5138.47	6804.96	7313.53	8178.78	8681.59	9293.79	10245.51
2035	103	5138.46	6804.94	7313.51	8178.81	8681.65	9293.77	10245.52
2035	103	5138.48	6804.98	7313.57	8178.82	8681.51	9293.80	10245.51
2035a	200	5139.26	6806.45	7314.93	8180.19	8682.57	9294.46	10245.62
2035a	200	5139.22	6806.47	7315.03	8180.26	8682.52	9294.35	10245.66
2035a	200	5139.21	6806.56	7314.92	8180.26	8682.61	9294.34	10245.68
2035a	200	5139.20	6806.56	7314.90	8180.23	8682.49	9294.31	10245.69
2035a	200	5139.19	6806.58	7314.95	8180.24	8682.64	9294.32	10245.67
2035a	200	5139.20	6806.50	7314.97	8180.21	8682.58	9294.27	10245.64
Y	201	5138.53	6804.82	7313.62	8178.78	8681.78	9293.77	10245.52
Y	201	5138.49	6804.87	7313.47	8178.75	8681.66	9293.74	10245.52
Y	201	5138.48	6805.00	7313.54	8178.85	8681.67	9293.75	10245.54
Y	201	5138.48	6804.97	7313.54	8178.82	8681.70	9293.79	10245.53
Y	201	5138.47	6804.96	7313.51	8178.77	8681.52	9293.85	10245.54
Y	201	5138.48	6804.97	7313.49	8178.84	8681.66	9293.87	10245.50

1. Standardise the peak positions for the 30 known samples (exclude samples Y).
2. Perform PCA on this data, retaining the first two PCs. Calculate the scores and eigenvalues. What will the sum of squares of the standardised data equal, and so what proportion of the variance is accounted for by the first two PCs?
3. Produce a scores plot of the first two PCs of this data, indicating the two groups using different symbols. Verify that there is a good discrimination using PCA.
4. Determine the origin of Y as follows. (a) For each variable, subtract the mean and divide by the standard deviation of the 30 known samples to give a 6×7 matrix $^{stand}X$. (b) Then, multiply this standardised data by the overall loadings, for the first PC to give $T = {}^{stand}XP'$ and predict the scores for these samples. (c) Superimpose the scores of Y onto the scores plot obtained in 3 and determine the origin of Y.
5. Why is it correct to calculate $T = {}^{stand}XP'$ rather than using the pseudo-inverse and calculate $T = {}^{stand}XP'(PP')^{-1}$?

4.4 Effect of centring on PCA

Section 4.3.1   Section 4.3.2.1   Section 4.6.3

The following data consist of simulations of a chromatogram sampled at 10 points in time (rows) and at eight wavelengths (columns).

0.131	0.069	0.001	0.364	0.436	0.428	0.419	0.089
0.311	0.293	0.221	0.512	1.005	0.981	0.503	0.427
0.439	0.421	0.713	1.085	1.590	1.595	1.120	0.386
0.602	0.521	0.937	1.462	2.056	2.214	1.610	0.587
1.039	0.689	0.913	1.843	2.339	2.169	1.584	0.815
1.083	1.138	1.539	2.006	2.336	2.011	1.349	0.769
1.510	1.458	1.958	1.812	2.041	1.565	1.075	0.545
1.304	1.236	1.687	1.925	1.821	1.217	0.910	0.341
0.981	1.034	1.336	1.411	1.233	0.721	0.637	0.334
0.531	0.628	0.688	0.812	0.598	0.634	0.385	0.138

1. Perform PCA on the data, both raw and centred. Calculate the first five PCs including scores and loadings.
2. Verify that the scores and loadings are all orthogonal, and the sum of squares of the loadings equals 1.
3. Calculate the eigenvalues (defined by sum of squares of the scores of the PCs) of the first five PCs for both raw and centred data.
4. For the raw data, verify that the sum of the eigenvalues approximately equals the sum of squares of the data.
5. The sum of the eigenvalues of the column centred data can be roughly related to the sum of the eigenvalues for the uncentred data as follows. Take the mean of each column, square it, multiply by the number of objects in each column (=10) and then add these values for all the eight columns together. This plus the sum of the eigenvalues of the column centred data matrix should be nearly equal to the sum of the eigenvalues of the raw data. Show this numerically and explain why.
6. How many components do you think are in the data? Explain why the mean centred data, in this case, gives answers that are easier to interpret.

4.5 Effect of pre-processing on PCA in LCMS

Section 4.3.1   Section 4.5   Section 4.6   Section 4.3.2.1

The intensity of the ion current at 20 masses (first row in the table) and 27 points in time of an LCMS chromatogram of two partially overlapping peaks is recorded as follows:

96	95	78	155	97	41	154	68	191	113	79	51	172	190	67	156	173	164	112	171
−5.22	21.25	7.06	−5.60	0.76	−1.31	−18.86	5.54	−1.11	9.07	4.78	5.24	−4.10	1.37	3.93	−0.09	−2.27	−0.01	0.47	−0.64
12.32	15.45	24.89	0.68	1.35	3.40	16.44	7.08	1.31	4.79	3.32	3.33	−1.78	−0.96	2.71	2.05	2.73	2.43	0.20	−0.84
115.15	88.47	51.13	28.10	20.01	19.24	15.58	45.09	1.12	1.57	−5.42	−5.08	−1.91	−0.83	4.04	6.03	−4.46	−1.31	−6.54	2.90
544.69	240.76	3.69	107.08	83.66	109.34	44.96	71.50	11.56	2.20	2.18	4.54	9.53	−0.90	28.94	18.46	10.98	5.09	2.72	2.43

1112.09	410.71	13.37	201.24	148.15	231.92	127.30	198.73	30.54	9.15	9.14	3.47	12.18	14.54	61.09	34.53	23.47	6.24	−2.42	6.37
1226.38	557.41	81.54	235.87	204.20	244.35	140.59	192.32	68.94	19.35	8.71	5.81	19.01	36.95	86.36	40.95	22.77	6.39	3.20	11.73
1490.17	622.00	156.30	240.23	164.59	263.74	138.90	113.47	113.68	11.16	19.63	14.14	39.38	58.75	60.70	37.19	21.58	6.83	7.45	11.69
1557.13	756.45	249.16	205.44	212.87	255.27	170.75	87.22	149.62	8.04	39.82	28.36	37.98	51.93	58.07	34.81	23.50	4.11	10.98	17.04
1349.73	640.86	359.48	244.02	186.86	204.96	115.28	64.56	147.56	27.79	43.44	24.52	51.96	83.34	51.66	31.79	22.40	8.07	7.82	33.33
1374.64	454.83	454.54	175.32	154.20	146.15	118.83	73.99	136.44	34.12	39.23	34.82	34.82	73.43	35.39	23.33	21.71	4.44	9.96	14.68
1207.78	528.04	508.69	207.57	172.53	140.39	112.46	21.15	127.40	28.48	46.53	41.21	35.42	75.82	33.93	26.14	25.39	15.19	14.47	15.84
1222.46	270.99	472.10	227.69	183.13	128.84	102.10	66.31	169.38	28.21	61.12	37.59	52.28	82.46	28.28	30.15	17.30	12.04	22.59	14.22
1240.02	531.42	468.20	229.35	166.19	155.55	130.11	78.76	153.54	24.97	41.98	30.96	45.26	74.40	25.23	28.80	23.96	11.88	11.33	15.65
1343.14	505.11	399.50	198.66	138.98	125.46	118.22	68.61	106.16	24.95	46.21	36.62	40.20	49.49	32.09	28.71	22.18	18.22	15.68	11.95
1239.64	620.82	310.26	207.63	136.69	168.99	118.21	61.13	116.28	7.25	54.27	40.56	37.36	49.02	29.62	23.40	21.02	13.04	15.36	9.99
1279.44	573.64	347.99	154.41	169.51	152.62	149.96	50.76	83.85	30.04	47.06	32.87	38.92	46.14	31.16	24.12	27.80	17.04	17.68	11.51
1146.30	380.67	374.81	169.75	138.29	135.18	158.56	32.72	80.89	39.74	38.38	31.24	24.59	35.64	28.30	23.28	18.39	21.93	23.61	11.28
1056.80	474.85	367.77	165.68	142.32	144.29	119.43	73.98	70.94	27.52	38.34	39.10	27.52	46.31	36.38	21.40	25.51	28.87	20.25	11.63
1076.39	433.11	309.50	189.47	141.80	123.69	118.59	74.54	62.70	37.77	36.19	29.90	33.93	39.34	35.33	25.92	13.83	24.03	17.15	10.10
1007.28	383.06	233.38	168.77	125.95	125.74	131.29	52.88	58.22	40.22	39.03	40.17	26.38	36.04	30.01	26.56	23.53	23.54	25.74	7.39
919.18	397.39	218.33	162.08	122.76	95.15	99.53	70.66	57.47	47.40	29.87	32.85	32.88	31.43	28.04	24.83	20.28	22.66	15.74	5.23
656.45	371.74	201.15	119.10	79.72	90.19	84.65	27.95	41.04	31.65	29.91	28.20	25.90	12.08	29.35	19.19	22.23	22.30	7.81	2.63
727.64	294.33	281.14	134.72	90.90	97.64	92.00	56.17	37.36	42.17	25.29	30.48	30.21	23.70	23.59	14.22	20.27	19.46	15.73	9.23
656.70	282.45	308.55	138.62	93.78	87.44	61.85	58.65	46.31	22.80	35.31	22.44	34.28	23.12	15.27	20.85	15.74	16.58	22.28	8.96
745.82	345.59	193.12	144.01	81.15	97.76	80.30	39.74	44.34	35.20	18.83	30.68	30.69	24.05	21.57	22.25	17.86	17.92	14.63	9.35
526.49	268.84	236.18	149.61	88.20	68.65	93.67	81.95	45.02	33.49	27.16	26.23	27.16	20.49	20.77	16.98	14.38	20.79	11.40	12.32
431.35	164.86	190.95	112.79	78.27	43.15	26.48	52.97	23.36	24.21	21.47	19.51	12.29	15.84	15.88	13.76	9.59	13.91	4.51	8.99

1. Produce a graph of the total ion current (the sum of intensity over the 20 masses) against time.
2. Perform PCA on the raw data, uncentred, calculating two PCs. Plot the scores of PC2 versus PC1. Are there any trends? Plot the scores of the first two PCs against elution time. Interpret the probable physical meaning of these two principal components. Obtain a loadings plot of PC2 versus PC1, labelling some of the points farthest from the origin. Interpret this graph with reference to the scores plot.
3. Scale the data along the rows, by making each row add up to 1. Perform PCA. Why is the resultant scores plot of little physical meaning?
4. Repeat the PCA in step 3, but remove the first three points in time. Compute the scores and loadings plots of PC2 versus PC1. Why has the scores plot dramatically changed in appearance compared with that obtained in question 2? Interpret this new plot.
5. Return to the raw data, retaining all 27 original points in time. Standardise the columns. Perform PCA on this data and produce graphs of PC2 versus PC1 for both the loadings and scores. Comment on the patterns in the plots.
6. What are the eigenvalues of the first two PCs of the standardised data? Comment on the size of the eigenvalues and how this relates to the appearance of the loadings plot in question 5.

4.6 Determining the number of significant components in a data set by cross-validation.
Section 4.3.2.2
The following data set represents six samples (rows) and seven measurements (columns):

62.68	52.17	49.50	62.53	56.68	64.08	59.78
113.71	63.27	94.06	99.50	62.90	98.08	79.61
159.72	115.51	128.46	124.03	76.09	168.02	120.16
109.92	81.11	72.57	72.55	42.82	106.65	87.80
89.42	47.73	68.24	73.68	49.10	78.73	59.86
145.95	96.16	105.36	107.76	48.91	139.58	96.75

The aim is to determine the number of significant factors in the data set.

1. Perform PCA on the raw data and calculate the eigenvalues for the six non-zero components. Verify that these eigenvalues add up to the sum of squares of the entire data set.

2. Plot a graph of eigenvalue against component number. Why is it not clear from this graph how many significant components are in the data? Change the vertical scale to a logarithmic one and produce a new graph. Comment on the difference in appearance.

3. Remove sample 1 from the data set and calculate the five non-zero PCs arising from samples 2 to 6. What are the loadings? Use these loadings to determine the predicted scores $\hat{t} = x \cdot p'$ for sample 1 using models based on 1, 2, 3, 4 and 5 PCs successively, and hence the predictions $\hat{x}$ for each model.

4. Repeat this procedure, leaving each of the samples out once. Hence, calculate the residual sum of squares over the entire data set (all six samples) for models based on one to five PCs, and so obtain PRESS values.

5. Using the eigenvalues obtained in question 1, calculate the residual sum of squares error for one to five PCs and auto-prediction.

6. List the RSS and PRESS values and calculate the ration $\text{PRESS}_a / \text{RSS}_{a-1}$. How many PCs do you think will characterise the data?

5

Classification and Supervised Pattern Recognition

5.1 Introduction

5.1.1 Background

Pattern recognition has been the fastest growing application of chemometrics in the past decade, especially in the application to metabolomics. The origins of chemometrics were primarily in analytical chemistry, with chemical engineering (primarily in the area of process control) also taking an important front seat in the 1980s and 1990s.

However, during the past few years, there has been substantial increase in applications primarily to biology and medicine, where the aim is not so much to measure an analyte to a high degree of accuracy, or determine kinetics or equilibria, or identify components of a mixture but to use laboratory-based data, primarily from spectroscopy or chromatography, to determine the provenance of a sample.

Most of these problems can be formulated as classification problems. That is, samples are assigned using their analytical signal into one or several predefined classes, or groups, or labels. Pattern recognition, as originally defined in the 1960s, was quite broadly based, for example, involving handwriting analysis and facial recognition; however, over the past few days, it has primarily involved classification. There are two main forms of pattern recognition, unsupervised, as discussed in Chapter 4, and supervised, the basis of this chapter.

The original applications of supervised pattern recognition in chemometrics were primarily concerned with classifying samples into two or more known classes. In classical analytical chemistry, the reference standards were known in advance; for example, we might know that an extract came from one of the two types of orange juice, one from Brazil and one from Greece. We are certain of the provenance of the orange, and we may be developing a method, for example, in near infrared (NIR) to assign the spectra into one of the classes. The higher the classification ability, the better the method. There is a huge and somewhat misleading literature in which algorithms are compared according to their classification ability.

Some of this classical literature is not directly applicable to many modern-day problems.

- We may not be certain that the analytical technique can separate the groups; for example, can we use the HPLC method in urine samples to unambiguously determine whether patients are diseased or not; the aim may be to hypothesise that groups can be separated; thus, the aim of pattern recognition is hypothesis testing.
- There may be additional groups that we do not know about, or subgroups, or mislabelled samples.
- There are often outliers, or atypical samples, for example, originating from patients with unusual genetics that are not typical of the background population.
- We may be interested in markers, or variables.

Some people use simulations to test the relative merits of algorithms, but this depends on the simulations being realistic – the characteristics of a population are rarely known perfectly in advance. However, simulations can be used to develop and understand how methods work.

Every method has its merits. A lot depends on the aim of the study, which is known in advance, either with certainty or hypothesised. In this chapter, we will first look at some of the main methods and then how to assess them. We will not discuss Bayesian enhancements (see Section 3.6.3) in this chapter. They are often used in the machine-learning community but much less common in the chemometrics community.

Chemometrics: Data Driven Extraction for Science, Second Edition. Richard G. Brereton.
© 2018 John Wiley & Sons Ltd. Published 2018 by John Wiley & Sons Ltd.
Companion website: http://booksupport.wiley.com

Table 5.1 Case study in Section 5.1.2: the data involve 20 samples in two classes (first 10 = class A, second 10 = class B) recorded using two variables.

Sample	x_1	x_2
1	0.847	1.322
2	0.929	0.977
3	1.020	1.547
4	0.956	0.842
5	1.045	1.687
6	1.059	1.363
7	0.860	2.876
8	0.973	2.383
9	0.680	0.201
10	1.171	1.964
11	1.428	1.226
12	1.372	0.982
13	1.118	0.616
14	0.900	0.266
15	1.766	1.453
16	1.273	0.580
17	1.298	0.929
18	1.478	1.018
19	1.388	0.858
20	1.081	0.584

5.1.2 Case Study

In order to illustrate the methods of this chapter, we will describe a simple case study. This involves 20 samples, 10 from one class (A) and 10 from another class (B), recorded using two variables. These data are presented in Table 5.1 and illustrated in Figure 5.1.

1. The matrix in Table 5.1 has dimensions
   (a) 2×20
   (b) 20×2
   (c) $10 \times 2 \times 2$
   (d) $2 \times 10 \times 2$

5.2 Two-Class Classifiers

Two-class classifiers aim to find a boundary between two groups. Samples at one side of the boundary are assigned to one group (which we will call A) and the other side to the other group (which we will call B). This is illustrated in Figure 5.2(a).

The boundary can be of varying complexity. Different classifiers are defined according to the nature of this boundary, as discussed below, and it does not have to be linear. There is no requirement that this boundary perfectly separates the classes, as shown in Figure 5.2(b). If there, for example, is no linear boundary, either the classifier should be changed or the data are really inseparable using commonly available methods.

Two-class classifiers force all samples to be assigned to one of the two groups and do not usually allow for outliers or ambiguous samples.

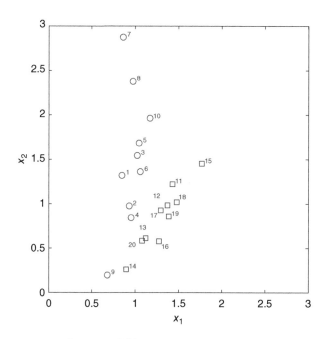

Figure 5.1 Data set in Table 5.1.

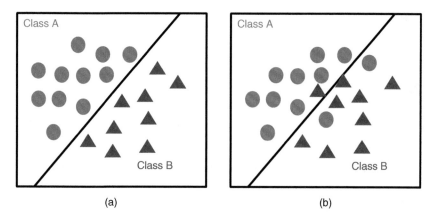

Figure 5.2 Two-class classifiers. (a) Linearly separable classes. (b) Linear inseparable classes.

1. There are 20 samples of polluted soils and 20 of unpolluted soils. A linear two-class classifier is constructed and classifies 18 of the polluted and 16 of the unpolluted correctly.

   (a) Two polluted are misclassified as unpolluted and four unpolluted as polluted.
   (b) Four polluted are misclassified as unpolluted and two unpolluted as polluted.
   (c) Six samples are misclassified, but we cannot assign these into any specific group.

5.2.1 Distance-Based Methods

Distance-based methods are conceptually the simplest and based on traditional statistics.

- For each group, the centroid is calculated. This is the average of all the samples in the group.
- Then, for each sample in the group, a distance to the centroid of each group is computed. We will discuss the different distance measures below.
- In the simplest situation, sample is assigned to the group whose centroid it is closest to.

In our example, the centroid of group A is $\begin{bmatrix} 0.955 & 1.517 \end{bmatrix}$ and group B is $\begin{bmatrix} 1.311 & 0.852 \end{bmatrix}$. Of course, the data can be described by any number of variables. Hence, the centroid of a group of dimensions 50×400, which might represent

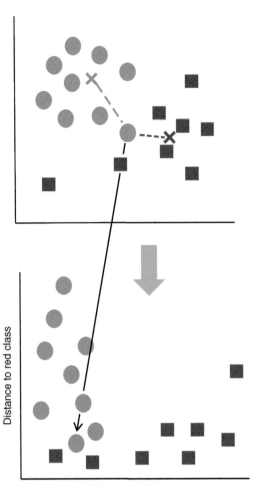

Figure 5.3 A class distance plot. Top illustrates two classes, with their centroids marked by crosses. A sample is indicated, with its distances to the centroids of the blue and red classes. Bottom projects onto a class distance plot, with the specific sample noted.

50 spectra recorded at 400 wavelengths, has dimensions 1×400. In our example, we are choosing two variables for simplicity.

There are many ways in which these distances can be interpreted.

- The first is simply to assign samples to groups using binary decision making.
- The second is as a class distance plot. This is illustrated in Figure 5.3. This can be used to graphically show the distribution of samples.
- The third is to determine whether any samples are ambiguous or outliers. Often, for two-class classifiers, this is not checked but will be discussed again in Section 5.3 in the context of one-class classifiers.

1. A small data set consists of six samples, the first two belong to class A and the last four belong to class B as follows:

$$\begin{bmatrix} 0.8 & 1.2 & 0.7 \\ 0.6 & 1.0 & 1.1 \\ 0.9 & 0.3 & 0.2 \\ 1.4 & 0.8 & 0.5 \\ 0.7 & 0.4 & 0.8 \\ 1.7 & 0.3 & 0.0 \end{bmatrix}.$$

What is the centroid of class A?

(a) [0.77 0.85 0.67]
(b) [1.02 0.67 0.55]
(c) [0.70 1.10 0.90]

5.2.1.1 Euclidean Distance to Centroids

The simplest type of distance is the Euclidean distance defined in vector terminology by

$$d_{iA}^2 = (\boldsymbol{x}_{iA} - \bar{\boldsymbol{x}}_A)(\boldsymbol{x}_{iA} - \bar{\boldsymbol{x}}_A)'$$

where

- A is the class and i the sample,
- $\boldsymbol{x}_{iA}$ is the row vector corresponding to the measurements for this sample and
- $\bar{\boldsymbol{x}}_A$ is the centroid of this class.

In our case, the square Euclidean distance from class A for sample 6 is calculated as follows.

- $\bar{\boldsymbol{x}}_A$ is [0.955 1.517] or the average of the first 10 readings.
- Hence, $\boldsymbol{x}_{6A} - \bar{\boldsymbol{x}}_A = [1.060 \quad 1.364] - [0.955 \quad 1.517] = [0.105 \quad -0.153]$.
- So $d_{6A}^2 = (\boldsymbol{x}_{6A} - \bar{\boldsymbol{x}}_A)(\boldsymbol{x}_{6A} - \bar{\boldsymbol{x}}_A)'$ can be calculated by the sum of squares of $\boldsymbol{x}_{6A} - \bar{\boldsymbol{x}}_A$ or $0.105^2 + (-0.153)^2 = 0.0345$.
- Therefore, the distance of sample 6 to the centroid of class A is the square root of 0.0345, or 0.186.
- Note that the units will be in the original measurement units. Sometimes, of course, the two variables measured will be in different units (e.g. pH and temperature), in which case either it is a combination or variables are transformed to similar units by standardisation (Section 4.6); however, more usually they are the same, for example, absorbances at different wavelengths.

The distance of this sample to class B is 0.754. Hence, it is closer to the centroid of class A and assigned to this class. This is the principle of the method of Euclidean distance to centroids (EDC).

---

1. The centroid of a group is [0.78 0.92 1.70 −1.33 0.68]. An individual sample has measurements [0.54 0.23 0.68 0.91 1.32]. What is the Euclidean distance of the sample to the centroid?

   (a) 0.93
   (b) 0.96
   (c) 7.00
   (d) 2.65

5.2.1.2 Linear Discriminant Analysis

However, different variables may be on quite different scales, consider one variable being a weight in kilogram and the other a length in metre. Should we use the same scale for each variable or weight them so that they are of equal significance, or variance? Another problem is that sometimes some variables are correlated. For example, we might be analysing the spectra of a set of compounds, and there could be one broad peak corresponding to several wavelengths, all varying in the same way, and they could unfairly weight the analysis.

To correct for this, we often calculate the Mahalanobis distance. The most straightforward method is called linear discriminant analysis (LDA). This involves weighting the variables by the inverse of the pooled variance–covariance matrix – so

$$d_{iA}^2 = (\boldsymbol{x}_{iA} - \bar{\boldsymbol{x}}_A)S_p^{-1}(\boldsymbol{x}_{iA} - \bar{\boldsymbol{x}}_A)'$$

(the variance and covariance are described in Section A.1). The pooled variance–covariance matrix is the average over both groups defined by

$$S_p = [(I_A - 1)S_A + (I_B - 1)S_B]/(I_A + I_B - 2)$$

where I_A is the number of samples in group A (in our case 10) and S_A its variance–covariance matrix (using the population variance) and similarly for group B. If both groups have equal number of samples, then it is simply the average of the variance–covariance matrices. Of course, if there are more than two groups, an equivalent equation can be found with as many groups as necessary.

For calculations in, for example, Excel or Matlab, you should use the population variance or covariance because we are primarily scaling variables rather than estimating population parameters.

- For our data set,

$$S_A = \begin{bmatrix} 0.0167 & 0.0381 \\ 0.0381 & 0.5367 \end{bmatrix} \quad \text{and} \quad S_B = \begin{bmatrix} 0.0519 & 0.0711 \\ 0.0711 & 0.1096 \end{bmatrix}$$

- Hence,

$$S_p = \begin{bmatrix} 0.0343 & 0.0546 \\ 0.0546 & 0.3232 \end{bmatrix} \quad \text{and} \quad S_p^{-1} = \begin{bmatrix} 39.857 & -6.729 \\ -6.729 & 4.230 \end{bmatrix}$$

- This means for the sixth sample, the distance from group A is given by

$$d_{6A}^2 = \begin{bmatrix} 0.105 & -0.153 \end{bmatrix} \begin{bmatrix} 39.857 & -6.729 \\ -6.729 & 4.230 \end{bmatrix} \begin{bmatrix} 0.105 \\ -0.153 \end{bmatrix}$$

$$= \begin{bmatrix} 5.216 & -1.355 \end{bmatrix} \begin{bmatrix} 0.105 \\ -0.153 \end{bmatrix} = (5.216) \times 0.105 + (-1.355) \times (-0.153)$$

$$= 0.755$$

- Hence, the distance is the square root of 0.755, or 0.869, distance units from the centre.

It is important to note that when there is more than one group, the pooled variance–covariance matrix is not the same as the overall variance–covariance matrix S, which in our case is $\begin{bmatrix} 0.0660 & -0.0047 \\ -0.0047 & 0.4337 \end{bmatrix}$. Of course, it is still possible to calculate the Mahalanobis distance from the centre of the entire data set if samples are considered to belong to a single group, and under such circumstance, we use the overall variance–covariance matrix as I_A is the same as I, but this is not strictly a classification problem and can be used in unsupervised pattern recognition (Section 4.8.1).

Unlike the Euclidean distance, the Mahalanobis distance is independent of the original measuring units. It is analogous to the number of standard deviations from the centre of a group. When there is only one variable, the number of standard deviations equals the Mahalanobis distance; thus, a sample 3.12 standard deviations from the centre of a group has a Mahalanobis distance of 3.12. When there is more than one variable, there is no positive or negative direction (rather like the distance from the centre to the circumference of a circle); thus, the Mahalanobis distance (which is directionless) must be used.

The main traditional limitation of the Mahalanobis distance is that the number of samples must never exceed the number of variables in the smallest group as the variance–covariance matrix has to have an inverse. In traditional multivariate statistics, this was not an issue, as variables were hard to measure; however, for modern chemometrics, often variables, such as mass spectral or chromatographic or NMR peaks/intensities, are plentiful and sample to variable ratios may be substantially less than 1. The traditional way of overcoming this was to select variables to be retained. However, it can be shown that the squared Mahalanobis distance is identical to the sum of squares of all the standardised scores of all non-zero principal components; hence, there is no problem performing principal component analysis (PCA) first and retaining all non-zero standardised PCs, even if the number of variables exceeds the number of samples.

1. A data set consists of 20 samples and 50 variables.

    (a) It is possible to calculate the Mahalanobis distance of each sample to the centroid, but only with variable selection.
    (b) It is possible to calculate the Mahalanobis distance of each sample to the centroid, using PCA as a final step.
    (c) It is not possible to calculate the Mahalanobis distance of each sample to the centroid.

2. The following data represent five samples and three variables considered from a single group: $\begin{bmatrix} 0.78 & 0.34 & 1.22 \\ 0.83 & 0.46 & 0.99 \\ 1.45 & 0.61 & 1.45 \\ 0.96 & 0.22 & 1.66 \\ 1.22 & 0.13 & 1.05 \end{bmatrix}$.

    What is the variance–covariance matrix as used for computing the Mahalanobis distance of each sample from the overall centroid?

    (a) $\begin{bmatrix} 0.0637 & 0.0114 & 0.0149 \\ 0.0114 & 0.0290 & 0.0028 \\ 0.0149 & 0.0028 & 0.0627 \end{bmatrix}$

    (b) $\begin{bmatrix} 0.0796 & 0.0142 & 0.0187 \\ 0.0142 & 0.0363 & 0.0035 \\ 0.0187 & 0.0035 & 0.0784 \end{bmatrix}$

$$
\text{(c)} \begin{bmatrix} 0.2821 & 0.1192 & 0.1366 \\ 0.1192 & 0.1904 & 0.0595 \\ 0.1366 & 0.0595 & 0.2821 \end{bmatrix}
$$

5.2.1.3 Quadratic Discriminant Analysis

LDA uses the pooled variance–covariance matrix of an entire data set, consisting of, in our example, two classes. It assumes that the variance of both\classes is the same. In some cases, this is an adequate assumption, and the main issue is that the variables themselves are correlated or on different scales.

However, sometimes each group has its own structure. Consider trying to separate elephants from mice, the variance of the elephants is very much larger than that of the mice. The result is that small elephants could be mistaken as large mice unless we modelled each group separately. To do this, a simple adjustment is involved. Instead of using the overall variance, calculate the variance of each group separately so that

$$
d_{iA}^2 = (\boldsymbol{x}_{iA} - \bar{\boldsymbol{x}}_A) \boldsymbol{S}_A^{-1} (\boldsymbol{x}_{iA} - \bar{\boldsymbol{x}}_A)'
$$

where $\boldsymbol{S}_A$ is the variance–covariance matrix of group A with a similar equation for group B, which is the basis of quadratic discriminant analysis (QDA).

- In the example in Table 5.1,

$$
\boldsymbol{S}_A = \begin{bmatrix} 0.0167 & 0.0381 \\ 0.0381 & 0.5367 \end{bmatrix} \quad \text{and} \quad \boldsymbol{S}_B = \begin{bmatrix} 0.0519 & 0.0711 \\ 0.0711 & 0.1096 \end{bmatrix}
$$

  as discussed above.
- Hence,

$$
\boldsymbol{S}_A^{-1} = \begin{bmatrix} 71.370 & -5.061 \\ -5.061 & 2.222 \end{bmatrix}
$$

- This means for the sixth sample, the distance from group A is given by

$$
d_{6A}^2 = \begin{bmatrix} 0.105 & -0.153 \end{bmatrix} \begin{bmatrix} 71.370 & -5.061 \\ -5.061 & 2.222 \end{bmatrix} \begin{bmatrix} 0.105 \\ -0.153 \end{bmatrix}
$$

$$
= \begin{bmatrix} 8.269 & -0.872 \end{bmatrix} \begin{bmatrix} 0.105 \\ -0.153 \end{bmatrix} = 8.292 \times 0.105 + (-0.875 \times -0.153)
$$

$$
= 1.002
$$

- Hence, the distance is the square root of 1.002, or 1.001, Mahalanobis distance units from the centre of group A.

QDA allows each group to have its own variance structure. In our example, the correlation between the two variables for the overall data set is −0.027; hence, they are virtually uncorrelated. If we consider just group A alone, it is 0.401, and group B alone 0.942; thus, each group on its own has quite different correlations. This is reflected in the variance–covariance matrix $\boldsymbol{S}$ for the overall data set where the off-diagonal elements are relatively small, compared with $\boldsymbol{S}_B$ where they are relatively large.

As usual, the number of variables should not exceed the number of samples in the smallest group. Although this can be overcome, as for LDA, PCA would have to be on each group separately and is rarely done for two-class classifiers. One-class classifiers overcome this and are described in the next section.

1. A data set consists of 50 samples, 20 in group A and 30 in group B, and seven variables. The dimensions of the variance–covariance matrix of group A are
    (a) 20×20
    (b) 2×2
    (c) 7×7

2. LDA uses the pooled variance–covariance matrix for the full data set, where there are two or more groups, in contrast to QDA, which uses the variance–covariance matrix of each group separately.

   (a) True
   (b) False

5.2.1.4 Comparing EDC, LDA and QDA

In order to understand the difference between the three methods, we compute the distances to the centroid of all 20 samples in Table 5.1. These are presented in Table 5.2. Samples are assigned to the class they are nearest to. For example, using QDA, we find that sample 6 has a distance of 1.001 to group A but 7.793 to class B; therefore, it is assigned to class A.

In our example, we find that there are three misclassified samples using EDC and one using QDA. All are correctly classified using LDA.

In order to better understand this, it is useful to visualise the boundaries between the classes graphically, as shown in Figure 5.4. The boundaries are the points where $d_A = d_B$ or the distance to the centroids of each class is equal, samples on either side of this are assigned into one of the classes. We also draw the equidistance contours for each class. We can see that for EDC, these are circular (or if there are more than two variables, hyper-spherical). For LDA, they form ellipses, the direction of the ellipses being given by the maximum variance of the pooled variance–covariance matrix: we see that for both classes, they are in the same direction as a 'common' matrix is computed. For QDA, they are in different directions, reflecting the different structure and orientation of the groups.

In addition, we see that the QDA boundaries are curved (or in our case parabolic). In some situations, we do not expect two groups to be linearly separable and as such these are more realistic boundaries. For the case study in this section, it is reasonable to use a linear boundary as LDA performs slightly better to QDA, but in other cases, groups are not linearly separable. It is also important to note that for the methods in this section, we just look at which side of the boundary a sample lies, but it is of course possible to look at how close a sample is to a boundary and form a probabilistic model, as discussed in Section 5.3.

Table 5.2 Class distances for the data in Table 5.1 using EDC, LDA and QDA together with the predicted class memberships.

Sample	True class	EDC			LDA			QDA		
		d_{iA}	d_{iB}	Predicted	d_{iA}	d_{iB}	Predicted	d_{iA}	d_{iB}	Predicted
1	A	0.222	0.660	A	0.580	3.525	A	0.831	10.167	A
2	A	0.540	0.401	B	1.036	2.550	A	0.744	6.075	A
3	A	0.073	0.754	A	0.388	2.850	A	0.541	9.936	A
4	A	0.674	0.354	B	1.393	2.226	A	1.012	4.563	A
5	A	0.194	0.877	A	0.494	2.957	A	0.706	10.870	A
6	A	0.186	0.570	A	0.869	2.314	A	1.001	7.793	A
7	A	1.363	2.074	A	3.146	6.139	A	2.456	23.890	A
8	A	0.867	1.568	A	1.724	4.627	A	1.236	18.031	A
9	A	1.343	0.905	B	2.337	3.479	A	2.357	3.367	A
10	A	0.498	1.121	A	1.191	2.845	A	1.680	11.765	A
11	B	0.556	0.393	B	3.341	0.744	B	4.197	1.992	B
12	B	0.678	0.145	B	3.343	0.340	B	3.920	0.497	B
13	B	0.915	0.304	B	2.547	1.048	B	2.284	0.881	B
14	B	1.251	0.715	B	2.412	2.219	B	1.731	1.813	A
15	B	0.814	0.755	B	5.195	2.474	B	6.898	2.011	B
16	B	0.989	0.274	B	3.433	0.480	B	3.498	1.992	B
17	B	0.681	0.079	B	2.982	0.211	B	3.355	0.854	B
18	B	0.723	0.237	B	3.938	0.929	B	4.774	0.930	B
19	B	0.788	0.078	B	3.631	0.484	B	4.159	0.962	B
20	B	0.941	0.352	B	2.432	1.254	B	2.069	1.091	B

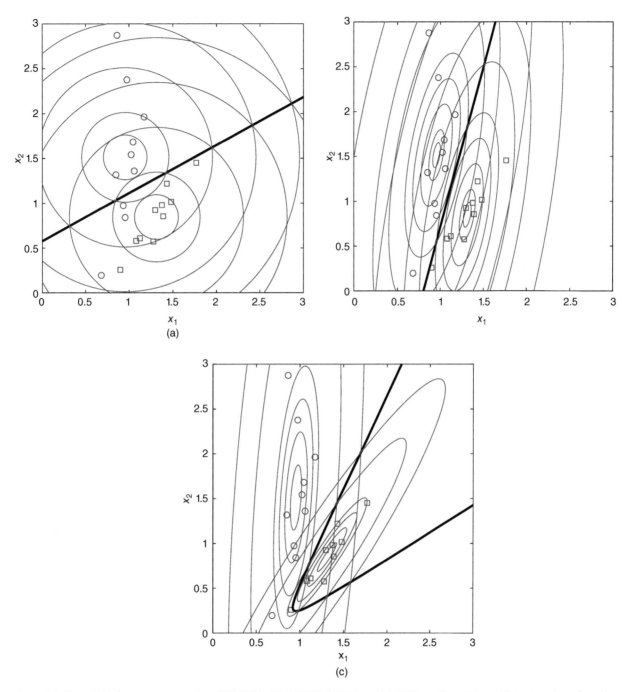

Figure 5.4 Boundaries between groups A and B in Table 5.1, (a) EDC, (b) LDA and (c) QDA together with equidistant contours from the centroids for each criterion.

An interesting property of the Mahalanobis distance is that the sum of squares of the distances of all samples to the centroid of its own group, or in-group, is equal to IJ, where I is the number in the group and J is the number of variables. In our example, this property only holds for QDA.

- Hence, the sum of squares for the distances of the samples in group A is given by $0.831^2 + 0.744^2 + \cdots + 1.680^2 = 20$ (see Table 5.2).
- And $IJ = 10 \times 2 = 20$.
- Similarly, the sum of squares $1.922^2 + 0.497^2 + \cdots + 1.091^2 = 20$ for group B.

Of course, the converse is not true for the sum of squares to another group, or out-group, as any fresh sample could be an outlying sample and will have no influence on the centroid or variance–covariance matrix of the in-group.

As an example, the sum of squares of the 10 samples of group A for the QDA distances to the centroid of group B is 1484.30.

Note that this property is not valid for LDA because the pooled variance–covariance matrix depends on both groups, whereas the centroid is unique to each group. For the purpose of brevity, we do not expand on these properties in this text.

1. A data set consists of 50 samples, 20 in group A and 30 in group B, and three variables. The sum of squares of the Mahalanobis distance using QDA for samples in group A to their centroid is

   (a) 20
   (b) 60
   (c) 3
   (d) 150

2. QDA will always classify more samples correctly to LDA as boundaries are quadratic and so extra terms can be added to better define the boundary.

   (a) True
   (b) False

5.2.2 Partial Least-Squares Discriminant Analysis

PLS-DA has become very topical over the past decade. In Chapter 6 and Section A.2.2, we discuss the algorithm that will not be reproduced here. The description in this section will exclusively relate to a two-class problem and PLS1.

The basis of PLS-DA is that the classifier is given a numerical value c for each sample, often $+1$ for class A and -1 for class B, and the X matrix is calibrated against this. Using a model

$$\hat{c} = tq$$

for each sample (see Section 6.5.1 for more details), an estimated value of c can be obtained. This predicts the class membership for each sample. In the simplest case, if it is >0, it is usually assigned to class A, otherwise assigned to class B. This is called a decision threshold, and in the case of, for example, unequal class sizes of differing variances, the decision threshold can be changed, but the simplest and default criterion is used below.

It is usual to centre X before performing PLS and compute the scores using centred X. For each sample, the vector t will be of dimensions $1 \times A$, where A is the number of PLS components retained and q will be of dimensions $A \times 1$. Usually, the X matrix is centred down the columns before performing PLS (as discussed in Chapter 6); however, if there are unequal class sizes, it is better to adjust the X matrix by the average of the means of the two classes so that

$$x_{inew} = x_i - (\bar{x}_{iA} + \bar{x}_{iB})/2$$

where x_{inew} is the transformed data for sample i and $\bar{x}_{iA}$ the mean or centroid of class A. In our example, both classes have equal sizes; thus, this transformation is the same as column mean centring.

- In our example, if we keep both PLS components, q is a vector $\begin{bmatrix} 1.061 \\ 1.984 \end{bmatrix}$.
- The PLS scores of both components of the centred data are given in Table 5.3.
- Hence, for sample 6, we have $\hat{c} = \begin{bmatrix} 0.193 & 0.061 \end{bmatrix} \begin{bmatrix} 1.061 \\ 1.984 \end{bmatrix} = 0.326$.
- If we keep only one component, it is $0.193 \times 1.061 = 0.204$.
- In both cases, the predicted value is >0 and the sample is assigned to class A according to both criteria.

The predicted values $\hat{c}$ are presented in Table 5.4. We can then assign the samples into their predicted groups, usually according to whether $\hat{c}$ is positive or negative. Note an important conclusion:

- If all non-zero PLS components are used, the position of the PLS boundary and assignment to classes is identical to that using LDA.

Table 5.3 PLS-DA components of data in Table 5.1.

Sample	Component 1	Component 2
1	0.257	0.295
2	−0.086	0.240
3	0.373	0.085
4	−0.218	0.224
5	0.485	0.044
6	0.193	0.061
7	1.620	0.126
9	1.132	0.053
9	−0.653	0.586
10	0.669	−0.119
11	−0.102	−0.324
12	−0.291	−0.239
13	−0.494	0.072
14	−0.699	0.342
15	−0.062	−0.712
16	−0.599	−0.092
17	−0.303	−0.154
18	−0.309	−0.358
19	−0.408	−0.244
20	−0.505	0.115

- If only one PLS component is used, the position of the PLS boundary and assignment to classes is identical to that using EDC.

The assignments can be compared between Tables 5.2 and 5.4. This property is only true if the X data are centred and if there are equal numbers in each class and the decision criterion is whether c is greater than or equal to 0; under different circumstances, the reader is referred to the more specialist literature.

For new samples (not part of the original training set), we often use a slightly different equation for prediction:

$$\hat{c} = \boldsymbol{xb}$$

where $\boldsymbol{x}$ is centred if appropriate and $\boldsymbol{b}$ is the first A coefficients of $(\boldsymbol{X'X})^{-1} \boldsymbol{X'c}$ or the product of the pseudo-inverse (see Section 2.2.3.3) of the original centred or transformed data matrix with $\boldsymbol{c}$. This can also be used as an alternative to computing $\hat{c} = \boldsymbol{tq}$ for the original training data as well.

- In our case, $\boldsymbol{b} = \begin{bmatrix} -2.646 \\ 0.738 \end{bmatrix}$.
- For sample 6, the centred value of $\boldsymbol{x}$ is $\begin{bmatrix} -0.073 & 0.179 \end{bmatrix}$.
- Hence, the estimated value using two PLS components $\hat{c} = (-2.646) \times (-0.073) + 0.738 \times 0.179 = 0.326$.
- This is the same result as mentioned above.

This approach, however, can also be employed for any future samples or those of unknown provenance.

Naturally, if there are many variables, PLS models can be obtained using any number of components as desired, representing an advantage over LDA or EDC. However, a disadvantage is that if the number of samples differs in each group, or if there are more than two groups, the choice of decision threshold (which value of c is used as a cut-off between classes and which column transformation) becomes complicated and there is no universal agreement. It is very common to use PLS-DA packages without explicitly understanding their basis, which is often unwise. As PLS methods can often give the same classification ability to more traditional statistical approaches using distance measures, their main advantage is predicting variables that are most discriminatory (e.g. marker compounds), as discussed in Section 5.6. It can also be used when the number of variables exceeds the number of samples, quite common, for

Table 5.4 PLS-DA predictions of c for one-component and two-component models for the centred data in Table 5.4.

Sample	True class	One component	Predicted	Two components	Predicted
1	A	0.272	A	0.856	A
2	A	−0.092	B	0.385	A
3	A	0.396	A	0.565	A
4	A	−0.231	B	0.214	A
5	A	0.514	A	0.602	A
6	A	0.204	A	0.326	A
7	A	1.718	A	1.969	A
8	A	1.201	A	1.306	A
9	A	−0.692	B	0.471	A
10	A	0.710	A	0.473	A
11	B	−0.109	B	−0.752	B
12	B	−0.309	B	−0.784	B
13	B	−0.524	B	−0.382	B
14	B	−0.742	B	−0.063	B
15	B	−0.066	B	−1.478	B
16	B	−0.635	B	−0.818	B
17	B	−0.321	B	−0.627	B
18	B	−0.328	B	−1.037	B
19	B	−0.433	B	−0.917	B
20	B	−0.535	B	−0.308	B

example using modern data-rich instrumentation, as there is no need to calculate the Mahalanobis distance (although performing PCA before the distance calculations is also possible).

1. A data set consists of 200 samples, 100 from class A and 100 from class B; there are 50 variables. The data matrix is centred and PLS-DA performed.

(a) A model with one PLS component will result in the same predictions as LDA.

(b) A model with 50 PLS components will result in the same prediction as LDA.

(c) We cannot say whether the same prediction ability is obtained as from LDA no matter how many components, the methods are unrelated.

5.2.3 K Nearest Neighbours

One of the oldest classifiers in chemometrics is k nearest neighbours (kNN). The basis of the method is simple.

- Choose k. This is usually an odd number from 3 onwards.
- Calculate the distance of the sample of interest from all the samples in the training set. Usually, the Euclidean distance is used. This should include itself (the reasons will be discussed below).
- Select the k samples that are closest.
- By a 'majority vote', assign the sample to the most popular class. For example, if $k = 5$, and the nearest samples (including itself) are members of classes A, A, B, A, A, then by a majority of 4 to 1, it is assigned to class A.
- There are various rules in the case of ties. When there are only two classes and k is an odd number, there will be no ambiguity.
- Note that in the algorithm described above, the closest sample will always be itself.

We list the five nearest samples for the data set in Table 5.1 and the assignments using $k = 3$ and $k = 5$ in Table 5.5. We see that when $k = 3$, only sample 9 is misclassified, whereas when $k = 5$, both samples 4 and 9 are misclassified.

Table 5.5 kNN for data in Table 5.1; the five nearest neighbours are listed and the assignments using $k = 3$ and $k = 5$.

Sample	Class	Nearest samples					Assignment	
		First	Second	Third	Fourth	Fifth	$k = 3$	$k = 5$
1	A	1	6	3	2	5	A	A
2	A	2	4	1	17	6	A	A
3	A	3	5	6	1	10	A	A
4	A	4	2	13	20	17	A	B
5	A	5	3	10	6	1	A	A
6	A	6	3	1	5	11	A	A
7	A	7	8	10	5	3	A	A
8	A	8	10	7	5	3	A	A
9	A	9	14	20	13	4	B	B
10	A	10	5	3	8	6	A	A
11	B	11	18	12	17	19	B	B
12	B	12	17	18	19	11	B	B
13	B	13	20	16	4	17	B	B
14	B	14	9	20	13	16	B	B
15	B	15	11	18	12	17	B	B
16	B	16	13	20	19	17	B	B
17	B	17	12	19	18	11	B	B
18	B	18	12	19	17	11	B	B
19	B	19	17	12	18	16	B	B
20	B	20	13	16	4	14	B	B

The kNN boundaries can be computed and are illustrated in Figure 5.5. Unlike for the methods in Sections 5.2.1 and 5.2.2, the boundaries are quite complicated. Note the position of sample 4, which changes from one side to the other according to the value of k.

A dilemma about kNN is whether to count the distance of a sample to itself as the nearest neighbour while computing models. In fact, if this is not done, there can be significant difficulties; sample 4 would be misclassified if $k = 3$, as shown in Table 5.5 and visualised in the graphs. The boundaries would, in practice, then be for the majority vote from the second, third and fourth nearest samples of the table. Sample 4 would now be misclassified, as its closest samples are 2, 13 and 20, and the boundary of Figure 5.5 would no longer be correct. However, modelling each sample, excluding itself, results in a very strange boundary, as illustrated in Figure 5.6 with certain rather unusual regions. This is a major dilemma of the kNN method. If we used $k = 1$, the method will always perfectly classify if we included itself, so the method is inherently biased.

Another problem of kNN results from unequal class sizes. If one class consists of many more samples than the other, it may weight results in its favour. Of course, there are ways of overcoming this, but the simplest and most straightforward implementation of kNN does not correct for relative class sizes.

Despite these limitations, kNN is a very simple approach and can cope with unusual distributions of data; thus, the boundaries do not need to be linear or quadratic.

1. A sample X belongs to class A. There are nine other samples in a data set, of which four belong to class A with distances from sample X of 0.8, 0.4, 0.1 and 0.9 and five belong to class B with distances from sample X of 0.5, 0.2, 0.7, 1.2 and 0.3. If all 10 samples in the data set are used to develop a kNN model

    (a) Sample X is predicted to belong to class A using models with $k = 3$, 5 and 7.
    (b) Sample X is predicted to belong to class A using models with $k = 3$ and 5, but erroneously predicted to belong to class B with $k = 7$.
    (c) Sample X is erroneously predicted to belong to class B using models with $k = 3$, 5 and 7.

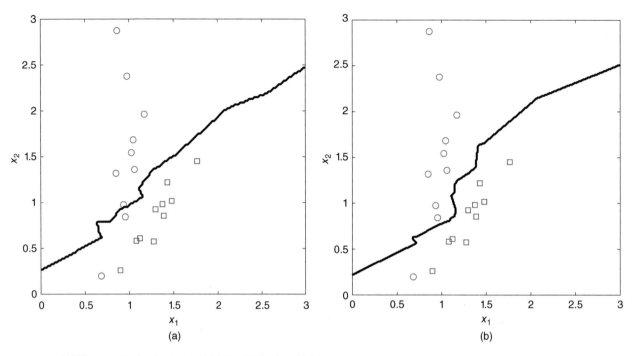

Figure 5.5 kNN boundaries for data set in Table 5.1; (a) $k = 3$ and (b) $k = 5$.

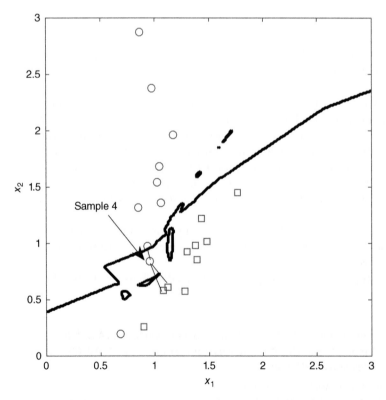

Figure 5.6 Appearance of kNN boundaries if the distance of a sample to itself is excluded for $k = 3$ and data in Table 5.1; sample 4 and its three neighbours marked.

5.3 One-Class Classifiers

In the case of two-class classifiers, we determined what is often called a hard boundary between two classes. We insist that each sample is unambiguously assigned to one or other of the classes according to which side of the boundary it falls. Another type of classifier is called a one-class classifier. The question asked is not 'Is a sample a member of either class A or B?' but 'Is a sample a member of class A or not?'.

This is equivalent to drawing a boundary around a class, as illustrated in the left-hand diagram of Figure 5.7, rather than between classes. The advantage of this is that each class can be modelled separately. In addition, more, new, classes can be considered without changing the original models. There are also several additional features of one-class classifiers. The first is that it is possible to have ambiguous samples, as they may appear to be a member of both classes simultaneously. This could occur because the analytical method is insufficient to distinguish between the classes or because it genuinely has both characteristics, an example being a compound that is both a ketone and an ester, it may have spectral characteristics of both types of functional group. The second is that there can be outliers, which are members of no prior class, they may be a mislabelled sample or genuinely from a new class. Hence, if we are modelling two classes, there can be four verdicts rather that two, for a two-class problem, namely class A, class B, ambiguous or outlier.

1. We are trying to model three classes. Using one-class classifiers, how many possible verdicts are there?
    (a) Three types of unambiguous samples, three types of ambiguous samples, three types of outliers.
    (b) Three types of unambiguous samples, four types of ambiguous samples, three types of outliers.
    (c) Three types of unambiguous samples, four types of ambiguous samples, one type of outliers.
    (d) Three types of unambiguous samples, three types of ambiguous samples, one type of outliers.

5.3.1 Quadratic Discriminant Analysis

The simplest and most straightforward method is QDA. This method can be used both as a two-class classifier (Section 5.2.1.3) or a one-class classifier. In its usual form, it is based on the assumption that the data are (multi)-normally distributed. As a two-class classifier, a sample is assigned to the class it is closest to. As a one-class classifier, a sample is assigned to a class if it is within a defined distance (often converted to a probability) from the centroid.

Figure 5.8 illustrates a typical data set in which the samples are multi-normally distributed, which is defined as follows.

- Each variable (in our case, two variables) in itself must be distributed normally.
- The sum of any combination of variables must also be distributed normally. In algebraic terms, if variable 1 is distributed normally, we call this x_1, then $ax_1 + bx_2 + \cdots + nx_j$ must also form a normal distribution, where a, b and so on can be any numbers.

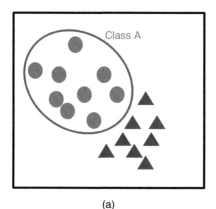
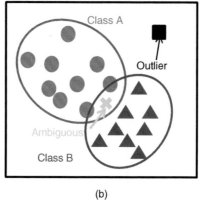

(a)   (b)

Figure 5.7 One-class classifiers: (a) separable class; (b) classes with ambiguous and outlying samples.

- In the vast majority of cases, if each variable on its own is normally distributed, the combination also forms a multi-normal distribution, although there are a few specialised exceptions.

The proportion of samples within a given distance from the centre of the distribution is calculated using the Mahalanobis distance. Assuming an underlying normal distribution, the proportion of samples m Mahalanobis distances or more from the centre is predicted using χ^2 with v degrees of freedom (see Section A.3.3), where v equals the number of variables, as follows.

- Determine what proportion of samples are expected to be 1.5 Mahalanobis distance units or less from the centre of a distribution as follows.
- Calculate the right-hand side of the χ^2 distribution for a value of 1.5^2 (=2.25) for two degrees of freedom, which is CHISQ.DIST.RT(1.5^2,2) in Excel.
- This is 0.345.
- Hence, $100 \times (1-0.345)$ or 67.5% are expected to be within 1.5 Mahalanobis distance units of the centre when there are two degrees of freedom or two variables.

Note that when there is only one degree of freedom, this provides the same result as for the normal distribution.

- Hence, when there is only one degree of freedom, then CHISQ.DIST.RT(1.5^2,1) = 0.133.
- However, because χ^2 is squared, it represents samples either side of the mean (i.e. samples more than 1.5 Mahalanobis distance on either side of the mean I).
- Hence, the proportion of samples more than 1.5 standard deviations greater than the mean is $(1-0.133)/2 = 0.067$.
- Hence, the proportion of samples less than 1.5 standard deviations more than the mean is $1-0.067$ or 0.933, which can be verified from the normal distribution presented in Table A.1.

In Figure 5.8, the predicted boundaries using the assumption of a multi-normal distribution are presented for various confidence limits. Hence, for example, the 0.95 boundary is that within which 95% of the samples should lie, 5% being outside. There are 100 samples, and we can see that five are outside the 95% boundary. Of course, although the underlying (population) distribution is normal, the observed (sample) distribution differs slightly; for example, there is no sample farther that the 99% confidence limit; however, these predicted contours do approximately predict the observed distribution of samples from the centre.

Using QDA as a one-class classifier involves choosing a specific boundary, often 90%, 95% or 99% ($p = 0.1, 0.05, 0.01$), confidence intervals or it could just involve a specified cut-off distance measure. Anything outside the chosen boundary has a low probability of belonging to the class being modelled (in-group). It is possible to calculate the probability that a sample is a member of a class based on its distance from the centroid, but in chemometrics, it is more usual to simply assign a sample to a class if it is within a specified boundary defined by a percentage confidence (e.g. 99%) or p value.

We calculate the Mahalanobis distances to the two centroids in Table 5.6 (which are the same as the corresponding columns of Table 5.2). If we use a 90% confidence limit for a classification threshold (this is also sometimes referred to

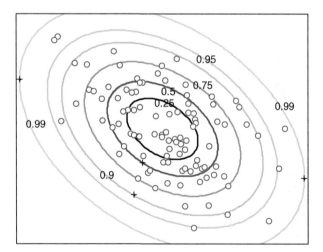

Figure 5.8 Typical Gaussian density estimator for a data set characterised by two variables, with contour lines at different levels of certainty indicated.

Table 5.6 QDA Mahalanobis distance to classes A and B for data in Table 5.1 together with the classification at a confidence limit of 90% (cut-off 2.146); shaded cells are outside the limits.

Sample	d_{iA}	d_{iB}	A model	B model	Decision
1	0.831	10.167	A	Not B	A
2	0.744	6.075	A	Not B	A
3	0.541	9.936	A	Not B	A
4	1.012	4.563	A	Not B	A
5	0.706	10.870	A	Not B	A
6	1.001	7.793	A	Not B	A
7	2.456	23.890	Not A	Not B	Outlier
8	1.236	18.031	A	Not B	A
9	2.357	3.367	Not A	Not B	Outlier
10	1.680	11.765	A	Not B	A
11	4.197	1.992	Not A	B	B
12	3.920	0.497	Not A	B	B
13	2.284	0.881	Not A	B	B
14	1.731	1.813	A	B	Ambiguous
15	6.898	2.011	Not A	B	B
16	3.498	1.992	Not A	B	B
17	3.355	0.854	Not A	B	B
18	4.774	0.930	Not A	B	B
19	4.159	0.962	Not A	B	B
20	2.069	1.091	A	B	Ambiguous

as $p = 0.1$, and statisticians like to say that it is a 10% probability of rejecting the null hypothesis that the sample is not part of the known group, although analytical chemists usually prefer to talk about confidence limits), we can calculate χ^2 for two degrees of freedom (as there are two variables) at a confidence level of 0.9. This is 4.605 as can be verified either in Excel or Matlab (use the CHISQ.INV function in Excel) or any other common statistically oriented package. This is the value of the squared Mahalanobis distance below which we expect to find 90% of the samples. Hence, the Mahalanobis distance threshold is $\sqrt{(4.605)} = 2.146$.

- Looking at our samples, sample 1 is 0.831 units from the centre of class A but 10.167 from class B. Hence, it is unambiguously assigned to class A.
- Samples 7 and 9 are classified as outliers, that is, members of no known class, using 90% confidence threshold.
- Samples 14 and 20 are ambiguous.

Note that the decisions change if the confidence limit or p value differs, for example, at a confidence threshold of 99% ($p = 0.01$), the limit is 3.035 when there are two variables. In our case study, this results in more ambiguous samples (number 13 is now ambiguous) but no outliers. Of course, we still expect a few samples to be outside the defined confidence limits even if the model and underlying population distribution are correct (e.g. if we set a confidence level of 99%, then 3 out of 300 samples are expected to be outside these limits), and it is, therefore, somewhat subjective to choose the confidence limit or p value as our threshold. Many users of automated packages do not appreciate this, and a large number of presentations have been shown where all training samples are well within 95% confidence limits, obviously some mistake or false assumption has been made in the calculation. Usually, a 90% confidence limit is considered too tight. The assumptions also suppose that the data are (multi)normally distributed, and this can be misleading in some cases. However, extreme samples, very far from centroids, and very ambiguous samples, often in regions of overlap, can usually be identified as the nature of the underlying distribution is not so critical.

For our case study, we can present the boundaries, graphically shown in Figure 5.9, for the 90% confidence limits. The conclusions presented in Table 5.6 can be verified by inspecting the positions of the data points relative to the 90% thresholds. In most realistic cases, of course, we will have far more data points and use much wider boundaries (higher confidence limits), but this is for illustrative purposes.

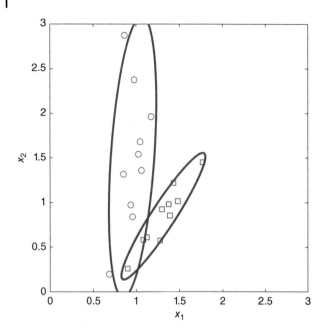

Figure 5.9 QDA one-class boundaries using 90% confidence ($p = 0.1$) for data in Table 5.1.

Some papers, instead of using a χ^2 criterion, use the Hotelling's T^2 as an alternative. This is related to the F distribution (see Sections A.3.4 and A.3.5), with a scaling factor, and the interested reader is referred to the detailed literature. Hotelling's T^2 is designed for very small sample sizes; however, both measures also assume underlying normality, which is rarely perfectly the case; hence, once there are more than half a dozen samples, the difference is only, in practice, useful for checking simulations.

It is also important to remember that the Mahalanobis distance cannot be calculated if there are more variables than samples, unless PCA is performed first on the data, as discussed in Section 5.2.1.2.

1. A data set is characterised by 10 variables; what is the Mahalanobis distance corresponding to a value of $p = 0.05$ assuming a χ^2 distribution?

   (a) 18.31
   (b) 16.92
   (c) 4.11
   (d) 4.28

2. A sample is at three Mahalanobis distance units from the centroid of class A and 2.5 from the centroid of class B. Four variables have been measured. If we use a 95% confidence limit ($p = 0.05$), using a χ^2 distribution,

   (a) The sample is assigned to class B.
   (b) The sample is assigned to neither class.
   (c) The sample is ambiguous.

5.3.2 Disjoint PCA and SIMCA

Self-independent modelling of class analogy (SIMCA) was probably the first method reported, in the 1970s, to be specifically aimed at chemometricians. It would be recognised nowadays as a one-class classifier.

When the number of variables is large, it often makes sense to reduce the variables by PCA first, before classification, as discussed in Chapter 4, and perform the classification (e.g. QDA) on the reduced PC model. Traditionally, if we perform PCA, we would take the entire data set consisting of all classes and reduce the variables using PCA, which is called a conjoint model. However, there is a conceptual problem. If the aim is to do one-class classification on group A, which we call the *in-group*, should we take group B into account as well? If PCA is performed on all the data together, in practice, group B is influencing group A in the PC model. There is also another disadvantage. One-class classification should be performed independently on each class; thus, it should not matter how many classes there are, and if new classes are introduced (we may have formed a model on esters and ketone and at a later stage decide to introduce amides), it should not influence the independent one-class model. However, if PCA is performed on all the data, all models have to be changed if one class (or outliers) classifiers is added to the data set, destroying an aim of one-class classifiers. Of course, many of the early examples in the machine-learning literature did not involve PCA and were on the raw data, but in chemometrics, we often have too many variables and need this first step.

Disjoint PCA involves performing PCA separately on each class. The basis of this is that each class is modelled independently, and additional classes and so on can be added at a future time without destroying the original model. The procedure is as follows:

- Select only those samples from the in-group to be modelled, for example, class A.
- Centre class A by its own mean.
- After any further suitable pre-processing (if any), perform PCA and select an appropriate number of components.
- Calculate the scores of all the samples from class A by $T_A = X_A P'_A$ on the centred X_A data.
- Then, take all other data and subtract the mean of class A to give X_{notA}.
- Calculate the estimated scores of X_{notA} by $\hat{T}_{notA} = X_{notA} P'_A$.

We illustrate the difference between disjoint and conjoint models shown in Figure 5.10, where there are two classes and two variables. The one-PC models are illustrated. The disjoint models are represented by the best-fit straight line through each class separately. The out group projects onto this line.

- For our data presented in Table 5.1, the mean of class A is $\begin{bmatrix} 0.954 & 1.516 \end{bmatrix}$.
- Hence, the scores of the first PC for class A after mean centring can be calculated, as given in Table 5.7. We use a one-PC model.
- The estimated mean centred data for this class can also be calculated as given in the table.
- We can now determine two measures of fit.
  - The first measure is how well individual samples fit the PC model. This is often called the Q statistic (or sometimes the squared error of prediction). It is defined, for sample i to the class A model, by $Q_{iA}^2 = \sum_{j=1}^{J} e_{ijA}^2$, where e is the difference between the observed and estimated values. For sample 6, looking at Table 5.7, we see

  $$\sum_{j=1}^{2} e_{6jA}^2 = (0.105 - (-0.011))^2 + (-0.153 - (-0.145))^2 = 0.0134$$

  for a one-PC model.
  - The second measure is how far individual samples are from the centre of the model called the D statistic in squared Mahalanobis distance units, usually scaled so that it can be interpreted statistically. For sample 6, in PC space, the sample has a PC1 score of -0.145 from the centroid, which can be interpreted as a distance along the best-fit

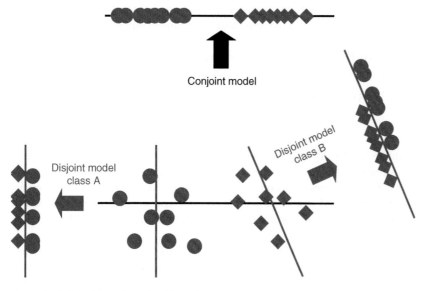

Conjoint model

Disjoint model class B

Disjoint model class A

Figure 5.10 Principles of disjoint PC models.

Table 5.7 Class A model using one PC (centred) for SIMCA and data in Table 5.1.

Class A

Samples	Scores	Variables	Loadings	Centred		Estimated centred	
1	−0.201	1	0.073	−0.107	−0.194	−0.015	−0.201
2	−0.540	2	0.997	−0.025	−0.539	−0.039	−0.538
3	0.036			0.066	0.031	0.003	0.035
4	−0.672			0.002	−0.674	−0.049	−0.671
5	0.177			0.091	0.171	0.013	0.176
6	−0.145			0.105	−0.153	−0.011	−0.145
7	1.349			−0.094	1.360	0.098	1.346
8	0.866			0.019	0.867	0.063	0.864
9	−1.332			−0.274	−1.315	−0.097	−1.328
10	0.462			0.217	0.448	0.034	0.461

Class B

Samples	Estimated scores	Variables	Loadings	Centred using class A		Estimated centred using class A	
11	−0.255	1	0.073	0.474	−0.290	−0.019	−0.254
12	−0.502	2	0.997	0.418	−0.534	−0.036	−0.501
13	−0.886			0.164	−0.900	−0.064	−0.884
14	−1.251			−0.054	−1.250	−0.091	−1.248
15	−0.004			0.812	−0.063	0.000	−0.004
16	−0.911			0.319	−0.936	−0.066	−0.908
17	−0.561			0.344	−0.587	−0.041	−0.559
18	−0.459			0.524	−0.498	−0.033	−0.458
19	−0.625			0.434	−0.658	−0.045	−0.623
20	−0.921			0.127	−0.932	−0.067	−0.918

straight line from the centre of the distribution; hence, we only need to rescale the units. If there is only one variable (or one PC), the mean squared Mahalanobis distance is simply the variance of the PC scores or 0.539. Therefore, the distance is $(−0.145)^2/0.539 = 0.039$ squared Mahalanobis distance units from the centre. As shown in Figure 5.1, this sample is quite central as expected.

- Similar calculations can be performed for class B using class A models. These are also presented in Table 5.7. Remember that the mean of class A must be used to transform the data of class B, so that for sample 13, for example

$$\begin{bmatrix} 0.164 & -0.900 \end{bmatrix} = \begin{bmatrix} 1.118 & 0.616 \end{bmatrix} - \begin{bmatrix} 0.954 & 1.516 \end{bmatrix}$$

In addition, we can also obtain a separate class B model, but for brevity, we only illustrate the calculations for a class A model.

The D statistic alone is not normally adequate to classify samples because the main variation between groups is often the main factor in conjoint PCA and is removed using a disjoint model. The Q statistic usually does provide good information about whether a sample is within a defined class or not. Usually, both measures are used to give a consensus. By suitable scaling, the D statistic can be viewed as a Mahalanobis distance and statistical tests are used to determine how well the sample fits into the model using QDA, as discussed in Section 5.3.1. Hence, SIMCA can be viewed as disjoint PCA followed by one-class QDA.

From Table 5.7, we can calculate these two measures, as presented in Table 5.8. In Figure 5.11, we show the first (disjoint) PC for class A. There are some interesting observations. For the Q statistic, mainly class A samples are close to the class A model and but not class B samples; however, a few class B samples are close to the class A model. Consider sample 14. It is close to the blue line, and therefore, its Q statistic is small, but it is quite far from the centroid (in this

Table 5.8 Q and D one-PC class A models for data in Table 5.1.

Distance from PC model (Q)				Distance from centroid (D)			
Sample				Sample			
1	0.009	11	0.244	1	0.075	11	0.121
2	0.000	12	0.208	2	0.540	12	0.468
3	0.004	13	0.052	3	0.002	13	1.455
4	0.003	14	0.001	4	0.838	14	2.900
5	0.006	15	0.663	5	0.058	15	0.000
6	0.013	16	0.149	6	0.039	16	1.537
7	0.037	17	0.149	7	3.375	17	0.583
8	0.002	18	0.312	8	1.390	18	0.390
9	0.032	19	0.231	9	3.287	19	0.724
10	0.034	20	0.038	10	0.396	20	1.571

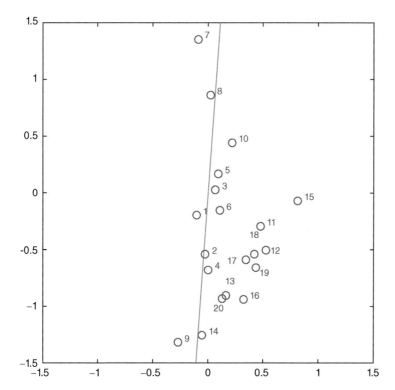

Figure 5.11 Class A disjoint model for PC1 for data set in Table 5.1, centred according to class A.

case, the origin) and hence has a high D statistic. For the D statistic, quite often samples from class B project into class A models, hence are not clearly separated in the new conjoint space; thus, both measures contain complementary information. In Figure 5.11, samples far from the line would exhibit high Q statistics even if they project close to its centre, whereas samples that project far from the centre of the line (or origin in our case) exhibit high D statistics even if they fall close to the line.

Detailed interpretation of the statistics and distance calculations are beyond the scope of this introductory text; however, it is important to understand the main steps of disjoint models, which are the basis of SIMCA, and it is usual to use both statistics or a combined measure. Statistical tests such as the χ^2, the F-test or Hotelling's T^2 can be used to further convert these distances into probabilities but depend on the underlying data being normally distributed, which is not always the case.

The advantage of disjoint models is that new classes can always be added to the data without disturbing the existing models. Another advantage is that PCA can be performed independently on any class, reducing the number of variables, which is often important in modern applications such as spectroscopy or chromatography where hundreds or indeed thousands of variables can be measured quite rapidly, and the principles in this section can easily be extended to multivariate situations. Of course, when there are several classes, it is not necessary to choose the same number of PCs for each class as they are independently modelled: in our case, the only way we can reduce the dimensionality is by going from two to one PC, but that is not of course the case when there are more than two variables; thus, for example, class A may be characterised by three PCs and class B by five PCs (in an imaginary example).

1. A data set consists of samples originating from three known classes.
    (a) We can form one type of conjoint PC model and one type of disjoint PC model.
    (b) We can form three types of conjoint PC models and three types of disjoint PC models.
    (c) We can form one type of conjoint PC model and three types of disjoint PC models.
    (d) We can form three types of conjoint PC models and one type of disjoint PC model.

5.4 Multi-Class Classifiers

When there are more than two classes, we have to adapt the methods discussed in Section 5.2 if we want to form hard boundaries between these classes, as illustrated in Figure 5.12.

For some types of classifiers, this procedure is quite straightforward. For distance-based methods discussed in Section 5.2.1, we simply take the distance (scaled as appropriate) to the nearest centroid. For kNN methods discussed in Section 5.2.3, assign to the class to which a sample has the most neighbours, with rules if there are ties (e.g. equally assign to more than one class).

However, for PLS-DA, the situation is significantly more difficult.

The simplest is to perform PLS-DA separately for each class using an approach called 'one versus all'. As we discuss in the next chapter, if there is only one c variable, we use the PLS1 algorithm by default. The c value is +1 for the in-group or class to be modelled and −1 for all the rest or out-group. Often, the X value is transformed separately for each calculation because the relative group sizes will differ, as discussed in Section 5.2.2. Hence, if there are three classes, three separate models are formed, namely (a) class A against all others, (b) class B against all others and (c) class C against all others. This is illustrated in Figure 5.13.

For each sample and each model, we will get an estimated value of c. For example, we may have three classes, and the estimated value of c for the class A model is −0.86, class B model +1.34 and class C model −1.20; therefore, it is assigned to class B. One problem, however, with these 'one versus all' models, is that there can frequently be ambiguity. For example, what happens if a sample has a positive estimated value of c for both class A and B models, and negative for class C, which do we assign it to? There are numerous rules, one might be that the sample is assigned to the class for which c is most positive, another for which it is closest to +1 – also remember that we could have a negative estimated value of c for all classes. There is no universal agreement or accepted rule, and there are often further complications if class sizes are unequal.

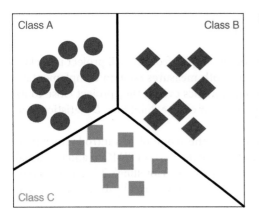

Figure 5.12 Multi-class classifiers.

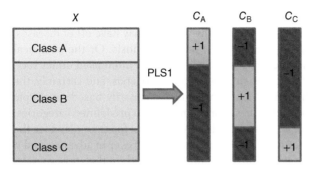

Figure 5.13 PLS1 multi-class models.

Some people instead use PLS2 (see Chapter 6), which models all classes at the same time. The C matrix consists of A columns, where A is the number of classes, each column corresponding to the c vector using PLS1. We will not discuss this method in detail, except to say that there are significant difficulties with transforming the X matrix, and often PLS2 results in quite misleading and hard to interpret results, hence is best avoided.

Many people using PLS-DA as a multi-class method are unaware of how the decisions are taken; for example, how columns are transformed, what is the decision criterion and whether PLS1 or PLS2 is used. For obvious groupings with clearly distinguished samples, this probably does not matter, but when there are tricky problems; for example, with overlapping groups, different group sizes, ambiguous or outlying samples, the method used is very critical to the model performance. The recommendation is not to use PLS-DA when there are more than two classes unless the problem is very straightforward or one is well aware of the steps in the algorithm employed. It is normally safer to use more straightforward approaches such as distance measures or one-class classifiers, which can of course be extended to any number of classes without difficulty.

1. PLS2 is always used as an alternative to PLS1 when there are more than two classes in a model for classification models.
   (a) True
   (b) False

5.5 Optimisation and Validation

Optimisation and validation are often confused. In the early days of chemometrics, and indeed multivariate classification literature, usually there was a known answer in advance. For example, there is classical work by the statistician R.A. Fisher who took 150 irises, divided into three classes each of difference species. There was no ambiguity about the desired answer. Hence, methods could be compared or optimised according to how well they predicted the known samples; that is, how well they were classified.

There is a huge chemometrics literature devoted to comparing and optimising classification methods. Investigators choose one or more favoured data sets and claim that method A (usually the one developed by the authors often in order to get a PhD or a grant) is better than all other methods. They often try hard to optimise their method, for example, by choosing an appropriate number of components (PCs or PLS components), tuning parameters (e.g. choosing the ones that give an answer that appears to lead to the highest correct classification rate), selecting variables and scaling and transforming data. An author may claim that their method results in 97% correctly classified, whereas a rival's method results in only 92%.

This mindset, in part, has its origins in analytical chemistry, where optimisation is an important aspect of most published methodological work. For example, chromatographic optimisation involves improving separations. Many laboratories report their performances against known reference standards, and the lower the error, the better the laboratory. In addition, much of traditional chemometrics also involved developing methods that tried to predict known answers – for example, to determine an NIR method that can predict HPLC quantification as accurately as possible and replace the definitive but slow HPLC method by a faster NIR technique. In all such cases, there is a well-established reference, and it is easy to say whether a chemometric method performs well or not.

However, chemometrics is now being employed in many other areas, for example, metabolomics or the study of human culture. In such situations, we may not be certain of the provenance of a sample even in the training set. As an example, are we certain that a patient has a known disease; hence, do we expect the GCMS of a urine sample to perfectly classify patients into two groups, diseased or healthy? Indeed, the healthy patients may have other diseases, or unusual genetics, and there may be a progression in disease (severe or mild) or misdiagnosis. Or the analytical technique may not be perfect. In mainstream statistics, Bayesian methods (see Chapter 3) are often used under such circumstances, but the majority of chemometrics literature and software prefers other approaches, and certainly, the vast majority of literature comparing and announcing new chemometrics techniques is primarily based on simple performance criteria – in our case, usually which method correctly classifies more samples into predefined categories. The problem here is that when doing pattern recognition, we are also often testing hypotheses; for example, does the NMR of serum extracts tell us whether a disease has progressed or not? We do not know the answer in advance, and it is actually very hard to find a fully controlled and representative set of samples. Therefore, a method that, for example, classifies the maximum proportion of samples into our preconceived groups may not be the most appropriate and may suffer from what is called overfitting.

Hence, we have a problem as are often trying to do two quite separate things. The first is optimisation; that is, get as high a success rate as possible, but the second is validation, that is test the underlying hypothesis, for example, that there really is a separation between groups. Traditionally, these two rather different aims have been mixed up and methods such as cross-validation (discussed in Chapter 4) have been used for both purposes. If we are certain of the underlying hypothesis as may happen in traditional analytical chemistry (e.g. we may want to separate unadulterated from adulterated orange juices and do some experiments in the laboratory we are really sure of before analysing the extracts), this is quite legitimate. But in other cases, where we cannot be sure of the provenance or representativeness of samples, we need to separate validation from optimisation.

1. Extracts of urine are obtained from 100 subjects, 50 of which have a disease and 50 of which are controls (do not have the disease). Method A classifies 87% correctly into one of the two categories, whereas Method B classifies 94% correctly.

   (a) Method B is definitely more appropriate and hence superior.
   (b) We cannot tell.

5.5.1 Validation

5.5.1.1 Test Sets
This is the process of determining whether a hypothesis is correct. The usual approach is to divide the overall data (often called auto-predictive) into two parts, a training set and a test set. Typically, 2/3 of the samples are assigned to the training set, although other splits are possible. This is illustrated in Figure 5.14.

The model is developed and optimised using the training set, but model is validated on the test set. Consider a simple illustration of a seating plan. We might be sociologists and want to study whether people of different genders sit next to each other. For example, for an audience in a family entertainment, there may be more or less random distribution between males and females. However, when giving a more formal presentation to an audience in a traditional society

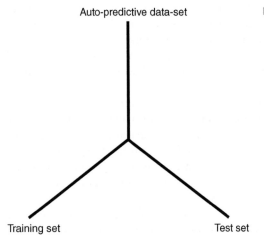

Figure 5.14 Division of data into training and test sets.

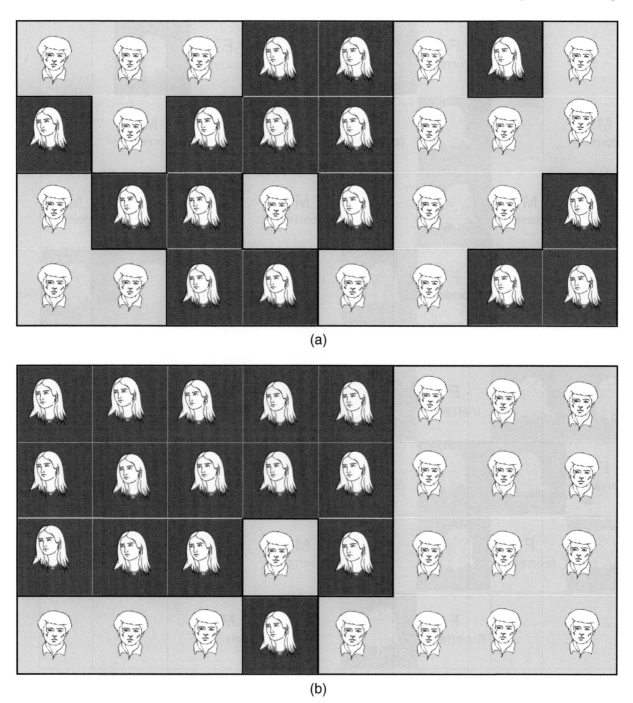

(a)

(b)

Figure 5.15 Two different seating plans.

on the whole males and females group separately, although there would be no rigid rules. Hence, we are interested in whether there is any gender-based clumping. Consider the two patterns shown in Figure 5.15.

Now consider a classifier, which can be formulated by a set of rules as follows.

- For each unknown place, look at the gender of the nearest neighbours both horizontally and vertically. For places in the middle of the seating plan, this will be four, for places at the edges 3, and in the corners just 2.
- Look at the gender of the nearest neighbours and assign the empty seat to the majority gender.

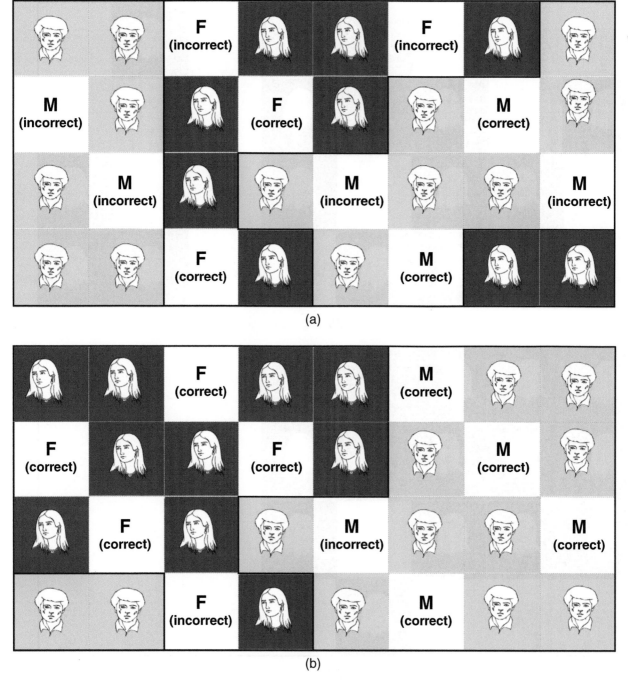

Figure 5.16 Dividing the data in Figure 5.15 into training and test sets.

Rather than checking these rules on the entire audience, we will remove approximately 1/3 of the samples as a test set and predict their gender as shown in Figure 5.16. Using the rules above, we can see

- For seating plan (a), 4/10 or 40% of the test set is correctly classified,
- Whereas for seating plan (b), 8/10 or 80% of the test set is correctly classified.

If the audience were completely randomly seated with respect to gender, we would expect around 50% correctly classified; however, the sample is small (10 members of the test set), and thus differs slightly, rather like an unbiased coin, if tossed 10 times will not always throw five Heads. However, the two results are consistent with the hypothesis that for seating plan (a), the audience is more or less randomly distributed according to gender, hence may be the family entertainment, whereas for seating plan (b), it is almost certainly not and may correspond to a traditional cultural event.

These conclusions from the test set do not tell us whether we have optimised our method or indeed anything about whether the method is good or not. The method provides quite different results according to the underlying data structure and primarily what we are doing is testing our hypothesis not finding out whether the method is better than any other method. Of course, different methods still do provide different answers, but this may be because each method has certain inherent assumptions about the data structure; thus, rather than answer whether one method is better than another, we are asking whether the data fit into a certain hypothesised structure.

Usually, the test set is predicted less well to the overall (auto-predictive) model or the training set.

We will illustrate the division into test and training set using the data given in Table 5.1.

- We will assign samples 1–3, and samples 11–13, to the test set and the remaining 14 samples to the training set.
- We will use the Euclidean distance to centroids method as an example, of course, any other method could be employed.

The division into test and training sets is illustrated in Table 5.9.

For the training set, we need to recalculate the means, for seven rather than the original 10 samples in each group, which are now $[0.963 \quad 1.617]$ and $[1.312 \quad 0.813]$ for the two classes A and B compared with the overall auto-predictive averages (involving all samples) of $[0.955 \quad 1.517]$ and $[1.310 \quad 0.851]$, which represent slight shifts in centroids. For methods such as LDA and QDA, it is also necessary to recalculate the variance–covariance matrix to take into account the smaller training set, rather than use the overall data matrix.

The calculated Euclidean distances are presented in Table 5.10. We can make a number of observations.

- Twelve out of 14 or 86% of the training set samples are correctly classified.
- Five out of 6 or 84% of the test set samples are correctly classified.

Table 5.9 Division into training and test set.

Sample		
	Training set	
4	0.956	0.842
5	1.045	1.687
6	1.059	1.363
7	0.860	2.876
8	0.973	2.383
9	0.680	0.201
10	1.171	1.964
14	0.900	0.266
15	1.766	1.453
16	1.273	0.580
17	1.298	0.929
18	1.478	1.018
19	1.388	0.858
20	1.081	0.584
	Test set	
1	0.847	1.322
2	0.929	0.977
3	1.020	1.547
11	1.428	1.226
12	1.372	0.982
13	1.118	0.616

Table 5.10 EDC model of data in Table 5.1 divided into training and test sets.

Sample	Training set				Test set		
	d_{iA}	d_{iB}			d_{iA}	d_{iB}	
4	0.775	0.357	B	1	0.317	0.690	A
5	0.108	0.914	A	2	0.640	0.417	B
6	0.271	0.606	A	3	0.090	0.790	A
7	1.264	2.112	A	11	0.607	0.429	B
8	0.766	1.607	A	12	0.755	0.180	B
9	1.444	0.879	B	13	1.012	0.276	B
10	0.405	1.160	A				
14	1.352	0.684	B				
15	0.819	0.785	B				
16	1.082	0.236	B				
17	0.765	0.117	B				
18	0.789	0.264	B				
19	0.869	0.089	B				
20	1.039	0.325	B				

- As these values are very similar, we deduce that the model was not particularly overfitted (otherwise the training set %CC (per cent correctly classified) would be much higher than the test set) and probably there is reasonable evidence of separation between the groups.
- If we calculate the average Euclidean distance from the centroids over the training and test sets, it is $[0.758 \quad 0.646]$ compared with $[0.715 \quad 0.631]$ for the overall (auto-predictive) result in Table 5.2, which is to be expected.
- Note that for this example, the average distance of the test set samples to the training set centroid is in fact less than that for the training set because we chose only three samples and would differ according to the samples chosen; if all the three possible combinations of samples were selected, then the average distance should be larger. Hence, there is some importance to how samples are selected for the test set.

In practice, in our example, the assumption that there is a differentiation between the classes is a reasonable one, although there are one or two samples in a region of overlap. Obviously, there will be other examples, and other methods, where this will not be so; however, for brevity, we are mainly illustrating the principles of the methods rather than searching out specific case studies. However, validation can protect against overfitting and allow us to determine whether we are justified in using our model or not.

1. The proportion of samples correctly classified using a test set is normally less than that for the training set, but this is not an infallible rule.

    (a) True
    (b) False

5.5.1.2 Performance Indicators and Contingency Tables

Of course, in most cases, quite sophisticated hypothesis tests could also be performed on the test set, but our aim is mainly to ask whether we can truly separate the groups and how well. Usually, a number of performance indicators are calculated, the most straightforward is the %CC or percentage correctly classified from the test set samples as mentioned above. A poor %CC does not necessarily indicate a poor method.

Sometimes, different classes are predicted with different efficiency, and the data can be presented in the form of a contingency table (also called a confusion matrix). This is illustrated in Table 5.11. The columns represent the known (or diagnosed) classes and the rows represent the predictions. From this table, it can be seen that classes A and B are somewhat less well distinguished compared with class C and are sometimes confused with each other: for example, they may represent two genetically similar organisms, whereas class C is quite distinct genetically.

Table 5.11 A simple contingency table.

	Known class		
Predicted class	A	B	C
A	40	6	0
B	10	44	1
C	0	0	49

Table 5.12 A 2 × 2 contingency table.

	A (positive)	B (negative)
A (positive)	42 (TP)	11 (FP)
B (negative)	8 (FN)	39 (TN)

There is quite a large literature on the use of contingency tables, which is beyond the scope of this chapter, but they are valuable ways of presenting results. We can calculate a number of statistics.

- The overall %CC is the percentage of samples correctly classified overall, which, in our case, is $100 \times (40 + 44 + 49)/150 = 89\%$.
- The %CC of each class can be calculated, for class A it is 80%, class B it is 88% and class C it is 98%. As all classes are of equal size, the overall %CC is the average for all classes. For unequal class sizes, this is not necessarily so, although Bayesian statistics could be used to scale each column to equal importance, reflecting equal prior probability of belonging to each class. In this chapter, we do not discuss Bayesian extensions, although these are introduced in Section 3.6.3 in the context of signal analysis. In machine learning, Bayesian adjustments are common but much less are used in chemometrics.

When there are only two classes, as in Table 5.12, the analysis becomes even simpler. One class is often called positive (e.g. with a known disease, or in forensics contaminated/guilty) and the other negative.

- True positive (TP), false positive (FP), true negative (TN) and false negative (FN) are self-evident and as defined in the table.
- Sensitivity is defined by $TP/(TP + FN)$ or $42/50 = 84\%$ in our case.
- Specificity is defined by $TN/(FP + TN)$ or $39/50 = 78\%$ in our case.
- In some areas, such as forensics, the likelihood ratio is used instead, defined by $LR^+ = \text{sensitivity}/(1 - \text{specificity})$ and $LR^- = (1 - \text{sensitivity})/\text{specificity}$.

Of course, contingency tables can also be obtained for training sets and auto-prediction (all samples) as well as test sets.

These indicators can, among others, be used to provide a quantitative assessment of model performance and therefore hypothesise whether the underlying assumptions about a data set are correct.

1. Test set results are in the following contingency table:

	Known class		
Predicted class	A	B	C
A	17	6	2
B	4	28	5
C	3	1	34

To the nearest percentage point.

(a) The per cent correctly classified for class B is 80%.

(b) The per cent correctly classified for class B is 76%.

(c) The per cent correctly classified for class B is 79%.

5.5.1.3 Iterations

When data sets are quite modest in size, however, the performance of a test set may well be influenced by one or two samples, which could, for example, be atypical or ambiguous or outliers; hence, a more sophisticated approach is to generate a test set many times over by selecting different samples in each occasion. Indicators of success can be averaged over each iteration. As computing power becomes more powerful, this repeated validation is less time consuming and, rather than taking hours on a typical desktop or laptop, can be done in minutes or seconds. The calculations can be performed, for example, 100 times over, using different test and training set combinations, and an average or consensus view is obtained.

Usually, the test sets are drawn to represent the same proportion of each class as the overall data set, so that, for example, if a data set of 60 samples consists of 30 each of classes A and B, the test set may consist of 10 samples randomly selected from each class, rather than 20 samples selected from the entire data set, which may be unevenly distributed between classes. This sometimes causes dilemmas when class sizes are unequal, the default is to keep to the original proportions. Hence, if there are 90 samples in class A and 30 in class B, the test set may consist of 30 from class A and 10 from class B, but this is certainly not a universal rule.

There are sometimes algorithms for repeated selection of test set samples, and methods in the literature include double cross-validation and iterative selection of test and training sets.

1. There is a suspected outlier in a data set.

    (a) It is best removed from the data set before validating pattern recognition methods.

    (b) By repeated generation of test and training sets, its influence can be averaged and this is the best approach unless we are certain that we would never encounter such a sample.

    (c) It should always be part of the test set and never the training set.

5.5.1.4 Permutation Methods

There are, however, other quite different approaches. Often, statisticians like to test against a null hypothesis that there is no significant discrimination between classes (in our case). A common approach involving permuting the classifier is the Monte Carlo method.

- The classification of each sample is randomly permuted. Hence, if there are 10 samples of class A and 10 of class B, membership of class A is randomly assigned to any 10 of the original samples and of class B to the other 10. This means that samples will be assigned to either class approximately half the time; obviously, analogously as we cannot guarantee that an unbiased coin will turn up Heads exactly half the time, sometimes as a sample will be chosen as class A slightly more frequently to class B.
- Then, the full classification procedure is performed, often involving a test and training set split, and an indicator of success such as the %CC is calculated on the permuted data, usually using the test set.
- This permutation is repeated many times (e.g. 100 times), each time generating a slightly different value of %CC. These form an ensemble of %CCs for the permuted (basically null) data set.
- The %CC (or any other quantitative indicator of success) of the unpermuted data set is compared with the null ensemble, and if it is much larger, it is considered significant.

The ensemble of results form a background and can be compared with the result on the unpermuted (real) data. Figure 5.17 illustrates typical results of Monte Carlo permutations. The bars represent that the frequency-specific %CC values are obtained, with the vertical line the result for a real data set. It is then possible to put an empirical probability onto this according to where the real result falls on the curve, in our case equivalent to the relative area right of the red line. An advantage of this is that the underlying probability that a sample is correctly classified may be less than 100%; this may be due to overlap, ambiguity, difficulties in finding representative samples or often in metabolomics due to differences between individuals. However, if the test set classification ability is significantly better than that of the null

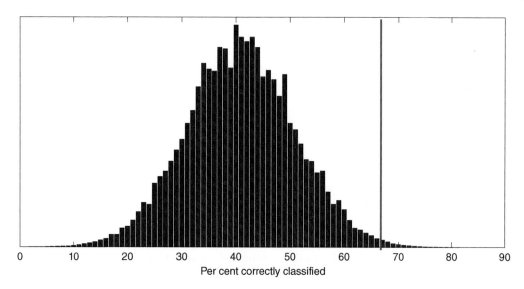

Figure 5.17 Monte Carlo methods: bars represent frequency of results of several permutations for the %CC, whereas the red line represents the unpermuted data.

or permuted data, it suggests that there is a significant difference between groups even if some individual samples are ambiguous.

1. Using PLS-DA, a sample has a c value of +1. When performing Monte Carlo methods
   (a) The c value is changed to −1.
   (b) The c value is changed to +1 or −1 with a 50% chance.
   (c) The c value is changed to +1 or −1 many times over, half the time +1 and half the time −1.
   (d) The c value is changed to +1 or −1 many times over, approximately half the time +1 and approximately half the time −1.

5.5.2 Optimisation

Optimisation involves obtaining a model that performs as well as possible. In most cases, this involves obtaining a value of %CC as high as possible and choosing the parameters that optimise this value.

In order to avoid overfitting (i.e. obtaining an unduly optimistic view), it is usually performed on the training set. There are many reasons for optimisation. Some of the most common are as follows.

- Choosing the optimum number of principal components.
- Choosing the optimum number of partial least-squares components.
- Choosing the best value of k when performing kNN.
- Choosing which variables (e.g. spectroscopic wavelengths or NMR peaks) to retain in the model.
- Reducing the number of variables before LDA.
- Scaling or transforming the data.

The first two are common in chemometrics and can be performed by several methods, some of which have been discussed already in Chapter 4 in the context of PCA. Cross-validation is well known, involving removing single or a group of samples, as a mini-test set, and forming the model on the remaining – the best model has the optimum cross-validated predictions. However, in the case of classification, the predictions are usually of the number of samples correctly classified rather than as a numerical fit to the data; thus, for example, if we had 50 samples in the original data, and found that using one PC or PLS component correctly classified 38 of the cross-validated samples, but two components predicted 42 correctly, and three predicted 46, then the three PC model would be the best of the three. A problem here is that for a single training set, the number correctly classified can be rather flat as the number of components is varied: to overcome this, iterative methods, as discussed in Section 5.5.1.3, can help, generating an average of many different training sets to give an average %CC (as appropriate).

Bootstrap training samples	Bootstrap test samples	Figure 5.18 Typical bootstrap sampling.
10	3	
7	5	
1	6	
9		
8		
7		
4		
1		
2		
9		

An alternative is the bootstrap, which is an iterative method. The main steps in the case of optimising the number of components are as follows.

- The training set is resampled as many times as there are samples.
- Each time, it is sampled repetitively; thus, some samples will be chosen more than once to give a bootstrap training set with some repetitions. Hence, if there are 50 samples in the training set, sample 1 may be chosen two times, sample 2 chosen three times and sample 3 never.
- Then, the ones never chosen are the bootstrap test set.
- Each time, the model is computed on the bootstrap training set: note that this can contain more than one sample of the same kind. It is performed using different numbers of components.
- Its performance is tested on the bootstrap test set for different numbers of components.
- The entire procedure is then repeated many times over, typically 200.
- The bootstrap test set performance for varying number of components is averaged over all these repetitions, and then an optimum number of components is determined.

The bootstrap could be used in other situations, such as optimising k for kNN.

A typical bootstrap sampling is illustrated in Figure 5.18.

- There are 10 samples in the original data set.
- In the bootstrap training set, samples 7, 1 and 9 are selected twice.
- Samples 3, 5 and 6 are never selected in the training set and therefore are part of the bootstrap test set.
- This partition of the data is repeated again many times over, typically 200; thus, the next time we might find samples 2 and 6 are selected twice and sample 9 three times, but samples 1, 4, 5 and 10 are never selected and hence form the bootstrap test set.

Naturally, some samples may be selected more than twice in the training set. The bootstrap test set has certain statistical properties that make this approach optimal.

Figure 5.19 illustrates a typical strategy including the bootstrap; the number of iterations/repetitions, of course, can differ. This also involves regenerating the test set many times over, as recommended in Section 5.5.1.3. Although computationally intense, with modern computer power, it is very feasible in realistic timescales.

There are, of course, other approaches to optimisation and this section is not exhaustive. However, it is important to separate optimisation from validation and also to establish a suitable criterion, usually the per cent correctly classified.

1. A data set consists of ten samples, numbered 1–10. When a bootstrap is used to select samples for the training set, the following 10 samples are selected: 2, 9, 3, 2, 10, 1, 4, 6, 9, 5. The bootstrap test set consists of

    (a) Samples 2 and 9.
    (b) Samples 7 and 8.
    (c) Samples 1, 3, 4, 5 and 6.

5.6 Significant Variables

A final and important topic quite specific to chemometrics relates to which variables are significant. This has an important role especially in metabolic profiling, where we may be interested to know compounds are best at discriminating between groups of samples. For example, which compounds are candidate biomarkers for a specific characteristic?

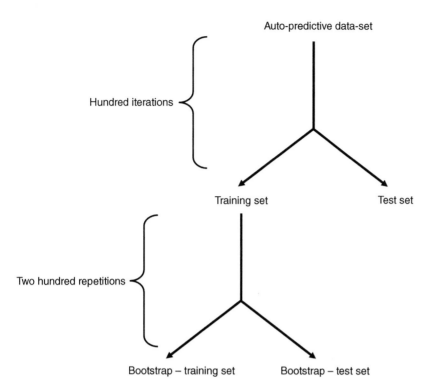

Figure 5.19 Division of data into test set and bootstrap test set and a typical iterative approach.

There are a large number of approaches. One important issue that many in the analytical chemistry community especially do not often appreciate is that often variable selection is an exploratory procedure. In spectroscopy, we often do place physical meaning to the most discriminatory variables, as discussed in Chapter 7, but in pattern recognition, it is more common just to determine which variables are most likely to be markers. There are often hundreds or thousands of variables, and as such, they may exceed the number of samples often by orders of magnitude. To illustrate the problem, let us consider tossing an unbiased coin 10 times (as an example of a small sample size). The distribution of Heads is presented in Figure 5.20 and can be obtained via the binomial theorem. The chance of obtaining eight Heads or more can be obtained by the binomial theorem and equals $\sum_{i=8}^{10} 0.5^i 0.5^{10-i} (10)!/((10-i)!i!)$ or around 5%. If we extend this calculation, we find that if we repeat this experiment 100 times, there is now only a 0.36% chance that we will never have a set tosses with eight Heads or more for an unbiased coin. Indeed, there is an approximately 50% chance that there will be six or more cases out of 100, where there are at least eight Heads out of 10 tosses. In other words, it is very likely that there will be several sets of tosses where the coin will turn up eight or more Heads.

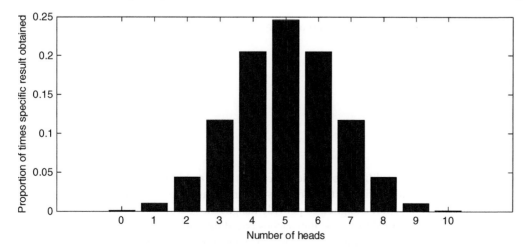

Figure 5.20 Distribution of Heads if an unbiased coin is tossed 10 times.

Table 5.13 Data set mentioned in Section 5.6.

Sample	Class	Variable									
		1	2	3	4	5	6	7	8	9	10
1	A	6.017	3.035	5.004	6.950	3.736	3.155	6.717	5.227	8.592	1.794
2	A	10.022	4.807	9.020	10.083	6.934	4.974	9.659	9.151	13.891	2.436
3	A	9.239	3.333	7.941	8.663	5.848	4.216	7.898	7.094	12.044	2.051
4	A	6.501	3.129	6.712	5.042	5.435	3.011	4.825	6.783	8.742	1.015
5	A	8.036	3.793	7.670	7.245	6.028	3.850	6.912	7.709	10.948	1.639
6	A	6.272	3.368	5.172	7.563	3.871	3.386	7.383	5.588	9.135	1.982
7	A	8.403	5.270	7.937	9.250	6.377	4.599	9.313	9.142	12.521	2.292
8	A	5.763	3.181	4.913	6.768	3.743	3.107	6.639	5.368	8.411	1.749
9	B	5.282	4.683	5.782	7.896	5.814	7.670	6.937	6.305	6.327	2.846
10	B	4.593	3.898	8.999	10.416	6.993	9.232	7.035	4.736	4.601	3.614
11	B	6.926	5.864	9.185	11.508	8.155	11.121	9.389	7.909	7.705	4.054
12	B	6.068	5.144	8.694	10.708	7.503	10.156	8.409	6.818	6.643	3.761
13	B	4.286	3.445	4.956	6.179	4.429	6.443	5.510	4.972	4.695	2.155
14	B	7.442	6.433	9.605	12.291	8.780	11.761	10.087	8.581	8.472	4.361
15	B	4.257	3.315	7.254	8.215	5.532	7.905	6.059	4.499	4.131	2.807
16	B	5.762	4.760	8.193	9.939	6.939	9.621	7.900	6.450	6.181	3.468

Extending this to real situations where we have thousands of variables and often quite limited sample sizes, it is quite possible to turn up potential markers whose distribution could be by chance. There is no really obvious solution to this except to increase the sample size, often by several orders of magnitude, which is usually impractical either in terms of cost (the grant body often has only limited funds), time and even the problem of finding representative samples. Hence, in many situations, the best that can be done is some exploratory analysis.

Nevertheless, chemometrics does play a very important role in pulling out potential markers. In areas such as spectroscopy, it can result in a definitive answer, but in many other areas, it is still also a useful tool for data mining.

In this section, we will illustrate the methods by a simple case study to show the main calculations and their pros and cons, but it is important to understand that the same principles can be extended to far larger real-world data sets. Table 5.13 presents a small 16×10 data set, in which the first eight samples are from class A and the second from class B. We are interested in variables that are most discriminatory. As this should ideally be regarded as an exploratory procedure, it can be done on the auto-predictive or overall data set, rather than a test set.

In this section, we describe three common approaches, but there are, of course, many more described in the literature.

1. An unbiased coin is tossed 12 times. What is the probability that there will be either more than nine or less than three Heads?
    (a) 0.0192
    (b) 0.0385
    (c) 0.0730
    (d) 0.1460

5.6.1 Partial Least-Squares Discriminant Loadings and Weights

Although PLS-DA holds little advantage over traditional statistical methods such as LDA as a classification method, it does hold a significant advantage in exploring variables that are most discriminatory between two or more classes. LDA and most other traditional methods were not designed to provide insight into variables, whereas PLS methods do provide this insight. Many traditional statistical tests use approaches based on the F-statistic, t statistic, ANOVA and so on, which we will discuss in Section 5.6.2, but this neglects interactions or correlations between variables and treat each variable in a univariate manner. Multivariate methods take these correlations into account.

Usually, the PLS-DA loadings or weights (see Chapter 6 and Section A.2.2) of the variables are calculated. Often, these are presented visually, but there are alternative numerical approaches as well. As an introductory text, we will primarily illustrate the visual approach.

- Usually, the first step is to standardise the data before PLS-DA. This means that all variables are on an equal scale. If not standardised, the data should be centred or column transformed, as discussed in Section 5.2.2.
- Then, PLS-DA is performed, coding the c values as appropriate. There is no requirement to split into test or training set.
- Using either visual inspection of the PLS scores T, or numerical tests (such as the t test), the most discriminatory PLS components can be determined. In most classification problems, these are the largest or first. Often, the first component alone is sufficient.
- The loadings P or weights for the corresponding discriminatory components are also compared with the scores. The more extreme the value, the more the variable influences this component; hence, variables with extreme loadings for discriminatory components correspond to markers that help discriminant different classes. The 'sign' of the loading (or weight) usually corresponds to the class it is associated with.

As an example, let us consider the data set given in Table 5.13. The scores and loadings of the first PLS component of our case study are shown in Figure 5.21, using the algorithm of this book. By comparing these, we see that

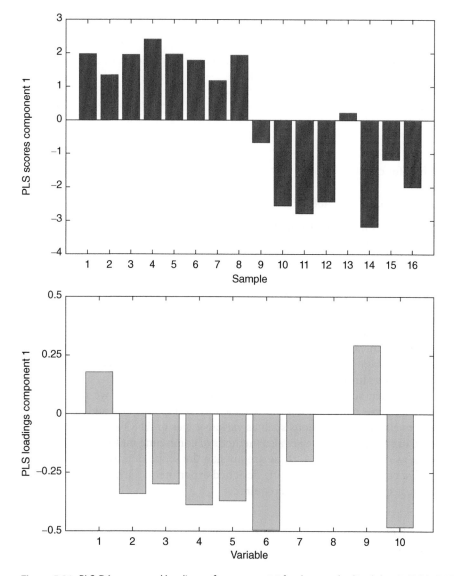

Figure 5.21 PLS-DA scores and loadings of component 1 for the standardised data in Table 5.13.

positive scores primarily relate to class A and negative to class B. The two variables with the most positive loadings are 1 and 9, whereas those with the most negative are 6 and 10, being the best markers for classes A and B respectively. Note that using most PLS1 algorithms, PLS loadings are not normalised, in contrast to PCA, or centred. For a rule of thumb, we select the extreme variables, which we visually identify as the markers for each group. The origin is not always significant, as the centre of gravity may lie one side of 0. Some workers prefer weights to loadings, as the latter are orthogonal; however, if the conclusions are visual, there is rarely a major difference in most practical situations.

There is a vast literature interpreting PLS-DA loadings, but the aim of this section is to introduce the usage as a method for determining variables that correspond to potential markers. PLS-DA is a very useful exploratory visual method, which can also be used to rank variables according to their significance.

1. PLS1-DA loadings for individual components where there are two classes usually
    (a) Have zero mean and a sum of squares of 1.
    (b) Have zero mean and a mean sum of squares of 1.
    (c) Have zero mean but do not necessarily have a sum (or mean sum) of squares of 1.
    (d) Do not necessarily have a zero mean or a sum (or mean sum) of squares of 1.

5.6.2 Univariate Statistical Indicators

It is also possible to determine univariate statistical indicators of whether variables are significant or not. There are various ways but they are mostly related. These can be done on the raw data.

One of the simplest is the t statistic defined as

$$t_j = (\bar{x}_{jA} - \bar{x}_{jB}) / (s_{jpooled} \sqrt{(1/I_A + 1/I_B)})$$

In our case for variable 3,

- $\bar{x}_{3A} = 3.740$ or the mean of the eight values for class A
- $\bar{x}_{3B} = 5.578$
- $s_{3pooled} = (s_{3A} + s_{3B})/2$ as both groups are equal sized or $(1.488 + 1.582)/2 = 1.535$
- The term $\sqrt{(1/I_A + 1/I_B)}$ is not strictly necessary if we are just interested in ranking variables, but it essential if the t value is to be converted to a probability, and in our case equals 0.5.
- Hence, the t value for variable is $(3.740 - 5.578)/(1.535 \times 0.5)$ or -1.352.
- This value could be converted into a p value using the t statistic, but this depends on the variables being normally distributed within each class, a condition that is rarely met. Although, it can also just be used to numerically rank and assign variables.

The t values for our data set are presented in Figure 5.22. For PLS-DA, we can calculate the loadings and weights for each PLS component, but for straightforward univariate statistics, there is only one value for each variable. In this example, we see that variables 6 and 10 are most diagnostic for class B and 1 and 9 for class A, a similar conclusion to PLS-DA; however by no means, all situations will be in agreement between the methods.

The univariate statistical tests are valuable when several factors result in separation. Although we may be able to separate two classes, say Male and Female, there may be other factors such as Young and Old and so the data could be grouped in different ways. There are elaborate extensions such as various types of ANOVA and multi-level approaches, often used in clinically designed experiments. There are also methods of combining multivariate methods with multi-level univariate approaches such as ASCA. However, these are outside the scope of a basic introduction and will only be necessary in certain specialist cases, normally involving designed experiments such as in clinical laboratories. In order to analyse such data, quite a specialist expertise is required.

1. Class A consists of 50 samples and class B of 100. For one of the variables, the mean for class A is 3.29 and class B 5.71. The standard deviation over class A is 1.28 and class B 2.79. What is the value of *t*, assuming class A is positive?

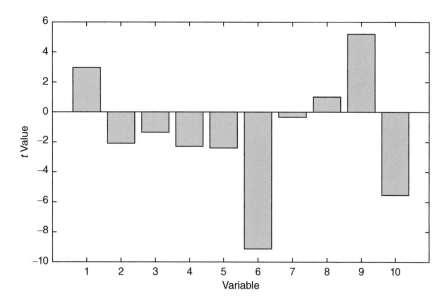

Figure 5.22 Values of *t* for the 10 variables in Table 5.13.

    (a) −1.057
    (b) −0.183
    (c) −1.189
    (d) −0.206

5.6.3 Variable Selection for SIMCA

The methods discussed in the last two sections are for two (or multi)-class classifiers. Strictly speaking, as SIMCA models each class separately, we should not be asking whether we can find variables that discriminate between classes. However, the classical SIMCA literature does define discriminatory power, so we report this for completion.

In order to determine this, it is necessary to fit each sample to both class (disjoint PC) models. For example, fit sample 1 to the PC model of both class A and class B. The residual matrices are then calculated, but there are now four such matrices.

- Samples in class A fitted to the model of class A.
- Samples in class A fitted to the model of class B.
- Samples in class B fitted to the model of class B.
- Samples in class B fitted to the model of class A.

We would expect matrices 2 and 4 to be a worse fit than matrices 1 and 3. The standard deviations for each variable are then calculated for these matrices to give

$$D_j = \sqrt{\frac{{}^{\text{class A}}_{}{}^{\text{model B}}s^2_{jresid} + {}^{\text{class B}}_{}{}^{\text{modelA}}s^2_{jresid}}{{}^{\text{class A}}_{}{}^{\text{model A}}s^2_{jresid} + {}^{\text{class B}}_{}{}^{\text{model B}}s^2_{jresid}}}$$

The bigger the value, the higher the discriminatory power and the more significant the variable. Discriminatory power can of course be calculated between any two classes.

1. Two groups are modelled. The standard deviation of variable 7 class A to class A is 12.35 and to class B is 17.12, whereas variable 7 class B to class A is 21.51 and to class B is 13.56. The discriminatory power is

    (a) 1.50
    (b) 0.67
    (c) 0.83

Problems

5.1 Classification Using Euclidean Distance and kNN
   Section 5.2.1.1 Section 5.2.3
   The following data represent three measurements, x, y and z, made on two classes of compound.

Object	Class	x	y	z
1	A	0.3	0.4	0.1
2	A	0.5	0.6	0.2
3	A	0.7	0.5	0.3
4	A	0.5	0.6	0.5
5	A	0.2	0.5	0.1
6	B	0.2	0.1	0.6
7	B	0.3	0.4	0.5
8	B	0.1	0.3	0.7
9	B	0.4	0.5	0.7

1. Calculate the centroids of each class (this is simply done by averaging the values of the three measurements over each class).
2. Calculate the Euclidean distance of all nine objects from the centroids of both classes A and B (you should obtain a table of 18 numbers). Verify that all objects do, indeed, belong to their respective classes.
3. An object of unknown origins has measurements (0.5, 0.3, 0.3). What is the distance from the centroids of each class and so which class is it more likely to belong to?
4. The k nearest neighbour criterion can also be used for classification. Find the distance of the object in question 3 from the nine objects in the table above. Which are the three closest objects, and does this confirm the conclusions in question 3?
5. Is there one object in the original data set that you might be slightly suspicious about?

5.2 Simple kNN Classification
   Section 5.2.3
   The following represents 5 measurements on 16 samples in two classes, A and B.

Sample						Class
1	37	3	56	32	66	A
2	91	84	64	37	50	A
3	27	34	68	28	63	A
4	44	25	71	25	60	A
5	46	60	45	23	53	A
6	25	32	45	21	43	A
7	36	53	99	42	92	A
8	56	53	92	37	82	A
9	95	58	59	35	33	B
10	29	25	30	13	21	B
11	96	91	55	31	32	B
12	60	34	29	19	15	B
13	43	74	44	21	34	B

Sample						Class
14	62	105	36	16	21	B
15	88	70	48	29	26	B
16	95	76	74	38	46	B

1. Calculate the 16×16 sample distance matrix, by computing the Euclidean distance between each sample.
2. For each sample, list the classes of the three and five nearest neighbours, using the distance matrix as a guide. Remember to include the sample itself in the assessment.
3. Verify that most samples belong to their proposed class. Is there a sample that is most probably misclassified?

5.3 Classification of Swedes into Fresh and Stored Using SIMCA.
   Section 4.3   Section 4.6.1   Section 4.6.4   Section 5.3.2
   The following consists of a training set of 14 swedes (the vegetable) divided into two groups, fresh and stored (indicated by F and S in the names), with the areas of eight GC peaks (A to H) from the extracts indicated. The aim is to set up a model to classify a swede into one of these two groups.

	A	B	C	D	E	F	G	H
FH	0.37	0.99	1.17	6.23	2.31	3.78	0.22	0.24
FA	0.84	0.78	2.02	5.47	5.41	2.8	0.45	0.46
FB	0.41	0.74	1.64	5.15	2.82	1.83	0.37	0.37
FI	0.26	0.45	1.5	4.35	3.08	2.01	0.52	0.49
FK	0.99	0.19	2.76	3.55	3.02	0.65	0.48	0.48
FN	0.7	0.46	2.51	2.79	2.83	1.68	0.24	0.25
FM	1.27	0.54	0.90	1.24	0.02	0.02	1.18	1.22
SI	1.53	0.83	3.49	2.76	10.3	1.92	0.89	0.86
SH	1.5	0.53	3.72	3.2	9.02	1.85	1.01	0.96
SA	1.55	0.82	3.25	3.23	7.69	1.99	0.85	0.87
SK	1.87	0.25	4.59	1.4	6.01	0.67	1.12	1.06
SB	0.8	0.46	3.58	3.95	4.7	2.05	0.75	0.75
SM	1.63	1.09	2.93	6.04	4.01	2.93	1.05	1.05
SN	3.45	1.09	5.56	3.3	3.47	1.52	1.74	1.71

In addition, two test set samples, X and Y, each belonging to one of the groups F and S have also been analysed by GC.

	A	B	C	D	E	F	G	H
FX	0.62	0.72	1.48	4.14	2.69	2.08	0.45	0.45
SY	1.55	0.78	3.32	3.2	5.75	1.77	1.04	1.02

1. Transform the data first by taking logarithms and then standardising over the 14 training set samples (use the population standard deviation). Why are these transformations used?
2. Perform PCA on the transformed PCs of the 14 objects in the training set and retain the first two PCs. What are the eigenvalues of these PCs and what percentage variability do they correspond to? Obtain a scores plot, indicating the objects from each class in different colours. Is there an outlier in the PC plot?
3. Remove this outlier, re-standardise the data over 13 objects and perform PCA again. Produce the scores plot of the first two PCs, indicating each class with different symbols. Comment on the improvement.
4. Rescale the data to provide two new data sets, one based on standardising over the first class and the other over the second class (minus the outlier) in all cases using logarithmically transformed data. Hence, data set (a)

involves subtracting the mean and dividing by the standard deviation for the fresh swedes and data set (b) for the stored swedes. Call these data sets ^FX and ^SX, each will be of dimensions 13×8, the superscript relating to the method of pre-processing.

5. For each data set (a) and (b), perform PCA over the objects belonging only to its own class (six or seven objects as appropriate) and keep the loadings of the first PC in each case. Call these loadings vectors ^Fp and ^Sp. Two row vectors, consisting of eight numbers, should be obtained.

6. For each data set, calculate the predicted scores for the first PC given by $^Ft = {}^FX{}^Fp'$ and $^St = {}^SX{}^Sp'$. Then, recalculate the predicted data sets using models (a) and (b) by multiplying the predicted scores by the appropriate loadings and call these $^F\hat{X}$ and $^S\hat{X}$.

7. For each of the 13 objects in the training set i, calculate the distance from the PC model of each class c by determining $d_{ic} = \sqrt{\sum_{j=1}^{J} ({}^c x_{ij} - {}^c \hat{x}_{ij})^2}$ where $J = 8$ and corresponds to the measurements, and the superscript c indicates a model for class c. For these objects, produce a class distance plot.

8. Extend the class distance plot to include the two samples in the test set using the method mentioned in steps 6 and 7 to determine the distance from the PC models. Are they predicted correctly?

5.4 Classification of Pottery from Pre-Classical Sites in Italy, Using Euclidean and Mahalanobis Distance (QDA) Measures.
Section 4.5 Section 4.6.4 Section 5.2.1.1 Section 5.2.1.3 Section 5.5.1.2
Measurements of elemental composition were performed on 58 samples of pottery from Southern Italy, divided into two groups: A (black carbon-containing bulks) and B (clayey ones). The data are as follows:

	Ti (%)	Sr (ppm)	Ba (ppm)	Mn (ppm)	Cr (ppm)	Ca (%)	Al (%)	Fe (%)	Mg (%)	Na (%)	K (%)	Class
A1	0.304	181	1007	642	60	1.640	8.342	3.542	0.458	0.548	1.799	A
A2	0.316	194	1246	792	64	2.017	8.592	3.696	0.509	0.537	1.816	A
A3	0.272	172	842	588	48	1.587	7.886	3.221	0.540	0.608	1.970	A
A4	0.301	147	843	526	62	1.032	8.547	3.455	0.546	0.664	1.908	A
A5	0.908	129	913	775	184	1.334	11.229	4.637	0.395	0.429	1.521	A
E1	0.394	105	1470	1377	90	1.370	10.344	4.543	0.408	0.411	2.025	A
E2	0.359	96	1188	839	86	1.396	9.537	4.099	0.427	0.482	1.929	A
E3	0.406	137	1485	1924	90	1.731	10.139	4.490	0.502	0.415	1.930	A
E4	0.418	133	1174	1325	91	1.432	10.501	4.641	0.548	0.500	2.081	A
L1	0.360	111	410	652	70	1.129	9.802	4.280	0.738	0.476	2.019	A
L2	0.280	112	1008	838	59	1.458	8.960	3.828	0.535	0.392	1.883	A
L3	0.271	117	1171	681	61	1.456	8.163	3.265	0.521	0.509	1.970	A
L4	0.288	103	915	558	60	1.268	8.465	3.437	0.572	0.479	1.893	A
L5	0.253	102	833	415	193	1.226	7.207	3.102	0.539	0.577	1.972	A
C1	0.303	131	601	1308	65	0.907	8.401	3.743	0.784	0.704	2.473	A
C2	0.264	121	878	921	69	1.164	7.926	3.431	0.636	0.523	2.032	A
C3	0.264	112	1622	1674	63	0.922	7.980	3.748	0.549	0.497	2.291	A
C4	0.252	111	793	750	53	1.171	8.070	3.536	0.599	0.551	2.282	A
C5	0.261	127	851	849	61	1.311	7.819	3.770	0.668	0.508	2.121	A
G8	0.397	177	582	939	61	1.260	8.694	4.146	0.656	0.579	1.941	A
G9	0.246	106	1121	795	53	1.332	8.744	3.669	0.571	0.477	1.803	A
G10	1.178	97	886	530	441	6.290	8.975	6.519	0.323	0.275	0.762	A
G11	0.428	457	1488	1138	85	1.525	9.822	4.367	0.504	0.422	2.055	A
P1	0.259	389	399	443	175	11.609	5.901	3.283	1.378	0.491	2.148	B
P2	0.185	233	456	601	144	11.043	4.674	2.743	0.711	0.464	0.909	B
P3	0.312	277	383	682	138	8.430	6.550	3.660	1.156	0.532	1.757	B

	Ti (%)	Sr (ppm)	Ba (ppm)	Mn (ppm)	Cr (ppm)	Ca (%)	Al (%)	Fe (%)	Mg (%)	Na (%)	K (%)	Class
P6	0.183	220	435	594	659	9.978	4.920	2.692	0.672	0.476	0.902	B
P7	0.271	392	427	410	125	12.009	5.997	3.245	1.378	0.527	2.173	B
P8	0.203	247	504	634	117	11.112	5.034	3.714	0.726	0.500	0.984	B
P9	0.182	217	474	520	92	12.922	4.573	2.330	0.590	0.547	0.746	B
P14	0.271	257	485	398	955	11.056	5.611	3.238	0.737	0.458	1.013	B
P15	0.236	228	203	592	83	9.061	6.795	3.514	0.750	0.506	1.574	B
P16	0.288	333	436	509	177	10.038	6.579	4.099	1.544	0.442	2.400	B
P17	0.331	309	460	530	97	9.952	6.267	3.344	1.123	0.519	1.746	B
P18	0.256	340	486	486	132	9.797	6.294	3.254	1.242	0.641	1.918	B
P19	0.292	289	426	531	143	8.372	6.874	3.360	1.055	0.592	1.598	B
P20	0.212	260	486	605	123	9.334	5.343	2.808	1.142	0.595	1.647	B
F1	0.301	320	475	556	142	8.819	6.914	3.597	1.067	0.584	1.635	B
F2	0.305	302	473	573	102	8.913	6.860	3.677	1.365	0.616	2.077	B
F3	0.300	204	192	575	79	7.422	7.663	3.476	1.060	0.521	2.324	B
F4	0.225	181	160	513	94	5.320	7.746	3.342	0.841	0.657	2.268	B
F5	0.306	209	109	536	285	7.866	7.210	3.528	0.971	0.534	1.851	B
F6	0.295	396	172	827	502	9.019	7.775	3.808	1.649	0.766	2.123	B
F7	0.279	230	99	760	129	5.344	7.781	3.535	1.200	0.827	2.305	B
D1	0.292	104	993	723	92	7.978	7.341	3.393	0.630	0.326	1.716	B
D2	0.338	232	687	683	108	4.988	8.617	3.985	1.035	0.697	2.215	B
D3	0.327	155	666	590	70	4.782	7.504	3.569	0.536	0.411	1.490	B
D4	0.233	98	560	678	73	8.936	5.831	2.748	0.542	0.282	1.248	B
M1	0.242	186	182	647	92	5.303	8.164	4.141	0.804	0.734	1.905	B
M2	0.271	473	198	459	89	10.205	6.547	3.035	1.157	0.951	0.828	B
M3	0.207	187	205	587	87	6.473	7.634	3.497	0.763	0.729	1.744	B
G1	0.271	195	472	587	104	5.119	7.657	3.949	0.836	0.671	1.845	B
G2	0.303	233	522	870	130	4.610	8.937	4.195	1.083	0.704	1.840	B
G3	0.166	193	322	498	80	7.633	6.443	3.196	0.743	0.460	1.390	B
G4	0.227	170	718	1384	87	3.491	7.833	3.971	0.783	0.707	1.949	B
G5	0.323	217	267	835	122	4.417	9.017	4.349	1.408	0.730	2.212	B
G6	0.291	272	197	613	86	6.055	7.384	3.343	1.214	0.762	2.056	B
G7	0.461	318	42	653	123	6.986	8.938	4.266	1.579	0.946	1.687	B

1. Standardise this matrix and explain why this transformation is important. Why is it usual to use the population rather than the sample standard deviation? All calculations below should be performed on this standardised data matrix.
2. Perform PCA, initially calculating 11 PCs, on the data of question 1. What is the total sum of the eigenvalues for all the 11 components, and what does this number relate to?
3. Plot the scores of PC2 versus PC1, using different symbols for classes A and B. Is there a good separation between classes? One object appears an outlier, which one?
4. Plot the loadings of PC2 versus PC1. Label these with the names of the elements.
5. Compare the loadings plot to the scores plot. Pick two elements that appear diagnostic of the two classes: these elements will appear in the loadings plot in the same direction of the classes (there may be more than one answer to this question). Plot the value of the standardised readings of these elements against each other, using different symbols, and show that reasonable (but not perfect) discrimination is possible.

6. From the loadings plots, choose a pair of elements that are very poor at discriminating (at right angles to the discriminating direction) and show that the resultant graph of the standardised readings of each element against the other is very poor and does not provide good discrimination.
7. Calculate the centroids of class A (excluding the outlier) and class B. Calculate the Euclidean distance of the 58 samples to both these centroids. Produce a class distance plot of distance to centroid of class A against class B, indicating the classes using different symbols, and comment.
8. Determine the variance–covariance matrix for the 11 elements and each of the classes (so that there should be two matrices of dimensions 11×11), remove the outlier first. Hence, calculate the Mahalanobis distance to each of the class centroids. What is the reason for using Mahalanobis distance rather than Euclidean distance? Produce a class distance plot for this new measure and comment.
9. Calculate the %correctly classified using the class distances in 8, using the lowest distance to indicate correct classification.

5.5 Linear Discriminant Analysis in QSAR to Study the Toxicity of Polyaromatic Hydrocarbons
Section 4.5   Section 5.2.1.2
Five molecular descriptors, A–E, have been calculated using molecular orbital computations for 32 PAHs, 10 of which have carcinogenic activity (A) and 22 not (I), as given below, the two groups being indicated.

		A	B	C	D	E
(1) Dibenzo[3,4;9,10]pyrene	A	−0.682	0.34	0.457	0.131	0.327
(2) Benzo[3,4]pyrene	A	−0.802	0.431	0.441	0.231	0.209
(3) Dibenzo[3,4;8,9]pyrene	A	−0.793	0.49	0.379	0.283	0.096
(4) Dibenzo[3,4;6,7]pyrene	A	−0.742	0.32	0.443	0.288	0.155
(5) Dibenzo[1,2;3,4]pyrene	A	−0.669	0.271	0.46	0.272	0.188
(6) Naphtho[2,3;3,4]pyrene	A	−0.648	0.345	0.356	0.186	0.17
(7) Dibenzo [1,2;5,6]anthracene	A	−0.684	0.21	0.548	0.403	0.146
(8) Tribenzo[3,4;6,7;8,9]pyrene	A	−0.671	0.333	0.426	0.135	0.292
(9) Dibenzo[1,2;3,4]phenanthrene	A	−0.711	0.179	0.784	0.351	0.434
(10) Tribenzo[3,4;6,7;8,9]pyrene	A	−0.68	0.284	0.34	0.648	−0.308
(11) Dibenzo[1,2;5,6]phenanthrene	I	−0.603	0.053	0.308	0.79	−0.482
(12) Benzo[1,2]anthracene	I	−0.715	0.263	0.542	0.593	−0.051
(13) Chrysene	I	−0.792	0.272	0.71	0.695	0.016
(14) Benzo[3,4]phenanthrene	I	−0.662	0.094	0.649	0.716	−0.067
(15) Dibenzo[1,2;7,8]anthracene	I	−0.618	0.126	0.519	0.5	0.019
(16) Dibenzo[1,2;3,4]anthracene	I	−0.714	0.215	0.672	0.342	0.33
(17) Benzo[1,2]pyrene	I	−0.718	0.221	0.541	0.308	0.233
(18) Phenanthrene	I	−0.769	0.164	0.917	0.551	0.366
(19) Triphenylene	I	−0.684	0	0.57	0.763	−0.193
(20) Benzo[1,2]naphthacene	I	−0.687	0.36	0.336	0.706	−0.37
(21) Dibenzo[3,4;5,6]phenanthrene	I	−0.657	0.121	0.598	0.452	0.147
(22) Picene	I	−0.68	0.178	0.564	0.393	0.171
(23) Tribenzo[1,2;3,4;5,6]anthracene	I	−0.637	0.115	0.37	0.456	−0.087
(24) Dibenzo[1,2;5,6]pyrene	I	−0.673	0.118	0.393	0.395	−0.001
(25) Phenanthra[2,3;1,2]anthracene	I	−0.555	0.126	0.554	0.25	0.304
(26) Benzo[1,2]pentacene	I	−0.618	0.374	0.226	0.581	−0.356
(27) Anthanthrene	I	−0.75	0.459	0.299	0.802	−0.503
(28) Benzene	I	−1	0	2	2	0
(29) Naphthalene	I	−1	0.382	1	1.333	−0.333

		A	B	C	D	E
(30) Pyrene	I	−0.879	0.434	0.457	0.654	−0.197
(31) Benzo[ghi]preylene	I	−0.684	0.245	0.42	0.492	−0.072
(32) Coronene	I	−0.539	0	0.431	0.45	−0.019

1. Perform PCA on the raw data and produce a scores plot of PC2 versus PC1. Two compounds appear outliers as evidenced by high scores on PC1. Distinguish the two groups using different symbols.
2. Remove these outliers, repeat the PCA calculation and produce a new scores plot for the first two PCs, distinguishing the groups. Perform all subsequent steps on the reduced data set of 30 compounds minus outliers using the raw data.
3. Calculate the variance–covariance matrix for each group (minus the outliers) separately and hence the pooled variance–covariance matrix C_{AB}.
4. Calculate the centroids for each class, and hence the linear discriminant function given by $(\bar{x}_A - \bar{x}_B) \cdot C_{AB}^{-1} \cdot x_i'$ for each object i. Represent this graphically. Suggest a cut-off value of this function which will discriminate most of the compounds. What is the percentage correctly classified?
5. One compound is poorly discriminated in 4, could this have been predicted at an earlier stage in the analysis?

5.6 Class Modelling Using PCA

Section 4.3   Section 4.5   Section 5.3.2

Two classes of compounds are studied. In each class, there are 10 samples, and eight variables have been measured. The data are as follows, with each column representing a variable and each row a sample.

Class A

−20.1	−13.8	−32.4	−12.1	8.0	−38.3	2.4	−21.0
38.2	3.6	−43.6	2.2	30.8	7.1	−6.2	−5.4
−19.2	1.4	39.3	−7.5	−24.1	−2.9	−0.4	−7.7
9.0	0.2	−15.1	3.0	10.3	2.0	−1.2	2.0
51.3	12.6	−13.3	7.5	20.6	36.7	−32.2	−14.5
−13.9	7.4	61.5	−11.6	−35	7.1	−3	−11.7
−18.9	−2.4	17.6	−8.5	−14.8	−13.5	9.9	−2.7
35.1	10.3	−0.4	6.5	9.9	31.2	−25.4	−9.4
16.6	6.0	5.8	−1.8	−6.4	19.6	−7.1	−1.2
7.1	−2.7	−24.8	7.1	14.9	1.1	−3.0	4.8

Class B

−2.9	−5.4	−12.0	−9.1	3.3	−13.3	−18.9	−30.5
30.7	8.3	−8.0	−39.1	3.8	−25.5	9.0	−47.2
15.1	7.1	10.9	−10.7	16.5	−17.2	−9.0	−34.6
−18.2	−13	−17	6.6	9.1	−9.6	−45.2	−34.6
12.2	2.8	−3.8	−5.2	4.0	1.2	−4.8	−11.2
19.8	19.8	55.0	−30	−26.3	0.3	33.2	−7.1
19.9	5.8	−3.1	−25.3	1.2	−15.6	9.5	−27.0
22.4	4.8	−9.1	−30.6	−3.2	−16.4	12.1	−28.9
5.5	0.6	−7.1	−11.7	−16.0	5.8	18.5	11.4
−36.2	−17.3	−14.1	32.3	2.75	11.2	−39.7	9.1

In all cases, perform uncentred PCA on the data. The exercises could be repeated with centred PCA, but only one set of answers is required.

1. Perform PCA on the overall data set involving all the 20 samples.
2. Verify that the overall data set is fully described by five PCs. Plot a graph of the scores of PC2 versus PC1 and show that there is no obvious distinction between the two classes.
3. Independent class modelling is common in chemometrics and is the basis of SIMCA. Perform uncentred PCA on classes A and B separately and verify that class A is described reasonably well using two PCs but class B by three PCs. Keep only these significant PCs in the data.
4. The predicted fit to a class can be computed as follows. To test the fit to class A, take the loadings of the PC model for class A, including two components (see question 3). Then, multiply the observed row vector for each sample in class A by the loadings model, to obtain two scores. Perform the same operation for each sample in class A. Calculate the sum of squares of the scores for each sample and compare this to the sum of squares of the original data for this sample. The closer these numbers are, the better. Repeat this for samples of class B, using the model of class A. Perform this operation as follows: (a) fitting all samples to the class A model and (b) fitting all samples to the class B model using three PCs this time.
5. A table consisting of 40 sums of squares (20 for the model of class A and 16 for the model of class B) should be obtained. Calculate the ratio of the sum of squares of the PC scores for a particular class model to the sum of squares of the original measurements for a given sample. The closer this is to 1, the better the model. A good result will involve a high value (>0.9) for the ratio using its own class model and a low value (<0.1) for the ratio using a different class model.
6. One class seems to be fit much better than the other. Which is it? Comment on the results and suggest whether there are any samples in the less good class that could be removed from the analysis.

5.7 Three-Class Linear Discriminant Analysis for the Metabolites of Wild-Type and Two Mutants of Arabidopsis
Section 4.5.1   Section 4.6.3   Section 5.2.1.2   Section 5.4

The following are the MS intensities of the 20 largest peaks of plant extracts, row scaled so that they sum to 100 for three groups of samples. This study investigated the metabolic network between wild-type and two mutant (methionine over accumulation 1 [mto1] and transparent test 4 [tt4]) plants regarding the alteration of metabolite accumulation in *Arabidopsis thaliana*. Samples 1–16 are wild type, samples 17–36 are tt4 and samples 37–49 are mto1.

Sample		61	72	73	74	75	76	77	79	89	100	103	117	129	131	133	147	148	156	204	217
																	m/z				
1	Wild	1.52	1.83	39.60	4.55	8.41	1.69	2.33	4.14	2.20	2.16	5.14	2.47	2.16	1.61	2.89	8.49	1.84	2.14	1.69	3.14
2	Wild	1.57	1.83	38.26	4.43	8.79	1.75	2.58	5.28	2.20	1.98	5.39	2.45	2.31	1.62	2.84	7.95	1.79	1.94	1.74	3.31
3	Wild	1.47	1.88	39.80	4.48	8.51	1.66	2.20	3.88	2.03	2.42	4.48	2.55	2.06	1.66	3.04	9.09	1.90	2.24	1.78	2.88
4	Wild	1.65	2.03	39.28	4.62	9.09	1.87	2.48	4.61	1.90	2.69	3.37	2.43	1.93	1.72	3.00	8.68	1.91	2.61	1.85	2.31
5	Wild	1.44	1.83	39.66	4.49	8.31	1.62	2.23	3.76	2.09	2.40	4.98	2.58	2.12	1.65	3.01	8.87	1.88	2.23	1.74	3.09
6	Wild	1.70	2.06	39.31	4.68	9.35	1.94	2.68	4.93	1.88	2.69	3.06	2.38	1.84	1.72	3.05	8.38	1.88	2.59	1.83	2.04
7	Wild	1.43	1.83	40.61	4.52	8.06	1.57	2.06	3.48	2.09	2.23	5.05	2.53	2.06	1.61	3.07	9.00	1.88	2.09	1.71	3.10
8	Wild	1.50	1.84	40.22	4.57	8.20	1.63	2.22	3.87	2.15	2.22	5.06	2.47	2.11	1.61	2.95	8.50	1.83	2.20	1.73	3.10
9	Wild	1.58	1.91	39.25	4.57	8.45	1.72	2.40	4.19	2.10	2.31	4.86	2.45	2.10	1.68	3.04	8.29	1.86	2.46	1.79	3.00
10	Wild	1.39	1.79	40.33	4.47	8.34	1.55	2.08	3.56	2.06	2.33	5.01	2.52	2.10	1.61	3.02	9.08	1.85	2.08	1.70	3.11
11	Wild	1.49	1.92	40.01	4.58	8.67	1.66	2.30	3.86	1.93	2.70	4.17	2.46	1.95	1.66	2.96	8.90	1.86	2.59	1.69	2.65
12	Wild	1.58	1.87	39.64	4.60	8.83	1.75	2.59	4.36	2.09	2.24	4.54	2.37	2.00	1.61	2.91	8.35	1.81	2.50	1.61	2.75
13	Wild	1.56	1.85	39.38	4.54	8.55	1.70	2.48	4.49	2.12	2.18	5.06	2.42	2.18	1.63	2.97	8.15	1.80	2.08	1.79	3.09
14	Wild	1.67	2.03	39.70	4.72	9.23	1.87	2.54	4.75	1.89	2.69	3.12	2.34	1.89	1.70	2.95	8.41	1.87	2.69	1.83	2.10
15	Wild	1.47	1.83	40.70	4.60	8.34	1.60	2.19	3.87	2.10	2.27	4.90	2.40	2.07	1.58	2.86	8.43	1.78	2.40	1.62	3.00
16	Wild	1.46	1.82	40.06	4.54	8.23	1.64	2.19	4.04	2.08	2.25	5.06	2.41	2.09	1.60	2.96	8.59	1.83	2.23	1.78	3.13
17	tt4	1.69	2.02	39.63	4.69	9.16	1.89	2.66	4.93	1.90	2.61	3.22	2.29	1.87	1.69	3.01	8.30	1.85	2.69	1.81	2.10
18	tt4	1.66	2.05	39.55	4.65	9.23	1.86	2.62	4.47	1.87	2.74	3.18	2.37	1.88	1.73	3.18	8.54	1.88	2.58	1.84	2.15
19	tt4	1.83	2.04	36.88	4.59	9.32	2.04	3.21	6.51	2.08	2.32	4.04	2.30	2.05	1.75	3.09	7.36	1.83	2.46	1.84	2.46
20	tt4	1.89	2.12	35.94	4.54	9.60	2.13	3.23	6.70	2.06	2.41	3.90	2.38	2.09	1.82	3.12	7.34	1.88	2.48	1.90	2.45

Sample		m/z																			
		61	72	73	74	75	76	77	79	89	100	103	117	129	131	133	147	148	156	204	217
21	tt4	1.70	2.06	39.69	4.71	9.13	1.93	2.71	4.74	1.86	2.72	3.15	2.29	1.82	1.72	3.18	8.28	1.86	2.56	1.84	2.06
22	tt4	1.51	1.86	40.11	4.59	8.37	1.69	2.41	4.16	2.08	2.27	4.93	2.35	2.02	1.61	3.01	8.27	1.80	2.31	1.70	2.94
23	tt4	1.52	1.90	39.50	4.54	8.45	1.71	2.39	4.13	1.97	2.49	4.60	2.40	2.05	1.68	3.14	8.57	1.86	2.43	1.78	2.89
24	tt4	2.02	2.23	35.65	4.66	9.81	2.28	3.57	7.18	2.06	2.57	2.93	2.28	1.99	1.85	3.02	7.22	1.92	2.91	1.92	1.93
25	tt4	1.73	2.06	38.54	4.64	9.40	1.98	2.91	5.04	1.88	2.74	3.04	2.34	1.85	1.77	3.22	8.35	1.90	2.63	1.92	2.04
26	tt4	1.56	1.94	39.60	4.61	8.41	1.76	2.35	4.14	2.01	2.54	4.52	2.41	1.99	1.68	3.07	8.46	1.88	2.50	1.80	2.79
27	tt4	1.77	2.07	39.17	4.74	9.35	2.00	2.80	5.10	1.92	2.65	3.07	2.29	1.84	1.72	3.03	8.09	1.87	2.77	1.79	1.97
28	tt4	1.71	2.00	36.78	4.44	9.28	1.95	2.95	5.84	2.02	2.30	4.43	2.42	2.15	1.77	3.19	7.78	1.86	2.29	2.02	2.82
29	tt4	1.74	2.09	38.49	4.70	9.25	2.00	2.69	4.95	1.93	2.82	3.12	2.33	1.89	1.78	3.04	8.30	1.93	2.94	1.93	2.09
30	tt4	1.89	2.13	37.46	4.71	9.81	2.14	3.10	5.82	1.97	2.65	3.03	2.28	1.88	1.79	3.06	7.68	1.89	2.85	1.92	1.94
31	tt4	1.61	1.91	39.12	4.60	8.57	1.81	2.61	4.74	2.10	2.29	4.69	2.33	2.05	1.66	2.97	8.03	1.82	2.53	1.73	2.83
32	tt4	1.70	2.00	36.19	4.38	9.31	1.94	2.91	5.96	2.14	2.27	4.82	2.51	2.22	1.76	3.02	7.96	1.90	2.26	1.78	2.97
33	tt4	1.70	2.02	39.69	4.74	9.15	1.89	2.71	4.75	1.88	2.66	3.36	2.26	1.84	1.70	3.04	8.11	1.83	2.78	1.78	2.12
34	tt4	1.58	1.93	38.88	4.54	8.94	1.82	2.52	4.73	1.98	2.55	4.09	2.38	1.98	1.69	3.05	8.65	1.88	2.44	1.77	2.60
35	tt4	1.79	2.07	37.90	4.68	9.20	2.00	2.97	5.52	2.03	2.54	3.74	2.26	1.99	1.76	2.98	7.61	1.85	2.91	1.87	2.33
36	tt4	1.51	1.89	39.40	4.56	8.53	1.74	2.44	4.22	2.03	2.47	4.79	2.36	2.02	1.66	3.11	8.41	1.83	2.38	1.73	2.91
37	mto1	1.58	1.87	41.05	4.69	8.47	1.61	2.15	3.54	1.94	2.56	4.40	2.33	2.03	1.63	2.96	8.65	1.82	2.17	1.72	2.83
38	mto1	1.50	1.81	40.30	4.55	8.28	1.57	2.06	3.51	2.04	2.40	5.07	2.48	2.20	1.64	3.04	8.88	1.85	1.88	1.70	3.22
39	mto1	1.82	2.06	39.61	4.78	9.19	1.92	2.63	4.55	1.85	2.76	3.00	2.30	1.93	1.75	3.05	8.45	1.92	2.48	1.83	2.11
40	mto1	1.72	2.01	39.96	4.70	9.22	1.84	2.55	4.41	1.80	2.77	3.02	2.30	1.93	1.72	3.14	8.75	1.89	2.33	1.79	2.17
41	mto1	1.60	1.84	40.44	4.61	8.55	1.64	2.16	3.60	2.04	2.38	4.74	2.41	2.10	1.62	2.96	8.69	1.84	2.05	1.72	2.99
42	mto1	1.79	1.98	40.39	4.76	9.25	1.81	2.50	4.17	1.92	2.55	3.43	2.31	1.96	1.67	2.97	8.41	1.85	2.28	1.73	2.29
43	mto1	1.55	1.86	41.14	4.65	8.60	1.57	2.22	3.60	1.84	2.61	4.18	2.33	2.00	1.64	3.11	8.75	1.80	2.13	1.69	2.73
44	mto1	1.59	1.82	40.03	4.58	8.46	1.65	2.21	3.95	2.06	2.35	4.90	2.41	2.14	1.63	2.95	8.57	1.83	2.06	1.74	3.08
45	mto1	1.82	2.00	39.81	4.80	9.23	1.88	2.73	4.60	1.90	2.63	3.31	2.22	1.96	1.70	2.95	8.17	1.84	2.49	1.77	2.19
46	mto1	1.81	1.98	40.69	4.82	9.25	1.83	2.68	4.36	1.82	2.61	3.09	2.22	1.88	1.67	3.02	8.28	1.81	2.43	1.68	2.05
47	mto1	1.70	1.82	39.11	4.55	8.85	1.77	2.64	3.90	2.20	2.12	5.07	2.37	2.11	1.61	2.99	8.43	1.84	2.21	1.60	3.10
48	mto1	1.65	1.91	39.96	4.63	9.01	1.72	2.58	4.06	1.87	2.67	3.62	2.37	2.00	1.68	3.01	8.93	1.88	2.24	1.76	2.44
49	mto1	1.61	1.82	39.84	4.57	8.58	1.64	2.58	3.77	1.98	2.47	4.66	2.39	2.09	1.64	3.04	8.79	1.85	1.99	1.75	2.94

The aim is to see whether the three groups can be distinguished. In this problem, we will only use auto-predictive approaches for simplicity, but a full analysis should also include independent test sets.

1. How many samples characterise each group?
2. Centre the data, perform PCA and plot a graph of the scores of the first two PCs against each other, distinguishing each class with different coloured symbols. Is there a promising separation?
3. Calculate the centroids of each of the three classes in the original space.
4. Determine the position of the centroids of the three classes in score space. To do this, first centre the original data, and then using the loadings from question 2, calculate $\bar{t}_A = \bar{x}_A P'$, where $\bar{x}_A$ is the centroid of A in the overall centred data set and P the overall loadings after centring. Plot these in the scores space of section 2, superimposed on the rest of the data.
5. Calculate the pooled variance–covariance matrix of the data set. Use the population statistics.
6. Using LDA (linear discriminant analysis) and the full variance–covariance matrix, calculate the 49×3 Mahalanobis distance matrix of each sample to the centroid of each group.
7. Calculate the predicted classification of each sample according to its Mahalanobis distance and comment.
8. QDA would be an alternative; however, in this case, it cannot be directly done using the 20 variables. Why is this, and what would be an alternative way of calculating the Mahalanobis distances for QDA?

5.8 Quadratic Discriminant Analysis for the Analysis of NMR in Urine Extracts of Diabetic and Control Donors: Training and Test Set Models
Section 4.5.1   Section 4.6.3   Section 5.3.1   Section 5.5.1.1   Section 5.5.1.2

The data below represent intensities of the NMR spectra of urine extracts at 20 frequencies for 54 female donors, 26 of which have diabetes (class D) and 28 do not (class C). The raw NMR spectra have been row scaled to 1000, and then 20 of the most diagnostic peaks selected, without further scaling. It is not the main aim of this problem to discuss scaling or variable selection issues.

		ppm																			
Sample	Class	7.84	7.76	7.64	7.55	7.51	7.47	7.42	7.37	7.32	7.27	7.16	7.09	2.67	2.17	1.53	1.49	1.41	1.15	1.06	0.88
1	D	14.747	0.331	7.042	13.532	1.741	3.643	9.967	13.561	3.169	6.843	3.023	3.188	13.704	15.001	5.781	5.796	7.307	4.141	6.341	9.691
2	D	24.612	0.775	11.107	25.175	1.298	2.167	8.263	13.651	3.04	7.616	2.794	1.822	8.925	13.257	5.633	4.468	7.383	4.301	5.735	8.562
3	D	25.826	0.635	11.466	26.008	1.393	2.001	11.005	17.827	2.982	8.665	5.276	0.588	6.474	12.948	4.984	5.161	5.944	2.665	5.384	6.54
4	D	42.823	0.837	17.791	39.209	0.876	2.438	9.923	16.686	3.248	6.827	2.261	3.666	44.077	14.097	5.469	6.937	7.559	4.593	5.502	6.245
5	D	42.623	0.925	18.411	41.482	1.284	2.912	8.632	13.905	3.251	7.521	2.472	3.336	30.335	15.165	6.861	6.076	7.813	4.061	4.959	7.803
6	D	12.869	0.739	5.886	13.186	0.778	2.167	7.294	12.871	3.833	8.102	2.741	4.979	24.775	17.353	5.853	6.546	7.294	2.68	5.927	7.526
7	D	38.674	1.256	16.479	36.443	1.183	1.642	10.125	16.697	2.616	5.817	1.516	3.043	60.505	16.46	4.977	6.216	6.675	2.683	5.007	6.362
8	D	31.248	0.94	12.985	29.332	1.257	2.248	9.243	16.351	3.335	7.124	2.437	4.008	22.096	13.499	5.073	5.653	6.087	2.372	5.09	5.85
9	D	16.029	0.012	5.627	12.757	1.202	3.47	11.432	13.704	2.605	4.617	2.221	3.542	46.43	14.528	5.742	7.525	6.578	2.774	6	6.452
10	D	37.297	1.123	16.778	38.038	1.269	2.151	4.9	8.04	4.137	5.281	2.925	2.041	10.133	11.311	6.263	5.144	7.544	4.072	4.964	5.799
11	D	49.435	1.329	22.256	50.142	1.739	2.168	9.753	14.867	4.732	8.445	3.288	1.42	10.107	9.696	4.813	4.509	6.207	2.652	4.481	4.968
12	D	32.211	1.075	14.556	32.756	0.963	2.223	7	11.811	3.663	5.403	2.629	1.008	6.666	12.35	5.68	5.24	6.985	2.35	4.818	5.663
13	D	25.134	0.833	9.483	21.063	0.9	1.763	7.732	13.291	3.188	5.61	2.497	5.634	11.22	13.656	5.981	6.24	6.943	18.847	5.707	5.688
14	D	18.158	0.838	7.985	18.456	0.796	1.412	5.581	10.303	2.877	4.76	2.408	2.256	8.747	13.997	7.07	6.533	8.4	5.401	5.38	6.919
15	D	26.053	0.937	11.672	26.53	1.51	2.439	9.063	14.688	4.045	7.272	6.234	0.82	8.82	12.73	5.681	5.937	7.26	4.226	4.828	5.359
16	D	21.044	0.733	7.649	15.848	1.021	1.929	7.77	13.182	3.299	5.978	2.218	4.042	24.165	13.28	5.739	8.458	6.919	37.637	6.669	5.331
17	D	21.417	1.17	9.515	21.929	1.535	2.547	9.458	15.373	4.017	7.745	3.85	0.964	18.284	13.463	6.479	6.698	7.941	5.1	5.426	6.114
18	D	53.276	2.198	23.62	53.946	1.613	2.731	7.217	11.582	4.294	7.41	3.633	0.829	16.137	18.765	5.24	6.203	7.41	4.493	6.063	6.22
19	D	16.177	1.819	7.252	16.803	1.293	2.569	8.515	14.726	3.889	7.907	4.937	0.744	12.507	17.667	6.689	5.876	6.927	4.695	5.161	6.397
20	D	17.941	0.695	6.645	13.641	0.935	1.693	5.515	10.242	3.263	4.973	2.315	3.877	24.405	16.174	6.757	8.086	7.537	31.226	6.012	7.333
21	D	17.515	0.632	7.349	16.624	1.025	2.053	6.808	11.515	3.033	6.77	4.734	1.954	5.192	12.556	5.677	4.82	7.134	3.222	5.776	6.928
22	D	45.416	0.765	19.613	44.437	1.174	1.788	7.457	11.969	3.827	7.122	2.368	5.057	8.164	10.321	4.794	4.163	6.093	4.209	5.679	5.5
23	D	23.28	0.704	10.384	20.025	1.648	3.349	7.666	10.228	3.103	5.954	3.33	2.939	13.026	12.232	5.719	5.507	7.08	6.842	6.175	5.98
24	D	28.64	0.866	12.736	28.403	1.342	2.161	7.691	12.311	3.435	6.965	3.269	0.913	8.725	11.697	5.63	5.491	7.105	5.31	5.336	6.456
25	D	17.202	2.225	6.547	14.218	1.213	2.188	9.234	14.882	3.765	9.371	4.962	1.854	10.971	15.2	5.347	5.03	6.969	5.136	5.482	6.553
26	D	25.782	0.78	10.889	24.76	0.975	1.919	6.499	10.813	3.452	7.266	3.908	1.377	5.439	11.57	5.223	4.169	6.707	2.993	5.592	6.476
27	C	17.29	0.756	8.033	18.027	1.637	3.55	8.829	11.741	3.489	6.506	10.798	1.235	14.816	14.66	6.158	8.182	7.34	4.341	6.738	5.986
28	C	23.188	0.77	11.577	23.613	2.895	4.724	15.92	24.559	3.295	8.142	3.84	1.628	31.918	14.141	5.165	5.737	6.497	4.463	5.827	6.043
29	C	32.104	0.571	14.286	31.27	1.893	2.06	15.902	27.15	2.694	9.602	3.01	2.255	20.223	14.242	5.742	5.422	7	3.362	5.428	6.773
30	C	34.779	3.614	16.185	36.197	2.126	2.647	16.383	25.011	4.911	12.295	8.746	1.319	11.646	11.643	5.078	4.701	6.361	2.7	4.961	5.281
31	C	27.195	1.634	12.404	24.998	2.718	3.949	18.92	27.569	4.65	13.153	3.947	3.213	28.033	12.296	5.082	5.992	6.288	4.246	5.947	5.317
32	C	24.316	1.008	11.285	25.266	2.003	2.341	17.006	28.208	3.645	11.747	3.996	3.887	16.884	12.57	5.741	5.666	7.169	3.057	5.498	6.451
33	C	18.677	0.805	8.52	17.006	1.496	1.913	5.256	11.195	3.183	5.703	3.32	1.758	48.073	23.002	5.201	5.452	7.077	5.354	5.152	6.898
34	C	49.637	1.127	21.715	49.409	2.005	3.054	11.553	18.901	3.879	7.521	2.911	0.87	16.509	12.922	4.639	4.125	5.668	2.876	4.548	5.062
35	C	20.125	1.006	8.751	19.351	1.832	3.369	10.109	16.653	4.353	8.371	5.32	2.485	8.201	12.019	5.667	6.72	7.081	2.746	5.574	7.104
36	C	19.39	0.315	7.983	16.055	1.624	3.326	8.59	14.203	3.958	6.577	2.8	3.039	32.034	12.462	5.595	7.561	6.706	6.1	6.848	7.187
37	C	15.357	0.45	5.93	13.871	1.009	1.542	6.936	12.787	2.999	5.64	2.478	2.916	21.454	12.935	5.922	7.923	7.317	4.136	7.007	7.458
38	C	8.91	0.404	3.808	9.155	0.986	1.14	9.714	17.184	6.165	19.067	3.462	0.593	19.277	14.619	10.444	7.805	13.749	2.848	13.155	10.112
39	C	39.818	2.038	18.05	40.635	0.97	1.557	4.767	8.047	2.911	6.162	7.768	0.883	13.484	10.511	5.08	4.765	6.795	2.414	4.394	6.126
40	C	20.148	0.373	9.059	18.089	0.968	1.633	6.092	9.835	2.189	4.784	2.241	2.618	35.452	12.318	5.492	6.369	6.741	4.113	5.845	6.213
41	C	73.806	1.245	32.199	73.48	1.797	2.595	10.488	14.514	4.246	9.054	3.116	1.202	9.427	9.075	4.205	3.959	5.706	3.858	3.956	3.79
42	C	5.878	1.304	2.747	6.514	1.2	2.6	5.142	9.515	4.787	4.852	6.606	0.66	30.569	14.802	6.523	14.779	7.575	3.044	7.077	7.958
43	C	7.539	1.177	2.42	5.036	2.054	14.03	7.39	29.377	12.359	7.36	22.272	4.133	40.679	66.384	5.948	15.129	6.295	2.536	6.074	6.965
44	C	4.411	4.37	1.987	4.519	1.544	2.531	6.547	12.154	4.132	7.069	2.323	0.772	56.857	13.157	5.083	10.155	7.173	3.018	7.401	6.574

		ppm																			
Sample	Class	7.84	7.76	7.64	7.55	7.51	7.47	7.42	7.37	7.32	7.27	7.16	7.09	2.67	2.17	1.53	1.49	1.41	1.15	1.06	0.88
45	C	8.007	1.747	3.779	8.708	2.269	5.411	10.216	13.007	5.229	7.044	3.603	4.029	33.459	13.999	6.322	16.342	6.617	2.654	7.645	7.085
46	C	14.045	1.28	6.026	14.043	1.031	3.99	4.946	10.64	5.759	4.625	2.792	6.647	30.275	17.834	10.045	12.137	8.596	4.115	5.707	7.833
47	C	9.118	2.857	3.464	7.96	0.732	1.87	3.534	8.105	4.461	4.355	2.74	1.357	31.533	16.911	8.252	9.734	8.049	3.103	5.894	8.132
48	C	22.562	0.693	10.43	23.228	1.456	2.93	15.061	24.705	3.696	8.557	3.033	1.369	33.071	15.119	6.606	9.879	8.183	2.9	5.313	6.663
49	C	17.14	0.802	7.5	17.303	1.032	2.04	8.482	18.528	18.33	43.603	2.63	0.374	7.519	8.531	28.564	4.969	31.701	2.089	34.271	21.094
50	C	7.341	0.634	2.928	6.823	1.199	1.676	10.114	20.766	15.185	37.447	3.164	0.673	5.89	8.987	24.015	4.838	24.554	2.514	28.198	14.837
51	C	12.766	0.523	5.784	13.409	1.477	1.493	8.451	15.009	3.406	7.259	3.187	0.747	8.49	19.029	9.482	7.507	10.12	4.959	6.256	9.576
52	C	12.816	0.806	6.072	13.302	1.837	2.311	11.705	19.693	3.482	9.061	8.955	1.135	6.624	15.234	7.091	6.934	8.707	8.1	5.917	7.634
53	C	32.625	0.89	14.753	33.662	1.694	7.578	6.943	25.369	8.272	6.844	19.868	1.029	7.319	51.042	4.794	4.141	6.59	3.416	4.239	6.067
54	C	9.195	0.593	4.369	10.078	1.778	4.557	9.002	19.381	5.371	7.574	9.287	0.979	7.856	25.331	6.367	5.353	8.645	4.693	6.227	10.473

1. Using centred data, perform PCA. Plot PCs 2, 3, 4 and 5 versus PC1 (four separate 2D graphs with PC1 on the vertical axis), distinguishing each group. Is there any likely discrimination?
2. Calculate the mean of each class, in the original space, and the variance–covariance matrix of each class.
3. Using the entire data set in auto-predictive mode, obtain a 54×2 distance matrix of the QDA Mahalanobis distance to the class centroids. Determine the predicted class membership for each sample, and the overall %CC. Comment.
4. Calculate the sum of squares of the Mahalanobis distance of all samples in group D to their centroid (the first 26 numbers of column 1 of the matrix in question 3) and the sum of squares of the Mahalanobis distance of all samples in group C to their centroid. Comment.
5. As a more realistic assessment, split the samples into training and test set. Use samples 1–22 as the training set for group D and 27–50 for group C. The remaining eight samples are the test set. Why, using the type of model above, on the raw data, is it necessary to have so many samples in the training set, and such small test sets, and how might this be overcome?
6. Calculate the Mahalanobis distances to the training sets (a 46×2 distance matrix) and the test sets (8×2). The means and variance–covariance matrices have to be recalculated for the training sets to reflect the different sample sizes compared with auto-prediction and used to estimate the new distances for the test sets. Calculate the overall %CCs for both training and test sets and comment.
7. Perform an auto-predictive model using PCA on centred data, and retaining five PCs. Calculate the Mahalanobis distance to both class centroids in the PC space of the scores of the first five PCs and the overall %CC.
8. Now split the data into training and test set as in question 5 and calculate the Mahalanobis distances for both training and test sets, and %CC for both, using five PC model on centred data. Note that to do this correctly, you must centre the test set according to the training set means and then estimate the test set scores using the training set loadings. Do this for a five PC model. Comment.
9. Repeat the calculation in 8 using 20 PCs and comment.

5.9 Partial Least-Squares Discriminant Analysis for the Distinction between Cancerous and Normal Lymph Nodes by Raman Spectroscopy: Training and Test Set Models
Section 4.5.1   Section 4.6.3   Section 5.2.2   Section 5.5.1.1   Section 5.5.1.2   Section 6.5.1
The following represents the Raman spectra of extracts from lymph nodes of 103 patients, either diagnosed benign (class 1) or cancerous (class 2). For illustrative purposes, the data have been digitised every $50\,cm^{-1}$, with the intensities averaged over a window; in practice, data are recorded 50 times better resolved, but the data have been simplified for illustration. Data are expressed in millions of counts.

		cm^{-1}																					
Sample	Class	675	725	775	825	875	925	975	1025	1075	1125	1175	1225	1275	1325	1375	1425	1475	1525	1575	1625	1675	1725
1	1	6.27	8.31	8.16	8.30	8.56	8.86	9.04	7.63	9.00	6.81	7.14	9.04	9.46	9.22	7.30	12.67	7.08	6.77	7.43	12.05	8.21	7.37
2	1	4.06	5.23	5.56	5.74	5.95	6.33	6.18	5.45	6.42	4.34	5.06	6.47	7.14	6.76	5.24	9.73	5.05	4.47	5.86	10.15	6.12	4.79
3	1	4.03	5.83	5.22	8.36	7.61	7.43	6.41	6.64	6.83	4.48	5.28	8.33	9.70	5.92	5.78	15.34	5.00	4.85	5.09	11.36	6.27	6.94
4	1	3.88	5.24	5.46	5.80	5.76	6.28	6.54	5.47	6.74	4.37	4.71	6.83	7.58	7.20	5.06	10.83	5.02	4.29	5.35	10.41	5.91	4.91

													cm⁻¹									
Sample	Class	675	725	775	825	875	925	975	1025	1075	1125	1175	1225	1275	1325	1375	1425	1475	1525	1575	1625	1675	1725
5	1	4.24	5.59	5.67	6.15	6.20	6.87	6.97	5.88	7.01	4.70	5.17	7.38	7.94	7.78	5.58	11.12	5.26	4.85	6.00	11.00	6.28	5.07
6	1	5.21	6.95	6.74	7.21	7.33	7.99	8.36	6.82	8.16	5.59	6.17	8.47	9.16	9.13	6.76	12.61	6.19	6.71	7.62	12.62	7.28	5.95
7	1	4.11	5.44	5.30	5.79	5.79	6.27	6.45	5.47	6.36	4.44	4.87	6.67	7.24	7.10	5.34	9.91	5.02	5.04	5.83	9.69	5.76	4.75
8	1	8.43	10.68	10.04	10.27	10.34	10.44	10.36	9.33	9.93	8.93	9.24	10.13	10.25	9.95	9.00	12.41	8.86	9.06	9.34	11.98	9.76	9.56
9	1	13.04	16.61	15.56	16.25	16.49	16.90	16.97	15.26	16.28	13.90	14.88	16.50	16.78	16.35	14.48	20.95	14.21	14.17	15.22	20.48	16.16	15.62
10	1	3.99	5.12	4.93	5.62	5.62	6.21	6.07	5.08	5.66	4.29	4.97	6.14	6.45	6.22	5.14	9.05	4.74	4.94	5.86	9.50	5.70	4.66
11	1	3.93	4.84	4.75	5.29	5.48	6.14	5.98	5.09	5.66	4.32	4.84	6.42	6.76	6.44	5.20	9.12	4.50	4.90	5.66	8.89	5.30	4.58
12	1	5.25	7.32	6.71	7.24	7.68	8.18	8.86	6.62	8.08	5.91	6.25	9.06	9.26	9.26	6.85	13.11	6.10	7.45	7.74	12.29	7.49	6.04
13	1	2.64	3.77	3.58	6.69	6.01	5.39	5.17	5.10	5.79	2.89	3.27	6.30	10.27	4.77	3.94	15.57	3.26	3.68	3.70	10.85	3.87	4.92
14	1	5.99	8.50	8.30	9.08	9.83	10.47	11.03	8.60	10.84	6.40	7.45	11.35	12.65	12.04	8.56	17.96	7.15	8.14	9.01	16.47	9.00	7.20
15	1	4.02	5.29	5.24	5.77	5.97	6.60	6.48	5.62	6.53	4.48	5.25	7.02	7.34	7.26	5.78	10.50	4.98	4.57	6.04	10.33	5.99	5.36
16	1	4.06	5.27	5.27	5.52	5.81	6.27	6.51	5.33	6.35	4.38	5.10	6.71	7.19	7.25	5.62	9.86	4.91	4.90	6.20	9.83	5.76	5.12
17	1	3.79	4.97	5.20	5.46	5.74	6.24	6.32	5.23	6.45	4.23	4.78	6.75	7.26	7.12	5.38	10.08	4.65	4.31	5.63	9.85	5.65	4.93
18	1	4.47	5.84	6.10	6.21	6.54	7.05	7.14	6.00	7.32	4.94	5.59	7.54	8.11	7.93	6.03	11.02	5.41	5.05	6.51	11.02	6.47	5.70
19	1	3.88	5.08	5.28	5.37	5.73	6.20	6.33	5.17	6.41	4.24	4.82	6.68	7.21	7.14	5.35	9.91	4.65	4.52	5.76	9.78	5.60	4.87
20	1	4.25	5.67	5.88	6.14	6.42	7.05	7.13	6.09	7.35	4.71	5.52	7.59	8.19	8.05	6.14	11.60	5.40	4.83	6.49	11.39	6.50	5.65
21	1	3.80	5.16	5.52	5.76	6.21	6.77	6.85	5.66	7.15	4.25	4.82	7.49	8.13	7.90	5.58	11.36	4.66	4.24	5.78	10.73	6.04	5.12
22	1	4.62	6.19	6.32	7.03	7.48	8.32	8.45	6.79	8.35	4.97	5.96	8.71	9.77	9.20	6.55	14.04	5.52	5.49	7.09	13.86	7.39	5.65
23	1	4.31	5.92	5.97	6.34	6.50	6.95	7.26	5.71	7.16	4.73	5.22	7.40	7.97	7.83	5.55	11.29	5.16	4.73	6.04	11.32	6.46	5.53
24	1	4.80	6.52	6.28	6.74	6.92	7.34	7.74	6.21	7.42	5.23	5.91	7.68	8.06	8.09	6.07	11.47	5.75	5.28	6.63	11.84	7.07	6.12
25	1	4.10	5.31	5.11	6.22	6.06	6.99	6.28	5.63	6.08	4.28	5.40	7.54	7.62	7.21	6.14	10.27	5.00	5.07	6.43	9.96	6.07	4.77
26	1	3.92	5.43	4.83	6.49	6.85	7.68	7.62	6.30	7.19	4.18	5.53	7.64	9.06	8.18	5.92	13.48	4.78	4.89	6.59	12.98	6.79	5.29
27	1	2.31	3.30	2.97	3.45	3.70	3.82	4.02	3.26	3.76	2.47	3.05	4.50	4.66	4.48	3.58	6.53	2.92	3.14	3.34	5.40	3.43	2.72
28	1	4.09	5.44	5.16	5.78	5.58	6.05	6.01	5.22	5.93	4.29	5.13	6.30	6.59	6.43	5.23	8.95	4.81	4.85	5.53	8.71	5.69	5.10
29	1	2.42	3.16	2.90	3.71	3.56	4.01	3.75	3.39	3.55	2.53	3.12	4.24	4.56	4.27	3.48	6.52	2.83	3.02	3.57	5.93	3.48	3.06
30	1	2.52	3.34	3.19	3.69	3.59	3.91	3.87	3.43	3.82	2.62	3.15	4.18	4.54	4.39	3.39	6.15	2.89	3.01	3.62	5.70	3.52	3.12
31	1	2.64	3.63	3.52	6.23	5.49	4.99	4.59	4.93	5.11	3.07	3.55	5.68	8.88	3.98	3.76	13.41	3.53	3.14	3.72	10.42	4.22	5.21
32	1	2.47	3.62	3.49	7.21	6.48	5.60	5.00	5.47	6.05	2.79	3.14	6.09	11.01	4.45	3.69	17.07	3.06	3.16	3.95	12.26	4.10	5.52
33	1	4.50	6.36	6.49	7.24	7.71	8.29	8.52	7.03	8.81	4.67	6.15	9.31	10.37	9.64	6.93	14.47	5.60	5.87	7.44	13.69	7.41	5.60
34	1	4.03	5.53	5.53	6.16	6.17	6.91	7.15	6.01	7.24	4.28	5.26	7.43	8.30	7.88	5.89	11.94	5.23	4.47	5.98	11.46	6.54	5.34
35	1	5.69	7.91	8.22	8.64	8.77	9.71	10.10	8.51	10.55	5.96	7.37	10.60	11.75	11.15	8.07	16.33	7.15	6.23	8.19	15.68	9.25	7.55
36	1	4.01	5.51	5.51	6.15	6.26	7.08	7.47	6.00	7.24	4.34	5.04	7.55	8.49	8.07	5.74	12.23	4.94	4.51	5.84	11.59	6.44	5.10
37	1	4.75	6.56	6.50	7.50	7.60	8.55	8.99	7.51	8.99	5.07	6.27	9.42	10.62	9.87	7.11	15.30	6.04	5.35	7.17	14.22	8.11	6.54
38	1	4.47	6.18	6.22	6.84	6.96	7.72	8.12	6.73	8.15	4.80	5.76	8.36	9.36	8.87	6.42	13.24	5.61	4.92	6.44	12.54	7.37	5.98
39	1	4.85	6.66	6.67	7.28	7.38	8.17	8.49	7.02	8.47	5.17	6.05	8.73	9.61	9.08	6.68	13.45	5.87	5.37	6.81	12.81	7.71	6.37
40	1	3.54	5.03	4.98	5.90	6.01	6.80	7.19	5.88	7.13	3.83	4.78	7.61	8.67	7.87	5.58	12.58	4.54	4.04	5.41	11.36	6.25	4.95
41	1	7.05	8.99	8.64	9.33	9.33	9.20	9.15	8.61	9.02	7.70	8.36	9.11	9.47	9.61	8.00	11.42	8.18	7.49	8.71	11.40	9.21	8.39
42	1	7.78	9.89	9.45	10.14	10.17	10.01	10.07	9.38	9.83	8.47	9.12	9.81	10.23	10.51	8.73	12.68	9.02	8.19	9.54	12.54	10.16	9.24
43	1	4.60	6.23	6.09	7.12	7.27	7.34	7.50	6.84	7.48	5.13	6.00	7.80	8.58	8.81	6.27	11.74	6.10	5.16	6.74	10.62	7.28	5.91
44	1	3.56	4.75	4.59	5.36	5.48	5.32	5.58	5.08	5.42	4.00	4.72	5.57	6.17	6.51	4.87	8.28	4.89	4.02	5.33	7.82	5.45	4.48
45	1	10.36	13.26	12.59	13.08	13.38	13.80	13.79	12.21	13.36	11.04	11.93	13.65	13.87	13.64	11.72	17.56	11.46	11.16	12.07	17.07	13.16	12.59
46	1	6.26	8.34	7.82	8.66	8.96	9.42	9.54	8.15	9.08	6.75	7.59	9.61	10.02	9.74	7.95	13.73	7.29	7.00	7.86	12.77	8.69	8.00
47	1	3.74	5.01	4.88	5.70	5.80	5.74	5.94	5.37	5.90	4.13	4.96	5.97	6.71	6.97	5.08	9.30	5.17	4.15	5.54	8.53	5.89	4.80
48	1	5.33	6.85	6.76	7.11	7.20	7.83	8.46	6.39	7.80	5.97	6.32	8.38	8.86	8.76	6.63	12.20	6.09	6.94	7.22	12.13	7.62	5.96
49	1	3.96	5.98	5.15	7.73	7.43	8.94	7.70	6.37	7.28	4.28	5.32	9.55	9.23	8.20	6.82	14.81	4.66	5.79	6.52	12.67	6.76	5.37
50	1	3.54	4.85	4.56	5.10	5.19	5.66	6.00	4.75	5.51	3.83	4.12	5.68	6.36	6.28	4.25	9.31	4.39	4.00	4.64	8.77	5.41	4.18
51	1	13.99	17.62	16.54	17.32	17.50	18.03	17.75	15.97	17.07	14.94	15.43	17.61	17.88	16.91	14.96	22.14	14.50	14.70	15.70	21.56	17.01	16.38
52	1	3.79	4.85	5.00	5.17	5.15	5.53	5.78	4.90	5.88	4.17	4.65	5.93	6.53	6.44	4.81	8.59	4.53	4.32	5.03	8.23	5.42	4.33
53	1	3.51	4.71	4.08	5.01	5.27	5.67	6.00	4.76	5.17	3.96	4.38	5.94	6.18	6.11	4.83	9.08	4.15	4.29	4.69	8.22	4.93	4.35
54	2	4.69	6.76	5.85	7.97	7.76	8.88	9.24	7.28	7.67	5.63	5.96	8.78	8.59	8.39	6.64	13.08	5.91	5.53	6.50	12.81	7.54	6.93
55	2	4.38	6.09	5.11	7.32	7.51	8.63	9.10	6.85	7.71	4.96	5.92	9.26	9.36	9.16	6.70	14.98	5.20	5.32	6.34	13.54	7.42	6.09

														cm^{-1}									
Sample	Class	675	725	775	825	875	925	975	1025	1075	1125	1175	1225	1275	1325	1375	1425	1475	1525	1575	1625	1675	1725
56	2	2.75	4.38	3.20	6.36	5.65	6.99	6.27	4.99	5.10	3.01	4.37	8.03	6.58	5.97	5.22	11.80	3.18	3.94	4.55	9.61	5.86	3.93
57	2	5.14	7.35	6.50	8.42	8.21	9.19	9.75	7.70	8.28	6.13	6.40	9.13	9.18	8.93	7.01	13.68	6.49	5.96	7.00	13.41	8.02	7.40
58	2	7.52	10.12	9.42	10.98	11.50	12.28	12.68	10.41	12.11	8.17	9.51	12.86	13.76	13.02	10.26	19.52	8.90	8.73	10.12	17.80	11.34	10.06
59	2	8.35	11.34	10.47	12.57	13.17	14.21	14.74	11.92	13.98	9.09	10.65	15.10	16.34	15.36	11.66	23.38	9.73	9.79	11.57	21.03	13.06	11.39
60	2	5.51	7.59	6.84	8.55	9.01	9.75	10.12	8.00	9.30	6.09	7.27	10.33	11.06	10.49	8.12	16.34	6.66	6.70	7.85	14.48	8.74	7.63
61	2	17.80	22.39	21.03	22.24	22.44	23.02	23.08	20.76	22.07	18.96	20.15	22.59	22.98	22.17	19.77	28.09	18.90	19.15	20.56	27.26	22.11	21.43
62	2	18.05	22.84	21.53	22.74	22.97	23.51	23.49	21.14	22.49	19.21	20.43	23.09	23.35	22.55	20.12	28.40	19.08	19.33	20.81	27.74	22.93	22.05
63	2	6.43	8.50	7.91	9.32	9.44	10.47	10.69	8.93	10.16	6.90	8.16	10.94	11.61	11.11	8.55	16.26	7.56	7.34	9.05	15.97	9.68	8.33
64	2	4.65	6.13	5.60	6.53	6.59	7.08	7.44	6.10	6.67	4.91	5.54	7.01	7.59	7.53	5.72	10.69	5.62	5.19	5.93	10.13	6.87	5.64
65	2	5.56	7.43	6.84	8.30	8.51	9.51	9.91	8.02	9.10	6.00	7.16	9.94	10.60	10.28	7.79	15.05	6.63	6.47	8.22	14.83	8.72	7.34
66	2	4.03	5.46	4.63	7.46	7.18	8.42	7.97	7.27	8.46	5.04	5.24	8.53	10.11	9.33	6.61	15.70	4.91	4.75	6.47	14.26	7.26	6.31
67	2	4.59	6.39	5.23	8.16	8.17	9.44	8.97	7.58	8.50	5.39	6.25	9.80	10.32	9.69	7.56	16.49	5.55	5.68	6.85	14.44	8.04	6.89
68	2	4.29	5.82	5.01	7.58	7.30	8.50	8.06	7.41	8.43	5.36	5.63	8.46	9.51	9.16	6.72	14.45	5.21	5.02	6.69	13.75	7.65	6.71
69	2	3.54	4.80	4.07	6.04	6.24	7.24	7.28	5.71	6.50	3.98	4.96	7.97	8.50	7.66	5.60	13.01	3.85	4.92	6.05	12.05	5.84	4.71
70	2	3.32	4.13	3.84	5.18	5.21	5.87	5.55	5.02	5.64	4.05	4.38	5.98	6.88	6.08	4.70	10.09	3.91	3.76	4.81	9.18	5.02	4.78
71	2	5.07	6.46	6.13	7.21	7.38	8.11	8.04	7.08	7.94	5.67	6.52	8.40	9.19	8.55	6.71	12.85	5.86	5.73	7.07	12.43	7.38	6.61
72	2	3.64	4.60	4.32	5.66	5.67	6.47	6.13	5.55	6.14	4.26	4.90	6.70	7.34	6.74	5.31	10.69	4.36	4.22	5.40	9.99	5.65	5.09
73	2	6.20	8.59	7.54	10.31	10.69	11.37	11.44	9.67	11.35	7.06	8.07	11.96	13.61	12.28	9.07	20.48	7.38	7.51	8.93	18.05	10.32	9.09
74	2	2.41	3.54	3.25	4.98	4.96	6.02	6.17	4.62	5.46	2.59	3.84	7.59	7.11	6.70	4.71	11.01	3.09	3.07	4.04	9.77	5.24	3.49
75	2	3.48	4.87	4.32	6.55	6.40	7.57	8.04	6.09	6.89	3.83	5.33	9.82	8.84	8.40	6.40	13.49	4.32	4.65	5.74	12.46	6.95	4.88
76	2	2.73	4.00	3.64	5.68	5.54	6.72	6.99	5.22	6.10	3.05	4.33	8.60	7.79	7.34	5.39	12.20	3.48	3.67	4.66	11.04	5.98	4.09
77	2	3.27	4.82	3.99	8.98	8.07	8.04	6.84	6.96	7.56	4.07	4.58	9.37	13.43	6.22	5.40	21.73	4.06	4.07	4.63	15.77	6.07	7.57
78	2	7.64	9.83	8.96	10.63	11.11	12.21	11.57	9.98	10.82	8.12	9.48	11.51	12.50	11.63	9.56	16.81	8.38	8.39	10.42	17.74	11.01	9.70
79	2	3.89	5.30	4.78	6.37	6.74	7.65	7.33	6.27	7.17	4.28	5.52	7.48	8.77	7.94	6.03	13.29	4.97	4.49	6.32	12.96	6.86	5.58
80	2	3.89	5.23	5.41	6.13	6.25	7.06	6.75	5.85	7.13	4.13	5.34	7.12	8.39	7.49	5.56	12.11	5.09	4.40	6.05	11.73	6.39	5.16
81	2	3.70	5.24	4.97	6.81	7.29	8.54	8.13	6.72	7.99	4.00	5.61	8.99	10.00	9.00	6.49	15.32	4.78	4.46	6.57	14.59	7.35	5.65
82	2	4.22	5.69	5.00	6.74	7.12	8.00	7.89	6.25	7.50	4.52	5.83	8.01	9.36	8.27	6.02	13.96	5.00	5.05	6.82	13.85	6.93	5.58
83	2	3.82	5.20	4.50	6.19	6.69	7.61	7.13	6.01	6.83	4.19	5.48	7.08	8.70	7.75	5.77	12.87	4.86	4.42	6.26	13.05	6.70	5.38
84	2	4.28	5.73	5.02	7.11	7.45	8.91	8.03	6.60	7.14	4.61	5.89	8.77	9.16	8.37	6.55	13.51	4.84	4.93	6.82	13.76	7.44	5.95
85	2	2.90	3.75	3.16	4.13	4.03	4.33	4.45	3.88	3.65	3.09	3.68	4.38	4.57	4.53	3.67	5.93	3.05	3.68	3.99	5.74	3.88	3.53
86	2	3.59	5.14	4.35	6.57	6.28	7.67	7.11	5.95	6.35	3.83	5.12	8.27	7.97	7.41	5.76	12.76	4.06	4.29	5.65	11.72	6.41	5.28
87	2	4.25	5.66	4.83	6.37	6.17	7.00	6.80	5.92	6.08	4.49	5.55	7.09	7.20	6.84	5.62	10.68	4.74	4.88	5.95	10.44	6.27	5.55
88	2	4.08	5.47	4.58	6.45	6.36	7.20	7.22	5.91	6.24	4.41	5.32	7.21	7.83	7.27	5.70	11.72	4.51	5.05	6.04	11.12	6.10	5.40
89	2	4.74	6.45	5.36	7.51	7.55	8.64	8.66	7.01	7.46	5.08	6.25	8.71	9.21	8.65	6.69	13.85	5.10	5.70	6.95	13.12	7.36	6.33
90	2	3.34	4.96	4.07	9.23	8.39	8.29	7.50	7.42	7.89	3.77	4.55	9.31	13.66	6.88	5.51	22.02	3.69	4.23	5.25	15.66	5.91	7.28
91	2	3.50	4.80	4.01	5.93	5.61	6.70	6.67	5.37	5.70	3.84	4.67	6.98	7.17	6.45	5.17	11.28	4.02	4.09	5.12	10.51	5.71	4.81
92	2	3.93	5.25	4.35	6.09	5.91	6.88	7.06	5.70	5.77	4.24	5.22	6.83	7.06	6.83	5.69	10.56	4.59	4.64	5.72	10.34	5.99	5.12
93	2	2.91	4.17	3.52	5.69	5.15	6.67	5.86	4.93	5.16	3.15	4.07	7.13	6.63	6.00	5.09	10.75	3.43	3.47	4.61	9.66	5.37	4.13
94	2	2.86	4.10	3.59	5.25	5.36	6.35	6.68	5.24	5.79	3.18	4.22	6.97	7.44	6.97	5.09	11.12	3.59	3.55	4.89	10.28	5.66	4.42
95	2	2.92	4.08	3.62	5.10	5.02	6.07	6.11	4.92	5.44	3.22	4.12	6.59	6.82	6.39	4.99	10.24	3.67	3.51	4.79	9.46	5.32	4.29
96	2	9.08	11.54	10.37	11.99	11.78	13.20	12.84	11.19	11.12	9.77	10.73	12.58	11.90	11.89	10.87	15.35	9.66	9.68	11.37	16.16	12.29	11.34
97	2	3.35	4.75	4.25	5.75	5.87	6.92	7.16	5.71	6.41	3.67	4.72	7.42	7.95	7.48	5.57	11.79	4.21	3.98	5.35	11.05	6.25	5.06
98	2	4.14	5.46	5.00	6.43	6.37	6.44	6.79	5.85	6.20	4.68	5.48	6.78	7.47	7.39	5.65	10.54	5.47	4.65	6.14	9.53	6.39	5.38
99	2	5.81	7.69	7.12	8.06	8.49	9.70	9.93	7.72	9.40	6.06	7.41	10.83	11.80	11.06	8.45	15.90	6.30	8.67	9.55	15.46	8.19	6.63
100	2	4.56	5.99	5.25	7.29	7.41	8.35	8.43	6.89	7.36	4.99	6.35	8.81	9.33	8.40	6.74	14.01	5.31	5.41	7.14	13.31	7.35	6.16
101	2	5.71	7.65	7.13	8.63	9.08	10.17	9.84	8.16	9.36	5.98	7.36	10.14	11.07	10.45	7.91	15.38	6.39	6.83	8.65	15.65	8.88	7.25
102	2	4.04	5.27	4.73	5.40	5.60	5.90	6.18	5.02	5.55	4.52	5.00	5.98	6.24	6.28	5.20	8.83	4.96	4.81	5.11	8.20	5.45	5.03
103	2	3.54	4.76	4.16	5.17	5.52	5.97	6.23	4.91	5.44	4.00	4.45	6.28	6.54	6.42	4.99	9.69	4.16	4.31	4.76	8.70	5.17	4.41

1. Centre the data and perform PCA. Plot the scores of PC2 versus PC1, distinguishing the two groups.
2. From this plot, identify four samples that appear to be outliers from their scores on PC1. Which are they? Reform the PC plot (centring the data again) without these samples.

3. The rows do not add up to a constant total. Although they are on comparable scales, it is safest to row scale these. This was not done before the outliers were removed because the former step was a very qualitative one, but before quantitative modelling, in the absence of other information, it is better to row scale. Scale each row to the total of 100 and then perform mean centred PCA on the remaining 99 samples. Plot the graph of scores of PC2 versus PC1 and of PC3 versus PC2, and comment.

4. Calculate the PLS scores on the normalised and then centred data. Use $c = +1$ for the benign class and $c = -1$ for the cancerous. Plot the scores of PLS components 2 versus 1 and of PLS components 3 versus 2. Comment.

5. Using the auto-predictive model in part 4, calculate the number of correctly classified for each class separately, and the overall %CC, using the criterion that an estimated value of c more than 0 is assigned to the benign class and less to the cancerous class. Determine models using 2, 5 and 10 PLS components. Comment.

6. Usually, auto-predictive PLS models are mainly used for exploratory analysis and looking at the variables. However, to get a true idea of predictive ability, the data should be split into test and training set. Remove samples 1–15 (from class 1) and 52–66 (for class 2) as a test set. Determine PLS models using the remaining samples. Now calculate the same statistics as in part 5 but both on the 69 training set and 30 test set samples separately. You should calculate nine numbers for each type of model, three of which are overall %CC for both training and test set. Comment.

7. What ways might one get a better view of the trends seen in the step above?

6

Calibration

6.1 Introduction

6.1.1 History, Usage and Terminology

Calibration involves connecting one (or more) sets of variables together. Usually, one set (often called a *block*) is a series of physical measurements, such as some spectra or molecular descriptors, and the other contains one or more parameters such as the concentrations of a number of compounds or biological activity. Can we predict the concentration of a compound in a mixture spectrum or the properties of a material from its known structural parameters? Calibration provides the answer. In its simplest form, calibration is a form of regression, as discussed in Chapter 2, in the context of experimental design. There are several aims of calibration.

- The first aim is to simply form a functional model between two sets of measurements, for example, concentration and absorbance.
- The second aim is to find out whether this functional model is a good one, for example, by using linear model, can we adequately predict the concentration of an analyte from its absorbance?
- The third aim is to predict the properties of a new, unknown sample from making a series of measurements on it, for example, spectral absorbances.

Multivariate calibration has historically been a major cornerstone of chemometrics. However, there are a large number of diverse schools of thought, mainly dependent on people's background and the software they are familiar with. Many mainstream statistical packages do not contain the partial least squares (PLS) algorithm, whereas most specialist chemometric software is based around this method. PLS is one of the most publicised algorithms for multivariate calibration that has been widely advocated by many in chemometrics, following the influence of S. Wold, whose father first proposed this in the context of economics. A mystique surrounding PLS has developed, a technique with its own terminology, conferences and establishment. A number of commercial packages on the marketplace perform PLS calibration and result in a variety of diagnostic statistics. It is, though, important to understand that a major historic (and economic) driving force was near infrared spectroscopy (NIR), primarily in the food industry and in process analytical chemistry. Each type of spectroscopy and chromatography has its own features and problems; thus, much software was developed to tackle specific situations that may not necessarily be very applicable to other techniques such as, for example, chromatography, NMR or MS. In many statistical circles, NIR and chemometrics are almost inseparably intertwined. As time moves on, instruments improve in quality; thus, many of the computational approaches developed several decades ago to deal with problems such as background correction, and noise distributions, are not so relevant nowadays. However, for historical reasons, it is often difficult to distinguish between these specialist methods required to prepare data in order to obtain meaningful information from the chemometrics, and the actual calibration steps themselves. Despite this, chemometric approaches to calibration have very wide potential applicability throughout all areas of quantitative laboratory science, but NIR spectroscopists will definitely form one important readership base.

There are very many circumstances in which multivariate calibration methods are appropriate. The difficulty is that to develop a comprehensive set of data, analytical data collection for a particular situation may take a huge investment in resources and time; thus, the applications of multivariate calibration in some areas of science are much less well established than in others. It is important to separate the methodology that has built up around a small number of spectroscopic methods such as NIR, from the general principles. Probably several hundred favourite diagnostics are available to the professional user of PLS, for example, in NIR spectroscopy, yet each one has been developed with a specific technique or problem in mind and is not necessarily generally applicable to all calibration problems.

There are a whole series of problems in chemistry for which multivariate calibration is appropriate, but each is different in nature. Many of the most successful applications have been in the spectroscopy or chromatography of mixtures. We will, for the sake of brevity, illustrate this chapter with a single case study in the area of mixture spectroscopy but anticipate that the readers will extend their understanding into other application areas.

- The simplest is calibration of the concentration of a single compound using a spectroscopic or chromatographic method, an example being determining the concentration of chlorophyll by electronic absorption spectroscopy (EAS) – sometimes called *UV/vis spectroscopy*. Instead of using one wavelength (as is conventional for the determination of molar absorptivity or extinction coefficients), multivariate calibration involves using all or several of the wavelengths. Each variable measures the same information, but better information is obtained by considering all the wavelengths, averaging out the noise.
- A more complex situation is a multi-component mixture where all pure standards are available. It is possible to control the concentration of the reference compounds, so that a number of carefully designed mixtures can be produced in the laboratory. Sometimes, the aim is to see whether a spectrum of a mixture can be employed to determine individual concentrations, and, if so, how reliably. The aim may be to replace a slow and expensive chromatographic method by a rapid spectroscopic approach. Another rather different aim might be impurity monitoring, how well the concentration of a small impurity be determined, for example, buried within a large chromatographic peak.
- A quite different approach is required if only the concentration of a portion of the components is known in a mixture, for example, polyaromatic hydrocarbons (PAHs) within coal tar pitch volatiles. In natural samples, there may be tens or hundreds of unknowns, but only a few can be quantified and calibrated. The unknown interferents cannot necessarily be determined, and it is not possible to design a set of samples in the laboratory containing all the potential components in real samples. Multivariate calibration is effective providing the range of samples used to develop the model is sufficiently representative of all future samples in the field (it is often stated that the correlation structure of the training set is representative of future samples). If it is not, the predictions from multivariate calibration could be dangerously inaccurate. In order to protect against samples not belonging to the original data set, a number of approaches for the determination of outliers and experimental design have been developed.
- In a final case, the aim of calibration is not so much to determine the concentration of a particular compound but to determine a bulk statistical parameter. Pure standards will no longer be available, but the training set must be a sufficiently representative group. An example is to determine the concentration of a class of compounds in food, such as protein in wheat. It is not possible (or desirable) to isolate each single type of protein, and we rely on the original samples being sufficiently representative, for example, of the variety of wheat likely to be analysed in a particular factory. This situation also occurs, for example, in quantitative structure–property relationships (QSPR) or quantitative structure–activity relationships (QSAR).

There are many pitfalls in the use of calibration models, perhaps the most serious being variability in instrument performance over time. Each measurement technique has different characteristics, and on each day and even hour, the response can vary. How serious this is for the stability of the calibration model should be assessed before investing a large effort. Sometimes, it is necessary to reform the calibration model on a regular basis, by running a standard set of samples, possibly on a daily or weekly basis. In other cases, multivariate calibration gives only a rough prediction, but if the quality of a product or the concentration of a pollutant appears to exceed a certain limit, other more detailed analyses can be used to investigate the quality of the sample. For example, online calibration in NIR can be used for screening whether a manufactured sample is of acceptable quality, and any dubious batches investigated in more detail using chromatography.

This chapter will describe the main algorithms and principles of calibration. We will concentrate on situations in which there is a direct linear relationship between blocks of variables. It is possible to extend the methods to include multi-linear (such as squared) terms simply by extending the X matrix, for example, in the case of spectroscopy at high concentrations or non-linear detection systems.

We will refer to physical measurements as the x block, for example, intensities of spectroscopic wavelengths or chromatographic peaks, and concentrations or descriptors as the c block. One area of confusion is that users of different techniques in chemometrics tend to employ incompatible notation. In the area of experimental design, it is usual to call the measured response 'y', for example, the absorbance in a spectrum and the concentration or any related parameter 'x'. In traditional multivariate calibration, this notation is swapped around. For the purpose of a coherent text, it would be confusing to use two opposite notations; however, some compatibility with the established literature is desirable. Figure 6.1 illustrates the notation used in this text.

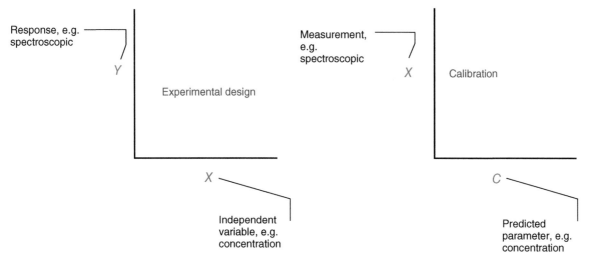

Figure 6.1 Different notations for calibration and experimental design as used in this book.

1. A series of 50 spectra are recorded at 120 wavelengths, and the concentrations of three compounds are independently known within these spectra.
   (a) The *x* block has dimensions 50×120 and the *c* block 3×50.
   (b) The *x* block has dimensions 50×120 and the *c* block 50×3.
   (c) The *x* block has dimensions 120×50 and the *c* block 50×3.
   (d) The *x* block has dimensions 120×50 and the *c* block 3×50.

6.1.2 Case Study

It is easiest to illustrate the methods in this chapter using a small case study, involving recording

- 25 EAS spectra at
- 27 wavelengths (220-350 nm at 5 nm intervals) and
- consisting of a mixture of 10 compounds (in our case, PAHs).

In reality, the spectra would usually be recorded at a higher digital resolution, but for illustrative purposes, we reduce the sampling rate. The aim is to predict the concentrations of individual PAHs from the mixture spectra. The spectroscopic data are presented in Table 6.1 and the corresponding concentrations of the compounds are presented in Table 6.2.

The methods in this chapter will be illustrated as applied to the spectroscopy of mixtures, as this is a common and highly successful application of calibration in chemistry. However, similar principles apply to a wide variety of calibration problems.

1. What are the dimensions of the *c* block in the case study of this section?
   (a) 25×10
   (b) 10×25
   (c) 25×27
   (d) 27×25

6.2 Univariate Calibration

Univariate calibration involves relating two single variables to each other and is often called *linear regression*. It is easy to perform using most data analysis packages.

Table 6.1 Case study consisting of 25 spectra recorded at 27 wavelengths in nanometre, absorbances in AU.

nm	220	225	230	235	240	245	250	255	260	265	270	275	280	285	290	295	300	305	310	315	320	325	330	335	340	345	350
1	0.771	0.714	0.658	0.537	0.587	0.671	0.768	0.837	0.673	0.678	0.741	0.755	0.682	0.633	0.706	0.290	0.208	0.161	0.135	0.137	0.162	0.130	0.127	0.165	0.110	0.075	0.053
2	0.951	0.826	0.737	0.638	0.738	0.911	1.121	1.162	0.869	0.870	0.965	1.050	0.993	0.934	1.008	0.405	0.254	0.185	0.157	0.159	0.180	0.155	0.150	0.178	0.140	0.105	0.077
3	0.912	0.847	0.689	0.514	0.504	0.622	0.805	0.892	0.697	0.728	0.790	0.728	0.692	0.639	0.717	0.292	0.224	0.168	0.134	0.123	0.136	0.115	0.095	0.102	0.089	0.068	0.048
4	0.688	0.679	0.662	0.558	0.655	0.738	0.838	0.883	0.670	0.656	0.704	0.668	0.562	0.516	0.573	0.295	0.212	0.172	0.138	0.144	0.179	0.130	0.134	0.191	0.107	0.060	0.046
5	0.873	0.801	0.732	0.640	0.750	0.820	0.907	0.955	0.692	0.681	0.775	0.866	0.798	0.749	0.827	0.311	0.213	0.170	0.151	0.162	0.201	0.158	0.170	0.239	0.146	0.094	0.067
6	0.953	0.850	0.732	0.612	0.724	0.882	1.013	1.087	0.860	0.859	0.925	0.938	0.869	0.810	0.868	0.390	0.241	0.175	0.138	0.141	0.167	0.129	0.135	0.178	0.115	0.078	0.056
7	0.613	0.577	0.547	0.448	0.508	0.463	0.473	0.549	0.510	0.560	0.604	0.493	0.369	0.343	0.363	0.215	0.178	0.165	0.136	0.142	0.183	0.122	0.129	0.193	0.089	0.041	0.030
8	0.927	0.866	0.799	0.605	0.618	0.671	0.771	0.842	0.646	0.658	0.716	0.829	0.783	0.731	0.794	0.292	0.220	0.155	0.120	0.123	0.144	0.122	0.127	0.164	0.113	0.078	0.056
9	0.585	0.577	0.543	0.450	0.515	0.614	0.734	0.815	0.624	0.642	0.724	0.716	0.680	0.649	0.713	0.298	0.213	0.167	0.136	0.126	0.137	0.109	0.104	0.129	0.098	0.074	0.057
10	0.835	0.866	0.803	0.550	0.563	0.577	0.642	0.732	0.675	0.713	0.836	0.938	0.911	0.847	0.910	0.358	0.226	0.173	0.153	0.160	0.186	0.156	0.157	0.193	0.134	0.093	0.066
11	0.477	0.454	0.450	0.433	0.538	0.593	0.661	0.724	0.554	0.561	0.616	0.469	0.355	0.322	0.345	0.211	0.154	0.139	0.114	0.118	0.154	0.100	0.100	0.154	0.071	0.030	0.016
12	0.496	0.450	0.402	0.331	0.389	0.504	0.613	0.599	0.383	0.344	0.353	0.366	0.338	0.309	0.334	0.194	0.123	0.094	0.076	0.070	0.077	0.065	0.056	0.065	0.053	0.036	0.025
13	0.594	0.512	0.441	0.392	0.489	0.562	0.637	0.647	0.415	0.387	0.421	0.461	0.396	0.361	0.424	0.161	0.103	0.076	0.062	0.076	0.110	0.082	0.094	0.144	0.078	0.043	0.028
14	0.512	0.478	0.409	0.319	0.375	0.395	0.436	0.487	0.439	0.449	0.474	0.438	0.362	0.331	0.360	0.204	0.158	0.122	0.089	0.087	0.107	0.075	0.079	0.114	0.064	0.040	0.028
15	0.662	0.583	0.586	0.564	0.687	0.708	0.748	0.757	0.611	0.581	0.641	0.727	0.638	0.596	0.658	0.277	0.171	0.133	0.113	0.128	0.168	0.125	0.143	0.211	0.114	0.067	0.044
16	0.768	0.588	0.501	0.386	0.429	0.539	0.671	0.740	0.607	0.627	0.693	0.772	0.751	0.710	0.781	0.287	0.174	0.116	0.093	0.089	0.095	0.087	0.081	0.087	0.081	0.069	0.047
17	0.635	0.557	0.487	0.377	0.404	0.488	0.597	0.669	0.603	0.625	0.647	0.576	0.515	0.477	0.524	0.260	0.188	0.139	0.105	0.096	0.104	0.084	0.071	0.077	0.061	0.045	0.031
18	0.575	0.489	0.432	0.375	0.408	0.449	0.525	0.569	0.442	0.451	0.501	0.519	0.474	0.458	0.500	0.219	0.151	0.119	0.095	0.092	0.107	0.085	0.081	0.106	0.072	0.047	0.032
19	0.768	0.713	0.644	0.528	0.599	0.807	1.009	1.035	0.750	0.722	0.790	0.907	0.921	0.866	0.924	0.375	0.219	0.144	0.118	0.116	0.123	0.120	0.114	0.119	0.115	0.096	0.070
20	0.811	0.655	0.593	0.496	0.601	0.694	0.810	0.835	0.669	0.671	0.718	0.641	0.566	0.524	0.573	0.290	0.182	0.148	0.123	0.119	0.141	0.103	0.098	0.130	0.080	0.051	0.036
21	0.827	0.714	0.660	0.535	0.601	0.660	0.729	0.775	0.619	0.601	0.640	0.667	0.571	0.525	0.602	0.242	0.160	0.120	0.103	0.122	0.164	0.127	0.133	0.182	0.105	0.059	0.037
22	0.673	0.492	0.427	0.447	0.584	0.751	0.908	0.931	0.656	0.611	0.623	0.552	0.466	0.418	0.443	0.275	0.199	0.146	0.104	0.094	0.106	0.074	0.070	0.095	0.064	0.042	0.030
23	0.949	0.852	0.711	0.559	0.586	0.701	0.855	0.907	0.737	0.731	0.801	0.956	0.934	0.881	0.934	0.380	0.239	0.158	0.127	0.125	0.138	0.126	0.124	0.138	0.118	0.093	0.063
24	0.939	0.835	0.723	0.622	0.716	0.791	0.908	1.031	0.894	0.948	1.036	1.079	0.948	0.897	0.983	0.386	0.260	0.189	0.150	0.160	0.199	0.155	0.163	0.219	0.145	0.101	0.070
25	1.055	0.989	0.894	0.681	0.663	0.726	0.837	0.914	0.813	0.840	0.916	0.892	0.837	0.785	0.846	0.359	0.237	0.179	0.151	0.150	0.175	0.145	0.128	0.147	0.116	0.086	0.058

Table 6.2 Concentrations of the 10 PAHs in the data in Table 6.1.

Spectrum	Polyarene concentration (mg/l)									
	Py	Ace	Anth	Acy	Chry	Benz	Fluora	Fluore	Nap	Phen
1	0.456	0.120	0.168	0.120	0.336	1.620	0.120	0.600	0.120	0.564
2	0.456	0.040	0.280	0.200	0.448	2.700	0.120	0.400	0.160	0.752
3	0.152	0.200	0.280	0.160	0.560	1.620	0.080	0.800	0.160	0.188
4	0.760	0.200	0.224	0.200	0.336	1.080	0.160	0.800	0.040	0.752
5	0.760	0.160	0.280	0.120	0.224	2.160	0.160	0.200	0.160	0.564
6	0.608	0.200	0.168	0.080	0.448	2.160	0.040	0.800	0.120	0.940
7	0.760	0.120	0.112	0.160	0.448	0.540	0.160	0.600	0.200	0.188
8	0.456	0.080	0.224	0.160	0.112	2.160	0.120	1.000	0.040	0.188
9	0.304	0.160	0.224	0.040	0.448	1.620	0.200	0.200	0.040	0.376
10	0.608	0.160	0.056	0.160	0.336	2.700	0.040	0.200	0.080	0.188
11	0.608	0.040	0.224	0.120	0.560	0.540	0.040	0.400	0.040	0.564
12	0.152	0.160	0.168	0.200	0.112	0.540	0.080	0.200	0.120	0.752
13	0.608	0.120	0.280	0.040	0.112	1.080	0.040	0.600	0.160	0.376
14	0.456	0.200	0.056	0.040	0.224	0.540	0.120	0.800	0.080	0.376
15	0.760	0.040	0.056	0.080	0.112	1.620	0.160	0.400	0.080	0.940
16	0.152	0.040	0.112	0.040	0.336	2.160	0.080	0.400	0.200	0.376
17	0.152	0.080	0.056	0.120	0.448	1.080	0.080	1.000	0.080	0.564
18	0.304	0.040	0.168	0.160	0.224	1.080	0.200	0.400	0.120	0.188
19	0.152	0.120	0.224	0.080	0.224	2.700	0.080	0.600	0.040	0.940
20	0.456	0.160	0.112	0.080	0.560	1.080	0.120	0.200	0.200	0.940
21	0.608	0.080	0.112	0.200	0.224	1.620	0.040	1.000	0.200	0.752
22	0.304	0.080	0.280	0.080	0.336	0.540	0.200	1.000	0.160	0.940
23	0.304	0.200	0.112	0.120	0.112	2.700	0.200	0.800	0.200	0.564
24	0.760	0.080	0.168	0.040	0.560	2.700	0.160	1.000	0.120	0.376
25	0.304	0.120	0.056	0.200	0.560	2.160	0.200	0.600	0.080	0.752

Py, pyrene; Ace, acenaphthene; Anth, anthracene; Acy, acenaphthylene; Chry, chrysene; Benz, benzanthracene; Fluora, fluoranthene; Fluore, fluorene; Nap, naphthalene; Phen, phenanthrene.

6.2.1 Classical Calibration

One of the simplest problems is to determine the concentration of a single compound using the response at a single detector, for example, a single spectroscopic wavelength or a chromatographic peak area.

Mathematically, a series of experiments can be performed to relate the concentration to spectroscopic measurements as follows:

$$x \approx cs$$

where, in the simplest case, x is a column vector consisting, for example, of absorbances at one wavelength for a number of samples and c is of the corresponding concentrations. Both vectors have I elements, equal to the number of samples. The scalar s relates these parameters and is determined by regression. Classically, most regression packages try to find s.

A simple method for solving this equation is to use the pseudo-inverse (see Section 2.2.3.3 for an introduction)

$$c'x \approx (c'c)s$$

so

$$(c'c)^{-1}c'x \approx (c'c)^{-1}(c'c)s$$

or

$$s \approx (c'c)^{-1}c'x = \frac{\sum_{i=1}^{I} x_i c_i}{\sum_{i=1}^{I} c_i^2}$$

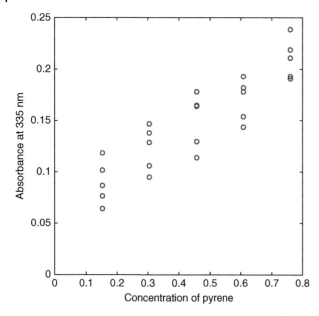

Figure 6.2 Absorbance at 335 nm for the PAH case study plotted against concentration of pyrene.

Many conventional texts formulate regression equations as summations rather than as matrices and vectors, but both approaches are equivalent; with modern spreadsheets and matrix-oriented programming environments, it is easier to build on the matrix-based equations and summations can become rather unwieldy if the problem is more complex. In Figure 6.2, the absorbance of the 25 spectra 335 nm is plotted against the concentration of pyrene. The graph is approximately linear and provides a best-fit slope calculated by

$$\sum_{i=1}^{I} x_i c_i = 1.916$$

and

$$\sum_{i=1}^{I} c_i^2 = 6.354$$

so that $\hat{x} = 0.301c$. The predictions are presented in Table 6.3. The spectra of the 10 pure standards are superimposed in Figure 6.3, with pyrene indicated in bold. It can be seen that pyrene has selective absorbances at higher wavelengths; hence, 335 nm will largely be characteristic of this compound. For most of the other compounds in these spectra, it would not be possible to obtain such good results from univariate calibration.

The quality of prediction can be determined by the residuals (or errors) often also called the *uncertainty*, that is, the difference between the observed and predicted, that is, $x - \hat{x}$, the smaller, the better. Generally, the root mean square error (RMSE) is calculated as

$$E = \sqrt{\sum_{i=1}^{I} \left(x_i - \hat{x}_i\right)^2 / d}$$

where d is called the *degrees of freedom*. In the case of univariate calibration, this equals the number of observations (N) minus the number of parameters in the model (P) or in our case, $25 - 1 = 24$ (see Section 2.2.1) so that

$$E = \sqrt{0.0279/24} = 0.0341$$

This error can be represented as a percentage of the mean $E_\% = 100(E/\bar{x}) = 24.1\%$ in this case. Sometimes, the percentage error is calculated relative to the standard deviation rather than mean: this is more appropriate if the data are mean centred (because the mean is 0), or if the data are all clustered at high values, in which case an apparently small error relative to the mean still may imply quite a large deviation. There are no hard and fast rules, and in this chapter, we will calculate errors relative to the mean unless otherwise stated. It is, however, always useful to check the

Table 6.3 Concentration of pyrene, absorbance at 335 nm and predictions of absorbance, using single-parameter classical calibration using method of Section 2.2.1.

Concentration	Absorbance at 335 nm	Predicted absorbance
0.456	0.165	0.137
0.456	0.178	0.137
0.152	0.102	0.046
0.760	0.191	0.229
0.760	0.239	0.229
0.608	0.178	0.183
0.760	0.193	0.229
0.456	0.164	0.137
0.304	0.129	0.092
0.608	0.193	0.183
0.608	0.154	0.183
0.152	0.065	0.046
0.608	0.144	0.183
0.456	0.114	0.137
0.760	0.211	0.229
0.152	0.087	0.046
0.152	0.077	0.046
0.304	0.106	0.092
0.152	0.119	0.046
0.456	0.130	0.137
0.608	0.182	0.183
0.304	0.095	0.092
0.304	0.138	0.092
0.760	0.219	0.229
0.304	0.147	0.092

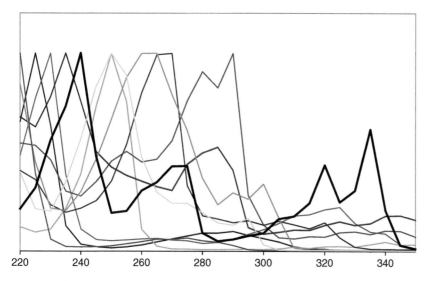

Figure 6.3 Spectra of pure standards, digitised at 5 nm intervals, pyrene indicated in bold.

original graph (Figure 6.2) just to be sure, and this percentage appears reasonable. Providing a consistent measure is used throughout, all percentage errors will be comparable. Several other indicators of goodness-of-fit are also available, such as the squared Pearson's correlation coefficient between measured and predicted often denoted by R^2, but this is a big topic that we only summarise in this section.

This approach to calibration, while widely used throughout most branches of science, is nevertheless not always appropriate in all applications. We may want to answer the question 'can the absorbance in a spectrum be employed to determine the concentration of a compound?' It is not always the most appropriate approach to use an equation that predicts the absorbance from the concentration when our experimental aim is the reverse. In other areas of science, the functional aim might, for example, be to predict an enzymic activity from its concentration. In the latter case, classical calibration, as outlined in this section, results in the correct functional model. Nevertheless, most chemists employ classical calibration, providing the errors are approximately normal, and there are no significant outliers, all the different univariate methods should result in approximately similar conclusions.

For a new or unknown sample, however, the concentration could be estimated (approximately) by using the inverse of the slope or

$$\hat{c} = 3.32\, x$$

1. For a set of observations, $\sum_{i=1}^{I} x_i c_i = 2.58$ and $\sum_{i=1}^{I} c_i^2 = 3.98$. What is the predicted value of c for a value of x of 0.75 using classical calibration and no intercept term?

    (a) 1.16
    (b) 0.49

6.2.2 Inverse Calibration

Although classical calibration is widely used, it is not always the most appropriate approach, for two main reasons. First of all, the ultimate aim is usually to predict the concentration (or independent variable) from the spectrum or chromatogram (response) rather than *vice versa*. The second relates to error distributions. The errors in the response are often due to instrumental performance. Over the years, instruments have become more reproducible. The independent variable (often concentration) is usually determined via a procedure involving weighing, dilution and so on and is often by far the largest source of uncertainty (error). The quality of volumetric flasks, syringes and so on has not improved dramatically over the years, whereas the sensitivity and reproducibility of instruments has increased many fold. Classical calibration fits a model so that all errors are considered to be in the response (Figure 6.4(a)), whereas a more appropriate assumption is that errors are primarily in the measurement of concentration (Figure 6.4(b)).

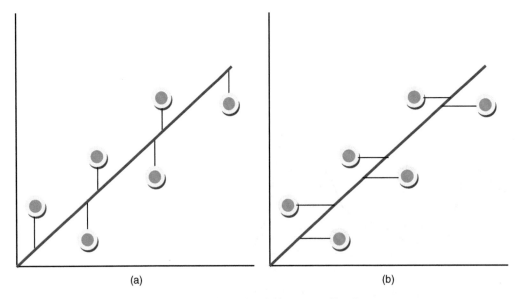

(a) (b)

Figure 6.4 Difference between errors in (a) classical and (b) inverse calibration.

Calibration can be performed by the inverse method whereby

$$c \approx xb$$

or

$$b \approx (x'x)^{-1}x'c = \frac{\sum_{i=1}^{I} x_i c_i}{\sum_{i=1}^{I} x_i^2}$$

giving for this example, $\hat{c} = 3.16x$, a *RMSE* of 0.110 or 24.2% relative to the mean. Note that b is only approximately the inverse of s (see above) because each model makes different assumptions about error distributions. The results are presented in Table 6.4. However, for good data, both models should provide fairly similar predictions, if not there could be some other factor that influences the data, such as an intercept, non-linearities, outliers or unexpected noise distributions. For heteroscedastic noise distributions, there are a variety of enhancements to linear calibration. However, these are rarely taken into consideration when extending the principles to the multivariate calibration.

The best-fit straight lines for both methods of calibration are given in Figure 6.5. At first, it looks as if these are quite a poor fit to the data, but an important feature is that the intercept is assumed to be zero. Both methods force the line through the point (0,0). Because of other compounds absorbing in the spectrum, this is a poor approximation, thus reducing the quality of regression. We will look at how to improve this model below.

Table 6.4 Concentration of pyrene, absorbance at 335 nm and predictions of absorbance, using single-parameter inverse calibration using method of Section 2.2.2.

Concentration	Absorbance at 335 nm	Predicted concentration
0.456	0.165	0.522
0.456	0.178	0.563
0.152	0.102	0.323
0.760	0.191	0.604
0.760	0.239	0.756
0.608	0.178	0.563
0.760	0.193	0.611
0.456	0.164	0.519
0.304	0.129	0.408
0.608	0.193	0.611
0.608	0.154	0.487
0.152	0.065	0.206
0.608	0.144	0.456
0.456	0.114	0.361
0.760	0.211	0.668
0.152	0.087	0.275
0.152	0.077	0.244
0.304	0.106	0.335
0.152	0.119	0.377
0.456	0.130	0.411
0.608	0.182	0.576
0.304	0.095	0.301
0.304	0.138	0.437
0.760	0.219	0.693
0.304	0.147	0.465

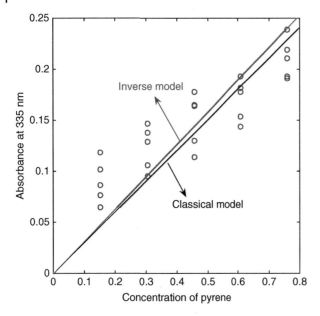

Figure 6.5 Best-fit straight lines for classical and inverse calibration, data for pyrene at 335 nm, no intercept, forcing the model through the origin.

1. For the example of question 1, Section 6.2.1, $\sum_{i=1}^{I} x_i^2 = 1.79$. What is the value of b in the relationship $c \approx xb$ for inverse calibration and no intercept term?

   (a) 1.44

   (b) 0.69

6.2.3 Intercept and Centring

In many situations, it is appropriate to include extra terms in the calibration model. Most commonly, an intercept (or baseline) term is included to give an inverse model of the form

$$c \approx b_0 + b_1 x$$

which can be expressed in matrix/vector notation by

$$c \approx Xb$$

for inverse calibration, where c is a column vector of concentrations and b is a column vector consisting of two numbers, the first equal to b_0 (the intercept) and the second equal to b_1 (the slope). X is now a matrix of two columns, the first of which is a column of 1's, the second the absorbances.

Exactly the same principles can be employed for calculating the coefficients, as discussed in Section 6.2.2, but in this case, b is a vector rather than scalar, and X is a matrix rather than a vector, so that using matrix notation,

$$b \approx (X'X)^{-1}X'c$$

or

$$\hat{c} = -0.17 + 4.23x$$

Note that the coefficient for b_1 is quite different to that for b, as calculated in Section 6.2.2. One reason is that there are still a number of interferents, from the other PAHs, in the spectrum at 335 nm, and these are modelled partly by the intercept term. The models of the previous sections force the best-fit straight line to pass through the origin. A better fit can be obtained if this condition is not required. The new best-fit straight line is presented in Figure 6.6 and results, visually, in a much better fit to the data.

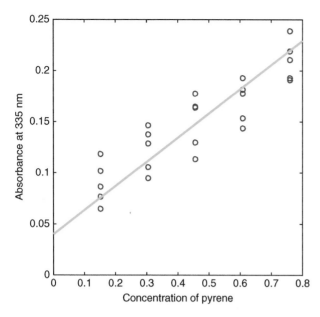

Figure 6.6 Best-fit straight line using inverse calibration and an intercept term.

The predicted concentrations are quite easy to obtain, the easiest approach involving the use of matrix-based methods, so that

$$\hat{c} = Xb$$

the RMSE being given by

$$E = \sqrt{0.229/(I-2)} = \sqrt{0.229/23} = 0.100\,\text{mg/l}$$

representing an $E_\%$ of 21.8% relative to the mean. Notice that the error term should be divided by 23 (number of degrees of freedom ($= N - P$) rather than 25) to reflect the two parameters used in the model.

One interesting and important consideration is that the apparent RMSE in Sections 6.2.2 and 6.2.3 is only reduced by a small amount, yet the best-fit straight line appears much worse if we neglect the intercept. The reason for this is that there is still a considerable replicate error, and this cannot readily be modelled using a single compound model. If this contribution were removed, the error would be reduced dramatically. Alternative approaches such as ANOVA could be employed to separate the replicate from the lack-of-fit error, as discussed in Chapter 2, and determine whether coefficients are significant, but the emphasis of this chapter is primarily on methods for calibration models, so we will not duplicate discussion.

An alternative, and common, method for including the intercept is to mean centre both the x and the c variables to fit the equation

$$c - \bar{c} \approx (x - \bar{x})b$$

or

$$^{cen}c \approx {}^{cen}xb$$

or

$$b \approx ({}^{cen}x'{}^{cen}x)^{-1\,cen}x'{}^{cen}c = \frac{\sum_{i=1}^{I}(x_i - \bar{x})(c_i - \bar{c})}{\sum_{i=1}^{I}(x_i - \bar{x})^2}$$

It is easy to show algebraically that

- the value of b when both variables have been centred is identical to the value of b_1 obtained when the uncentred data are modelled including an intercept term (=4.23 in this example),
- whereas the value of b_0 (intercept term for uncentred data) is given by $\bar{c} - b\bar{x} = 0.47 - 4.23 \times 0.15 = -0.17$, so the two methods are related.

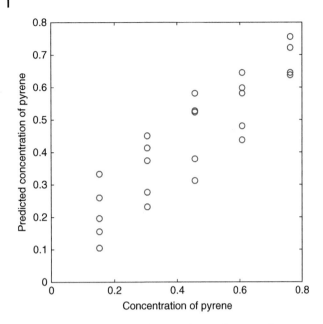

Figure 6.7 Predicted (vertical) versus known (horizontal) concentrations using methods of Section 6.2.3.

It is quite common to centre both sets of variables for this reason, the calculations being mathematically simpler than including an intercept term. Note that both blocks must be centred, and the predictions are of the concentrations minus their mean; hence, the mean concentration must be added back to return to the original values.

In calibration, it is common to plot a graph of predicted versus observed concentrations as presented in Figure 6.7. This looks superficially similar to that of the previous figure, but the vertical scale is different and the graph crosses the origin (providing the data have been mean centred). There is a variety of potential graphical output and it is important not to get confused between each type of graph, but to distinguish each type of information carefully.

It is important to realise that the predictions for the method described in this section differ to those obtained for the uncentred data. It is also useful to realise that similar methods can be applied to classical calibration, the details omitted for brevity, as it is recommended that inverse calibration is performed in normal circumstances in chemometrics.

1. Forty unique univariate measurements are recorded on a series of samples. A linear model including an intercept is used to fit the data.
    (a) There are 40 degrees of freedom for assessing the error.
    (b) There are 39 degrees of freedom for assessing the error.
    (c) There are 38 degrees of freedom for assessing the error.

6.3 Multiple Linear Regression

6.3.1 Multi-detector Advantage

Multiple linear regression (MLR) is an extension when more than one response is employed. There are three principal reasons for this. The first is that there may be more than one component in a mixture. Under such circumstances, it is usual to employ more than one response (the exception being if the concentrations of some of the components are correlated): for N components, at least N wavelengths should normally be used. The second is that each detector contains extra, and often complementary, information: some individual wavelengths in a spectrum may be influenced by noise or unknown interferants. Using, for example, a 100 wavelengths averages out the information and will often provide a better result than relying on a single wavelength. The third is that there may be an unknown number of factors that influence a response; thus, using very few variables will result in these factors being confounded and problems with finding unique and robust solutions.

When dealing with mixture spectra or chromatograms, if all the concentrations of all compounds are to be predicted, the number of variables or wavelengths must at least equal the number of compounds. However, if there are correlations either between the concentration profiles or the spectral profiles, it may not always be possible to resolve all the individual compounds. In our case study, the experiments are designed so that the concentrations are uncorrelated (using methods in Section 2.3.4) and the spectra all have unique characteristics.

1. A series of mixtures is known to contain 10 unique compounds and is studied spectroscopically.

   (a) Exactly 10 wavelengths must be recorded and will always result in a unique solution for each compound.
   (b) More than 10 wavelengths are always required for a unique solution.
   (c) The number of wavelengths required to obtain a unique solution for all compounds depends on whether there is correlation between spectra and selective wavelengths.
   (d) There may not be a unique solution for all the compounds in the mixture, no matter how many wavelengths, dependent on the experimental design and nature of the spectra.

6.3.2 Multi-wavelength Equations

In certain applications, equations can be developed that are used to predict the concentrations of compounds by monitoring at a finite number of wavelengths. A classical area is in pigment analysis by EAS, for example, in the area of chlorophyll chemistry. In order to determine the concentration of four pigments in a mixture, investigators recommend monitoring at four different wavelengths, and use an equation that links absorbance at each wavelength to concentration of the pigments.

In the PAH case study, only certain compounds absorb above 330 nm, the main ones being pyrene, fluoranthene, acenaphthalene and benzo[a]anthracene (note that the small absorbance due to a fifth component may be modelled as an interferant, although adding this to the model will, of course, result in better predictions but a more complicated model). It is possible to choose four wavelengths, preferably ones in which the absorbance ratios of these four compounds differ. The absorbance at wavelengths 330, 335, 340 and 345 nm of the pure compounds are illustrated in Figure 6.8. Of course, it is not necessary to select four sequential wavelengths, any four wavelengths would be sufficient, providing the four compounds are the main ones represented by these variables and their relative intensities differ, to give an X matrix with 25 rows and four columns.

Calibration equations can be obtained, as follows, using inverse methods.

- First, select the absorbances of the 25 spectra at these four wavelengths.
- Second, obtain the corresponding C matrix consisting of the relevant concentrations. These new (reduced) matrices are presented in Table 6.5.
- The aim is to find coefficients B relating X and C by $C \approx X B$, where B is a 4×4 matrix, each *column* representing a compound and each *row* a wavelength. This equation can be solved using regression methods of Section 6.2.2, changing vectors and scalars to matrices, so that $B = (X'X)^{-1}X'C$ giving the matrix in Table 6.6.
- If desired, represent in equation form, for example, the first column of B suggests that

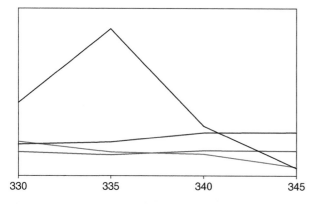

Figure 6.8 Absorbances of Pyr, Fluor, Benz and Ace between 330 and 345 nm.

Table 6.5 Matrices for four components.

	X				C		
330	335	340	345	Py	Ace	Benz	Fluora
0.127	0.165	0.110	0.075	0.456	0.120	1.620	0.120
0.150	0.178	0.140	0.105	0.456	0.040	2.700	0.120
0.095	0.102	0.089	0.068	0.152	0.200	1.620	0.080
0.134	0.191	0.107	0.060	0.760	0.200	1.080	0.160
0.170	0.239	0.146	0.094	0.760	0.160	2.160	0.160
0.135	0.178	0.115	0.078	0.608	0.200	2.160	0.040
0.129	0.193	0.089	0.041	0.760	0.120	0.540	0.160
0.127	0.164	0.113	0.078	0.456	0.080	2.160	0.120
0.104	0.129	0.098	0.074	0.304	0.160	1.620	0.200
0.157	0.193	0.134	0.093	0.608	0.160	2.700	0.040
0.100	0.154	0.071	0.030	0.608	0.040	0.540	0.040
0.056	0.065	0.053	0.036	0.152	0.160	0.540	0.080
0.094	0.144	0.078	0.043	0.608	0.120	1.080	0.040
0.079	0.114	0.064	0.040	0.456	0.200	0.540	0.120
0.143	0.211	0.114	0.067	0.760	0.040	1.620	0.160
0.081	0.087	0.081	0.069	0.152	0.040	2.160	0.080
0.071	0.077	0.061	0.045	0.152	0.080	1.080	0.080
0.081	0.106	0.072	0.047	0.304	0.040	1.080	0.200
0.114	0.119	0.115	0.096	0.152	0.120	2.700	0.080
0.098	0.130	0.080	0.051	0.456	0.160	1.080	0.120
0.133	0.182	0.105	0.059	0.608	0.080	1.620	0.040
0.070	0.095	0.064	0.042	0.304	0.080	0.540	0.200
0.124	0.138	0.118	0.093	0.304	0.200	2.700	0.200
0.163	0.219	0.145	0.101	0.760	0.080	2.700	0.160
0.128	0.147	0.116	0.086	0.304	0.120	2.160	0.200

Table 6.6 Matrix *B* for Section 6.3.2.

	Py	Ace	Benz	Fluor
330	−3.870	2.697	14.812	−4.192
335	8.609	−2.391	3.033	0.489
340	−5.098	4.594	−49.076	7.221
345	1.848	−4.404	65.255	−2.910

$$\text{Estimated}\,[\text{pyrene}] = -3.870\,A_{330} + 8.609\,A_{335} - 5.098\,A_{340} + 1.848\,A_{345}$$

In many areas of optical spectroscopy, these types of equations are very common. Note although that changing the wavelengths monitored can have a radical influence on the coefficients, and slight wavelength irreproducibility between spectrometers can lead to equations that are not easily transferred.

- Finally, estimate the concentrations by

$$\hat{C} = XB$$

as indicated in Table 6.7.

The estimates by this approach are very much better to those from the univariate approaches in this particular example. Figure 6.9 is of the predicted versus known concentrations for pyrene and should be compared with Figure 6.7.

Table 6.7 Estimated concentration for four components as described in Section 6.3.2.

Py	Ace	Benz	Fluor
0.507	0.123	1.877	0.124
0.432	0.160	2.743	0.164
0.182	0.122	1.786	0.096
0.691	0.132	1.228	0.130
0.829	0.144	2.212	0.185
0.568	0.123	1.986	0.125
0.784	0.115	0.804	0.077
0.488	0.126	1.923	0.137
0.345	0.096	1.951	0.119
0.543	0.168	2.403	0.133
0.632	0.096	0.421	0.081
0.139	0.081	0.775	0.075
0.558	0.078	0.807	0.114
0.423	0.058	0.985	0.070
0.806	0.110	1.535	0.132
0.150	0.079	1.991	0.087
0.160	0.089	1.228	0.050
0.319	0.089	1.055	0.095
0.174	0.128	2.670	0.131
0.426	0.096	1.248	0.082
0.626	0.146	1.219	0.118
0.298	0.071	0.925	0.093
0.278	0.137	2.533	0.129
0.702	0.137	2.553	0.177
0.338	0.148	2.261	0.123

The three other compounds in the model are now no longer lumped together as background noise and so we can take their systematic variation into account. The RMSE of prediction is now

$$E = \sqrt{\sum_{i=1}^{I} \left(c_i - \hat{c}_i \right)^2 / 21}$$

(note that the divisor is 21 not 25, as 4 degrees of freedom are lost because there are four compounds in the model), equal to 0.042% or 9.13%, of the average concentration, a significant improvement. Further improvement could be obtained by including the intercept (usually performed by centring the data) and including the concentrations of more compounds. However, the number of wavelengths must be increased if the more compounds are used in the model.

It is also possible to employ classical methods. For the single detector, single wavelength model of Section 2.1.1

$$\hat{c} = x(1/s)$$

where s is a scalar, and x and c are vectors corresponding to the concentrations and absorbances for each of the I samples. Where there are several components in the mixture, this becomes

$$\hat{C} = XS'(SS')^{-1}$$

and the trick is to estimate S, which can be done in one of the two ways:

- by knowledge of the true spectra or
- by regression as $CS \approx X$, so $\hat{S} = (C'C)^{-1}C'X$.

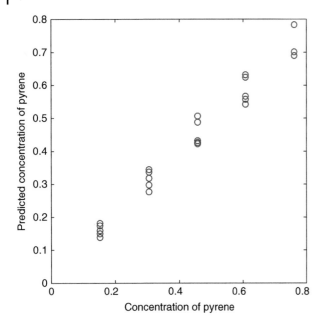

Figure 6.9 Predicted versus known concentration of pyrene, using a four-component model and the wavelengths 330, 335, 340 and 345 nm (uncentred).

However, as in univariate calibration, the coefficients obtained using both approaches may not be exactly equivalent, as each method makes different assumptions about the error structure. However, if the models from both methods differ substantially, this could be an indication of outliers, additional significant components and so on.

Note that the number of wavelengths must be more than or equal to the number of compounds in the mixture (or factors as appropriate). Mathematically, this is because the matrix SS' must be invertible. In addition, the concentrations and spectra of compounds must not be correlated, again chemical sense and mathematical sense come to the same conclusion – when there are complex multi-component mixtures, careful thought must be given to the design.

Equations where the number of wavelengths equals the number of analytes make assumptions that the concentrations of the significant analytes are all known and work well only if this is true. Applying to mixtures where there are significant unknown interferants can result in serious estimation errors.

1. A mixture consists of three components, recorded at three wavelengths and four samples. The X matrix is
$$\begin{bmatrix} 0.571 & 0.960 & 0.763 \\ 1.090 & 1.230 & 1.048 \\ 0.863 & 1.230 & 0.445 \\ 0.635 & 1.019 & 0.437 \end{bmatrix}, \text{ where the columns represent the three wavelengths and the } C \text{ matrix is } \begin{bmatrix} 0.78 & 0.35 & 1.04 \\ 0.52 & 0.90 & 1.76 \\ 0.39 & 1.21 & 0.50 \\ 0.44 & 0.88 & 0.23 \end{bmatrix}'$$
where the columns represent the three compounds.

   (a) The concentration of compound 1 can be approximated by

$$-1.43A_1 + 1.01A_2 + 0.80A_3$$

   (b) The concentration of compound 1 can be approximated by

$$-1.43A_1 + 1.40A_2 + 1.10A_3$$

6.3.3 Multivariate Approaches

The methods in Section 6.3.2 could be extended to all 10 PAHs, and with appropriate choice of 10 wavelengths may give reasonable estimates of concentrations. However, all the recorded wavelengths contain some information and there is no reason to restrict to only 10 selected wavelengths.

There is a fairly confusing literature on the use of MLR for calibration in chemometrics, primarily because many workers present their arguments in a very formalised manner. However, the choice and applicability of any method depends on three main factors.

- The number of compounds in the mixture ($N = 10$ in this case) or responses to be estimated.
- The number of samples ($I = 25$ in this case) often spectra or chromatograms.
- The number of variables ($J = 27$ wavelengths in this case).

In order to have a sensible model, the number of compounds must be less than or equal to the smaller of the number of samples or number of variables. In certain specialised cases, this limitation can be infringed if it is known that there are correlations between different compounds and this is taken into account in the model (e.g. by modelling less factors than the known or estimated number of compounds). This may happen, for example, in environmental chemistry where there could be tens or hundreds of compounds in a sample, but the presence of one (e.g. a homologous series) indicates the presence of another; hence, in practice, there are only a few independent factors or groups of compounds. In addition, correlations can be built into a design. In many real-world situations, definitely there will be correlations in complex multi-component mixtures. However, the methods described below are for the case where the number of compounds is smaller than the number of samples or number of detectors, and each compound in the mixture can be quantified independently: for brevity, we illustrate the methods in this chapter using just one case study.

The X data matrix is ideally related to the concentration and spectral matrices by

$$X \approx CS$$

where X is a 25×27 matrix, C a 25×10 matrix and S a 10×27 matrix in the example discussed here. In calibration, it is assumed that a series of experiments are performed in which C is known (e.g. a set of mixtures of compounds with known concentrations are recorded spectroscopically). An estimate of S can then be obtained by

$$\hat{S} = (C'C)^{-1}C'X$$

and then the concentrations can be predicted

$$\hat{C} = X\hat{S}'(\widehat{SS}')^{-1}$$

exactly as above. This can be extended to estimating the concentrations in any spectrum by

$$\hat{c} = x\hat{S}'(\widehat{SS}')^{-1} = xB$$

where $\hat{c}$ is a row vector with the 10 predicted concentrations.

Unless the number of experiments is exactly equal to the number of compounds, the prediction will not completely model the data. This approach works because the matrices $(C'C)$ and $(\widehat{SS}')$ are square matrices whose dimensions equal the number of compounds in the mixture (10×10) and have inverses, providing the experiments have been well designed. The predicted concentrations, using this approach, are given in Table 6.8, together with the percentage root mean square calibration error: note that there are only 15 degrees of freedom (=25 experiments – 10 compounds). Had the data been centred the number of degrees of freedom would be reduced further. The predicted concentrations are reasonably good for most compounds apart from acenaphthylene.

The predicted spectra are presented in Figure 6.10 and are not nearly as well predicted as the concentrations. In fact, it would be remarkable that for such a complex mixture, it is possible to reconstruct all 10 spectra well, given that there is a great deal of overlap. Pyrene, which is emphasised in bold, exhibits most of the main peak maxima of the known pure spectra (compare with Figure 6.3). Often, other knowledge of the system is required to produce better reconstructions of individual spectra, and more elaborate approaches involving constraints will be discussed in Section 7.4.3. The prime aim of the methods in this chapter is quantitative estimates of concentration rather than resolution of spectra. The reason why concentration predictions appear to work significantly better than spectral reconstruction is that, for most compounds, there are characteristic regions of the spectrum containing prominent features. These parts of the spectra for individual compounds will be predicted well and will disproportionately influence the effectiveness of the method for determining concentrations, whereas other regions that are less significant for predicting concentrations may not be resolved so well.

MLR predicts concentrations well in this case because all significant compounds are included in the model. If we knew only a few compounds, there would be much poorer predictions. Consider the situation in which only pyrene,

Table 6.8 Estimated concentrations for the case study using uncentred MLR and all wavelengths.

Spectrum	Polyarene concentration (mg/l)									
	Py	Ace	Anth	Acy	Chry	Benz	Fluora	Fluore	Nap	Phen
1	0.509	0.092	0.200	0.151	0.369	1.731	0.121	0.654	0.090	0.433
2	0.438	0.100	0.297	0.095	0.488	2.688	0.148	0.276	0.151	0.744
3	0.177	0.150	0.303	0.217	0.540	1.667	0.068	0.896	0.174	0.128
4	0.685	0.177	0.234	0.150	0.369	1.099	0.128	0.691	0.026	0.728
5	0.836	0.137	0.304	0.155	0.224	2.146	0.159	0.272	0.194	0.453
6	0.593	0.232	0.154	0.042	0.435	2.185	0.071	0.883	0.146	1.030
7	0.777	0.164	0.107	0.129	0.497	0.439	0.189	0.390	0.158	0.206
8	0.419	0.040	0.198	0.284	0.044	2.251	0.143	1.280	0.088	0.299
9	0.323	0.141	0.247	0.037	0.462	1.621	0.196	0.101	−0.003	0.298
10	0.578	0.236	0.020	0.107	0.358	2.659	0.093	0.036	0.070	0.305
11	0.621	0.051	0.214	0.111	0.571	0.458	0.062	0.428	0.022	0.587
12	0.166	0.187	0.170	0.142	0.087	0.542	0.100	0.343	0.103	0.748
13	0.580	0.077	0.248	0.133	0.051	1.120	−0.042	0.689	0.176	0.447
14	0.468	0.248	0.057	−0.006	0.237	0.558	0.157	0.712	0.103	0.351
15	0.770	0.016	0.066	0.119	0.094	1.680	0.187	0.450	0.080	0.920
16	0.101	0.026	0.100	0.041	0.338	2.230	0.102	0.401	0.201	0.381
17	0.169	0.115	0.063	0.069	0.478	1.054	0.125	0.829	0.068	0.523
18	0.271	0.079	0.142	0.106	0.222	1.086	0.211	0.254	0.151	0.261
19	0.171	0.152	0.216	0.059	0.274	2.587	0.081	0.285	0.013	0.925
20	0.399	0.116	0.095	0.170	0.514	1.133	0.101	0.321	0.243	1.023
21	0.651	0.025	0.146	0.232	0.230	1.610	−0.013	0.940	0.184	0.616
22	0.295	0.135	0.256	0.052	0.349	0.502	0.237	0.970	0.161	1.037
23	0.296	0.214	0.116	0.069	0.144	2.589	0.202	0.785	0.162	0.588
24	0.774	0.085	0.187	−0.026	0.547	2.671	0.128	1.107	0.108	0.329
25	0.324	0.035	0.036	0.361	0.472	2.217	0.094	0.918	0.128	0.779
RMSE$_\%$	9.79	44.87	15.58	69.43	13.67	4.71	40.82	31.38	29.22	16.26

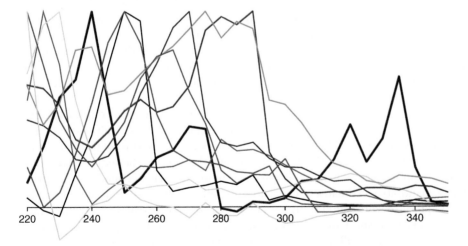

Figure 6.10 Spectra of the 10 PAHs estimated by MLR, with pyrene indicated in bold.

Table 6.9 Estimates for three PAHs using the full data set and MLR but including only three compounds in the model.

Spectrum	Polyarene concentration (mg/l)		
	Py	Ace	Anth
1	0.539	0.146	0.156
2	0.403	0.173	0.345
3	0.199	0.270	0.138
4	0.749	0.015	0.231
5	0.747	0.103	0.211
6	0.489	0.165	0.282
7	0.865	0.060	−0.004
8	0.459	0.259	0.080
9	0.362	0.121	0.211
10	0.512	0.351	−0.049
11	0.742	−0.082	0.230
12	0.209	0.023	0.218
13	0.441	0.006	0.202
14	0.419	0.095	0.051
15	0.822	0.010	0.192
16	0.040	0.255	0.151
17	0.259	0.162	0.122
18	0.323	0.117	0.104
19	0.122	0.179	0.346
20	0.502	0.085	0.219
21	0.639	0.109	0.130
22	0.375	−0.062	0.412
23	0.196	0.316	0.147
24	0.638	0.218	0.179
25	0.545	0.317	0.048
$RMSE_{\%}$	22.04986	105.7827	52.40897

acenaphthene and anthracene are known. The C matrix now has only three columns, and the predicted concentrations are given in Table 6.9. The errors are, as expected, much larger than those in Table 6.8. The absorbances of the remaining seven compounds are mixed up with those of the three modelled components. This problem could be overcome if some characteristic wavelengths or regions of the spectrum at which the selected compounds absorb most strongly are identified, or if the experiments were designed so that there are correlations in the data, or even by a number of methods for weighted regression, but the need to provide information about all significant compounds within a spectral region (in our case) is a major limitation of MLR.

1. In the equation $\hat{c} = xB$, what are the dimensions of B in the case there are 20 samples, 15 wavelengths and five compounds?

   (a) 15×1
   (b) 15×5
   (c) 5×20
   (d) 1×5

2. So long as the number of compounds of interest is always less than or equal to the number of wavelengths monitored, it is always possible to form an MLR model to predict the concentration well of these compounds in a series of mixtures?

    (a) True

    (b) False

6.4 Principal Components Regression

MLR-based methods have the disadvantage that all significant components (or compounds in our case) in the region of interest must be known. Principal components analysis (PCA)-based methods do not require details about the spectra or concentrations of all the compounds in a mixture, although it is important to make a sensible estimate how many significant components characterise a mixture, but not necessarily their chemical characteristics.

Principal components are primarily abstract mathematical entities and further details are described in Chapter 4. In multivariate calibration, the aim is to convert these to compound concentrations. Principal components regression (PCR) uses regression (sometimes also called *transformation* or *rotation*) to convert PC scores to concentrations. This process is often loosely called *factor analysis*, although terminology differs according to author and discipline. Note that although the chosen example in this chapter involves calibrating concentrations to spectral absorbances, it is equally possible, for example, to calibrate the property of a material to its structural features, or the activity of a drug to molecular parameters.

6.4.1 Regression

If the vector $\hat{c}_n$ contains the estimated concentrations of compound n in a series of spectra (25 in this instance), it can be obtained from the PC scores matrix, T, as follows:

$$\hat{c}_n = T r_n$$

where r_n is a column vector whose length equals the number of PCs retained in the model, sometimes called a *rotation* or *transformation vector*. Ideally, the length of r_n should be equal to the number of compounds in the mixture (=10 in this case). However, noise, spectral similarities and correlations between concentrations and spectra, sometimes, make it hard to provide an exact estimate of the number of significant components: this topic has been discussed in Section 4.3.2. We will assume, for the purpose of this section, that 10 PCs are retained in the model.

The scores of the first 10 PCs are presented in Table 6.10, using raw data. Naturally, data could be pre-processed, including centring if desired.

The transformation vector can be obtained by using the pseudo-inverse of T.

$$r_n = (T'T)^{-1}T'c_n$$

where c_n are the concentrations of compound n. Note that the matrix $(T'T)$ is actually a diagonal matrix, whose elements consist 10 eigenvalues of the PCs and so each element of r could also be expressed as a summation

$$r_{na} = \frac{\displaystyle\sum_{i=1}^{I} t_{ia} c_{in}}{\displaystyle\sum_{a=1}^{A} g_a}$$

or even as a product of vectors

$$r_{na} = \frac{t'_a c_n}{\displaystyle\sum_{a=1}^{A} g_a}$$

We remind the reader of the main notation:

Table 6.10 Scores of the first 10 PCs for PAH case study.

PC1	PC2	PC3	PC4	PC5	PC6	PC7	PC8	PC9	PC10
2.757	0.008	0.038	0.008	0.026	0.016	0.012	−0.004	0.006	−0.006
3.652	−0.063	−0.238	−0.006	0.021	0.000	0.018	0.005	0.009	0.013
2.855	−0.022	0.113	0.049	−0.187	0.039	0.053	0.004	0.007	−0.003
2.666	0.267	0.040	−0.007	0.073	0.067	−0.002	−0.002	−0.013	−0.006
3.140	0.029	0.006	−0.153	0.111	−0.015	0.030	0.022	0.014	0.006
3.437	0.041	−0.090	0.034	−0.027	−0.018	−0.010	−0.014	−0.032	0.006
1.974	0.161	0.296	0.107	0.090	−0.010	0.003	0.037	0.004	0.008
2.966	−0.129	0.161	−0.147	−0.043	0.016	−0.006	−0.010	0.013	−0.035
2.545	−0.054	−0.143	0.080	0.074	0.073	0.013	0.008	0.025	−0.006
3.017	−0.425	0.159	0.002	0.096	0.049	−0.010	0.013	−0.018	0.022
2.005	0.371	0.003	0.120	0.093	0.032	0.015	−0.025	0.003	0.003
1.648	0.239	−0.020	−0.123	−0.090	0.051	−0.017	0.021	−0.009	0.007
1.884	0.215	0.020	−0.167	−0.024	−0.041	0.041	−0.007	−0.017	0.001
1.666	0.065	0.126	0.070	−0.007	−0.005	−0.016	0.036	−0.025	−0.009
2.572	0.085	−0.028	−0.095	0.184	−0.046	−0.045	−0.016	0.013	−0.006
2.532	−0.262	−0.126	0.047	−0.084	−0.076	0.004	0.005	0.017	0.010
2.171	0.014	0.028	0.166	−0.080	0.008	−0.018	−0.007	−0.003	−0.007
1.900	−0.020	0.027	−0.015	−0.006	−0.018	−0.005	0.030	0.029	−0.002
3.174	−0.114	−0.312	−0.059	−0.014	0.066	−0.009	−0.016	−0.011	0.007
2.610	0.204	0.037	0.036	−0.069	−0.041	−0.020	−0.005	0.012	0.033
2.567	0.119	0.155	−0.090	−0.017	−0.050	0.017	−0.023	−0.013	0.005
2.389	0.445	−0.190	0.045	−0.091	−0.026	−0.026	0.021	0.006	−0.015
3.201	−0.282	−0.043	−0.062	−0.066	−0.015	−0.026	0.032	−0.015	−0.009
3.537	−0.182	−0.071	0.166	0.094	−0.069	0.026	−0.013	−0.016	−0.021
3.343	−0.113	0.252	0.012	−0.086	0.019	−0.031	−0.048	0.018	0.004

Table 6.11 Vector r for pyrene.

Pyrene
0.166
0.470
0.624
−0.168
1.899
−1.307
1.121
0.964
−3.106
−0.020

- n refers to compound number (e.g. pyrene = 1),
- a to PC number (e.g. 10 significant components, not necessarily equal to the number of compounds in a series of samples) and
- i to sample number (=1–25 in this case).

This vector for pyrene using 10 PCs is presented in Table 6.11. If the concentration of some or all the compounds of interest are known, PCR can be extended simply by replacing the vector c_k with a matrix C each column corresponding to a compound in the mixture, so that

$$\hat{C} = TR$$

where

$$R = (T'T)^{-1}T'C$$

The number of PCs must be at least equal to the number of compounds of interest in the mixture. R has dimensions $A \times N$.

If the number of PCs used in the model and number of significant compounds are equal, so that, in this example, T and C are 25×10 matrices, then R is a square matrix of dimensions $N \times N$ and

$$\hat{X} = TP = TRR^{-1}P = \hat{C}\hat{S}$$

hence, by calculating $R^{-1}P$, it is possible to estimate the spectra of each individual component without knowing this information in advance, and by calculating TR concentration estimates can be obtained. These principles are discussed further in Chapter 7.

Table 6.12 presents the concentration estimates using PCR with 10 significant components. The percentage mean square error of prediction (equalling the square root sum of squares of the errors of prediction divided by 15 – the number of degrees of freedom that equals 25–10 to account for the number of components in the model, and not by 25) for all 10 compounds is also presented. In most cases, it is slightly better to that using MLR, there are certainly less very large errors. However, the big advantage is that the prediction using PCR is the same if only 1 or all the 10 compounds are included in the model. In this, it differs from MLR, the estimates of Table 6.9 are much worse than Table 6.8, for example. The main first task when using PCR is to determine how many significant components

Table 6.12 Concentration estimates of the PAHs using PCR and 10 components (uncentred).

Spectrum	Polyarene concentration (mg/l)									
	Py	Ace	Anth	Acy	Chry	Benz	Fluora	Fluore	Nap	Phen
1	0.505	0.113	0.198	0.131	0.375	1.716	0.128	0.618	0.094	0.445
2	0.467	0.120	0.286	0.113	0.455	2.686	0.137	0.381	0.168	0.782
3	0.161	0.178	0.296	0.174	0.558	1.647	0.094	0.836	0.162	0.161
4	0.682	0.177	0.231	0.165	0.354	1.119	0.123	0.720	0.049	0.740
5	0.810	0.128	0.297	0.156	0.221	2.154	0.159	0.316	0.189	0.482
6	0.575	0.170	0.159	0.107	0.428	2.240	0.072	0.942	0.146	1.000
7	0.782	0.152	0.104	0.152	0.470	0.454	0.162	0.477	0.169	0.220
8	0.401	0.111	0.192	0.170	0.097	2.153	0.182	1.014	0.062	0.322
9	0.284	0.084	0.237	0.106	0.429	1.668	0.166	0.241	0.022	0.331
10	0.578	0.197	0.023	0.157	0.321	2.700	0.077	0.194	0.090	0.300
11	0.609	0.075	0.194	0.103	0.550	0.460	0.080	0.472	0.038	0.656
12	0.185	0.172	0.183	0.147	0.083	0.558	0.086	0.381	0.101	0.701
13	0.555	0.103	0.241	0.092	0.084	1.104	0.007	0.576	0.156	0.475
14	0.461	0.167	0.067	0.089	0.212	0.624	0.111	0.812	0.114	0.304
15	0.770	0.019	0.076	0.089	0.115	1.669	0.178	0.393	0.068	0.884
16	0.109	0.033	0.101	0.040	0.349	2.189	0.108	0.352	0.190	0.376
17	0.178	0.102	0.073	0.086	0.481	1.057	0.112	0.805	0.073	0.486
18	0.271	0.067	0.145	0.104	0.221	1.077	0.183	0.273	0.142	0.250
19	0.186	0.135	0.217	0.101	0.253	2.618	0.071	0.369	0.036	0.919
20	0.406	0.109	0.111	0.145	0.534	1.126	0.111	0.306	0.220	0.973
21	0.665	0.110	0.152	0.130	0.284	1.541	0.044	0.720	0.165	0.614
22	0.315	0.112	0.258	0.092	0.336	0.501	0.205	0.981	0.162	1.009
23	0.327	0.179	0.126	0.115	0.126	2.610	0.160	0.847	0.161	0.537
24	0.766	0.075	0.168	0.029	0.525	2.692	0.121	1.139	0.124	0.383
25	0.333	0.110	0.053	0.210	0.539	2.151	0.135	0.709	0.086	0.738
RMSE$_\%$	10.27	36.24	15.76	42.06	9.05	4.24	31.99	24.77	21.11	16.19

are necessary to model the data using methods such as cross-validation of percentage error, but after that, not every compound in the mixture needs to be modelled individually.

For a new sample, not necessarily part of the training set, we can predict the concentration of all analytes in the model as follows, first by estimating the scores using the loadings matrix, and then transforming these estimated scores

$$\widehat{t} = xP' \text{ hence } \widehat{c} = \widehat{t}r$$

or in, one step, $\widehat{c} = xB$ where $B = P'r$. Note that the matrix B, which is used for estimation, differs according to the various methods described in this chapter.

1. A set of 30 mixture spectra recorded at 50 wavelengths consisting of eight main compounds and four PCs are retained for PCR. For each individual compound, a vector r is calculated. Its dimensions are

   (a) 8×1
   (b) 1×8
   (c) 4×1
   (d) 1×4

6.4.2 Quality of Prediction

A key issue in calibration is to determine how well the data have been modelled. In this section, we will discuss the training set or calibration error; Section 6.6 provides a further overview of how the approaches below can be used. We can now look at how the model changes for PCR as we change the number of principal components to help us decide how many PCs are suitable for the model.

Most look at how well the concentration is predicted, or the c (or according to some authors y) block of data.

The simplest method is to determine the sum of square of residuals between the true and predicted concentrations

$$S_c = \sum_{i=1}^{I} \left(c_{in} - \widehat{c}_{in} \right)^2$$

where

$$\widehat{c}_{in} = \sum_{a=1}^{A} t_{ia} r_{an}$$

for compound n using a principal components. The larger this error, the worse the prediction, hence the error reduces as more components are calculated.

Often, the error is reported as a RMSE, divided by the number of degrees of freedom (where A equals the number of PCs in the model),

$$RMSE = \sqrt{\frac{\sum_{i=1}^{I} (c_{in} - \widehat{c}_{in})^2}{I - A}}$$

If the data are centred, a further degree of freedom is lost; hence, the sum of square residual is divided by $I - A - 1$. This error can also be presented as a percentage error,

$$RMSE_\% = 100 E / \overline{c}_n$$

where $\overline{c}_n$ is the mean concentration in the original units. Sometimes, the percentage of the standard deviation is calculated instead, but in this section, we will compute errors as a percentage of the mean unless specifically stated.

It is also possible to report errors in terms of quality of modelling of spectra (or chromatograms), often called the x block error.

The quality of modelling of the spectra using PCA (the x variance) can likewise be calculated as follows:

$$S_x = \sum_{i=1}^{I} \sum_{j=1}^{J} \left(x_{ij} - \widehat{x}_{ij} \right)^2$$

where

$$\widehat{x}_{ij} = \sum_{a=1}^{A} t_{ia} p_{aj}$$

This error can also be expressed in terms of eigenvalues or scores, so that

$$S_x = \sum_{i=1}^{I} \sum_{j=1}^{J} x_{ij}^2 - \sum_{a=1}^{A} g_a = \sum_{i=1}^{I} \sum_{j=1}^{J} x_{ij}^2 - \sum_{a=1}^{A} \sum_{i=1}^{I} t_{ia}^2$$

for A principal components.

These can be converted to RMSEs as above,

$$RMSE_x = \sqrt{S_x/IJ}$$

Note that many people divide by IJ (=25 × 27 = 675 in our case) rather than the more strictly correct $IJ - A$ (adjusting for degrees of freedom) because IJ is very large relative to A, and we will adopt this convention.

The percentage RMSE may be defined by (for uncentred data)

$$RMSE_{x\%} = 100RMSE/\bar{x}$$

Note that if x is centred, the divisor is often given by $\sqrt{\sum_{i=1}^{I} \sum_{j=1}^{J} \frac{(x_{ij}-\bar{x}_j)^2}{IJ}}$, where $\bar{x}_j$ is the average of all the measurements for the samples for variable j: obviously, there are several other ways of defining this error, if you try to follow a paper or a package, read very carefully the documents provided by the authors, and if there is no documentation, do not trust the answers.

Note that the x error depends only on the number of PCs, no matter how many compounds are being modelled, but the error in concentration estimates also depends on which specific compound is being estimated, there being a different percentage error for each compound in the mixture. For zero PCs, the estimates of the PCs and concentrations are simply 0, or, if mean centred, the mean. The graphs of RMSEs for both the concentration estimates of pyrene and the spectra as increasing numbers of PCs are calculated are given in Figure 6.11, using a logarithmic scale for the error. Although the x error graph declines steeply, which might falsely suggest that only a small number of PCs are required for the model, the c error graph exhibits a much gentler decline. Sometimes, these graphs are presented either as

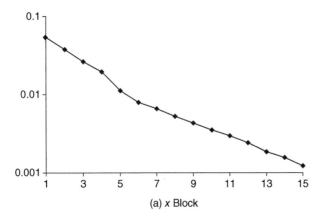

(a) *x* Block

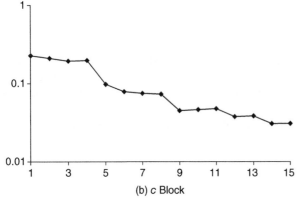

(b) *c* Block

Figure 6.11 Root mean square errors of estimation of pyrene using uncentred PCR between 1 and 15 PCs.

percentage variance remaining (or explained by each PC) or eigenvalues. Note a very small rise in the RMSE c error at four components: this is because the number of degrees of freedom changes more rapidly than the absolute error.

It is most common to use the c block error as this relates to how well the concentrations rather than the spectra are modelled, but sometimes comparing both gives a good idea of how many PCs are best to use in the model.

1. If a centred data set consists of 30 samples, 20 variables and is modelled by five PCs, how many degrees of freedom are there to assess the error in the predicted values of c?

   (a) 30
   (b) 25
   (c) 24

2. A data set consists of 100 samples and 30 variables. A four PC model gives eigenvalues of 103, 70, 16 and 4 and the total sum of squares of the data is 215, there is no pre-processing. What is the $RMSE_x$?

   (a) 0.073
   (b) 0.271
   (c) 22
   (d) 4.69

6.5 Partial Least Squares Regression

PLS is often presented as the main regression technique used when there is multivariate data. In fact, its use is not always justified, and the originators of the method were well aware of this, but that being said, in some applications, PLS has been spectacularly successful. In some areas such as QSAR, or even metabolomics and psychometrics, PLS is an invaluable tool because the underlying factors have little or no physical meaning; hence, a linearly additive model in which each underlying factor can be interpreted chemically is not anticipated. In spectroscopy or chromatography, we usually expect linear additivity, and under such circumstances, simpler methods such as MLR are often adequate, providing there is fairly full knowledge of the system. However, PLS is always an important tool when there is partial knowledge of the data, a typical example being the measurement of protein in wheat by NIR spectroscopy. A model can be obtained from a series of wheat samples, and PLS will use typical features in this data set to establish a relationship to the known amount of protein. The basis is that if a training set is well designed, it will encompass the sort of samples that are likely to be encountered in the future. PLS models can be very robust, providing future samples do indeed contain similar features to the original data. Another example is the determination of vitamin C in orange juices using spectroscopy: a very reliable PLS model could be obtained using orange juices from a particular region of Spain, but what if some Brazilian orange juice is included? There is no guarantee that the model will perform well on the new data, as there may be different spectral features; hence, it is always important to be aware of the limitations of the method, particularly to remember that the use of PLS cannot compensate for poorly designed experiments or inadequate experimental data. Of course, it is possible to look for outliers that do not have the same correlation structure as the calibration or training set, but this is another layer of data analysis that many neglect.

An important feature of PLS is that it takes into account errors both in the concentration estimates and in the spectra. A method such as PCR assumes that the concentration estimates are error free. Much traditional statistics rests on this assumption that all errors are in the variables (spectra). If in medicine it is decided to determine the concentration of a compound in the urine of patients as a function of age, it is assumed that age can be estimated exactly, the statistical variation being in the concentration of a compound and the nature of the urine sample. Yet, in chemistry, there are often significant errors in sample preparation, for example, accuracy of weighings and dilutions and so the independent variable in itself also contains errors. Classical and inverse calibration force the user to choose which variable contains the error, whereas PLS assumes that it is equally distributed in both the x and c blocks and differs in this key assumption from the methods described above.

6.5.1 PLS1

The most widespread approach is often called *PLS1*. Although there are several algorithms, the main ones due to Wold and Martens, the overall principles are quite straightforward.

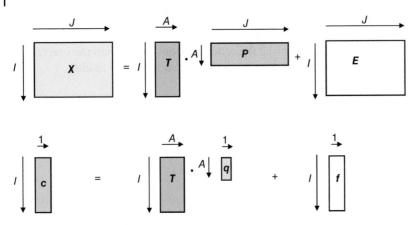

Figure 6.12 Principles of PLS1.

Instead of modelling exclusively the x variables, two sets of models are obtained as follows:

$$X = TP + E$$
$$c = Tq + f$$

where q has analogies to a loadings vector, although it is not normalised. These matrices are represented in Figure 6.12. The product of T and P approximates to the spectral data and the product of T and q to the true concentrations; the common link is T. An important feature of PLS is that it is possible to obtain a scores matrix that is common to both the concentrations (c) and measurements (x). Note that T and P for PLS are different to T and P obtained in PCA, and unique sets of scores and loadings are obtained when modelling each compound in the data set. Hence, if there are 10 compounds of interest, there will be 10 sets of T, P and q. In this way, PLS differs from PCR in which there is only one set of T and P, no matter how many compounds are being modelled, the PCA step taking no account of the c block. It is important to recognise that there are several algorithms for PLS available in the literature, and although the predictions of c are the same in each case, the scores and loadings are not. In this book and the associated Excel software, we use the algorithm in Section A.2.2. Although the scores are orthogonal (as in PCA), the loadings are not (which is an important difference to PCA), and, furthermore, the loadings are not normalised, so the sum of squares of each p vector does not equal 1, using our algorithm. Another matrix called the *weights matrix* can also be generated for which the rows (corresponding to components) are orthogonal, as described in the appendix, but for simplicity, we will primarily discuss the loadings in this chapter. If you are using a commercial software package, it is important to check exactly what constraints and assumptions the authors make about the scores and loadings in PLS, as there are different definitions according to author.

Additionally, the analogy to g_a or the eigenvalue of a PC involves multiplying the sum of squares of both t_a and p_a together, so we define the magnitude of a PLS component as

$$g_a = \left(\sum_{i=1}^{I} t_{ia}^2 \right) \left(\sum_{j=1}^{J} p_{aj}^2 \right)$$

This will have the property that the sum of values of g_a for all non-zero components adds up to the sum of squares of the original (pre-processed) data. Notice that in contrast to PCA, the size of successive values of g_a does not necessarily decrease as each component is calculated. This is because PLS does not only model the x data and is a compromise between x and c block regression.

There are a number of alternative ways of presenting the PLS regression equations in the literature, all mathematically equivalent. In the models above, we obtain three arrays T, P and q. Some authors calculate a normalised vector, w, proportional to q, and the second equation becomes

$$c = TBw + f$$

where B is a diagonal matrix. Analogously, it is also possible to the define the PCA decomposition of x as a product of three arrays (the singular value decomposition (SVD) method which is used in Matlab is a common alternative to NIPALS), but the models used in this chapter have the simplicity of using a single scores matrix for both blocks of data and model each block using just two matrices.

For a data set consisting of 25 spectra observed at 27 wavelengths, for which eight PLS components are calculated, there will be

- a T matrix of dimensions 25×8,
- a P matrix of dimensions 8×27,
- an E matrix of dimensions 25×27,
- a q vector of dimensions 8×1 and
- an f vector of dimensions 25×1.

Note that there will be 10 separate sets of these arrays, in the case discussed in this chapter, one for each compound in the mixture, and that the T matrix will be dependent on compound, which differs from PCR.

Each successive PLS component approximates both the concentration and spectral data better. For each PLS component, there will be a

- spectral scores vector t,
- spectral loadings vector p and
- concentration loadings scalar q.

In most implementations of PLS, it is conventional to centre both the x and c data, by subtracting the mean of each column, before analysis. In fact, there is no general scientific need to do this. Many spectroscopists and chromatographers perform PCA uncentred, but many early applications of PLS (e.g. outside chemistry) were of such a nature that centring the data was appropriate. Many of the historical developments in PLS as used for multivariate calibration in chemistry relate to applications in NIR spectroscopy, where there are specific spectroscopic problems, such due to baselines, which, in turn, would favour centring. Below, we use the most widespread implementation (involving centring) for the sake of compatibility with the most common computational implementations of the method.

For a given compound, the remaining percentage error in the x matrix for a PLS components can be expressed in various ways, as discussed in Section 6.4.2. Note that there are slight differences according to authors that take into account the number of degrees of freedom left in the model. The predicted measurements simply involve calculating $\hat{X} = TP$ and adding back the column means where appropriate, and error indicators in the x block can be defined similar to those used in PCR, see Section 6.4.2. The only difference is that each compound generates a separate scores matrix, unlike PCR where there is a single scores matrix for all compounds in the mixture, and so there will be a different behaviour in the x block residuals according to compound when using PLS.

The concentration of compound n in the training set or original data is estimated by

$$\hat{c}_{in} = \sum_{a=1}^{A} t_{ian} q_{an} + \bar{c}_n$$

or, in matrix terms

$$c_n - \bar{c}_n = T_n q_n$$

where $\bar{c}_n$ is a vector of the average concentration. Hence, the scores of each PLS component are proportional to the contribution of the component to the concentration estimate. The method of the concentration estimation for two PLS components and pyrene is presented in Table 6.13.

For estimating new samples that were not included in the training set, we can use the equation $\hat{c}_n = xb_n = \hat{t}_n q_n$. The calculation of b and $\hat{t}$ is somewhat complicated and described in detail in Section A.2.2. Unlike PCA, the product of the scores and loadings matrices does not approximate the best fit to the X matrix using least squares criteria.

The mean square error in the concentration estimate can be defined just as in PCR. It is also possible to define this error in various different ways using t and q. In the case of the c block estimates, it is usual to divide the sum of squares by $I - A - 1$. These error terms have been discussed in greater detail in Section 6.4.2. The x block is usually mean centred; hence, to obtain a percentage error, most people divide by the standard deviation, whereas for the c block, the estimates are generally expressed in the original concentration units, thus we will retain the convention of dividing by the mean unless there is a specific reason for another approach. As in all areas of chemometrics, each group and software developer has their own favourite way of calculating parameters; hence, it is essential never to accept output from a package blindly.

The calculation of x block error is presented for the case of pyrene. Table 6.14 is of the magnitudes of the first 15 PLS components, defined as the product of the sum of squares for t and p of each component. The total sum of squares of the mean centred spectra is 10.313; hence, the first two components account for

Table 6.13 Calculation of concentration estimates for pyrene using two PLS components.

Component 1 $q = 0.222$		Component 2 $q = 0.779$		Estimated concentration	
Scores	$(t_{i1}q)$	Scores	$(t_{i2}q)$	$(t_{i1}q + t_{i2}q)$	$(t_{i1}q + t_{i2}q) + \bar{c}$
0.088	0.020	0.052	0.039	0.058	0.514
0.532	0.118	−0.139	−0.104	0.014	0.470
0.041	0.009	−0.169	−0.126	−0.117	0.339
0.143	0.032	0.334	0.249	0.281	0.737
0.391	0.087	0.226	0.168	0.255	0.711
0.457	0.102	−0.002	−0.002	0.100	0.556
−0.232	−0.052	0.388	0.289	0.238	0.694
0.191	0.042	−0.008	−0.006	0.037	0.493
−0.117	−0.026	−0.148	−0.111	−0.137	0.319
0.189	0.042	−0.136	−0.101	−0.059	0.397
−0.250	−0.055	0.333	0.249	0.193	0.649
−0.621	−0.138	−0.046	−0.034	−0.173	0.283
−0.412	−0.092	0.105	0.078	−0.013	0.443
−0.575	−0.128	0.004	0.003	−0.125	0.331
0.076	0.017	0.264	0.197	0.214	0.670
−0.264	−0.059	−0.485	−0.362	−0.420	0.036
−0.358	−0.080	−0.173	−0.129	−0.209	0.247
−0.485	−0.108	−0.117	−0.087	−0.195	0.261
0.162	0.036	−0.356	−0.265	−0.229	0.227
0.008	0.002	0.105	0.078	0.080	0.536
0.038	0.008	0.209	0.156	0.164	0.620
−0.148	−0.033	0.080	0.059	0.026	0.482
0.197	0.044	−0.329	−0.245	−0.201	0.255
0.518	0.115	−0.041	−0.031	0.085	0.541
0.432	0.096	0.050	0.037	0.133	0.589

Table 6.14 Magnitudes of first 15 PLS1 components (centred data) for pyrene.

7.944
1.178
0.484
0.405
0.048
0.158
0.066
0.010
0.004
0.007
0.001
0.002
0.002
0.003
0.001

$100 \times (7.944 + 1.178)/10.313 = 88.4\%$ of the overall variance; hence, the *RMSE* after two PLS components have been calculated is $\sqrt{1.191/(27 \times 25)} = 0.042$ (as 1.191 is the residual error) or, expressed as a percentage of the mean centred data $\text{RMSE}_\% = 0.042/\sqrt{10.313/(27 \times 25)} = 40.0\%$. This could be expressed as a percentage of the mean of the raw data $= 0.042/0.430 = 9.76\%$. The latter appears much lower and is a consequence that the mean of the data is considerably higher than the standard deviation of the mean centred data. It is probably better to simply determine the percentage residual sum of square error $(=100 - 88.4 = 11.6\%)$ as more components are computed, but it is important to be aware that there are several approaches for the determination of errors.

The error in concentration predictions for pyrene using two PLS components can be computed from Table 6.13.

- The sum of squares of the errors is 0.385.
- Dividing this by 22 and taking the square root leads to a *RMSE* of 0.128 mg/l.
- The average concentration of pyrene is 0.456 mg/l.
- Hence, the *RMSE*$_\%$ (compared with the raw data) is 28.25%.

Relative to the standard deviation of the centred data, it is even higher. Hence, the quality of reconstruction of the *x* and *c* blocks is different; it is important to recognise that the percentage error of prediction in concentration may diverge considerably from the percentage error of prediction of the spectra. It is sometimes possible to reconstruct spectral blocks quite well but still not predict concentrations very effectively. It is best practice to look at errors in both blocks simultaneously to gain an understanding of the quality of predictions.

The *RMSE* for modelling both blocks of data as successive number of PLS components are calculated for pyrene are illustrated in Figure 6.13, and those for acenaphthene in Figure 6.14. Several observations can be made. First of all, the shape of the graph of residuals for the two blocks is often very different, see especially acenaphthene. Second, the graph of *c* residuals tends to change much more dramatically than that for *x* residuals, according to compound, as might be expected. Third, the estimate of the number of significant PLS components might give different answers according to which block is used for the test.

The errors using 10 PLS components are summarised in Table 6.15 and are better than PCR in this case. It is, however, important not to get too excited about the improved quality of predictions. The *c* or concentration variables may

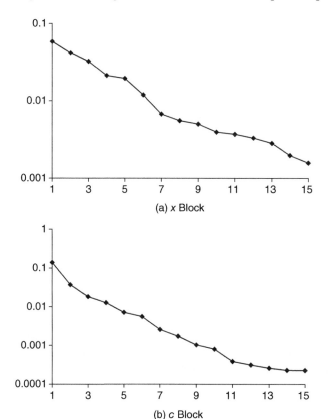

(a) *x* Block

(b) *c* Block

Figure 6.13 Root mean square errors in *x* and *c* blocks, PLS1 centred and pyrene using between 1 and 15 PCs.

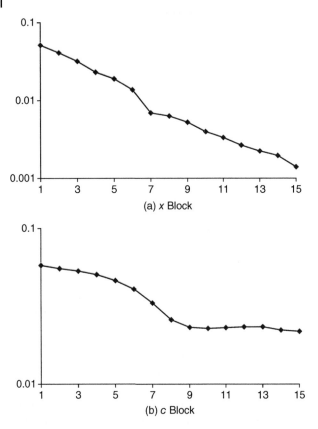

Figure 6.14 Residual errors in *x* and *c* blocks, PLS1 centred and acenaphthene.

contain errors in themselves, and what has been shown is that PLS forces the solution to model the *c* block better; it does not necessarily imply that the other methods are worse at discovering the truth. If, however, we have a lot of confidence in the experimental procedure to determine *c* (e.g. weighing and dilution), PLS will result in a more faithful reconstruction.

1. PLS1 loadings, like PC loadings, are always normalised and orthogonal.
   - (a) True
   - (b) False

2. Using a two-component model on a mean centred training set, the estimated PLS scores for a sample for compound A are [1.27 3.51] and the values of *q* for the first two components are 0.31 and 0.12. The mean concentration is 0.56 ml. What is its estimated concentration in the sample?
   - (a) 1.37 ml
   - (b) 0.25 ml
   - (c) 0.81 ml

6.5.2 PLS2

An extension to PLS1 was proposed in the mid-1980s, called *PLS2*. In fact, there is little conceptual difference with PLS1, except that the latter uses a concentration matrix, C, rather than concentration vectors for each individual compound in a mixture, and the algorithm is iterative. The equations above alter slightly in that Q is now a matrix of dimensions $A \times N$ and not a vector, so that

$$X = TP + E$$
$$C = TQ + F$$

The number of columns in C and Q are equal to the number of compounds (in our case) of interest. In PLS1, one compound is modelled at a time, whereas in PLS2, all known compounds can be included in the model simultaneously.

Table 6.15 Concentration estimates of the PAHs using PLS1 and 10 components (centred).

Spectrum	Polyarene concentration (mg/l)									
	Py	Ace	Anth	Acy	Chry	Benz	Fluora	Fluore	Nap	Phen
1	0.462	0.112	0.170	0.147	0.341	1.697	0.130	0.718	0.110	0.553
2	0.445	0.065	0.280	0.175	0.440	2.758	0.138	0.408	0.177	0.772
3	0.147	0.199	0.285	0.162	0.562	1.635	0.111	0.784	0.159	0.188
4	0.700	0.174	0.212	0.199	0.333	1.097	0.132	0.812	0.054	0.785
5	0.791	0.167	0.285	0.111	0.223	2.118	0.171	0.211	0.176	0.519
6	0.616	0.226	0.176	0.040	0.467	2.172	0.068	0.752	0.116	0.928
7	0.767	0.119	0.108	0.180	0.452	0.522	0.153	0.577	0.177	0.202
8	0.476	0.085	0.228	0.157	0.109	2.155	0.129	0.967	0.046	0.184
9	0.317	0.145	0.232	0.042	0.440	1.576	0.171	0.187	0.009	0.367
10	0.614	0.178	0.046	0.154	0.334	2.702	0.039	0.174	0.084	0.219
11	0.625	0.029	0.237	0.121	0.574	0.543	0.042	0.423	0.039	0.516
12	0.179	0.161	0.185	0.175	0.091	0.560	0.098	0.363	0.110	0.709
13	0.579	0.119	0.262	0.061	0.118	1.074	0.012	0.522	0.149	0.428
14	0.463	0.198	0.067	0.054	0.226	0.561	0.134	0.788	0.110	0.330
15	0.752	0.041	0.062	0.075	0.113	1.646	0.193	0.401	0.072	0.943
16	0.149	0.017	0.115	0.037	0.338	2.186	0.062	0.474	0.196	0.349
17	0.148	0.106	0.050	0.096	0.453	1.044	0.112	0.974	0.092	0.585
18	0.274	0.075	0.149	0.119	0.223	1.098	0.199	0.280	0.147	0.256
19	0.151	0.119	0.213	0.109	0.236	2.664	0.075	0.536	0.050	0.953
20	0.458	0.140	0.114	0.095	0.555	1.067	0.100	0.220	0.198	0.944
21	0.615	0.080	0.120	0.189	0.226	1.581	0.040	1.024	0.198	0.738
22	0.318	0.091	0.267	0.097	0.329	0.523	0.187	0.942	0.157	0.967
23	0.295	0.160	0.124	0.160	0.122	2.669	0.182	0.826	0.171	0.531
24	0.761	0.072	0.167	0.047	0.541	2.687	0.153	1.049	0.122	0.378
25	0.296	0.120	0.047	0.197	0.555	2.166	0.170	0.590	0.082	0.758
$RMSE_{\%}$	5.47	19.06	7.85	22.48	3.55	2.46	21.96	12.96	16.48	7.02

This is illustrated in Figure 6.15. There is a single scores and loadings matrix for all the compounds that are modelled, rather than separate scores matrices for each compound in the mixture.

It is a simple extension to predict all the concentrations simultaneously, the PLS2 predictions, together with *RMSEs* being given in Table 6.16. Note that there is now only a single set of scores and loadings for the *x* (spectroscopic) data set and one set of g_a common to all 10 compounds. However, the concentration estimates are different between

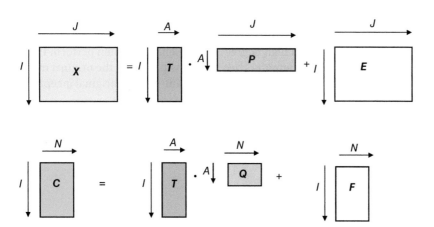

Figure 6.15 Principles of PLS2.

Table 6.16 Concentration estimates of the PAHs using PLS2 and 10 components (centred).

Spectrum	Polyarene concentration (mg/l)									
	Py	Ace	Anth	Acy	Chry	Benz	Fluora	Fluore	Nap	Phen
1	0.505	0.110	0.193	0.132	0.365	1.725	0.125	0.665	0.089	0.459
2	0.460	0.116	0.285	0.105	0.453	2.693	0.144	0.363	0.150	0.760
3	0.162	0.180	0.294	0.173	0.563	1.647	0.094	0.787	0.161	0.157
4	0.679	0.173	0.224	0.164	0.343	1.134	0.123	0.752	0.038	0.748
5	0.811	0.135	0.294	0.149	0.230	2.152	0.162	0.221	0.183	0.475
6	0.575	0.182	0.156	0.108	0.442	2.228	0.077	0.827	0.153	1.002
7	0.779	0.151	0.107	0.143	0.469	0.453	0.167	0.484	0.156	0.199
8	0.397	0.100	0.198	0.183	0.093	2.165	0.181	1.035	0.070	0.306
9	0.295	0.089	0.238	0.108	0.433	1.665	0.158	0.238	0.032	0.341
10	0.581	0.203	0.029	0.148	0.327	2.690	0.079	0.191	0.088	0.287
11	0.609	0.070	0.207	0.108	0.559	0.453	0.079	0.484	0.049	0.636
12	0.190	0.176	0.186	0.144	0.086	0.549	0.083	0.411	0.105	0.709
13	0.565	0.107	0.249	0.095	0.092	1.088	0.000	0.595	0.173	0.478
14	0.468	0.173	0.067	0.089	0.214	0.610	0.108	0.830	0.124	0.322
15	0.771	0.018	0.073	0.096	0.112	1.668	0.175	0.415	0.077	0.906
16	0.119	0.030	0.110	0.037	0.345	2.189	0.101	0.442	0.192	0.369
17	0.181	0.098	0.070	0.090	0.468	1.061	0.106	0.903	0.076	0.510
18	0.278	0.067	0.151	0.102	0.226	1.073	0.178	0.292	0.147	0.249
19	0.184	0.131	0.218	0.102	0.245	2.617	0.071	0.434	0.034	0.925
20	0.410	0.120	0.111	0.134	0.543	1.117	0.115	0.243	0.215	0.963
21	0.663	0.100	0.147	0.129	0.262	1.558	0.040	0.845	0.152	0.630
22	0.308	0.108	0.257	0.093	0.335	0.509	0.209	0.954	0.156	0.998
23	0.320	0.179	0.123	0.114	0.129	2.610	0.164	0.817	0.157	0.537
24	0.763	0.072	0.165	0.038	0.524	2.696	0.120	1.123	0.130	0.390
25	0.327	0.110	0.049	0.216	0.544	2.150	0.139	0.650	0.091	0.746
$RMSE_\%$	10.25	34.11	13.66	44.56	6.99	4.26	33.41	18.62	25.83	14.77

PLS2 and PLS1. In this way, PLS differs to PCR where it does not matter if each compound is modelled separately or altogether. The reasons are rather complex but relate to the fact that for PCR, the PCs are calculated independently of the c variables, whereas the PLS components are influenced by both blocks of variables.

In some cases, PLS2 is helpful, especially as it is easier to perform computationally if there are several c variables compared with PLS1. Instead of obtaining 10 independent models, one for each PAH, in this example, we can analyse all the data in step. However, in many situations, PLS2 estimates are, in fact, worse than PLS1 estimates, so a good strategy might be to perform PLS2 as a first step, which could provide further information such as which wavelengths are significant and which concentrations can be determined with a high degree of confidence, and then perform PLS1 individually for the most appropriate compounds. In practice, as computers become more powerful, the original motivation of PLS2, which is to save time as only one overall model is required, is less important since its original inception; however, this method is described for completion.

1. For PLS2, are the dimensions of $\mathbf{Q}$ if the number of samples is I, variables J, PLS components A and compounds in the model N?

    (a) $N \times A$
    (b) $A \times N$
    (c) $A \times J$

6.5.3 Multi-way PLS

Two-way data such as HPLC–DAD, LC–MS, LC–NMR and fluorescence excitation–emission spectroscopy are increasingly common in chemistry. Conventionally, either a univariate parameter (e.g. a peak area at a given wavelength) (methods in Section 6.2) or a chromatographic elution profile at a single wavelength (methods in Sections 6.3-6.5.2) is used for calibration, allowing the use of the regression techniques described above. However, additional information has been recorded for each sample, often involving both an elution profile and a spectrum. For example, a series of two-way chromatograms may be available and can be organised into a three-way array often visualised as a box, sometimes denoted by $\underline{X}$ where the line underneath the array name indicates a third dimension. Each level of the box consists of a single chromatogram. Sometimes, these three-way arrays are called *tensors*, but tensors often have special properties in physics, which are unnecessarily complex and confusing to the chemometrician. We will use the notation of tensors only where it helps understand the existing methods.

Enhancements of the standard methods for multivariate calibration are required. Although it is possible to use methods such as three-way MLR, most chemometricians have concentrated on developing approaches based on PLS, which will be the focus below. Theoreticians have extended these methods to cases where there are several dimensions in both the x and c blocks, but the most complex practical case in chemistry is where there are three dimensions in the x block, as happens for a series of coupled chromatograms or in fluorescence excitation–emission spectroscopy, for example. A simple simulated numerical example is presented in Table 6.17, in which the x block consists of 4 two-way chromatograms, each of dimensions 5×6. There are three components in the mixture, the c block consisting of a 4×3 matrix. We will restrict the discussion for the case where each column of c is to be estimated independently (analogous to PLS1) rather than all in one go. Note that although PLS is by far the most popular approach for multi-way calibration, it is possible to envisage methods analogous to MLR or PCR, but they are rarely used.

6.5.3.1 Unfolding

One of the simplest methods is to create a single, long data matrix from the original three-way tensor. In the case of Table 6.17, we have four samples, which could be arranged as a $4 \times 5 \times 6$ tensor (or 'box'). The three dimensions will be denoted as I, J and K. It is possible to change the shape so that any binary combination of variables is converted to a new variable, for example, the intensity of the variable at $J = 2$ and $K = 3$ and the data can now be represented by $5 \times 6 = 30$ new variables and is the unfolded form of the original data matrix. This operation is illustrated in Figure 6.16.

It is now a simple task to perform PLS (or indeed any other multivariate approach), as discussed above. The 30 variables are centred and the predictions of the concentrations performed while increasing number of components are used (note that 3 is the maximum permitted for column centred data in this case, so this example is somewhat simple). All the methods described above can be applied.

An important aspect of three-way calibration involves centring, which can be rather complex. There are four different ways in which the data can be treated.

- No centring.
- Centre the columns in each $J \times K$ plane and then unfold with no further centring; hence, for example, $x_{1,1,1}$ becomes $390 - (390 + 635 + 300 + 65 + 835)/5 = -55$.
- Unfold the raw data and centre afterwards; hence, for example, $x_{1,1,1}$ becomes $390 - (390 + 488 + 186 + 205)/4 = 72.75$.
- Combine methods 2 and 3, start with centring as in step 2, then unfold and re-centre a second time.

These four methods are illustrated in Table 6.18 for the case of the $x_{i,1,1}$, the variables in the top left-hand corner of each of the 4 two-way data sets. Note that methods 3 and 4 yield radically different answers; for example, sample 2 has the highest value (=170.75) using method 3, but the lowest using method 4 (=−87.85).

Standardisation is also sometimes employed, but must be done before unfolding for meaningful results, an example might be in the GC–MS of a series of samples, each mass being of different absolute intensity. A sensible three-step strategy might be as follows.

- Standardise each mass in each individual chromatogram to provide I standardised matrices of dimensions $J \times K$
- Unfold
- Then, centre each of the variables.

Standardising at the wrong stage of the analysis can result in meaningless data; hence, it is always essential to carefully think of the physical (and numerical) consequences of any pre-processing which is far more complex and has far more options than for simple two-way data.

Table 6.17 Three-way calibration data set.

(a) x Block, each of the 4 (=I) samples gives a two-way 5 × 6 (=J × K) matrix

Sample 1

390	421	871	940	610	525
635	357	952	710	910	380
300	334	694	700	460	390
65	125	234	238	102	134
835	308	1003	630	1180	325

Sample 2

488	433	971	870	722	479
1015	633	1682	928	1382	484
564	538	1234	804	772	434
269	317	708	364	342	194
1041	380	1253	734	1460	375

Sample 3

186	276	540	546	288	306
420	396	930	498	552	264
328	396	860	552	440	300
228	264	594	294	288	156
222	120	330	216	312	114

Sample 4

205	231	479	481	314	268
400	282	713	427	548	226
240	264	576	424	336	232
120	150	327	189	156	102
385	153	482	298	542	154

(b) c Block, the concentrations of three compounds in each of the four samples

Compound concentrations

1	9	10
7	11	8
6	2	6
3	4	5

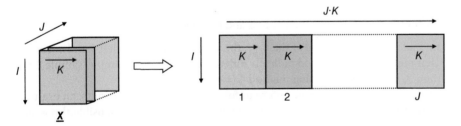

Figure 6.16 Unfolding a data matrix.

Table 6.18 Four methods of mean centring the data in Table 6.17, illustrated by the variable $x_{i,1,1}$ as discussed in Section 6.5.3.1.

Method 1	Method 2	Method 3	Method 4
390	−55	72.75	44.55
488	−187.4	170.75	−87.85
186	−90.8	−131.25	8.75
205	−65	−112.25	34.55

After this pre-processing, all the normal multivariate calibration methods can be employed.

1. A data set consists of 20 samples, recorded using excitation emission fluorescence consisting of 40 excitation and 50 emission wavelengths. The unfolded data set has dimensions
   - (a) 20×40
   - (b) 20×50
   - (c) 20×200
   - (d) 20×2000

6.5.3.2   Tri-linear PLS1

Some of the most interesting theoretical developments in chemometrics over the past few years have been in so-called *multi-way* or *multi-mode* data analysis. Many such methods have been available for some years, especially in the area of psychometrics, but a few do have relevance to chemistry and allied laboratory science. We will restrict the discussion in this chapter to tri-linear PLS1, involving a three-way x block and a single c variable. If there are several known calibrants, the simplest approach is to perform tri-linear PLS1 individually on each variable.

As centring can be quite complex for three-way data, and there is no inherent reason to do this, for simplicity, so for in the example of this section, we use uncentred data for illustrative purposes. The data in Table 6.17 can be considered to be arranged in the form of a cube, with three dimensions, I for the number of samples and J and K for the measurements.

Tri-linear PLS1 attempts to model both the x and c blocks simultaneously. In this review, we will illustrate the use with the algorithm of Section A.2.4, based on the methods proposed by de Jong and Bro.

Superficially, tri-linear PLS1 has many of the same objectives to normal PLS1, and the method as applied to the x block is often represented diagrammatically as in Figure 6.17, replacing 'squares' or matrices by 'boxes' or tensors, and replacing, where necessary, the dot product ('·') by something called a *tensor product* ('⊗'). The c block decomposition can be represented as per PLS1 and is omitted from the diagram for brevity. In fact, as we shall see, this is an oversimplification and is not an entirely accurate description of the method.

In tri-linear PLS1, for each component, it is possible to determine

- a scores vector (t), of length I or 4 in this example,
- a weight vector, which has analogy to a loadings vector ($^j p$) of length J or 5 in this example, referring to one of the dimensions (e.g. time), whose sum of squares equals 1,
- another weight vector, which also has analogy to a loadings vector ($^k p$) of length K or 6 in this example, referring to the other one of the dimensions (e.g. wavelength) whose sum of squares also equals 1.

Superficially, these vectors are related to scores and loadings in normal PLS, but in practice, they are somewhat different concepts, a key reason being that more of these vectors are not orthogonal in tri-linear PLS1. In this chapter,

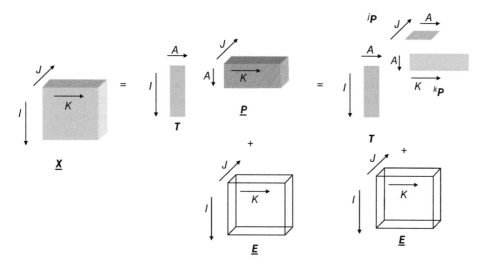

Figure 6.17 Representation of tri-linear PLS1.

we keep the notation scores and loadings, simply for the purpose of retaining familiarity with terminology usually used in two-way data analysis.

In addition, a vector q is determined after each new component is included in the model, by

$$q = (T'T)^{-1}T'c$$

so that

$$\hat{c} = Tq$$

or

$$c = Tq + f$$

where T is the scores matrix, whose columns consist of the individual scores vectors for each component and has dimensions $I \times A$ or 4×3 in this example if three PLS components are computed, and q is a column vector of dimensions $A \times 1$ or 3×1 in our example.

A key difference from bilinear PLS1 as described in Section 6.5.1 is that the elements of q have to be recalculated afresh as new components are computed, whereas for two-way PLS, the first element of q, for example, is the same, no matter how many components are calculated and does not need to be recalculated. This limitation is a consequence of non-orthogonality of individual columns of matrix T.

The x block residuals after each component are often computed conventionally by

$$^{resid,a}x_{ijk} = {}^{resid,a-1}x - t_i^j p^k p$$

where $^{resid,a}x_{ijk}$ is the residual after a components are calculated, which would lead to a model

$$\hat{x}_{ijk} = \sum_{a=1}^{A} t_i^j p_j^k p_k$$

Sometimes, these equations are written as tensor products, but there are quite a large number of ways of multiplying tensors together; hence, this notation can be confusing and it is often conceptually more convenient to deal directly with vectors and matrices, just as in Section 6.5.3.1 by unfolding the data. This procedure is called *matricisation*.

In mathematical terms, we can state that

$$^{unfolded}\hat{X} = \sum_{a=1}^{A} t_a {}^{unfolded}p_a$$

where $^{unfolded}p_a$ is simply a row vector of length JK. Where tri-linear PLS1 differs from unfolded PLS described in Section 6.5.3.1, is that a matrix P_a of dimensions $J \times K$ can be obtained for each PLS component is given by

$$P_a = {}^j p_a^{\ k} p_a$$

and P_a is unfolded to give $^{unfolded}p_a$.

Figure 6.18 represents this procedure, avoiding tensor multiplication, using conventional matrices and vectors together with unfolding. A key problem with the common implementation of tri-linear PLS1 is that, as the scores and loadings of successive components are not orthogonal, the methods for determining residual errors are simply an approximation. Hence, the x block residual is not modelled very well, and the error matrices (or tensors) do not have an easily interpretable physical meaning and cannot be directly related to errors or instrumental noise. It also means that there are no obvious analogies to eigenvalues. This means that it is not easy to determine the size of the components or the modelling power using the x scores and loadings, but, nevertheless, the main aim is to predict the concentration (or c block); hence, we are not always worried about this limitation, so tri-linear PLS1 is an acceptable method, providing care is taken to interpret the output and not to expect the residuals to have a physical meaning.

In order to further understand this method, tri-linear PLS1 is performed on the first compound (column 1 in Table 6.17(b)). The main results are given in Table 6.19 (using uncentred data), with the residual errors for sample 1. It can be seen that three components provide an exact model of the concentration, but there is still an apparent residual error in the x matrix: this error has no real physical or statistical meaning, except that it is quite small. Despite this limitation, it is essential to recognise that the concentration has been modelled correctly.

A beauty of multi-mode methods is that the dimensions of c (or indeed $\underline{X}$) can be changed; for example, a matrix C consisting of several different compounds can be employed, exactly as in PLS2, or even, a tensor. It is possible to define

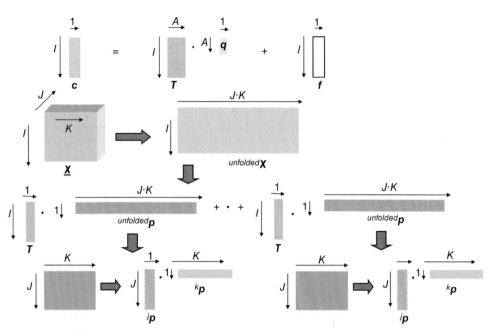

Figure 6.18 Matricisation in three-way calibration (*x* block only illustrated).

Table 6.19 Calculation of three tri-linear PLS1 components for the data in Table 6.17 and residuals for sample 1.

t	^jp	^kp	q	$\hat{c}$	RMS concentration residuals	RMS of x 'residuals'
Component 1						
3135.35	0.398	0.339	0.00140	4.38	2.28	190.69
4427.31	0.601	0.253		6.19		
2194.17	0.461	0.624		3.07		
1930.02	0.250	0.405		2.70		
	0.452	0.470				
		0.216				
Component 2						
−757.35	−0.252	0.381	0.00177	1.65	0.58	131.56
−313.41	0.211	0.259	0.00513	6.21		
511.73	0.392	0.692		6.50		
−45.268	0.549	0.243		3.18		
	−0.661	0.485				
		0.119				
Component 3						
−480.37	−0.875	−0.070	0.00201	1	0.00	97.93
−107.11	−0.073	0.263	0.00508	7		
−335.17	−0.467	0.302	0.00305	6		
−215.76	−0.087	0.789		3		
	0.058	0.004				
		0.461				

the number of dimensions in both the x and c blocks; for example, a three-way x block and a two-way c block may consist of a series of two-way chromatograms, each containing several compounds. However, unless one has a good grip of the theory or there is a real need from the nature of the data, it is often more straightforward to reduce the problem to one of tri-linear PLS1: for example, a concentration matrix C can be treated as several concentration vectors, in the same way that a calibration problem that might appear to need PLS2 can be reduced to several calculations using PLS1.

Although there has been a huge interest in multi-mode calibration in the theoretical chemometrics literature, it is important to recognise that there are limitations to the applicability of such techniques. Good, very high order data are rare in chemistry. Even three-way calibration, such as in DAD–HPLC, has to be used cautiously, as there are frequent experimental difficulties with alignments of chromatograms in addition to interpretation of the numerical results. However, there have been some significant successes in areas such as sensory research and psychometrics and certain techniques such as fluorescence excitation emission spectroscopy, where the wavelengths are very stable, have demonstrated significant promise.

Using genuine three-way methods (even when redefined in terms of matrices) differs from unfolding in that the connection between different variables is retained; in an unfolded data matrix, there is no indication if two variables, for example, share the same elution time or spectroscopic wavelength for each joint variable such as a wavelength elution time pair (e.g. the absorbance at 300 nm and 20.2 min) is independent.

1. A c or concentration vector is estimated via PLS using $\hat{c} = Tq$.
    (a) Using two-way and three-way PLS, the first element of q is the same, no matter how many components.
    (b) Using two-way PLS, the first element of q is the same, no matter how many components, but differs according to the number of components for three-way PLS.
    (c) Using two-way and three-way PLS, the first element of q differs according to the number of components.

6.6 Model Validation and Optimisation

Unquestionably, one of the most important aspects of all calibration methods is model validation. Numerous questions need to be answered.

- How many significant components are needed to characterise a data set to optimise the model?
- How well is an unknown predicted?
- How representative are the data used to produce a model?

Using auto-predictive or training set models, it is possible to obtain a fit as close as desired using more and more PLS or PCA components, until the raw data are fitted exactly; however, the later terms are unlikely to be physically meaningful and may model noise. There is a large literature on how to decide what model to adapt, which requires an appreciation of model validation, experimental design and how to measure errors. Most methods aim to guide the experimenter as to how many significant components to retain. The methods are illustrated below with reference to both PCR and PLS1, but similar principles apply to all calibration methods, including MLR, PLS2 and tri-linear PLS1. The most common question asked is how many components should be retained, and then once the optimum has been determined, how well an independent test set is predicted.

6.6.1 Auto-prediction

The simplest approach to determining number of significant components is by measuring the auto-prediction error. This is also called the root mean square error of calibration (*RMSEC*). Usually (but not exclusively), the error is calculated on the concentration data matrix (c), and we will restrict the discussion below to errors in concentration for brevity: it is important to understand that similar equations can be obtained for the x block.

As more components are calculated, the residual error reduces. It is normally calculated by

$$RMSEC = \sqrt{\frac{\sum_{i=1}^{I} (c_i - \hat{c}_i)^2}{I - A - 1}}$$

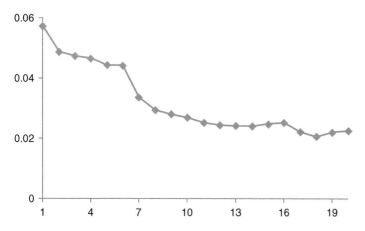

Figure 6.19 *RMSEC* auto-predictive errors for acenaphthylene using PLS1.

where A components have been used for the model and the data have been centred.

Note that these errors can easily be converted to a percentage variance or mean square errors as described in Sections 6.4 and 6.5. The value of *RMSEC* will usually decline in value as more components are calculated.

Note that there can sometimes be very slight increases, especially when the number of components is large, as although $\sum_{i=1}^{I}(c_i - \hat{c}_i)^2$ will never increase, the denominator will get smaller as the number of components increases.

The auto-predictive error for acenaphthylene is presented in Figure 6.19, using PLS1. A similar calculation can be done for PCR.

The auto-predictive error can be used to determine how many components to use in the model, in a number of ways.

- A standard cut-off percentage error can be used, for example, 1%. Once the error has reduced to this cut-off, ignore later PLS (or PCA) components.
- Sometimes, an independent measure of the noise level is possible. Once the error has declined to the noise level, ignore later PLS (or PCA) components.
- Occasionally, the error can reach a plateau. Take the PLS components up to this plateau.

By plotting the magnitudes of each successive component (or errors in modelling the x block), it is also possible to determine prediction errors for the x block. However, the main aim of calibration is to predict concentrations rather than spectra; thus, this information, although quite useful, is less frequently employed in calibration. More details have been discussed in the context of PCA in Section 4.3.2 and also Section 6.4.2, the ideas for PLS are similar.

Many statistically oriented chemometricians do not like to use auto-prediction determining the number of significant components, as it is always possible to fit data perfectly simply by increasing the number of terms (or components) in the model. There is, though, a difference between statistical and chemical thinking. A chemist might know (or have a good intuitive feel) for parameters such as noise levels and, therefore, in some circumstances, be able to successfully interpret the auto-predictive errors in a perfectly legitimate physically meaningful manner.

1. The root mean square error of calibration may not always decrease when more components are added to the model, using the equation presented in this section.
    (a) True
    (b) False

6.6.2 Cross-validation

An important chemometric tool is called *cross-validation*. The basis of the method is that the predictive ability of a model formed on part of a data set can be tested by how well it predicts the remainder of the data and has been introduced previously in other contexts (see Sections 4.3.2.2 and 5.5.2).

It is possible to determine a model using $I - 1$ (=24) samples leaving out one sample (i). How well does this model fit the original data? Below we describe the method when the data are centred, the most common approach.

For PCR, the following steps are used to determine cross-validation errors using leave one out (LOO).

- The procedure below is repeated using successively more PCs, and a root mean square cross-validation error over all samples is obtained each time the number of PCs in the model is increased.
- Using the commonest leave-one-out method, select one sample each time to leave out from the data for a given number of PCs, so that in our case, each of the 25 samples are left out once.
- If centring the data before PCA, centre both the remaining $I - 1$ (=24 in this example) concentrations and spectra each time, remembering to recalculate the means $\bar{c}_{noti}$ and $\bar{x}_{noti}$ removing sample i, and subtracting these means from the original data. This process needs repeating I times, each time a sample is removed.
- Determine the prediction error for the sample left out as follows.
  - Perform PCA to give loadings and scores matrices T_{noti} and P_{noti} for the remaining $I - 1$ samples of the x data, then obtain a vector r_{noti} for the c data using standard regression techniques. Note that these arrays will differ according to which sample is removed from the analysis.
  - Call the spectrum of sample i x_i (a row vector).
  - Subtract the mean of the $I - 1$ (=24 in our case) samples from this to give $x_i - \bar{x}_{noti}$, where $\bar{x}_{noti}$ is the mean spectrum excluding sample i, if mean centring, and similarly for the c block.
  - Estimate $\hat{t}_i = (x_i - \bar{x}_{noti})P'_{noti}$, where P_{noti} are the loadings obtained from the PCA model using $I - 1$ samples excluding sample i or $\hat{t}_i = x_i P'_{noti}$ if not for the defined number of PCs.
  - Then, calculate $^{cv}\hat{c}_i = \hat{t}_i r_{noti} + \bar{c}_{noti}$ (centred) or $^{cv}\hat{c}_i = \hat{t}_i r_{noti}$, which is the estimated concentration of sample i using the model based on the remaining $(I - 1)$ (=24 samples), remembering to add on the mean of these samples again.
- After all samples have been left out once, calculate the mean cross-validation error, defined by $RMSECV = \sqrt{\frac{\sum_{i=1}^{I}(c_i - ^{cv}\hat{c}_i)^2}{I}}$. Note that this is divided by I because each sample in the original data set represents an additional degree of freedom; however, many PCA components have been calculated and however the data have been pre-processed.
- This is repeated as successive PCs are calculated.

The *RMSECV* can then be plotted again the number of PCs in the model, as per PLS and illustrated below.

Cross-validation in PLS is slightly more complicated. The reason is that the scores are obtained using both the c and x blocks simultaneously; thus, predicting the concentration of the left out sample above is slightly more elaborate than above. The product of T and P is no longer the best least squares approximation for the x block; hence, it is not possible to obtain an estimate $\hat{t}_i$ using just P and x_i. Many people use what is called a *weights vector*, which can be employed in prediction. The method is illustrated in more detail in Problem 6.8, but the interested reader should first look at Section A.2.2 for a description of the PLS1 algorithm. Below we will illustrate the use of cross-validation for PLS predictions. However, the steps in the calculation of *RMSECV* are the same as PCR with the only difference in the computation of the estimated value of c and so the prediction error for the left out sample as detailed in the appendix.

For acenaphthylene using PLS1, the cross-validated error is presented in Figure 6.20. An immediate difference between auto-prediction and cross-validation is evident. In the former case, the data will be better modelled as more components are employed in the calculation; hence, the error will reduce. However, cross-validated errors normally reach a minimum as the correct number of components are found and then increase afterwards. This is because later components really represent noise and not systematic information in the data. In this case, the cross-validation error

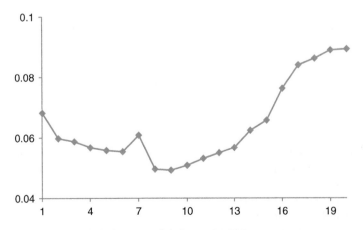

Figure 6.20 *RMSECV* for acenaphthylene using PLS1.

has a minimum after nine components are included in the model and then increases steadily afterwards. By inspecting the minimum, we can estimate that there are nine significant PLS components in the model. There are no hard and fast rules though, and it is important not to be carried away by automated approaches. Note that a different number of significant components may be found for each compound in the mixture if using PLS1. In addition, different conclusions might be drawn from the x block.

Cross-validation has two main purposes.

- It can be employed as a method for determining how many significant components characterise the data. From Figure 6.20, it appears that nine components are necessary to obtain an optimal model for acenaphthylene. This number will not always equal the number of compounds in the mixture, as spectral similarity and correlations between concentrations will often reduce this, whereas impurities or interferents may increase it; our design has been carefully chosen so that each compound's concentration profile is orthogonal, but in most real-world situations, this will not be so. Later components probably model noise and it would be unwise to include them if using to predict concentrations of unknowns.
- It can be employed as an estimate of predictive ability. The minimum cross-validated prediction error for acenaphthylene of 0.0493 mg/l equals 41.1% of the mean. This compares with an auto-predictive error of 0.0281 mg/l or 23.44% using nine components and PLS1, which is an over-optimistic estimate. The cross-validated error is almost always higher than the auto-predictive (or calibration) error and provides an indicative of what the error would be if we were to estimate the concentration of an unknown sample.

Many refinements to cross-validation have been proposed in the literature, which have been discussed in Chapter 4; it is equally possible to apply these to calibration as well as pattern recognition.

---

1. Calculating the *RMSEC* and *RMSECV* for an eight-component model using centred data and 40 samples involves
   (a) 40 degrees of freedom for *RMSEC* and 40 for *RMSECV*.
   (b) 31 degrees of freedom for *RMSEC* and 31 for *RMSEC*.
   (c) 31 degrees of freedom for *RMSEC* and 40 for *RMSECV*.
   (d) 40 degrees of freedom for *RMSEC* and 31 for *RMSECV*.

6.6.3 Independent Test Sets

A significant weakness of cross-validation is that the results depend on the design and scope of the original data set used to form the model. This data set is often called a *training set* (see Chapter 5), although some authors prefer the term *calibration set*. Consider a situation in which a series of mixture spectra are recorded, but it happens that the concentration of two compounds are correlated, so that the concentration of compound A is high when compound B likewise is high, and *vice versa*. A calibration model can be obtained from analytical data, which predicts both concentrations quite well. Even cross-validation might suggest the model is good. However, if asked to predict a spectrum where compound A is in a high concentration and compound B in a low concentration, it is likely to give very poor results, as it has not been trained to cope with this new situation. Cross-validation is very useful for removing the influence of internal factors such as instrumental noise or dilution errors or instrumental noise but cannot help very much if there are correlations in the concentration of compounds in the training set. If new data have a different correlation structure to the training (or calibration) set, predictions may be very poor. This is a huge dilemma in chemometrics: we cannot predict the composition of all possible future samples, so should the training set be very broad in scope, in which case the model may be quite poor or the sample size has to be very large, or should it be quite narrow and so predicted effectively and on a modest number of samples but not robust to future samples with a different structure. There is no obvious answer.

In some cases, there will be inevitable correlations in the concentration data because it is not easy to find samples without this. Examples routinely occur in environmental monitoring. Several compounds often arise from a single source. For example, PAHs are well-known pollutants; hence, if one or two PAHs are present in high concentrations, it is a fair bet that others will be too. There may be some correlations, for example, in the occurrence of compounds of different molecular weights if a homologous series occurs, for example, as the by-product of a biological pathway, there may be an optimum chain length, which is most abundant in samples from a certain source. It would be hard to find a series of field samples in which there are no correlations between the concentrations of compounds. Consider, for example, setting up a model of PAHs coming from rivers close to several specific sources of pollution. The model may behave well on this training set, but can it be safely used to predict the concentrations of PAHs in an unknown sample from a very different source? Another serious problem occurs in process control. Consider trying to set up a calibration model using NIR to determine the concentration of chemicals in a manufacturing process. If the process is behaving well, the predictions may be quite good, but it is precisely to detect problems in the process that the calibration model is effective: is it possible to rely on the predictions if data have a completely different structure? Some chemometricians do look at the structure of the data and samples that do not fit into the structure of the calibration set are often called *outliers*. It is beyond the scope of this text to provide extensive discussion about how to spot outliers, this depends on the software package and often the type of data, although they can sometimes be spotted visually from scores plots, but it is important to understand how design of training sets influences model validation.

Instead of validating the predictions internally, it is possible to test the predictions against an independent data set, often called a *test set*. Computationally, the procedure is similar to cross-validation. For example, a model is obtained using I samples, and then the predictions are calculated using an independent test set of L samples to give

$$RMSEP = \sqrt{\frac{\sum_{l=1}^{L} (c_l - {}^{test}\widehat{c}_l)^2}{L}}$$

The value of $\widehat{c}_l$ is determined in the same manner as per cross-validation (Section 6.6.2), but only one calibration model is formed, from all the I samples of the training set.

We will use the data in Table 6.1 for the training set, but test the predictions using the spectra obtained from a new data set presented in Table 6.20. In this case, each data set has the same number of samples, but this is not a requirement. The graph of the root mean square test set error (often called the root mean square error of prediction or root mean square error of prediction, *RMSEP*) for acenaphthylene is presented in Figure 6.21 and shows a different trend to *RMSECV*, which mainly increases as the number of PLS components increase. This suggests that the correlation structure is quite different and additional components, in our case, merely over-fit the data, although this may certainly not always be the case. The minimum error is 72.1%, much higher than that for cross-validation. Normally, the minimum test set error is higher than that for cross-validation, but if the structure of the test set is encompassed in the training set, these two errors will be very similar.

If, however, we use the data in Table 6.20 as the training set and that in Table 6.1 as the test set, a very different story emerges, as shown in Figure 6.22, for acenaphthylene. The minimum error is 55.0% and the trends are quite different. Neither auto-prediction nor cross-validation will be especially useful for determining the test set behaviour on this occasion.

It is important to recognise that cross-validation can, therefore, sometimes give a misleading and over-optimistic answer. However, whether cross-validation is useful depends, in part, on the practical aim of the measurements. If, for example, data containing all the possible features in Table 6.1 is unlikely ever to occur, it may be safe to use the model obtained from Table 6.20 for future predictions. For example, if it is desired to determine the amount of vitamin C in orange juices from a specific region of Spain, it might be sufficient to develop a calibration method only on these juices. It could be expensive and time-consuming to find a more representative calibration set. Brazilian orange juice may exhibit different features, but is it necessary to go to all the trouble of setting up a calibration experiment that includes Brazilian orange juice? Is it really necessary or practicable to develop a method to measure vitamin C in all conceivable orange juices or foodstuffs? The answer is no; hence, in some circumstances, living within the limitations of the original data set is entirely acceptable and finding a test set that is wildly unlikely to occur in real situations represents an artificial experiment; remember we want to save time (and money) while setting up the calibration model. If at some future date extra orange juice from a new region is to be analysed, the first step is to set up a data set from this new source of information as a test set and determine whether the new data fits into the structure of the existing database or whether the calibration method must be developed afresh. It is, though, very important to recognise the limitations and calibration models especially if they are to be applied to situations that are wider than those represented by the initial training sets.

There are a number of variations on the theme of test sets, one being simply to take a few samples from a large training set and assign them to a test set, for example, take five out of the 25 samples from the case study and assign them to a test set, using the remaining 20 samples for the training set, so splitting the data as discussed in a different context in Section 5.5.1 and use *RMSEP* as an indicator of quality of predictions.

The optimum size and representativeness of training and test sets for calibration modelling are a big subject. Some chemometricians recommend using hundreds or thousands of samples, but this can be expensive and time consuming. In some cases, a completely orthogonal data set is unlikely ever to occur and field samples with these features cannot

Table 6.20 Independent test set.

(a) Spectra ('x' block)

nm	220	225	230	235	240	245	250	255	260	265	270	275	280	285	290	295	300	305	310	315	320	325	330	335	340	345	350
1	0.687	0.706	0.660	0.476	0.490	0.591	0.723	0.783	0.589	0.611	0.670	0.663	0.646	0.624	0.678	0.280	0.185	0.144	0.127	0.122	0.139	0.120	0.102	0.106	0.097	0.075	0.053
2	0.710	0.644	0.603	0.483	0.552	0.646	0.808	0.885	0.722	0.759	0.821	0.808	0.737	0.698	0.771	0.286	0.199	0.143	0.117	0.127	0.161	0.133	0.130	0.160	0.119	0.087	0.062
3	0.682	0.698	0.655	0.466	0.477	0.575	0.707	0.768	0.578	0.601	0.661	0.654	0.636	0.616	0.666	0.273	0.178	0.138	0.120	0.115	0.131	0.113	0.095	0.099	0.091	0.069	0.047
4	0.790	0.803	0.713	0.545	0.608	0.705	0.829	0.917	0.705	0.738	0.852	0.944	0.922	0.873	0.943	0.363	0.233	0.177	0.152	0.156	0.185	0.154	0.156	0.196	0.142	0.100	0.070
5	0.521	0.522	0.507	0.434	0.539	0.593	0.661	0.718	0.558	0.571	0.624	0.476	0.364	0.335	0.353	0.218	0.159	0.140	0.112	0.115	0.150	0.093	0.097	0.151	0.068	0.030	0.018
6	0.677	0.631	0.606	0.464	0.507	0.470	0.486	0.478	0.320	0.298	0.337	0.344	0.259	0.236	0.261	0.119	0.079	0.076	0.072	0.091	0.129	0.090	0.103	0.161	0.072	0.030	0.017
7	0.609	0.536	0.522	0.486	0.632	0.748	0.847	0.844	0.555	0.504	0.539	0.574	0.489	0.448	0.500	0.241	0.155	0.118	0.095	0.105	0.133	0.096	0.111	0.167	0.088	0.049	0.034
8	0.787	0.612	0.541	0.464	0.556	0.604	0.659	0.655	0.518	0.486	0.510	0.479	0.392	0.355	0.370	0.224	0.147	0.116	0.088	0.090	0.106	0.070	0.079	0.118	0.056	0.029	0.019
9	0.812	0.634	0.613	0.563	0.671	0.704	0.752	0.794	0.675	0.656	0.703	0.777	0.665	0.611	0.683	0.284	0.207	0.154	0.122	0.133	0.168	0.124	0.140	0.206	0.109	0.062	0.040
10	0.682	0.509	0.420	0.415	0.519	0.657	0.820	0.847	0.657	0.639	0.641	0.612	0.558	0.513	0.564	0.278	0.184	0.126	0.092	0.086	0.091	0.076	0.070	0.078	0.068	0.056	0.042
11	0.920	0.795	0.703	0.559	0.594	0.679	0.792	0.858	0.749	0.768	0.831	0.940	0.902	0.869	0.918	0.368	0.233	0.161	0.134	0.140	0.164	0.147	0.146	0.170	0.134	0.103	0.073
12	0.955	0.781	0.681	0.587	0.680	0.848	1.064	1.128	0.924	0.941	1.006	1.058	0.980	0.926	0.984	0.394	0.256	0.175	0.137	0.137	0.153	0.135	0.128	0.144	0.125	0.101	0.073
13	0.825	0.777	0.672	0.471	0.484	0.584	0.725	0.817	0.720	0.765	0.831	0.821	0.798	0.767	0.820	0.305	0.220	0.161	0.133	0.128	0.145	0.128	0.107	0.105	0.102	0.081	0.055
14	0.842	0.792	0.722	0.573	0.631	0.775	0.949	1.039	0.812	0.844	0.924	0.984	0.929	0.888	0.936	0.361	0.244	0.173	0.144	0.149	0.176	0.150	0.146	0.174	0.136	0.099	0.070
15	0.731	0.718	0.632	0.474	0.527	0.621	0.776	0.841	0.666	0.711	0.790	0.742	0.695	0.665	0.707	0.306	0.233	0.183	0.145	0.136	0.152	0.117	0.105	0.126	0.094	0.068	0.048
16	0.744	0.778	0.721	0.560	0.629	0.685	0.766	0.819	0.620	0.626	0.710	0.766	0.698	0.664	0.742	0.265	0.166	0.129	0.116	0.135	0.182	0.144	0.152	0.205	0.128	0.081	0.056
17	0.610	0.618	0.620	0.528	0.629	0.687	0.760	0.796	0.588	0.598	0.667	0.602	0.492	0.472	0.506	0.251	0.177	0.158	0.136	0.145	0.192	0.134	0.139	0.207	0.109	0.057	0.040
18	0.606	0.591	0.547	0.400	0.460	0.474	0.522	0.563	0.449	0.463	0.527	0.529	0.453	0.426	0.476	0.218	0.156	0.128	0.103	0.110	0.145	0.100	0.108	0.162	0.082	0.040	0.024
19	0.520	0.490	0.514	0.491	0.618	0.651	0.701	0.709	0.510	0.482	0.519	0.481	0.364	0.332	0.352	0.211	0.154	0.134	0.107	0.119	0.158	0.104	0.120	0.187	0.083	0.038	0.026
20	0.718	0.625	0.560	0.452	0.529	0.595	0.662	0.658	0.487	0.453	0.497	0.549	0.504	0.472	0.519	0.240	0.146	0.111	0.091	0.093	0.114	0.087	0.090	0.123	0.075	0.046	0.030
21	0.761	0.624	0.579	0.526	0.646	0.736	0.834	0.864	0.694	0.677	0.729	0.770	0.665	0.622	0.690	0.293	0.205	0.152	0.117	0.123	0.148	0.110	0.119	0.173	0.097	0.058	0.038
22	0.646	0.471	0.373	0.361	0.442	0.527	0.619	0.647	0.546	0.534	0.545	0.477	0.397	0.362	0.379	0.232	0.164	0.124	0.090	0.084	0.094	0.068	0.063	0.084	0.053	0.035	0.023
23	0.801	0.650	0.604	0.488	0.551	0.659	0.810	0.840	0.687	0.689	0.729	0.837	0.800	0.766	0.811	0.319	0.229	0.156	0.120	0.118	0.122	0.113	0.109	0.125	0.101	0.077	0.052
24	0.826	0.657	0.589	0.521	0.610	0.809	1.008	1.063	0.832	0.830	0.875	0.905	0.866	0.827	0.879	0.374	0.236	0.159	0.122	0.113	0.117	0.105	0.095	0.100	0.098	0.084	0.060
25	0.986	0.887	0.783	0.548	0.536	0.582	0.685	0.785	0.745	0.798	0.890	0.933	0.911	0.861	0.890	0.325	0.234	0.172	0.144	0.145	0.162	0.145	0.135	0.151	0.122	0.094	0.064

(Continued)

Table 6.20 (Continued)

(b) Concentrations ('c' block)

Spectrum	Py	Ace	Anth	Acy	Chry	Benz	Fluora	Fluore	Nap	Phen
					Polyarene concentration (mg/l)					
1	0.456	0.120	0.168	0.120	0.336	1.620	0.120	0.600	0.120	0.564
2	0.456	0.040	0.224	0.160	0.560	2.160	0.120	1.000	0.040	0.188
3	0.152	0.160	0.224	0.200	0.448	1.620	0.200	0.200	0.040	0.376
4	0.608	0.160	0.280	0.160	0.336	2.700	0.040	0.200	0.080	0.188
5	0.608	0.200	0.224	0.120	0.560	0.540	0.040	0.400	0.040	0.564
6	0.760	0.160	0.168	0.200	0.112	0.540	0.080	0.200	0.120	0.376
7	0.608	0.120	0.280	0.040	0.112	1.080	0.040	0.600	0.080	0.940
8	0.456	0.200	0.056	0.040	0.224	0.540	0.120	0.400	0.200	0.940
9	0.760	0.040	0.056	0.080	0.112	1.620	0.080	1.000	0.200	0.752
10	0.152	0.040	0.112	0.040	0.336	1.080	0.200	1.000	0.160	0.940
11	0.152	0.080	0.056	0.120	0.224	2.700	0.200	0.800	0.200	0.564
12	0.304	0.040	0.168	0.080	0.560	2.700	0.160	1.000	0.120	0.752
13	0.152	0.120	0.112	0.200	0.560	2.160	0.200	0.600	0.160	0.376
14	0.456	0.080	0.280	0.200	0.448	2.700	0.120	0.800	0.080	0.376
15	0.304	0.200	0.280	0.160	0.560	1.620	0.160	0.400	0.080	0.188
16	0.760	0.200	0.224	0.200	0.336	2.160	0.080	0.400	0.040	0.376
17	0.760	0.160	0.280	0.120	0.448	1.080	0.080	0.200	0.080	0.564
18	0.608	0.200	0.168	0.160	0.224	1.080	0.040	0.400	0.120	0.188
19	0.760	0.120	0.224	0.080	0.224	0.540	0.080	0.600	0.040	0.752
20	0.456	0.160	0.112	0.080	0.112	1.080	0.120	0.200	0.160	0.752
21	0.608	0.080	0.112	0.040	0.224	1.620	0.040	0.800	0.160	0.940
22	0.304	0.080	0.056	0.080	0.336	0.540	0.160	0.800	0.200	0.752
23	0.304	0.040	0.112	0.120	0.112	2.160	0.160	1.000	0.160	0.564
24	0.152	0.080	0.168	0.040	0.448	2.160	0.200	0.800	0.120	0.940
25	0.304	0.120	0.056	0.160	0.448	2.700	0.160	0.600	0.200	0.188

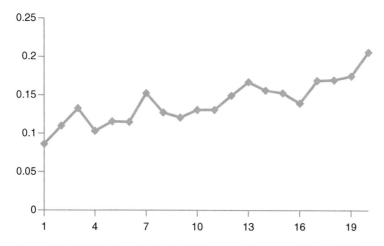

Figure 6.21 *RMSEP* using data in Table 6.1 as a training set and data in Table 6.20 as a test set, PLS1 (centred) and acenaphthylene.

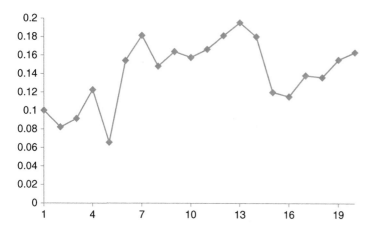

Figure 6.22 *RMSEP* using data in Table 6.20 as a training set and data in Table 6.1 as a test set, PLS1 (centred) and acenaphthylene.

be found. Hence, there is no 'perfect' way of validating calibration models, but it is essential to have some insight into the aims of an experiment before deciding how important the structure of the training set is to the success of the model in foreseeable future circumstances. Sometimes, it can take so long to produce good calibration models that new instruments and measurement techniques become available, superseding the original data sets.

1. Usually, *RMSEC* < *RMSECV* < *RMSEP*

    (a) True

    (b) False

Problems

6.1 Quantitative retention–property relationships of benzodiazepines
Section 6.5   Section 6.6.2
QSAR, QSPR and similar techniques are used to relate one block of properties to another. In this data set, six chromatographic retention parameters of 13 compounds are used to predict a biological property (A1).

Compound	Property	Retention parameters					
	A1	C18	Ph	CN-R	NH$_2$	CN-N	Si
1	−0.39	2.90	2.19	1.49	0.58	−0.76	−0.41
2	−1.58	3.17	2.67	1.62	0.11	−0.82	−0.52
3	−1.13	3.20	2.69	1.55	−0.31	−0.96	−0.33
4	−1.18	3.25	2.78	1.78	−0.56	−0.99	−0.55
5	−0.71	3.26	2.77	1.83	−0.53	−0.91	−0.45
6	−1.58	3.16	2.71	1.66	0.10	−0.80	−0.51
7	−0.43	3.26	2.74	1.68	0.62	−0.71	−0.39
8	−2.79	3.29	2.96	1.67	−0.35	−1.19	−0.71
9	−1.15	3.59	3.12	1.97	−0.62	−0.93	−0.56
10	−0.39	3.68	3.16	1.93	−0.54	−0.82	−0.50
11	−0.64	4.17	3.46	2.12	−0.56	−0.97	−0.55
12	−2.14	4.77	3.72	2.29	−0.82	−1.37	−0.80
13	−3.57	5.04	4.04	2.44	−1.14	−1.40	−0.86

1. Perform standardised cross-validated PLS on data (note that the property parameter should be mean centred, but there is no need to standardise) calculating five PLS components. If you are not using the Excel add-in, the following steps are required, as described in more detail in Problem 6.8.
   - Remove one sample.
   - Calculate the standard deviation and mean of the c and x block parameters of the remaining 12 samples (note that it is not strictly necessary to standardise the c block parameter).
   - Standardise these according to the parameters calculated above and then perform PLS.
   - Use this model to predict the property of the 13th sample. Remember to correct this for the standard deviation and mean of the 12 samples used to form the model. This step is a tricky one as it is necessary to use a weight vector.
   - Continue for all 13 samples.
   - When calculating the room mean square error, divide by the number of samples (13) rather than the number of degrees of freedom.

   If this is your first experience of cross-validation, you are recommended to use the Excel add-in, or first attempt Problem 6.8.

   Produce a graph of root mean square error, against component number. What appears to be the optimum number of components in the model?

2. Using the optimum number of PLS components obtained in question 1, perform PLS (standardising the retention parameters) on the overall data set and obtain a table of predictions for the parameter A1. What is the root mean square error?

3. Plot a graph of predicted versus observed values of parameter A1 from the model in question 2.

6.2 Classical and inverse univariate regression
Section 6.2
The following are some data. A response (x) is recorded at a number of concentrations (c).

c	x
1	0.082
2	0.174
3	0.320
4	0.412
5	0.531

c	x
6	0.588
7	0.732
8	0.792
9	0.891
10	0.975

1. There are two possible regression models, namely the inverse $\hat{c} = b_0 + b_1 x$ and the classical $\hat{x} = a_0 + a_1 c$. Show how the coefficients on the right of each equation would relate to each other algebraically if there were no errors.
2. The regression models can be expression in matrix form, $\hat{c} = Xb$ and $\hat{x} = Ca$. What are the dimensions of the six matrices/vectors in these equations? Using this approach, calculate the four coefficients from question 1.
3. Show that the coefficients for each model are approximately related as in question 1 and explain why this is not exact.
4. What different assumptions are made by both models?

6.3 Multivariate calibration with several x and c variables, factors that influence the taste and quality of cocoa
Section 4.3 Section 6.5.1 Section 6.5.2 Section 6.6.1
Sometimes, there are several variables both in the 'x' and 'c' blocks, and in many applications outside analytical chemistry of mixtures, it is difficult even to define which variables belong to which block.
The following data represent eight blends of cocoa, together with scores obtained from a taste panel of various qualities.

	ingredients			Assessments					
Sample	%Cocoa	%Sugar	%Milk	Lightness	Colour	Cocoa-odour	Smooth-text	Milk-taste	Sweetness
1	20.00	30.00	50.00	44.89	1.67	6.06	8.59	6.89	8.48
2	20.00	43.30	36.70	42.77	3.22	6.30	9.09	5.17	9.76
3	20.00	50.00	30.00	41.64	4.82	7.09	8.61	4.62	10.50
4	26.70	30.00	43.30	42.37	4.90	7.57	5.96	3.26	6.69
5	26.70	36.70	36.70	41.04	7.20	8.25	6.09	2.94	7.05
6	26.70	36.70	36.70	41.04	6.86	7.66	6.74	2.58	7.04
7	33.30	36.70	30.00	39.14	10.60	10.24	4.55	1.51	5.48
8	40.00	30.00	30.00	38.31	11.11	11.31	3.42	0.86	3.91

Can we predict the assessments from the ingredients? To translate into terminology of this chapter, refer to the ingredients as x and the assessments as c.

1. Standardise using the population standard deviation, all the variables. Perform PCA separately on the 8×3 X matrix and on the 8×6 C matrix. Retain the first two PCs, plot the scores and loadings of PC2 versus PC1 of each of the blocks, labelling the points, to give four graphs. Comment.
2. After standardising both the C and X matrices, perform PLS1 on the six c block variables against the three x block variables. Retain two PLS components for each variable (note that it is not necessary to retain the same number of components in each case) and calculate the predictions using this model. Convert this matrix (which is standardised) back to the original non-standardised matrix and present these predictions as a table.
3. Calculate the percentage root mean square prediction errors for each of the six variables as follows. (i) Calculate residuals between predicted and observed. (ii) Calculate the root mean square of these residuals, taking care to divide by 5 rather than 8 to account for the loss of three degrees of freedom due to the PLS components and the centring. (iii) Divide by the sample standard deviation for each parameter and multiply by 100. (Note that it is probably more relevant to use the standard deviation than the average in this case.)

4. Calculate the six correlation coefficients between the observed and predicted variables and plot a graph of the percentage root mean square error obtained in question 3 against the correlation coefficient and comment.
5. It is possible to perform PLS2 rather than PLS1. To do this, you must either produce an algorithm as presented in Section A.2.3 in Matlab or use the VBA Add-in. Use a six-variable C block and calculate the percentage root mean square errors as in question 3. Are they better or worse than for PLS1?
6. Instead of using the ingredients to predict taste/texture, it is possible to use the sensory variables to predict ingredients. How would you do this (you are not required to perform the calculations in full)?

6.4 Univariate and multivariate regression
Section 6.2   Section 6.3.3   Section 6.5.1

The following represent 10 spectra, recorded at eight wavelengths, of two compounds A and B, whose concentrations are given by the two vectors.

Spectrum								
1	0.227	0.206	0.217	0.221	0.242	0.226	0.323	0.175
2	0.221	0.412	0.45	0.333	0.426	0.595	0.639	0.465
3	0.11	0.166	0.315	0.341	0.51	0.602	0.537	0.246
4	0.194	0.36	0.494	0.588	0.7	0.831	0.703	0.411
5	0.254	0.384	0.419	0.288	0.257	0.52	0.412	0.35
6	0.203	0.246	0.432	0.425	0.483	0.597	0.553	0.272
7	0.255	0.326	0.378	0.451	0.556	0.628	0.462	0.339
8	0.47	0.72	0.888	0.785	1.029	1.233	1.17	0.702
9	0.238	0.255	0.318	0.289	0.294	0.41	0.444	0.299
10	0.238	0.305	0.394	0.415	0.537	0.585	0.566	0.253

Spectrum	Concentration A	Concentration B
1	1	3
2	3	5
3	5	1
4	7	2
5	2	5
6	4	3
7	6	1
8	9	6
9	2	4
10	5	2

Most calculations will be on compound A, although in certain cases, we may utilise the information about compound B in some cases.

1. After centring the data matrix down the columns, perform univariate calibration at each wavelength for compound A (only), using a method that assumes all errors are in the response (spectral) direction. Eight slopes should be obtained, one for each wavelength.
2. Predict the concentration vector for compound A using the results of regression at each wavelength, giving eight predicted concentration vectors. From these vectors, calculate the root mean square error of predicted − true concentrations at each wavelength and indicate which wavelengths are best for prediction.
3. Perform multi-linear regression using all the wavelengths at the same time for compound A, as follows: (i) On the uncentred data, find, assuming $X \approx cs$ where c is the concentration vector for compound A. (ii) Predict the concentrations as $\hat{c} = Xs'(ss')^{-1}$. What is the root mean square error of prediction?

4. Repeat the calculations of question 3, but this time include both concentration vectors in the calculations, replacing the vector c by the matrix C. Comment on the errors.
5. After centring the data both in the concentration and spectral directions, calculate the first and second components for the data and compound A using PLS1.
6. Calculate the estimated concentration vector for component A (i) using one PLS and (ii) using two PLS components. What is the root mean square error for prediction in each case?
7. Explain why information on only compounds A is necessary for good prediction using PLS, but both information on both compounds are needed for good prediction using MLR.

6.5 Principal components regression
Section 4.3   Section 6.4
A series of 10 spectra of two components is recorded at eight wavelengths. The following is the data.

Spectrum								
1	0.070	0.124	0.164	0.171	0.184	0.208	0.211	0.193
2	0.349	0.418	0.449	0.485	0.514	0.482	0.519	0.584
3	0.630	0.732	0.826	0.835	0.852	0.848	0.877	0.947
4	0.225	0.316	0.417	0.525	0.586	0.614	0.649	0.598
5	0.533	0.714	0.750	0.835	0.884	0.930	0.965	0.988
6	0.806	0.979	1.077	1.159	1.249	1.238	1.344	1.322
7	0.448	0.545	0.725	0.874	1.005	1.023	1.064	1.041
8	0.548	0.684	0.883	0.992	1.166	1.258	1.239	1.203
9	0.800	0.973	1.209	1.369	1.477	1.589	1.623	1.593
10	0.763	1.019	1.233	1.384	1.523	1.628	1.661	1.625

Spectrum	Concentration A	Concentration B
1	1	1
2	2	5
3	3	9
4	4	2
5	5	6
6	6	10
7	7	3
8	8	4
9	9	8
10	10	7

1. Perform PCA on the centred data, calculating loadings and scores for the first two PCs. How many PCs are needed to model the data almost perfectly (you are not asked to do cross-validation)?
2. Perform principal components regression as follows. Take each compound in turn and regress the concentrations onto the scores of the two PCs; you should centre the concentration matrices first. You should obtain two coefficients of the form $\hat{c} = t_1 r_1 + t_2 r_2$ for each compound. Verify that the results of PCR provide a good estimate of the concentrations for each compound. Note that you will have to add the mean concentration back to the results.
3. Using the coefficients obtained in question 2, give the 2×2 rotation or transformation matrix that rotates the scores of the first two PCs onto the centred estimates of concentrations.
4. Calculate the inverse of the matrix obtained in question 3 and, hence, determine the spectra of both components in the mixture from the loadings.

5. A different design can be used as follows:

Spectrum	Concentration A	Concentration B
1	1	10
2	2	9
3	3	8
4	4	7
5	5	6
6	6	5
7	7	4
8	8	3
9	9	2
10	10	1

Using the spectra obtained in question 4, and assuming no noise, calculate the data matrix that would be obtained. Perform PCA on this matrix and explain why there are considerable differences between the results using this design and the earlier design, and hence why this design is not satisfactory.

6.6 Multivariate calibration and prediction in spectroscopic monitoring of reactions
Section 4.3   Section 6.3.3   Section 6.5.1
The aim is to monitor reactions using a technique called *flow injection analysis*, which is used to record a UV/vis spectrum of a sample. These spectra are reported as summed intensities over the FIA trace below. The reaction is of the form $A + B \rightarrow C$ and so the aim is to quantitate the three compounds and produce a profile with time. A series of 25 three-component mixtures are available, their spectra and concentrations of each component are presented below, with spectral intensities and wavelengths recorded.

	234.39	240.29	246.20	252.12	258.03	263.96	269.89	275.82	281.76	287.70	293.65	299.60	305.56	311.52	317.48	323.45	329.43	335.41	341.39	347.37	353.36	358.86
1	12.268	14.149	12.321	6.863	2.479	1.559	1.695	2.261	3.262	4.622	6.283	8.199	10.199	12.041	13.617	14.453	14.068	13.014	11.167	8.496	6.139	4.456
2	7.624	9.817	8.933	4.888	1.346	0.621	0.738	1.101	1.683	2.452	3.372	4.409	5.417	6.292	6.857	6.976	6.487	5.575	4.396	3.005	1.829	1.113
3	10.259	11.537	9.759	5.170	1.914	1.317	1.447	1.848	2.544	3.516	4.770	6.155	7.820	9.289	10.953	12.315	12.990	13.131	12.597	11.171	9.428	7.551
4	8.057	8.464	7.115	4.239	2.056	1.534	1.587	2.033	2.922	4.145	5.610	7.391	9.198	11.007	12.527	13.257	12.597	11.428	9.346	6.380	4.086	2.749
5	18.465	21.359	18.855	10.408	3.448	2.012	2.208	2.957	4.235	5.981	8.165	10.601	13.244	15.543	17.655	18.938	18.925	17.988	16.159	13.274	10.336	7.820
6	11.687	12.672	10.783	6.238	2.739	1.943	2.020	2.584	3.678	5.187	7.023	9.207	11.469	13.649	15.530	16.517	15.939	14.688	12.395	9.012	6.225	4.407
7	14.424	17.271	15.422	8.688	2.886	1.619	1.753	2.386	3.497	4.996	6.816	8.912	11.003	12.896	14.285	14.784	13.965	12.361	10.026	7.018	4.496	2.927
8	10.096	12.184	10.821	6.017	2.001	1.152	1.256	1.701	2.477	3.533	4.822	6.304	7.820	9.196	10.285	10.782	10.351	9.365	7.840	5.797	4.020	2.816
9	11.158	12.751	10.981	6.012	2.229	1.463	1.573	2.027	2.844	3.977	5.405	7.024	8.833	10.472	12.079	13.172	13.329	12.889	11.718	9.715	7.700	5.948
10	14.251	15.483	13.098	7.348	3.112	2.207	2.310	2.914	4.066	5.675	7.684	10.009	12.567	14.950	17.251	18.727	18.759	18.015	16.180	13.035	10.071	7.652
11	16.518	18.466	16.008	9.102	3.572	2.343	2.460	3.185	4.545	6.417	8.709	11.376	14.161	16.741	18.919	20.045	19.508	18.069	15.515	11.681	8.338	5.998
12	16.513	19.051	16.833	9.579	3.459	2.103	2.227	2.935	4.265	5.983	8.144	10.635	13.176	15.491	17.325	18.131	17.368	15.745	13.166	9.591	6.523	4.501
13	17.159	19.509	16.987	9.452	3.451	2.187	2.333	3.024	4.265	5.983	8.131	10.566	13.214	15.593	17.782	19.153	19.080	18.155	16.233	13.157	10.150	7.670
14	16.913	18.425	15.741	8.950	3.759	2.614	2.717	3.432	4.804	6.713	9.080	11.827	14.780	17.523	19.979	21.421	21.154	20.033	17.698	13.915	10.441	7.795
15	14.437	16.934	15.067	8.745	3.224	1.940	2.027	2.700	3.948	5.651	7.680	10.087	12.406	14.592	16.101	16.467	15.159	13.023	9.955	6.080	3.100	1.555
16	16.117	19.285	17.162	9.433	2.960	1.623	1.788	2.401	3.430	4.837	6.602	8.558	10.665	12.468	14.060	14.994	14.949	14.082	12.572	10.341	8.033	6.058
17	5.991	6.551	5.665	3.476	1.684	1.233	1.248	1.566	2.225	3.139	4.228	5.574	6.900	8.244	9.291	9.696	9.006	7.927	6.177	3.837	2.105	1.215
18	12.738	15.538	13.856	7.603	2.386	1.307	1.436	1.928	2.754	3.886	5.298	6.873	8.547	9.990	11.210	11.873	11.710	10.889	9.556	7.676	5.820	4.320
19	10.106	10.967	9.247	5.196	2.273	1.654	1.723	2.138	2.952	4.096	5.534	7.205	9.075	10.832	12.582	13.774	13.873	13.427	12.118	9.853	7.705	5.928
20	14.632	16.343	14.095	7.981	3.205	2.158	2.254	2.869	4.030	5.644	7.641	9.961	12.433	14.737	16.796	17.974	17.675	16.590	14.485	11.230	8.311	6.137
21	13.294	15.525	13.701	7.851	2.932	1.819	1.907	2.494	3.585	5.081	6.893	9.023	11.163	13.166	14.698	15.292	14.439	12.847	10.395	7.150	4.512	2.930
22	10.969	13.628	12.320	6.932	2.199	1.160	1.254	1.720	2.533	3.637	4.965	6.503	7.994	9.336	10.222	10.415	9.628	8.244	6.388	4.166	2.357	1.327
23	7.356	9.084	8.112	4.465	1.461	0.844	0.921	1.221	1.739	2.450	3.332	4.336	5.397	6.340	7.139	7.577	7.442	6.913	6.023	4.772	3.584	2.663
24	7.748	8.738	7.523	4.228	1.742	1.224	1.278	1.586	2.174	3.007	4.050	5.263	6.602	7.852	9.060	9.863	9.896	9.512	8.551	6.946	5.430	4.187
25	7.501	8.829	7.837	4.589	1.822	1.181	1.220	1.568	2.243	3.173	4.289	5.629	6.949	8.223	9.164	9.485	8.808	7.691	6.010	3.843	2.176	1.272

The concentrations of the three compounds in the 25 calibration samples are as follows:

	Concentration A (mM)	Concentration B (mM)	Concentration C (mM)
1	0.276	0.090	0.069
2	0.276	0.026	0.013
3	0.128	0.026	0.126
4	0.128	0.153	0.041
5	0.434	0.058	0.126
6	0.200	0.153	0.069
7	0.434	0.090	0.041
8	0.276	0.058	0.041
9	0.200	0.058	0.098
10	0.200	0.121	0.126
11	0.357	0.153	0.098
12	0.434	0.121	0.069
13	0.357	0.090	0.126
14	0.276	0.153	0.126
15	0.434	0.153	0.013
16	0.434	0.026	0.098
17	0.128	0.121	0.013
18	0.357	0.026	0.069
19	0.128	0.090	0.098
20	0.276	0.121	0.098
21	0.357	0.121	0.041
22	0.357	0.058	0.013
23	0.200	0.026	0.041
24	0.128	0.058	0.069
25	0.200	0.090	0.013

The spectra recorded with time (minutes) along the first column are presented below. The aim is to estimate the concentrations of each compound in the mixture.

	234.39	240.29	246.20	252.12	258.03	263.96	269.89	275.82	281.76	287.70	293.65	299.60	305.56	311.52	317.48	323.45	329.43	335.41	341.39	347.37	353.36	358.86
1.08	11.657	12.457	10.788	6.372	2.609	1.700	1.747	2.278	3.312	4.721	6.391	8.417	10.358	12.262	13.583	13.905	12.656	10.765	7.999	4.499	1.927	0.722
3.67	11.821	12.617	10.921	6.461	2.664	1.747	1.793	2.330	3.377	4.805	6.500	8.554	10.526	12.462	13.812	14.142	12.873	10.962	8.157	4.602	1.993	0.765
7.22	11.922	12.725	11.015	6.518	2.692	1.767	1.815	2.355	3.407	4.842	6.547	8.614	10.600	12.541	13.903	14.241	12.977	11.068	8.254	4.686	2.065	0.822
9.48	11.886	12.688	10.986	6.497	2.685	1.768	1.817	2.356	3.401	4.828	6.520	8.571	10.541	12.466	13.816	14.160	12.927	11.042	8.268	4.745	2.140	0.896
13.07	11.755	12.561	10.866	6.401	2.633	1.735	1.792	2.327	3.355	4.757	6.420	8.429	10.373	12.261	13.605	13.969	12.790	10.980	8.290	4.864	2.315	1.071
15.45	11.478	12.260	10.596	6.218	2.553	1.694	1.759	2.281	3.273	4.621	6.222	8.149	10.023	11.832	13.141	13.521	12.441	10.750	8.226	4.984	2.541	1.303
19.17	11.769	12.561	10.823	6.309	2.588	1.738	1.824	2.352	3.339	4.675	6.268	8.162	10.029	11.815	13.148	13.603	12.664	11.112	8.746	5.647	3.242	1.920
22.06	11.459	12.232	10.520	6.076	2.465	1.664	1.760	2.272	3.207	4.474	5.990	7.777	9.564	11.253	12.560	13.076	12.303	10.945	8.819	5.983	3.721	2.387
25.01	11.848	12.632	10.831	6.220	2.524	1.719	1.827	2.344	3.282	4.553	6.081	7.865	9.680	11.378	12.737	13.354	12.706	11.471	9.470	6.723	4.456	3.012
28.02	12.151	12.936	11.063	6.310	2.551	1.747	1.862	2.379	3.312	4.576	6.109	7.888	9.722	11.424	12.848	13.539	13.034	11.922	10.038	7.389	5.127	3.592
31.04	11.844	12.609	10.764	6.108	2.463	1.692	1.807	2.305	3.196	4.409	5.884	7.586	9.365	11.001	12.406	13.131	12.737	11.746	10.028	7.545	5.380	3.843
34.09	11.882	12.640	10.770	6.077	2.427	1.663	1.783	2.278	3.165	4.369	5.840	7.526	9.304	10.934	12.360	13.144	12.815	11.902	10.254	7.835	5.686	4.109
37.15	12.065	12.813	10.901	6.126	2.443	1.680	1.805	2.301	3.185	4.387	5.860	7.544	9.337	10.972	12.429	13.263	13.022	12.173	10.604	8.241	6.092	4.460
40.36	12.152	12.902	10.960	6.140	2.451	1.695	1.819	2.313	3.187	4.376	5.842	7.506	9.298	10.920	12.400	13.291	13.128	12.372	10.885	8.604	6.474	4.794
43.49	12.055	12.799	10.855	6.052	2.406	1.669	1.798	2.278	3.131	4.290	5.725	7.347	9.114	10.702	12.189	13.113	13.043	12.387	11.011	8.834	6.761	5.056
47.18	12.113	12.843	10.872	6.030	2.379	1.648	1.778	2.260	3.102	4.254	5.684	7.292	9.059	10.641	12.148	13.124	13.122	12.535	11.235	9.118	7.057	5.315
50.22	11.908	12.622	10.677	5.907	2.331	1.620	1.752	2.223	3.045	4.169	5.566	7.133	8.866	10.411	11.907	12.896	12.950	12.426	11.201	9.168	7.154	5.409
55.2	11.655	12.358	10.441	5.762	2.272	1.586	1.716	2.173	2.969	4.056	5.411	6.927	8.616	10.114	11.590	12.583	12.690	12.228	11.091	9.155	7.206	5.482
60.06	12.194	12.913	10.892	5.982	2.341	1.630	1.768	2.242	3.067	4.191	5.601	7.168	8.928	10.486	12.041	13.117	13.290	12.876	11.747	9.783	7.764	5.935

(Continued)

64.59	12.662	13.389	11.277	6.179	2.417	1.687	1.833	2.319	3.164	4.315	5.762	7.365	9.179	10.775	12.385	13.528	13.762	13.391	12.289	10.319	8.251	6.333
70.34	12.372	13.083	11.008	6.014	2.353	1.648	1.794	2.265	3.078	4.190	5.592	7.136	8.899	10.445	12.035	13.185	13.481	13.192	12.193	10.331	8.331	6.425
75.41	12.389	13.095	11.007	5.995	2.332	1.634	1.781	2.250	3.059	4.165	5.562	7.097	8.862	10.400	12.001	13.176	13.524	13.276	12.327	10.508	8.524	6.599
81.09	11.939	12.629	10.611	5.767	2.239	1.570	1.715	2.169	2.946	4.012	5.355	6.830	8.532	10.011	11.566	12.714	13.081	12.859	11.968	10.233	8.316	6.442
86.49	12.120	12.809	10.758	5.839	2.269	1.595	1.742	2.201	2.985	4.059	5.418	6.905	8.628	10.122	11.704	12.880	13.264	13.075	12.194	10.465	8.525	6.614
92.43	11.463	12.129	10.185	5.523	2.147	1.514	1.654	2.087	2.825	3.836	5.117	6.518	8.145	9.554	11.057	12.193	12.589	12.434	11.628	10.009	8.182	6.362
99.09	12.277	12.969	10.871	5.876	2.269	1.596	1.747	2.210	2.998	4.077	5.443	6.935	8.673	10.177	11.794	13.023	13.472	13.345	12.516	10.818	8.869	6.904
105.32	12.211	12.894	10.808	5.837	2.259	1.593	1.745	2.202	2.978	4.046	5.398	6.871	8.594	10.083	11.692	12.924	13.409	13.308	12.519	10.864	8.943	6.976
110.39	11.263	11.923	9.997	5.396	2.090	1.479	1.622	2.044	2.759	3.740	4.988	6.346	7.938	9.311	10.804	11.967	12.431	12.361	11.651	10.131	8.357	6.526
123.41	11.438	12.089	10.123	5.451	2.111	1.498	1.644	2.066	2.786	3.771	5.028	6.386	7.996	9.375	10.899	12.098	12.615	12.585	11.919	10.430	8.644	6.763
144.31	11.871	12.527	10.472	5.606	2.153	1.526	1.681	2.121	2.861	3.875	5.172	6.565	8.233	9.653	11.249	12.525	13.131	13.165	12.537	11.047	9.215	7.237

1. Perform PCA (uncentred) on the 25 calibration spectra, calculating the first two PCs. Plot the scores of PC2 versus PC1 and label the points. Perform PCA (uncentred) on the 25×3 concentration matrix of *A*, *B* and *C*, calculating the first two PCs, likewise plotting the scores of PC2 versus PC1 and labelling the points. Comment.

2. Predict the concentrations of A in the calibration set using MLR and assuming only compound A can be calibrated, as follows. (i) Determine the vector $\hat{s} = \frac{c'X}{\sum c^2}$ where *X* is the 25×22 spectral calibration matrix and *c* a 25×1 vector. (ii) Determine the predicted concentration vector $\hat{c} = \frac{X\hat{s}'}{\sum \hat{s}^2}$ (notice that the denominator is simply the sum of squares when only one compound is used).

3. In the model of question 2, plot a graph of predicted against true concentrations. Determine the root mean square error both in mM and as a percentage of the average. Comment.

4. Repeat the predictions using MLR, but this time for all three compounds simultaneously as follows. (i) Determine the matrix $\hat{S} = (C'C)^{-1}C'X$, where *X* is the 25×22 spectral calibration matrix and *c* a 25×1 vector. (ii) Determine the predicted concentration matrix $\hat{C} = X\hat{S}'(\hat{S}\hat{S}')^{-1}$.

5. In the model of question 3, plot a graph of predicted against true concentration for compound A. Determine the root mean square error both in mM and as a percentage of the average for all the three compounds. Comment.

6. Repeat questions 2 and 3, but instead of MLR, use PLS1 (centred) for the prediction of the concentration of A retaining the first three PLS components. Note that to obtain a root mean square error, it is best to divide by 21 rather than 25 if three components are retained. You are not asked to cross-validate the models. Why are the predictions much better?

7. Use the 25×22 calibration set as a training set, obtain a PLS1 (centred) model for all the three compounds retaining three components in each case and centring the spectroscopic data. Use this model to predict the concentrations of compounds A–C in the 30 reaction spectra. Plot a graph of estimated concentrations of each compound against time.

6.7 PLS1 algorithm

Section 6.5.1   Section A.2.2

The PLS1 algorithm is quite simple and described in detail in Section A.2.2. However, it can be easily set up in Excel or programmed into Matlab in a few lines, and the aim of this problem is to set up the matrix-based calculations for PLS.

The following is a description of the steps you are required to perform.

1. Centre both the *X* and *c* blocks by subtracting the column means.

2. Calculate the scores of the first PLS component by

$$h = X'c$$

and then

$$t = \frac{Xh}{\sqrt{\sum h^2}}$$

3. Calculate the *x* loadings of the first PLS component by

$$p = t'X / \sum t^2$$

Note that the denominator is simply the sum of squares of the scores.

4. Calculate the *c* loadings (a scalar in PLS1) by

$$q = c't/\sum t^2$$

5. Calculate the contribution to the concentration estimate by tq and the contribution to the *x* estimate by tp.
6. Subtract the contributions in steps (d) and (e) from the current *c* vector and *X* matrix and use these residuals for the calculation of the next PLS component by returning to step (b).
7. To obtain the overall concentration estimate, simply multiply Tq, where T is a scores matrix with *A* columns corresponding to the PLS components, and q a column vector of size *A*. Add back the mean value of the concentrations to produce real estimates.

The method will be illustrated by a small simulated data set, consisting of four samples, six measurements and one *c* parameter, which is exactly characterised by three PLS components.

X					c
10.1	6.6	8.9	8.2	3.8	0.5
12.6	6.3	7.1	10.9	5.3	0.2
11.3	6.7	10.0	9.3	2.9	0.5
15.1	8.7	7.8	12.9	9.3	0.3

1. Calculate the loadings and scores of the first three PLS components, laying out the calculations in full.
2. What are the residual sum of squares for the '*x*' and '*c*' blocks as each successive component is computed (Hint: Start from the centred data matrix and simply sum the squares of each block, repeat for the residuals)? What percentage of the overall variance is accounted for by each component?
3. How many components are needed to describe the data exactly? Why does this answer not say much about the underlying structure of the data?
4. Provide a table of true concentrations and of predicted concentrations as one, two and three PLS components are calculated.
5. If only two PLS components are used, what is the root mean square error of prediction of concentrations over all four samples? Remember to divide by 1 and not 4. (Why is this?)

6.8 Cross-validation in PLS
Section 5.5.1 Section 6.6.2 Section A.2.2
The following consists of 10 samples, whose spectra are recorded at six wavelengths. The concentration of a component in the samples is given by as a *c* vector. This data set has been simulated to give an exact fit for two components as an example of how cross-validation works.

Sample	Spectra						c
1	0.10	0.22	0.20	0.06	0.29	0.10	1
2	0.20	0.60	0.40	0.20	0.75	0.30	5
3	0.12	0.68	0.24	0.28	0.79	0.38	9
4	0.27	0.61	0.54	0.17	0.80	0.28	3
5	0.33	0.87	0.66	0.27	1.11	0.42	6
6	0.14	0.66	0.28	0.26	0.78	0.36	8
7	0.14	0.34	0.28	0.10	0.44	0.16	2
8	0.25	0.79	0.50	0.27	0.98	0.40	7
9	0.10	0.22	0.20	0.06	0.29	0.10	1
10	0.19	0.53	0.38	0.17	0.67	0.26	4

1. Select samples 1–9 and calculate their means. Mean centre both the *x* and *c* variables over these samples.
2. Perform PLS1, calculate two components, on the first nine samples, centred as in question 1. Calculate t, p, h and the contribution to the *c* values for each PLS component (given by qt), and verify that the samples can be

exactly modelled using two PLS components. (Note you will have to add on the mean of samples 1–9 to c after prediction.) You should use the algorithm of Problem 6.7 or Section A.2.2 and will need to find the vector h to answer question 3.

3. Cross-validation is to be performed on sample 10, using the model of samples 1–9 as follows:
   (a) Subtract the means of samples 1–9 from sample 10 to produce a new x vector, similarly for the c value.
   (b) Then, calculate the predicted score for the first PLS component and sample 10 by $\hat{t}_{10,1} = x_{10} h_1 / \sqrt{\sum h_1^2}$, where h_1 has been calculated above on samples 1–9 for the first PLS component and calculate the new residual spectral vector $x_{10} - \hat{t}_{10,1} p_1$.
   (c) Calculate the contribution to the mean centred concentration for sample 10 as $\hat{t}_{10,1} q_1$, where q_1 is the value of q for the first PLS component using samples 1–9 and calculate the residual concentration $c_{10} - \hat{t}_{10,1} q_1$.
   (d) Find $\hat{t}_{10,2}$ for the second component using the residual vectors above using the vector h determined for the second component using the prediction set of nine samples.
   (e) Calculate the contribution to predicting c and x from the second component.

4. Demonstrate that, for this particular set, cross-validation results in an exact prediction of concentration for sample 10 – remember to add the mean of samples 1–9 back after prediction.

5. Unlike for PCA, it is not possible to determine the predicted scores by xp', but it is necessary to use a vector h. Why is this?

6.9 Multivariate calibration in three-way diode array HPLC

Section 6.5.3

The aim of this problem is to perform a variety of methods of calibration on a three-way data set. Ten chromatograms are recorded of 3-hydroxypyridine impurity within a main peak 2-hydroxypyridine. The aim is to employ PLS to determine the concentration of the minor component.

For each concentration, a 20×10 chromatogram is presented, taken over 20 s in time (1 s digital resolution) and in this data set, for simplicity absorbances every 12 nm starting at 230 are presented.

Five concentrations are used, replicated twice. The 10 concentrations (mM) in the table below are presented in the following arrangement.

0.0158	0.0158
0.0315	0.0315
0.0473	0.0473
0.0631	0.0631
0.0789	0.0789

0.089	0.011	0.020	0.048	0.097	0.132	0.116	0.055	0.005	0.000	0.087	0.011	0.020	0.048	0.096	0.130	0.114	0.054	0.005	0.000
0.150	0.015	0.032	0.080	0.161	0.223	0.198	0.092	0.008	0.001	0.148	0.015	0.031	0.079	0.159	0.220	0.196	0.090	0.008	0.001
0.224	0.020	0.046	0.118	0.239	0.334	0.299	0.137	0.011	0.001	0.221	0.019	0.045	0.117	0.236	0.329	0.295	0.135	0.010	0.001
0.300	0.024	0.060	0.158	0.320	0.449	0.402	0.184	0.013	0.001	0.296	0.024	0.059	0.156	0.315	0.442	0.396	0.181	0.013	0.001
0.366	0.028	0.073	0.192	0.389	0.548	0.492	0.224	0.016	0.001	0.360	0.027	0.071	0.189	0.383	0.539	0.484	0.220	0.015	0.001
0.412	0.030	0.081	0.216	0.438	0.618	0.555	0.252	0.017	0.001	0.405	0.030	0.080	0.213	0.430	0.607	0.546	0.247	0.017	0.001
0.436	0.032	0.085	0.229	0.463	0.654	0.588	0.267	0.018	0.001	0.428	0.031	0.084	0.225	0.455	0.642	0.578	0.262	0.017	0.001
0.439	0.031	0.086	0.231	0.467	0.659	0.594	0.269	0.018	0.001	0.431	0.031	0.084	0.226	0.458	0.647	0.582	0.264	0.017	0.001
0.428	0.030	0.083	0.224	0.453	0.641	0.577	0.261	0.017	0.001	0.419	0.030	0.082	0.220	0.445	0.629	0.566	0.256	0.017	0.001
0.405	0.028	0.079	0.212	0.429	0.607	0.547	0.247	0.016	0.001	0.397	0.028	0.077	0.208	0.421	0.596	0.536	0.243	0.016	0.001
0.377	0.026	0.073	0.197	0.398	0.564	0.508	0.230	0.015	0.001	0.370	0.026	0.072	0.193	0.391	0.554	0.499	0.225	0.015	0.001
0.346	0.024	0.067	0.180	0.365	0.517	0.466	0.211	0.014	0.001	0.340	0.023	0.066	0.177	0.359	0.508	0.458	0.207	0.013	0.001
0.315	0.022	0.061	0.164	0.332	0.470	0.423	0.191	0.013	0.001	0.309	0.021	0.060	0.161	0.326	0.462	0.416	0.188	0.012	0.001
0.285	0.020	0.055	0.148	0.300	0.424	0.382	0.173	0.011	0.000	0.280	0.019	0.054	0.146	0.295	0.418	0.376	0.170	0.011	0.000
0.257	0.018	0.049	0.133	0.270	0.382	0.344	0.156	0.010	0.000	0.253	0.017	0.049	0.131	0.266	0.377	0.339	0.153	0.010	0.000
0.231	0.016	0.044	0.120	0.243	0.344	0.310	0.140	0.009	0.000	0.228	0.016	0.044	0.118	0.240	0.339	0.306	0.138	0.009	0.000

0.208	0.014	0.040	0.108	0.219	0.309	0.279	0.126	0.008	0.000	0.206	0.014	0.039	0.107	0.216	0.306	0.275	0.124	0.008	0.000
0.188	0.013	0.036	0.097	0.197	0.279	0.251	0.113	0.007	0.000	0.186	0.013	0.036	0.096	0.195	0.276	0.248	0.112	0.007	0.000
0.170	0.012	0.033	0.088	0.178	0.252	0.227	0.102	0.007	0.000	0.168	0.011	0.032	0.087	0.176	0.249	0.224	0.101	0.006	0.000
0.154	0.010	0.029	0.080	0.161	0.228	0.205	0.093	0.006	0.000	0.152	0.010	0.029	0.079	0.159	0.226	0.203	0.092	0.006	0.000
0.055	0.011	0.015	0.032	0.063	0.081	0.069	0.035	0.005	0.001	0.067	0.013	0.018	0.038	0.076	0.099	0.085	0.042	0.006	0.001
0.100	0.014	0.024	0.056	0.111	0.149	0.130	0.062	0.007	0.001	0.120	0.016	0.028	0.066	0.132	0.178	0.156	0.074	0.008	0.001
0.163	0.019	0.036	0.088	0.177	0.243	0.215	0.101	0.010	0.001	0.191	0.021	0.042	0.103	0.207	0.284	0.252	0.118	0.011	0.001
0.237	0.023	0.050	0.126	0.255	0.353	0.315	0.145	0.012	0.001	0.271	0.026	0.057	0.144	0.291	0.404	0.360	0.166	0.014	0.001
0.309	0.027	0.063	0.164	0.331	0.462	0.413	0.190	0.015	0.001	0.347	0.030	0.071	0.183	0.370	0.518	0.464	0.213	0.016	0.001
0.369	0.030	0.074	0.195	0.394	0.552	0.495	0.226	0.017	0.001	0.406	0.033	0.082	0.214	0.433	0.608	0.545	0.249	0.018	0.001
0.409	0.032	0.081	0.215	0.435	0.612	0.550	0.250	0.018	0.001	0.443	0.034	0.088	0.233	0.472	0.664	0.596	0.272	0.019	0.001
0.426	0.032	0.084	0.224	0.454	0.639	0.575	0.261	0.018	0.001	0.457	0.035	0.091	0.240	0.486	0.685	0.616	0.280	0.020	0.001
0.426	0.032	0.084	0.224	0.452	0.638	0.574	0.261	0.018	0.001	0.452	0.034	0.089	0.237	0.480	0.677	0.609	0.277	0.019	0.001
0.411	0.030	0.081	0.216	0.436	0.615	0.554	0.251	0.017	0.001	0.433	0.032	0.085	0.227	0.460	0.649	0.584	0.265	0.018	0.001
0.387	0.028	0.076	0.203	0.410	0.579	0.522	0.237	0.016	0.001	0.406	0.029	0.079	0.213	0.430	0.607	0.547	0.248	0.017	0.001
0.359	0.026	0.070	0.188	0.379	0.536	0.483	0.219	0.015	0.001	0.375	0.027	0.073	0.196	0.396	0.560	0.504	0.229	0.015	0.001
0.328	0.023	0.064	0.171	0.347	0.490	0.441	0.200	0.013	0.001	0.342	0.024	0.066	0.178	0.361	0.510	0.459	0.208	0.014	0.001
0.298	0.021	0.058	0.155	0.314	0.444	0.400	0.181	0.012	0.001	0.310	0.022	0.060	0.161	0.326	0.461	0.415	0.188	0.013	0.001
0.269	0.019	0.052	0.140	0.284	0.401	0.361	0.164	0.011	0.001	0.279	0.020	0.054	0.145	0.294	0.416	0.374	0.170	0.011	0.000
0.243	0.017	0.047	0.126	0.255	0.361	0.325	0.147	0.010	0.001	0.252	0.018	0.049	0.131	0.264	0.374	0.336	0.152	0.010	0.000
0.218	0.015	0.042	0.114	0.230	0.324	0.292	0.132	0.009	0.000	0.226	0.016	0.044	0.117	0.238	0.336	0.302	0.137	0.009	0.000
0.197	0.014	0.038	0.102	0.207	0.292	0.263	0.119	0.008	0.000	0.204	0.014	0.039	0.106	0.214	0.302	0.272	0.123	0.008	0.000
0.177	0.013	0.034	0.092	0.186	0.263	0.237	0.107	0.007	0.000	0.184	0.013	0.036	0.095	0.193	0.272	0.245	0.111	0.007	0.000
0.160	0.011	0.031	0.083	0.168	0.237	0.214	0.097	0.006	0.000	0.166	0.012	0.032	0.086	0.174	0.246	0.221	0.100	0.007	0.000
0.074	0.016	0.021	0.044	0.086	0.110	0.093	0.047	0.008	0.001	0.080	0.017	0.023	0.047	0.093	0.119	0.102	0.051	0.008	0.001
0.130	0.020	0.032	0.073	0.145	0.193	0.168	0.081	0.010	0.001	0.137	0.021	0.033	0.077	0.153	0.204	0.178	0.086	0.010	0.001
0.203	0.025	0.046	0.111	0.222	0.302	0.267	0.126	0.013	0.001	0.210	0.025	0.047	0.114	0.229	0.313	0.276	0.130	0.013	0.001
0.283	0.029	0.061	0.152	0.306	0.422	0.376	0.175	0.016	0.002	0.288	0.029	0.061	0.155	0.311	0.430	0.383	0.177	0.015	0.001
0.358	0.033	0.074	0.190	0.384	0.534	0.477	0.220	0.018	0.002	0.358	0.033	0.074	0.191	0.385	0.536	0.479	0.220	0.018	0.001
0.414	0.036	0.085	0.220	0.443	0.619	0.555	0.254	0.020	0.002	0.411	0.035	0.084	0.218	0.440	0.616	0.552	0.252	0.019	0.001
0.447	0.037	0.090	0.237	0.478	0.670	0.601	0.275	0.020	0.002	0.441	0.036	0.089	0.233	0.471	0.662	0.593	0.271	0.020	0.001
0.457	0.036	0.092	0.242	0.488	0.686	0.616	0.281	0.020	0.001	0.449	0.035	0.090	0.237	0.480	0.674	0.605	0.275	0.020	0.001
0.450	0.035	0.090	0.237	0.479	0.674	0.606	0.276	0.020	0.001	0.440	0.034	0.088	0.232	0.469	0.661	0.594	0.270	0.019	0.001
0.429	0.033	0.085	0.226	0.456	0.642	0.578	0.263	0.018	0.001	0.419	0.032	0.083	0.221	0.446	0.629	0.565	0.257	0.018	0.001
0.401	0.030	0.079	0.210	0.425	0.599	0.539	0.245	0.017	0.001	0.391	0.029	0.077	0.206	0.416	0.586	0.527	0.239	0.016	0.001
0.369	0.027	0.072	0.193	0.390	0.550	0.495	0.225	0.016	0.001	0.360	0.027	0.071	0.189	0.382	0.538	0.484	0.220	0.015	0.001
0.336	0.025	0.066	0.175	0.355	0.500	0.450	0.205	0.014	0.001	0.328	0.024	0.064	0.172	0.347	0.490	0.440	0.200	0.014	0.001
0.304	0.022	0.059	0.158	0.320	0.452	0.407	0.185	0.013	0.001	0.297	0.022	0.058	0.155	0.314	0.442	0.398	0.180	0.012	0.001
0.273	0.020	0.053	0.143	0.288	0.407	0.366	0.166	0.011	0.001	0.267	0.020	0.052	0.140	0.282	0.398	0.358	0.162	0.011	0.001
0.246	0.018	0.048	0.128	0.259	0.365	0.329	0.149	0.010	0.001	0.240	0.018	0.047	0.126	0.254	0.358	0.322	0.146	0.010	0.000
0.221	0.016	0.043	0.115	0.233	0.328	0.295	0.134	0.009	0.000	0.216	0.016	0.042	0.113	0.228	0.321	0.289	0.131	0.009	0.000
0.199	0.015	0.039	0.104	0.209	0.295	0.265	0.120	0.008	0.000	0.195	0.014	0.038	0.102	0.205	0.289	0.260	0.118	0.008	0.000
0.179	0.013	0.035	0.093	0.188	0.265	0.239	0.108	0.007	0.000	0.175	0.013	0.035	0.092	0.185	0.260	0.234	0.106	0.007	0.000
0.162	0.012	0.032	0.085	0.170	0.240	0.215	0.098	0.007	0.000	0.159	0.012	0.031	0.083	0.167	0.235	0.211	0.096	0.006	0.000
0.061	0.018	0.021	0.038	0.075	0.090	0.074	0.040	0.008	0.001	0.067	0.019	0.022	0.042	0.081	0.099	0.082	0.043	0.009	0.001
0.108	0.022	0.030	0.063	0.125	0.161	0.137	0.068	0.010	0.001	0.118	0.023	0.032	0.068	0.135	0.175	0.151	0.075	0.011	0.002

(Continued)

0.173 0.026 0.042 0.097 0.193 0.258 0.225 0.108 0.013 0.002 | 0.188 0.027 0.045 0.104 0.208 0.278 0.244 0.117 0.014 0.002
0.249 0.030 0.056 0.136 0.273 0.371 0.328 0.154 0.015 0.002 | 0.266 0.031 0.059 0.144 0.290 0.396 0.351 0.164 0.016 0.002
0.324 0.034 0.070 0.174 0.351 0.484 0.430 0.200 0.018 0.002 | 0.341 0.034 0.073 0.183 0.368 0.508 0.453 0.210 0.018 0.002
0.385 0.036 0.081 0.206 0.415 0.576 0.515 0.237 0.019 0.002 | 0.399 0.037 0.083 0.213 0.428 0.596 0.533 0.245 0.020 0.002
0.425 0.037 0.087 0.226 0.456 0.637 0.570 0.261 0.020 0.002 | 0.435 0.038 0.089 0.231 0.465 0.651 0.583 0.267 0.021 0.002
0.443 0.037 0.090 0.235 0.474 0.664 0.595 0.272 0.020 0.002 | 0.448 0.037 0.091 0.237 0.478 0.670 0.601 0.275 0.021 0.002
0.441 0.036 0.089 0.233 0.471 0.661 0.594 0.271 0.020 0.001 | 0.442 0.036 0.089 0.233 0.471 0.662 0.594 0.271 0.020 0.002
0.425 0.034 0.085 0.225 0.453 0.637 0.572 0.261 0.019 0.001 | 0.423 0.033 0.085 0.223 0.450 0.633 0.569 0.259 0.019 0.001
0.400 0.031 0.080 0.211 0.426 0.599 0.538 0.245 0.017 0.001 | 0.396 0.031 0.079 0.208 0.421 0.592 0.532 0.242 0.017 0.001
0.370 0.029 0.073 0.194 0.393 0.553 0.497 0.226 0.016 0.001 | 0.365 0.028 0.072 0.191 0.387 0.544 0.489 0.223 0.016 0.001
0.338 0.026 0.067 0.177 0.358 0.504 0.453 0.206 0.015 0.001 | 0.332 0.025 0.066 0.174 0.352 0.495 0.445 0.203 0.014 0.001
0.306 0.023 0.060 0.160 0.324 0.457 0.410 0.187 0.013 0.001 | 0.301 0.023 0.059 0.157 0.318 0.448 0.403 0.183 0.013 0.001
0.276 0.021 0.054 0.145 0.292 0.411 0.370 0.168 0.012 0.001 | 0.271 0.021 0.053 0.142 0.286 0.403 0.362 0.165 0.012 0.001
0.249 0.019 0.049 0.130 0.262 0.370 0.332 0.151 0.011 0.001 | 0.244 0.018 0.048 0.127 0.257 0.362 0.325 0.148 0.010 0.001
0.223 0.017 0.044 0.117 0.236 0.332 0.298 0.135 0.009 0.000 | 0.219 0.017 0.043 0.114 0.231 0.325 0.292 0.133 0.009 0.001
0.201 0.015 0.040 0.105 0.212 0.298 0.268 0.122 0.008 0.000 | 0.197 0.015 0.039 0.103 0.207 0.292 0.262 0.119 0.008 0.000
0.181 0.014 0.036 0.095 0.191 0.268 0.241 0.109 0.008 0.000 | 0.177 0.014 0.035 0.093 0.187 0.263 0.236 0.107 0.007 0.000
0.163 0.013 0.033 0.086 0.172 0.242 0.217 0.099 0.007 0.000 | 0.160 0.012 0.032 0.084 0.169 0.237 0.213 0.097 0.007 0.000

0.082 0.023 0.027 0.051 0.100 0.122 0.102 0.054 0.011 0.002 | 0.081 0.023 0.027 0.051 0.098 0.120 0.100 0.053 0.011 0.002
0.140 0.028 0.038 0.081 0.160 0.207 0.178 0.088 0.013 0.002 | 0.137 0.027 0.038 0.080 0.158 0.204 0.175 0.087 0.013 0.002
0.213 0.032 0.052 0.119 0.237 0.317 0.277 0.133 0.016 0.002 | 0.210 0.032 0.051 0.118 0.234 0.313 0.273 0.131 0.016 0.002
0.293 0.036 0.066 0.160 0.320 0.436 0.386 0.182 0.019 0.002 | 0.290 0.035 0.066 0.158 0.317 0.432 0.382 0.180 0.018 0.002
0.366 0.039 0.079 0.197 0.396 0.546 0.485 0.226 0.021 0.002 | 0.363 0.038 0.079 0.196 0.393 0.542 0.482 0.224 0.020 0.002
0.420 0.040 0.089 0.225 0.452 0.628 0.561 0.259 0.022 0.002 | 0.418 0.040 0.088 0.224 0.451 0.626 0.558 0.258 0.022 0.002
0.451 0.041 0.093 0.240 0.484 0.675 0.604 0.278 0.022 0.002 | 0.450 0.040 0.093 0.240 0.483 0.674 0.603 0.277 0.022 0.002
0.459 0.040 0.094 0.244 0.492 0.688 0.617 0.283 0.022 0.002 | 0.459 0.039 0.094 0.244 0.492 0.688 0.617 0.282 0.022 0.002
0.450 0.038 0.091 0.238 0.481 0.674 0.604 0.276 0.021 0.002 | 0.450 0.038 0.091 0.239 0.482 0.675 0.605 0.277 0.021 0.001
0.428 0.035 0.086 0.226 0.457 0.640 0.575 0.263 0.020 0.002 | 0.429 0.035 0.086 0.227 0.458 0.642 0.576 0.263 0.019 0.001
0.399 0.032 0.080 0.210 0.425 0.596 0.535 0.244 0.018 0.001 | 0.400 0.032 0.080 0.211 0.426 0.598 0.537 0.245 0.018 0.001
0.366 0.029 0.073 0.193 0.389 0.547 0.491 0.224 0.016 0.001 | 0.367 0.029 0.073 0.193 0.390 0.548 0.493 0.224 0.016 0.001
0.333 0.026 0.066 0.175 0.353 0.496 0.446 0.203 0.015 0.001 | 0.334 0.026 0.066 0.175 0.354 0.498 0.447 0.204 0.015 0.001
0.301 0.024 0.060 0.158 0.319 0.448 0.402 0.183 0.013 0.001 | 0.301 0.024 0.060 0.158 0.319 0.449 0.403 0.184 0.013 0.001
0.271 0.021 0.054 0.142 0.286 0.402 0.362 0.165 0.012 0.001 | 0.271 0.021 0.054 0.142 0.287 0.403 0.362 0.165 0.012 0.001
0.243 0.019 0.048 0.128 0.257 0.361 0.324 0.148 0.011 0.001 | 0.243 0.019 0.048 0.128 0.257 0.362 0.325 0.148 0.010 0.001
0.218 0.017 0.044 0.115 0.231 0.324 0.291 0.132 0.010 0.001 | 0.219 0.017 0.044 0.115 0.231 0.324 0.291 0.133 0.009 0.000
0.196 0.016 0.040 0.103 0.207 0.291 0.261 0.119 0.009 0.001 | 0.196 0.016 0.039 0.103 0.208 0.291 0.261 0.119 0.008 0.000
0.177 0.014 0.036 0.093 0.187 0.262 0.235 0.107 0.008 0.000 | 0.177 0.014 0.036 0.093 0.187 0.262 0.235 0.107 0.007 0.000
0.160 0.013 0.033 0.084 0.169 0.236 0.212 0.096 0.007 0.000 | 0.159 0.013 0.032 0.084 0.169 0.236 0.212 0.096 0.007 0.000

1. One approach to calibration is to use one-way PLS. This can be in either the spectroscopic or time directions. In fact, the spectroscopic dimension is often more useful. Produce a 10×10 table of summed intensities over the 20 chromatographic points in time at each wavelength for each sample.

2. Standardise the data and perform auto-predictive PLS1 calculating three PLS components. Why is it useful to standardise the measurements?

3. Plot graphs of predicted versus known concentrations for one, two and three PLS components and calculate the root mean square errors in mM.

4. Perform PLS1 cross-validation on the c values for the first eight components and plot a graph of cross-validated error against component number.

5. Unfold the original data matrix to give a 10×200 data matrix.

6. It is desired to perform PLS calibration on this data set, but first to standardise the data. Explain why there may be problems with this approach. Why is it desirable to reduce the number of variables from 200, and why was this variable selection less important in the PLS1 calculations?

7. Why is the standard deviation a good measure of variable significance? Reduce the data set to 100 significant variables with the highest standard deviations to give a 10×100 data matrix.

8. Perform auto-predictive PLS1 on the standardised reduced unfolded data of question 7 and calculate the errors as one, two and three components are computed.

9. How might you improve the model of question 8 still further?

7

Evolutionary Multivariate Signals

7.1 Introduction

Some of the classical applications of chemometrics are to evolutionary data. Such type of information is increasingly common and often involves simultaneously recording spectra whilst a physical parameter such as time or pH is changed; signals evolve during the change in this parameter. Multivariate methods are useful when there are several channels or variables that are each ordered sequentially; thus, we are looking for common trends in all or some of these variables against time or any other sequential variable. The data can be organised into a multivariate data matrix, usually with the prime sequential variable represented by rows and the other variable by columns. Although the second variable may sometimes also be sequentially related (e.g. wavelength), our primary interest is to know how the process evolves over a primary variable (e.g. time), and we will call this variable (usually representing successive rows) as the *sequential variable* (Figure 7.1).

In the modern laboratory, one of the most widespread applications is in the area of coupled chromatography, such as HPLC-DAD (high-performance liquid chromatography-diode array detector), LC-MS (liquid chromatography mass spectrometry) and LC-NMR (liquid chromatography nuclear magnetic resonance). A chromatogram is recorded whilst a UV/vis, MS (mass spectrometry) or NMR (nuclear magnetic resonance) spectrum is recorded. The information can be presented in a matrix form, with time along the rows and wavelength, mass number or frequency along the columns, as already introduced in Chapter 4. Multivariate approaches can be employed to analyse these data. However, there are a wide variety of other applications ranging from pH titrations to processes that change in a systematic way with time as in kinetics. Many of the approaches in this chapter have quite a wide applicability, for example, baseline correction, data scaling and 3D PC plots; however, for brevity, we illustrate the chapter primarily with case studies from coupled chromatography, as this has been the source of a huge literature over the past decades.

With modern laboratory computers, it is possible to obtain large quantities of information very rapidly. For example, spectra sampled at 1 nm intervals over a 200 nm region can be obtained every second using modern chromatography; hence, in an hour, 3600 spectra in time $\times$ 200 spectral frequencies or 720 000 pieces of information can be produced from a single chromatogram. A typical medium to large site may contain a hundred or more coupled chromatographs, meaning the potential of acquiring 72 million data points per hour of this type of information. Add on all the other instruments, and it is not difficult to see how billions of numbers are generated on a daily basis.

In Chapters 4–6, we have discussed a number of methods for multivariate data analysis, but most of the methods described did not take into account the sequential nature of the information. When performing principal components analysis (PCA) on a data matrix, the order of the rows and columns is irrelevant. Figure 7.2 represents three cross sections through a data matrix. The first could correspond to a chromatographic peak, whereas the others not. However, as PCA and most other classical multivariate methods would not distinguish these sequences, clearly other enhancements are useful. We will primarily illustrate the methods in this chapter with reference to coupled chromatography to avoid repetition.

In many cases, underlying factors corresponding to individual compounds in a chromatographic mixture are unimodal in time, where they have one maximum. The aim is to deconvolute the experimentally observed sum into individual components and determine the features of each component. The change in spectral characteristics across the chromatographic peak can be used to provide this information.

Chemometrics: Data Driven Extraction for Science, Second Edition. Richard G. Brereton.
© 2018 John Wiley & Sons Ltd. Published 2018 by John Wiley & Sons Ltd.
Companion website: http://booksupport.wiley.com

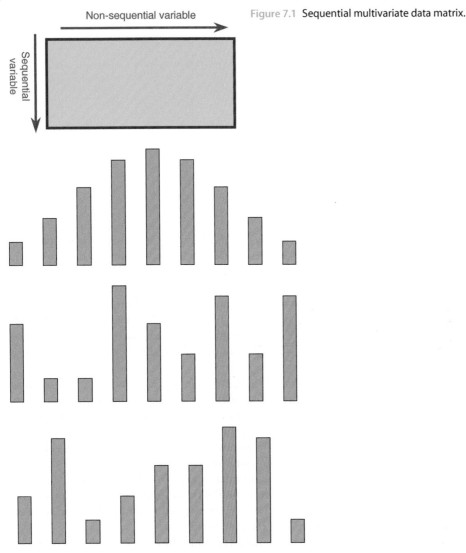

Figure 7.1 Sequential multivariate data matrix.

Figure 7.2 Three possible sequential patterns that would be treated identically using standard multivariate techniques.

To practicing chemists, there are three main questions that can be answered by applying chemometric techniques to coupled chromatography, of increasing difficulty.

- *How many peaks in a cluster?* Can we detect small impurities? Can we detect metabolites against a background? Can we determine whether there are embedded peaks?
- *What are the characteristics of each pure compound?* What are the spectra? Can we obtain mass spectra or NMR spectra of embedded chromatographic peaks at low levels of sufficient quality that we can be confident of their identities?
- *What are the quantities of each component?* Can we quantitate small impurities? Could we use chromatography of mixtures for reaction monitoring and kinetics? Can we say with certainty the level of dope or a potential environmental pollutant when it is detected in low quantities buried within a major peak?

There are a large number of 'named' methods in the literature, but they are mainly based around certain main principles, of evolutionary factor analysis, by which factors corresponding to individual compounds evolve in time (or any other sequential parameter such as pH). We will not discuss all such methods comprehensively and primarily outline the main principles.

Such methods are not only applicable to coupled chromatography but also in areas such as pH dependence of equilibria, by which the spectra of a mixture of chemical species can be followed with change in pH. It would be possible to record 20 spectra and then treat each independently. Sometimes, this can lead to good quantification, including the information that each component will be unimodal or monotonic over the course of a pH titration results in further insight. Another important application is in industrial process control where concentrations of compounds or levels of various factors may have a specific evolution over time.

A few of the methods discussed in this chapter, such as 3D PC plots and variable selection, have significant roles in most applications of chemometrics; thus, the interest in the techniques is by no means restricted to chromatographic applications; however, in order to reduce repetition, most methods are introduced in one main context. In reality, there are hundreds of bespoke methods often differing for each type of instrumental technique and problem: a full description would fill up several books. Many are very specialist or advocated by one group for one application but can be very successful and are incorporated into instrumental software. In this chapter, we will only introduce a few, but the general principles of how to approach a problem are universal. Some of the most widespread techniques are available as public domain software and as such are well publicised, though this does not always mean they are the most appropriate, simply their advocates have been best in making user-friendly software available usually at no cost.

7.2   Exploratory Data Analysis and Pre-processing

7.2.1   Baseline Correction

A first step before applying most methods in this chapter is often baseline correction. The reason for this is that most chemometric approaches look at variation above a baseline, and if this is not done, artefacts can be introduced.

Baseline correction is best performed on each variable (such as mass or wavelength) independently. There are many ways of doing this, but it is first important to identify regions of baseline and peaks, as shown in Figure 7.3, which is of an LC-MS data set. Note that the right-hand side of this tailing peak is not used, only take regions that you are confident in. Then, normally, a function is fitted to the baseline regions. This can simply involve the average or else a linear or polynomial best fit to the baseline regions. Sometimes, identifying baseline regions both before and after a peak cluster is useful; however, if the cluster is quite sharp, this is not essential, it would be hard in the case illustrated. Sometimes, the baseline is identified over the entire chromatogram (or spectrum), in other cases separately for each pack cluster or each region or window. After that, obtain a simple mathematical model that is usually multi-linear and then subtract the baseline from the entire region of interest, separately for each non-sequential variable such as wavelength. In the examples in this chapter, it is assumed that either there are no baseline problems or correction has already been performed.

1.   The baseline is usually fitted to a multi-linear model to regions where there are suspected to be no peaks.
    (a) True
    (b) False

7.2.2   Principal Component-Based Plots

These are very useful for exploratory visualisation of data; for example, how many significant components are there, and which regions of the sequential data (e.g. elution time or pH) are most characteristic? However, it is important to

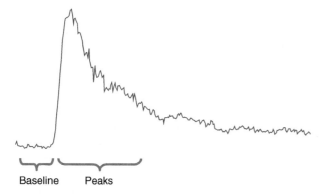

Baseline   Peaks

Figure 7.3 Dividing data into regions before baseline correction.

Table 7.1 Data set A.

	A	B	C	D	E	F	G	H	I	J	K	L
1	0.1102	−0.0694	−0.0886	0.0622	−0.0079	−0.0336	0.0518	−0.0459	−0.032	0.0645	0.0174	−0.0558
2	−0.0487	0.0001	0.0507	−0.0014	0.072	−0.0377	0.0123	0.1377	−0.0034	−0.0015	0.0355	0.0608
3	0.036	0.0277	0.1005	−0.0009	0.0386	0.0528	−0.0612	−0.0259	0.0293	−0.0246	0.0283	0.048
4	0.2104	0.1564	0.1828	0.1073	0.0185	0.1912	0.0499	−0.0587	0.0669	0.1275	0.1371	0.1521
5	0.1713	0.3206	0.4304	0.3531	0.2383	0.1575	−0.0015	0.1367	0.1024	0.1143	0.2227	0.2164
6	0.497	0.6192	0.7367	0.7042	0.3234	0.293	0.1919	0.1325	0.341	0.4269	0.4225	0.2212
7	0.6753	1.1198	1.3239	1.0167	0.6054	0.3783	0.3703	0.3343	0.529	0.5496	0.5986	0.5579
8	1.0412	1.5129	1.6344	1.3823	1.0843	0.6825	0.3584	0.334	0.5212	0.8334	0.9435	0.8741
9	1.0946	1.5543	1.9253	1.5951	1.1767	0.7215	0.5764	0.5695	0.7138	0.8645	0.9545	0.9038
10	0.9955	1.4794	1.5299	1.5679	1.1986	0.793	0.7043	0.5333	0.7661	0.9224	0.9744	0.8434
11	0.672	1.1315	1.2793	1.254	1.0619	0.9552	0.907	0.7855	0.78	0.7912	0.8432	0.6739
12	0.469	0.7531	0.8139	1.0496	1.094	1.1321	1.1164	1.0237	0.7796	0.6313	0.6549	0.5869
13	0.3113	0.3894	0.5844	0.7349	0.9656	1.2339	1.3362	1.2283	0.959	0.5641	0.4393	0.4386
14	0.0891	0.2121	0.3344	0.5837	0.9758	1.3175	1.3713	1.2238	0.9459	0.6646	0.4327	0.3938
15	0.0567	0.1408	0.169	0.4609	0.7807	1.1592	1.3094	1.1237	0.7724	0.4457	0.3217	0.3639
16	0.0391	−0.0211	0.1684	0.3332	0.5427	0.8509	0.9616	0.7876	0.5951	0.3343	0.2212	0.2178
17	0.0895	−0.0086	0.079	0.1721	0.2747	0.4634	0.582	0.5677	0.3231	0.1546	0.0379	0.1021
18	0.007	−0.024	0.0842	0.1622	0.1922	0.2974	0.3571	0.2925	0.1289	0.0491	0.0518	0.0979
19	0.0146	−0.0567	−0.0672	−0.0239	−0.0113	0.2454	0.1721	0.1047	0.1577	0.0129	0.0458	0.0307
20	0.0012	−0.0043	−0.0362	−0.0564	0.0693	0.0468	0.0213	0.1182	0.0152	−0.0342	−0.014	0.0308
21	−0.0937	0.0324	0.0371	−0.0405	−0.0648	−0.0053	0.0218	0.0975	−0.0222	−0.0138	−0.0065	0.0017
22	−0.0031	0.0127	0.0323	−0.0533	0.067	0.0716	0.0479	−0.0383	0.0038	−0.0186	−0.0026	−0.0653
23	−0.0387	−0.0041	0.0175	0.0052	0.0199	−0.0507	0.0263	0.0342	0.0072	0.0242	0.0579	−0.0072
24	−0.0449	0.0076	−0.0191	0.0046	0.0572	0.0946	−0.0018	0.0182	−0.0368	−0.0236	0.0619	0.0853
25	−0.0986	0.0244	0.0185	0.0395	−0.0291	0.0236	−0.0137	−0.0263	0.0156	0.003	0.0237	0.027

understand that the appearance of such graphs can change dramatically according to how the data are transformed or pre-processed, and in this section, we will look at how such scaling can influence these graphs.

Scores and loadings plots have been introduced in Section 4.5. In this chapter, we will explore some further properties especially useful where one or both of the variables are related in sequence. Table 7.1 represents a two-way data set, corresponding to a HPLC-DAD, each elution time is represented by a row and each measurement (such as successive wavelengths) by a column, giving a 25 × 12 data matrix, which will be called *data set A*. The data represent two partially overlapping chromatographic peaks. The profile (sum of intensity over all 12 spectral wavelengths at each elution time) is presented in Figure 7.4.

The simplest plots are the scores and loadings of the raw data, see Figure 7.5. These would suggest that there are two components, with a region of overlap between times 9 and 14, with wavelengths H and G most strongly associated with the slowest eluting compound and wavelengths A, B, C, L and K with the fastest eluting compound. For further discussion of the interpretation of these types of graph, see Section 4.5.

The data set of Table 7.2 is of the same size but represents three partially overlapping peaks. The profile (Figure 7.6) appears to be slightly more complex than that for data set A, and the PC scores plot presented in Figure 7.7 definitely appears to contain more features. Each turning point represents a pure compound; thus, it appears that there are three compounds, centred at times 9, 13 and 17. In addition, the spectral characteristics of the compounds centred at times 9 and 17 are probably quite similar compared with that centred at time 13. Comparing the loadings plot suggests that wavelengths F, G and H are strongly associated with the middle eluting compound, whereas A, B, J, K and L with the other two compounds. There is some distinction, in that wavelengths A, K and L appear most associated with the slowest eluting compound (centred at time 17) and B and J with the fastest. The loadings and scores could also be combined into a biplot (Section 4.7.1).

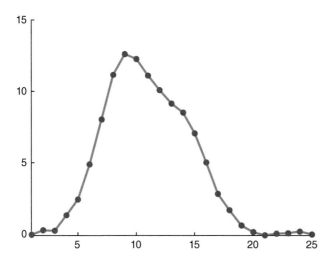

Figure 7.4 Profile of data from data set A.

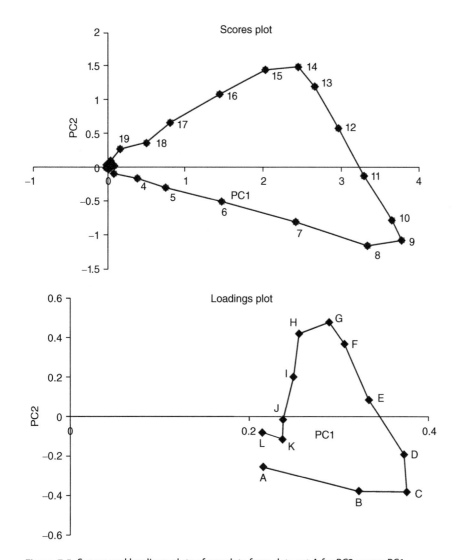

Figure 7.5 Scores and loadings plots of raw data from data set A for PC2 versus PC1.

Table 7.2 Data set B.

	A	B	C	D	E	F	G	H	I	J	K	L
1	−0.1214	0.0097	−0.0059	0.0136	0.0399	0.0404	−0.0530	−0.0066	0.0078	−0.0257	0.0641	0.0784
2	0.0750	−0.0200	0.0183	−0.0251	−0.1072	0.0218	0.0074	0.0876	−0.0341	−0.0295	−0.0715	−0.0746
3	−0.0256	0.1103	−0.0246	−0.0229	−0.0347	0.0102	0.0365	−0.1114	−0.0079	−0.0271	−0.0654	−0.0437
4	0.0838	0.0486	−0.0155	−0.0142	0.0045	−0.0213	0.0396	−0.0499	0.0421	0.0396	0.1128	0.0614
5	0.1956	0.2059	0.1601	0.1567	0.1594	0.0886	0.0310	0.0657	0.0898	0.0472	0.1688	0.1335
6	0.4605	0.5753	0.5696	0.3477	0.1596	0.0725	0.1147	0.0382	0.1049	0.3195	0.3498	0.2801
7	0.9441	1.1101	0.9926	0.6515	0.3413	0.2480	0.1340	0.1802	0.3163	0.4346	0.6187	0.7223
8	1.3161	1.6053	1.3641	0.9179	0.5933	0.3124	0.2991	0.3650	0.6141	0.8600	0.8351	0.8289
9	1.5698	1.8485	1.5372	1.1441	0.6591	0.3880	0.4163	0.4519	0.7866	1.0155	1.0880	0.8584
10	1.3576	1.6975	1.5099	1.1872	0.7743	0.6088	0.6598	0.6486	0.7726	1.0264	0.9573	0.8324
11	1.0215	1.1341	1.0579	0.9695	0.8587	0.9031	0.9594	0.9789	0.9984	0.9810	0.7305	0.8146
12	0.5267	0.6154	0.7074	0.8352	0.9716	1.2316	1.3408	1.1761	0.9874	0.7335	0.6364	0.5917
13	0.3936	0.3650	0.5143	0.6098	1.0398	1.2065	1.3799	1.3346	0.9871	0.7738	0.7354	0.6865
14	0.4351	0.3077	0.3630	0.5386	0.8882	1.1305	1.2870	1.2373	0.9289	0.8656	0.8967	0.9762
15	0.7120	0.4754	0.2429	0.3367	0.5563	0.7994	0.9244	0.8403	0.9415	0.9667	1.2267	1.2778
16	1.0076	0.5493	0.3146	0.2574	0.2568	0.4061	0.5775	0.6494	0.8331	1.1937	1.5703	1.7106
17	1.2155	0.5669	0.3203	0.1255	0.1279	0.1431	0.2725	0.3421	0.6828	1.1376	1.6209	1.8048
18	1.1392	0.4750	0.1603	0.0936	0.1297	0.0917	0.1154	0.2421	0.5582	0.9945	1.4607	1.6280
19	0.6988	0.4000	0.1668	0.0071	−0.0290	0.0012	0.0062	0.1631	0.3016	0.6982	1.0083	1.1998
20	0.3291	0.1766	0.0825	0.0714	−0.0180	−0.0078	−0.0121	0.0176	0.2330	0.2982	0.4253	0.6602
21	0.2183	0.1892	0.0436	−0.0689	0.0239	0.0410	0.0367	0.0322	0.1284	0.2204	0.2760	0.1539
22	0.1135	0.0517	0.0418	−0.0307	0.0017	0.0023	0.1068	−0.0517	0.0411	0.0370	0.0782	0.0798
23	−0.0442	0.0156	0.0520	−0.0867	0.0020	−0.1177	0.0374	−0.0282	−0.0036	0.0217	−0.0521	0.0518
24	−0.0013	−0.1103	−0.0536	−0.0875	−0.0212	−0.0066	0.0199	−0.0258	−0.0119	−0.1066	0.0664	−0.0489
25	0.0697	0.0827	0.0093	0.0298	−0.0511	0.0637	−0.1094	0.0358	−0.0279	0.0319	0.0480	−0.0383

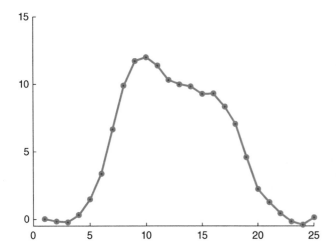

Figure 7.6 Profile of data set B.

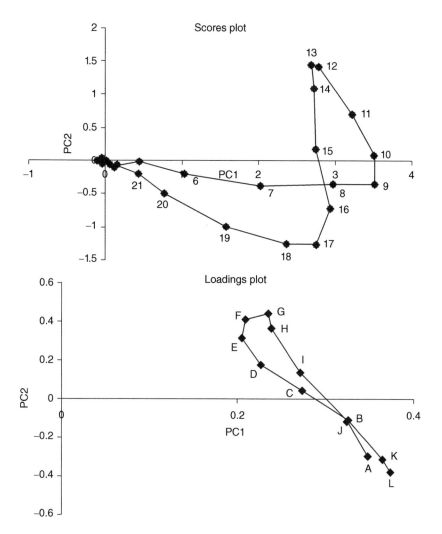

Figure 7.7 Scores and loadings plots of data set B from Table 7.2 for PC2 versus PC1.

It is sometimes clearer to present these graphs in three dimensions as shown in Figure 7.8 and Figure 7.9. Notice that the three-dimensional scores plot for data set A is not particularly informative and the two-dimensional plot shows the main trends more clearly. The reason for this is that there are only two main components in the system; hence, the third dimension consists primarily of noise and thus degrades the information. If the three dimensions were scaled according to the size of the PCs (or their eigenvalues), the graphs in Figure 7.8 would be quite flat. However, for data set B, the directions are much clearer than in the two-dimensional projections; thus, adding an extra PC can be beneficial if there are more than two significant components.

1. There is always an advantage in increasing the number of PCs from 2 to 3 in a scores plot of a hyphenated chromatogram.

   (a) True
   (b) False

7.2.3   Scaling the Data after PCA

Occasionally it is useful to scale the data after PCA, if circumstances are appropriate.

A useful trick is to normalise the scores of each PC. This involves calculating

$$^{norm}t_{ia} = \frac{t_{ia}}{\sqrt{\sum_{a=1}^{A} t_{ia}^2}}$$

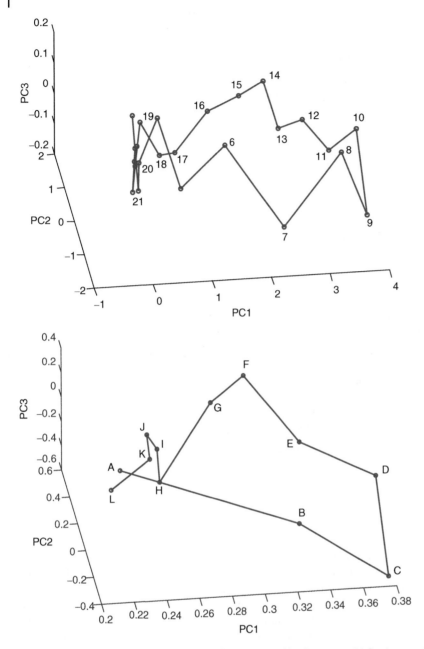

Figure 7.8 Three-dimensional projections of scores (a) and loadings - top (b) for data set A - bottom.

Note that the word normalisation can mean quite different things in other contexts. If only two PCs are used, this will project the scores onto a circle, whereas if three PCs are used, the projection will be onto a sphere. It is best to set *A* according to the suspected number of compounds in the region of the chromatogram being studied (or chemical species in, for example, equilibrium studies).

Figure 7.10 illustrates the scores of data set A normalised over two PCs. Between times 3 and 21, the points in the chromatogram are in sequence on the arc of a circle. The extremes (3 and 21) could represent the purest elution times, but points influenced primarily by noise might lie anywhere on the circle. Hence, time 25, which is clearly not representative of the fastest eluting component, is close to time 3 (this is entirely fortuitous and depends on the noise distribution). As elution times 4–9 are closely clustered, they probably better represent the faster eluting compound. Notice how points on a straight line (Figure 7.5), for example, points 4–8 in the raw scores plot, project onto clusters

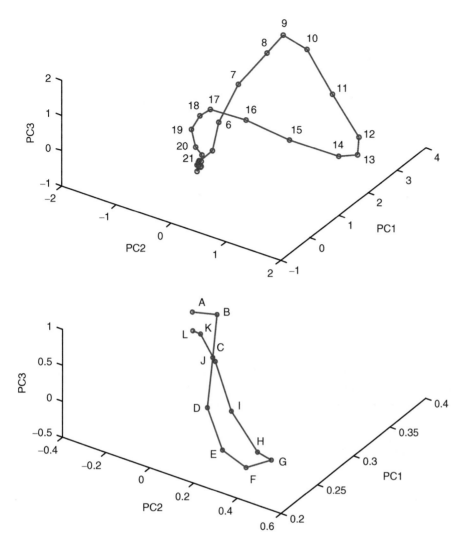

Figure 7.9 Three-dimensional projections of scores (a) and loadings - top (b) for data set B - bottom.

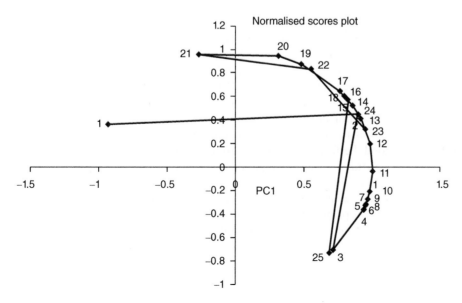

Figure 7.10 Scores plots of data set A with each PC normalised.

in the normalised scores plot. Hence, lines are transformed to close clusters via this procedure. Obviously, if there are, for example, three significant components, it is best to transform into a sphere, or to take a small region for which it is suspected that there are only two significant components. Removing areas of noise can also be of advantage as they could appear almost anywhere.

Similar observations could be made, for example, about studying a pH titration where the different species are observed using UV/vis spectra, pH ranges consisting primarily of one species will cluster in the normalised PC scores plots and is a good way of determining what ranges correspond to just one species.

1. Several sequential points in a 2D PC scores plot fall approximately on a straight line from the origin. The PC scores are then normalised.
    (a) These points will fall on an extended arc on the normalised plot in sequence.
    (b) These points will fall on a straight line on the normalised plot.
    (c) These points will cluster closely apart from those of low intensity.

2. If scores are normalised over two PCs, the original points will fall on an arc and be in sequence of time (or pH as appropriate).
    (a) True
    (b) False

7.2.4   Scaling the Data before PCA

It is also possible to scale the raw data before performing PCA. Some such techniques have been discussed in Chapter 4, but we will especially describe the application to evolutionary data such as coupled chromatography where we expect the sequential variable to consist of a series of unimodal peaks and will illustrate this by case studies A and B, discussed above, as appropriate. It is important to realise that similar principles apply, for example, in the UV/vis spectroscopy of processes such as equilibria where peaks, corresponding, for example, to the appearance and disappearance of different chemical species also have characteristic profiles along this dimension.

7.2.4.1   Scaling the Rows

Each successive row in a data matrix formed from a coupled chromatogram corresponds to a spectrum taken at a given elution time. One of the simplest methods of scaling involves summing each row to a constant total. Put mathematically

$$^{const}x_{ij} = \frac{x_{ij}}{\sum\limits_{j=1}^{J} x_{ij}}$$

Note that some people also call this *normalisation*, but we will avoid that terminology, to avoid confusion with other uses of the word. The influence on PC scores plots has already been introduced (Section 4.6.2) but will be examined in more detail in this chapter. This method introduces what is called *closure*, and we are really now just looking at how the ratio between absorbance at each wavelength changes with time. Figure 7.11(a) shows what happens if the rows of data set A are first scaled to a constant total and then PCA performed on this data. At first glance, this appears rather discouraging because the noise points have a disproportionate influence. These points contain largely nonsensical data, which are emphasised when scaling each point in time to the same total. An expansion of points 5–19 is slightly more encouraging (Figure 7.11(b)), but still not very good. Performing PCA only on points 5–19 (after scaling the rows as described above), however, provides a very clear picture of what is happening, all the points fall roughly on a straight line, with the purest points at the ends (Figure 7.11(c)). Unlike normalising the scores after PCA (Section 7.2.2), where the data must fall exactly on a geometric figure such as a circle or a sphere (depending on the number of PCs chosen), the straight line is only approximate and depends on primarily two components in the region of the data that has been chosen. It is a valuable technique for looking at mixtures spectroscopically with the position along the straight line relating to the proportion of each compound in the mixture. A little noise will result in small deviations from the straight line.

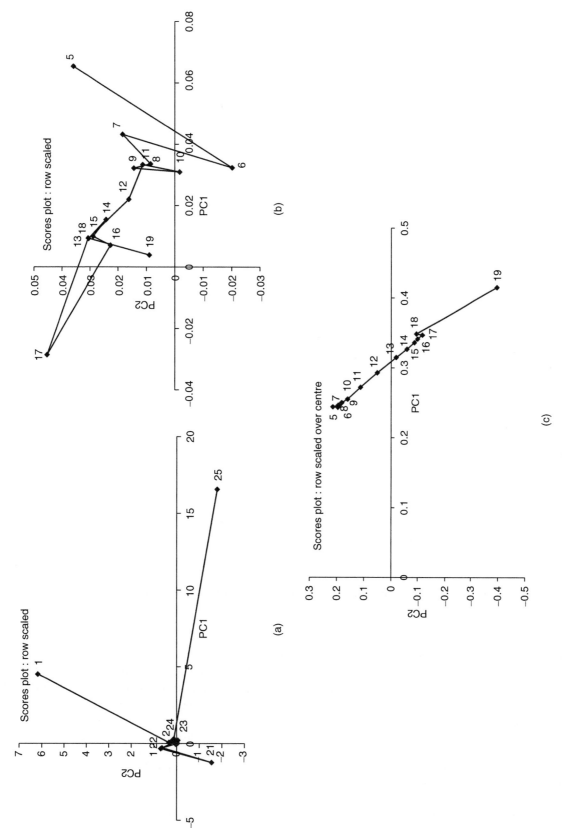

Figure 7.11 Scores plots of data set A, each row summed to a constant total. (a) Entire data set, (b) expansion of region data points 5–19 and (c) performing the scaling and then PCA exclusively over points 5–19.

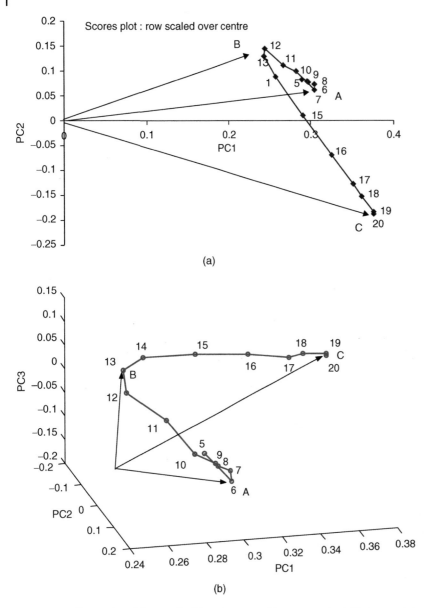

Figure 7.12 Scores plot of data set B with rows summed to a constant total between data points 5 and 20 and three main directions indicated. (a) Two PCs and (b) three PCs.

The corresponding scores plot for the first two PCs of data set B, using the points 5–20, is presented in Figure 7.12(a). There are now two linear regions, one between compounds A (fastest) and B and another between compounds B and C (slowest). Some important features are of interest. The first is that there are now three main directions in the graph, but the direction due to B is unlikely to represent the pure compound, probably the line would need to be extended further along the top right-hand corner. However, it appears likely that there is only a small or negligible region where the three components co-elute, otherwise the graph could not easily be characterised by two straight lines. The trends are clearer in three dimensions (Figure 7.12(b)). Note that the point 5 is primarily influenced.

Summing each row to a constant total is not the only method of dealing with individual rows or spectra. Two variations described below can be employed.

- *Selective summation to constant total.* This allows each portion of a row to be scaled to a constant total; for example, it might be interesting to scale the wavelengths 300–400, 400–500 and 500–600 nm each to 1. Or perhaps the wavelengths 300–400 nm are more diagnostic than the others; hence, why not scale these to a total of 5 and the others to a total of 1? Sometimes, more than one type of measurement can be used to study an evolutionary process, such as UV/vis and MS, each data block could be scaled to a constant total. When doing selective summation, it is important to very carefully consider the consequences of pre-processing.

- *Scaling to a base peak*. In some forms of measurement, such as mass spectrometry (e.g. LC-MS or GC-MS), it is possible to select a base peak and scale to this; for example, if the aim is to analyse the LC-MS of two isomers, rationing to the molecular ion can be performed, so that

$$^{scaled}x_{ij} = \frac{x_{ij}}{x_{i(\text{molecular ion})}}$$

In certain cases, the molecular ion can then be discarded. This method of pre-processing can be used to investigate how the ratio of fragment ions varies across a cluster.

1. A series of spectra correspond primarily to a mixture of two components.

   (a) If the PCs are normalised after PCA, those spectra primarily corresponding to pure compounds will fall on the extreme ends of a straight line in the scores plot.
   (b) If the data are row scaled before PCA, those spectra primarily corresponding to pure compounds will fall on the extreme ends of a straight line in the scores plot.

7.2.4.2 Scaling the Columns

In many cases, it is useful to transform the columns, for example, each wavelength or mass number or spectral frequency. This can be used to put all the variables on a similar scale.

Mean centring, involving subtracting the mean of each column, is the simplest method. Many PC packages do this automatically, but may be inappropriate in the case of signal analysis because the interest is about variability above the baseline rather that around an average.

Standardisation is a common technique that has already been discussed (Section 4.6.4) and is sometimes called *auto-scaling*. It can be mathematically described by

$$^{stand}x_{ij} = \frac{x_{ij} - \bar{x}_j}{\sqrt{\sum_{i=1}^{I}(x_{ij} - \bar{x}_j)^2/I}}$$

where there are I points in time (or length of a column), and $\bar{x}_j$ is the average of variable j. Notice that it is conventional to divide by I rather than $I-1$ in this application, if doing the calculations using many statistical packages, use the 'population' rather than 'sample' standard deviation. Matlab users should be careful when performing this scaling to ensure that the population standardisation is performed. This can be quite useful, for example, in mass spectrometry where the variation of an intense peak (such as a molecular ion of isomers) is no more significant than that of a much less intense peak, such as a significant fragment ion. However, standardisation will also emphasise variables that are pure noise, and if there are, for example, 200 mass numbers of which 180 correspond to noise, this could substantially degrade the analysis.

The most dramatic change is normally to the loadings plot. Figure 7.13 illustrates this for data set B. The scores plot hardly changes in appearance, although the centre is now shifted around the origin compared with Figure 7.7. The loadings plot, however, has changed considerably in appearance and is much clearer and more spread out.

Standardisation is most useful if the magnitudes of the variables are very different, as might occur in LC-MS. Table 7.3 is of data set C, which consists of 25 points in time and 8 variables, making a 25 × 8 matrix. As can be seen, the magnitude of the variables is quite different, with variable H having a maximum of 100, but others being much smaller. We assume that the variables are not in a particular sequence, or are not best represented sequentially; hence, the loadings graphs will consist of a series of points that are not joined up. Figure 7.14 is of the raw profile together with scores and loadings plots. The scores plot suggests that there are two components in the mixture, but the loadings are not very well distinguished and are dominated by variable H. Standardisation (Figure 7.15) largely retains the pattern in the scores plot, but the loadings change radically in appearance, and in this case fall approximately on a circle because there are two main components in the mixture. The variables most corresponding to each pure component fall at the

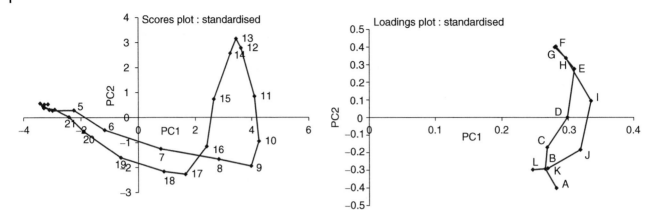

Figure 7.13 Scores and loadings after data set B has been standardised.

Table 7.3 Data set C.

	A	B	C	D	E	F	G	H
1	0.407	0.149	0.121	0.552	−0.464	0.970	0.389	−0.629
2	0.093	−0.062	0.084	−0.015	−0.049	0.178	0.478	1.073
3	0.044	0.809	0.874	0.138	0.529	−1.180	0.040	1.454
4	−0.073	0.307	−0.205	0.518	1.314	2.053	0.658	7.371
5	1.461	1.359	−0.272	1.087	2.801	0.321	0.080	20.763
6	1.591	4.580	0.207	2.381	5.736	3.334	2.155	41.393
7	4.058	7.030	0.280	2.016	9.001	4.651	3.663	67.949
8	4.082	8.492	0.304	4.180	11.916	5.705	4.360	92.152
9	5.839	10.469	0.529	3.764	12.184	6.808	3.739	105.228
10	5.688	10.525	1.573	5.193	12.100	5.720	5.621	106.111
11	3.883	10.111	2.936	4.802	10.026	5.292	7.061	99.404
12	3.630	9.139	2.356	4.739	9.257	4.478	7.530	92.409
13	2.279	8.052	3.196	3.777	9.926	3.228	10.012	92.727
14	2.206	7.952	4.229	5.118	8.629	1.869	9.403	86.828
15	1.403	5.906	2.867	4.229	7.804	1.234	8.774	73.230
16	1.380	5.523	1.720	2.529	4.845	2.249	6.621	52.831
17	0.991	2.820	0.825	1.986	2.790	1.229	3.571	31.438
18	0.160	0.993	0.715	0.591	1.594	0.880	1.662	15.701
19	0.562	−0.018	−0.348	−0.290	0.567	0.070	1.257	6.528
20	0.590	−0.308	−0.715	0.490	0.384	0.595	0.409	2.657
21	0.309	0.371	−0.394	0.077	−0.517	0.434	−0.250	0.551
22	−0.132	−0.081	−0.861	−0.279	−0.622	−0.640	1.166	0.079
23	0.371	0.342	−0.226	0.374	−0.284	0.177	−0.751	−0.197
24	−0.215	−0.577	−0.297	0.834	0.720	−0.248	0.470	−1.053
25	−0.051	0.608	−0.070	−0.087	−0.068	−0.537	−0.208	0.601

ends of the circle. It is important to recognise that this pattern is an approximation and will only happen if there are two main components, otherwise the loadings will fall onto the surface of a sphere (if three PCs are employed and there are three compounds in the mixture) and so on. However, standardisation can have a remarkable influence on the appearance of loadings plots.

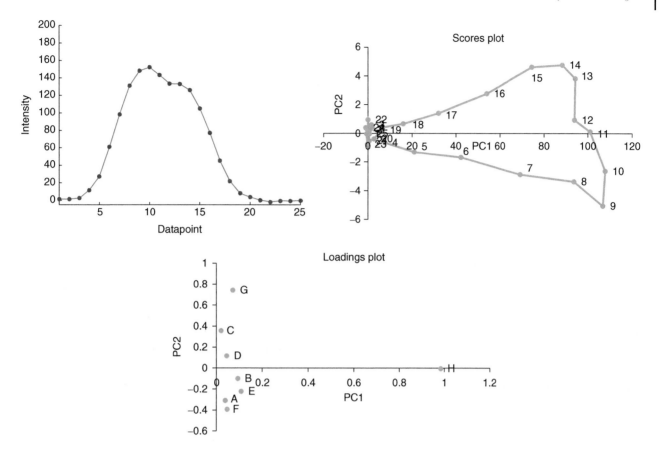

Figure 7.14 Intensity profile and unscaled scores and loadings from data set C in Table 7.3.

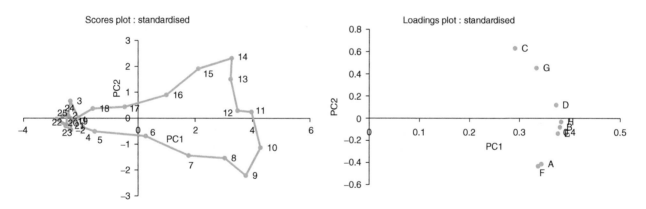

Figure 7.15 Scores and loadings after the data set C in Table 7.3 has been standardised.

Sometimes, weighting by the standard deviation can be performed without centring so that

$$^{scaled}x_{ij} = \frac{x_{ij}}{\sqrt{\sum_{i=1}^{I}(x_{ij} - \bar{x}_j)^2/I}}$$

It is, of course, possible to use any weighting criterion for the columns, so that

$$^{scaled}x_{ij} = {}^jw x_{ij}$$

Table 7.4 Method for ranking variables using data set C.

Time	A	B	C	D	E	F	G	H
(a) Data between times 4 and 19, each row summed to a total of 1								
4	−0.006	0.026	−0.017	0.043	0.110	0.172	0.055	0.617
5	0.053	0.049	−0.010	0.039	0.101	0.012	0.003	0.752
6	0.026	0.075	0.003	0.039	0.093	0.054	0.035	0.674
7	0.041	0.071	0.003	0.020	0.091	0.047	0.037	0.689
8	0.031	0.065	0.002	0.032	0.091	0.043	0.033	0.702
9	0.039	0.070	0.004	0.025	0.082	0.046	0.025	0.708
10	0.037	0.069	0.010	0.034	0.079	0.038	0.037	0.696
11	0.027	0.070	0.020	0.033	0.070	0.037	0.049	0.693
12	0.027	0.068	0.018	0.035	0.069	0.034	0.056	0.692
13	0.017	0.060	0.024	0.028	0.075	0.024	0.075	0.696
14	0.017	0.063	0.034	0.041	0.068	0.015	0.074	0.688
15	0.013	0.056	0.027	0.040	0.074	0.012	0.083	0.694
16	0.018	0.071	0.022	0.033	0.062	0.029	0.085	0.680
17	0.022	0.062	0.018	0.044	0.061	0.027	0.078	0.689
18	0.007	0.045	0.032	0.027	0.071	0.039	0.075	0.704
19	0.067	−0.002	−0.042	−0.035	0.068	0.008	0.151	0.784
(b) Ranked data over these times (1 = least, 16 = most)								
4	1	2	2	15	16	16	8	1
5	15	4	3	12	15	2	1	15
6	8	16	6	11	14	15	4	2
7	14	15	5	2	13	14	6	6
8	11	9	4	6	12	12	3	12
9	13	13	7	3	11	13	2	14
10	12	11	8	9	10	10	5	11
11	9	12	11	8	6	9	7	8
12	10	10	9	10	5	8	9	7
13	4	6	13	5	9	5	12	11
14	5	8	16	14	4	4	10	4
15	3	5	14	13	8	3	14	9
16	6	14	12	7	2	7	15	3
17	7	7	10	16	1	6	13	6
18	2	3	15	4	7	11	11	13
19	16	1	1	1	3	1	16	16

where w is a weighting factor. The weights may correspond to noise content or standard deviations or significance of a variable. Quite complex criteria can be employed. In the extreme, if $w = 0$, this becomes a form of variable selection, which will be discussed in Section 7.2.5.

In rare and interesting cases, it is possible to rank the size of the variables along each column. The suitability depends on the type of pre-processing performed first on the rows. However, a common method is to give the most intense reading in any column a value of I and the least intense 1. If the absolute values of each variable are not very meaningful, this procedure is an alternative that takes into account relative intensities. This procedure is exemplified with reference to the data set C, and illustrated in Table 7.4.

- Choose a region where the peaks elute, in this case from time 4 to 19 as suggested by the scores plot.
- Scale the data in this region, so that each row is of a constant total.
- Rank the data in each column, from 1 (low) to 16 (high).

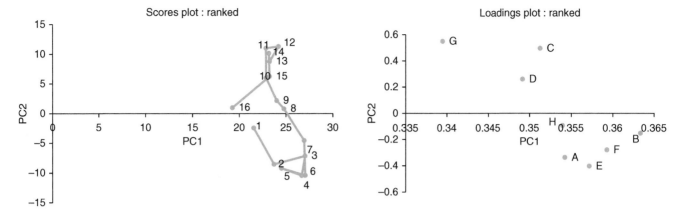

Figure 7.16 Scores and loadings of the ranked data in Table 7.4.

The PC scores and loadings plots are presented in Figure 7.16. Many similar conclusions can be deduced, as shown in Figure 7.15. For example, the loadings arising from variable *C* are close to the slowest eluting peak centred on times 14–16, whereas variables *A* to *F* correspond mainly to the fastest eluting peak. When ranking variables, it is unlikely that the resultant scores and loadings plots will fall onto a smooth geometric figure such as a circle or a line. However, this procedure can be useful for exploratory graphical analysis, especially if the data set is fairly complex with several different compounds and also many measurements on different intensity scale.

It is, of course, possible to scale both the rows and columns simultaneously, first by scaling the rows and then the columns. Note that the reverse (scaling the columns first) is rarely useful and standardisation followed by summing to a constant total has no physical meaning, as standardisation introduces negative numbers, which make row scaling inappropriate.

1. It is desired to row scale to a constant total and standardise the columns.
    (a) Row scaling must be done first.
    (b) Standardisation must be done first.
    (c) It does not matter.

7.2.5 Variable Selection

Variable selection has an important role throughout chemometrics but will be described below primarily in the context of coupled chromatography. This involves keeping only a portion of the original measurements, selecting only those such as wavelengths or masses that are most relevant to the underlying problem. There are a huge number of combinations of approaches limited only by imagination. In this section, we give only a brief summary of some of the main methods. Often, several steps are combined.

Variable selection is particularly important in techniques such as LC-MS or GC-MS. Raw data form is sometimes called a *sparse data matrix*, in which the majority of data points are zero or represent noise. In fact, only a small percentage (perhaps 5% or less) of the measurements are of any interest. The trouble with this is that if multivariate methods are applied to the raw data, often the results are nonsense, dominated by noise. Consider the case of recording the LC-MS of two closely eluting isomers, whose fragment ions are of principal interest. The most intense peak might be the molecular ion; however, in order to study the fragmentation ions, a method such as standardisation described above is required to place equal significance on all the ions. Unfortunately, not only are perhaps 20 or so fragment ions increased in importance, but so are 200 or so ions that represent pure noise, thus the data get worse not better. Typically, out of 200 or 300 masses, there may be around 20 significant ones, and the aim of variable selection is to find these. However, too much variable reduction has the disadvantage that the dimensions of the multivariate matrices are reduced. It is important to find an optimum size as illustrated in Figure 7.17. What tricks can we use to remove irrelevant variables?

Some simple methods, often used as an initial filter of irrelevant variables, are as follows. Note that it is often important to first perform baseline correction (Section 7.2.1).

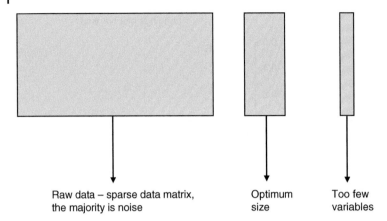

Raw data – sparse data matrix, the majority is noise

Optimum size

Too few variables

Figure 7.17 Optimum size for variable reduction.

- Remove variables outside a given region; for example, in mass spectrometry, these may be at low or high m/z values; in UV/vis spectroscopy, there may be a wavelength range where there is no significant absorbance.
- Sometimes, it is possible to measure the noise content for each variable, simply by looking at the standard deviation of the noise region. The higher the noise, the less significant the mass. This technique is useful in combination with other methods often as a first step.

Many other methods then use simple functions of the data. The first step is to obtain a numerical value corresponding to each variable, and rank them.

- The very simplest is to order the variables according to their mean, for example, $\bar{x}_j$, which is the average of column j, and reject those with low means as primarily representing noise rather than signal. If all the measurements are on approximately the same scale such as in many forms of spectroscopic detection, this is quite a good approach, but is less useful if there remain significant background peaks or if there are dominant high-intensity peaks such as molecular ions that are not necessarily very diagnostic.
- A variant is to employ the variance v_j (or standard deviation). Large peaks that do not vary much and are probably uninteresting may have a small standard deviation and thus can be rejected as they do not change within the region of interest. However, this depends crucially on determining a region of the data where compounds are present, and if noise regions are included, this method will often fail.
- A compromise is to select peaks according to a criterion of variance over mean, $v_j/\bar{x}_j$, and select those with the highest ratio. This may pick some less intense measurements that vary through interesting regions of the data. Intense peaks may still have a big absolute variance, but this might not be particularly significant relative to the average intensity. The problem with this approach, however, is that some measurements that are primarily noise could have a mean close to 0; hence, the ratio becomes large and they will be accidentally selected. To prevent this, first remove the noisy variables by another method and then from the remaining select those with highest relative variance. Variables can have low noise but still be uninteresting if they correspond, for example, to solvent or base peaks.
- A modification of the method described above is to select peaks using a criterion of $v_j/(\bar{x}_j + e)$, where e is an offset that relates to noise level. The advantage of this is that variables with low means are not accidentally selected. Of course, the value of e must be carefully chosen.

After ordering the variables according to one of the functions above, it is then important to choose how many to retain or what the cut-off criterion for rejection is. There are no general guidelines as to how many variables should be selected, some people use statistical tests, others cut-off the selection according to what appears sensible or manageable. The optimal method depends on the technique employed and the general features of a particular source of data.

There are yet further approaches, for example, to look at smoothness of variables, correlations between successive points and so on. In some cases after selecting variables, contiguous variables can then be combined into a smaller number of very significant (and less noisy) variables: this could be valuable in the case of LC-NMR or GC-IR where neighbouring variables often correspond to peaks in a spectrum of a single compound in a mixture, but is unlikely be valuable in HPLC-DAD where there are often contiguous regions of a spectrum that correspond to different compounds. Sometimes, features of variables, such as the change in relative intensity over a peak cluster, can also be taken into account; variables diagnostic for an individual compound are quite likely to vary in a smooth and predictable way, whereas those due to noise will vary in a random manner.

For each type of coupled chromatography (and indeed for any technique where chemometric methods are employed), there are often specific methods for variable selection. In some cases, such as LC-MS, this is usually a crucial first step before further analysis, whereas in the case of HPLC-DAD, it is often omitted and less essential.

1. It is intended to select variables according to the ratio of variance over mean. This often does not work well, why.

   (a) Noisy variables may have small means, so the variance over mean may be fortuitously large.
   (b) Variables that are present in high intensity throughout may have low relative variance and be rejected by mistake.

7.3 Determining Composition

After exploring data via PC plots, baseline correction, pre-processing, scaling and variable selection, as required, the next step is to normally look at the composition of different regions. Many chemometric techniques try to identify pure variables that are associated with one specific component in a mixture. In chromatography, these are usually regions in time where a single compound elutes, although measurements such as an m/z value can also be characteristic of a single compound or a peak in IR spectroscopy. Below we will concentrate primarily on methods for determining pure variables in the chromatographic direction, but many can be modified quite easily for spectroscopy. There is an enormous battery of techniques, but below we summarise the main groups of approaches.

7.3.1 Composition

The concept of composition is an important one. There are many alternative ways of expressing the same idea, that of rank being popular also, which derives from matrices: ideally, the rank of a matrix equals the number of independent components or non-zero eigenvectors.

A region of composition 0 contains no detectable compounds, one of composition 1, one detectable compound and so on. Composition 1 regions are also called *selective* or *pure regions*. A complexity arises in that because of noise, a matrix over a region of composition 1 will not necessarily be described by only one PC, and it is important to try to identify how many PCs are significant and correspond to real information. The concept of composition is an approximate idea, and a composition 1 region could better be defined as a region of a chromatogram where only one compound is observable – very small tails of interferences are often not detected once they are less than a few per cent of the main peak, as their intensity is no more than the typical noise level.

There are many cases of varying difficulty. Figure 7.18 illustrates four cases as exemplified by chromatography. Case (a) is the most studied and easiest, in which each peak has a composition 1 or selective region. Although not the hardest

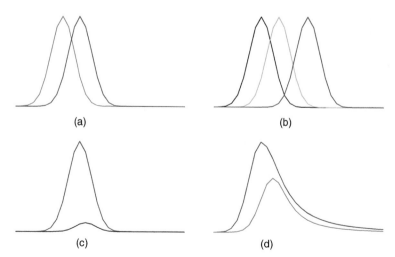

Figure 7.18 Different types of problems in chromatography.

of problems, there is often considerable value in the application of chemometric techniques in such a situation. For example, there may be a requirement for quantification in which the complete peak profile is required, including the area of each peak in the region of overlap. The spectra of the compounds might not be very clear and chemometrics can improve the quality. In complex peak clusters, it might simply be important to identify how many compounds are present, which regions are pure, so what the spectra in the selective regions are and whether it is necessary to improve the chromatography. Finally, this has a potential in the area of automation of pulling out spectra from chromatograms containing several compounds where there is some overlap. Case (b) involves a peak cluster where one or more do not have a selective region. Case (c) is of an embedded impurity peak and is surprisingly common. Many modern separations involve asymmetric peaks such as in case (d) and a lot of conventional chemometric methods fail under such circumstances.

To understand the problem, it is possible to produce a graph of ratios of intensities between the various components. For ideal peak shapes corresponding to the four cases above, these are presented in Figure 7.19. Notice the use of a logarithmic scale, as the ideal ratios will vary over a large range. Case (a) corresponds to two Gaussian peaks (for more information about peak shapes, see Section 3.2.1) and is quite straightforward. In case (b), the ratios of the first to second, and of second to third, peaks are superimposed; it can be seen that the rate of change is different for each pair of peaks; this relates to the different separation of the maxima. Note that there is a huge dynamic range, which is due to noise-free simulations being used. Case (c) is typical of an embedded peak, showing a purity maximum for the smaller component in the centre of its elution. Finally, the ratio arising from case (d) is typical of tailing peaks; many multivariate methods cannot cope easily with this type of data. However, these graphs are of ideal situations, and, in practice, it is only practicable to observe small effects if data are of an appropriate quality. In reality, measurement noise and spectral similarity limit data quality. In practice, it is only realistic to detect two (or more components) if the ratios of intensities of the two peaks are within a certain range, for example no more than 50:1, as indicated by region a in Figure 7.20. Outside these limits, it is unlikely that a second component will be detected. In addition, when the intensity of signal is sufficiently low (say 1% of the maximum, outside region b in Figure 7.20), the signal will be swamped by noise, and thus no signal is detected. Region a would appear to be of composition 2, the area of region b where region a has ended of composition 1 and the chromatogram outside region b of composition 0. If noise levels are higher, these regions become narrower.

Below we indicate a number of approaches for the determination of composition.

1. Peak B is completely embedded within a larger peak A. Over the detectable area including peak A:
    (a) There will be no composition 1 region.
    (b) There will be one contiguous composition 1 region.
    (c) There will be two contiguous composition 1 regions.

7.3.2 Univariate Methods

So far, the simplest are univariate approaches. It is important not to overcomplicate a problem if not justified by the data. Most conventional chromatography software contains methods for estimating ratios between peak intensities at given wavelengths (or mass numbers or any non-sequential marker variable). If two spectra are sufficiently dissimilar, then this method can work well. The two most diagnostic wavelengths can be chosen by a number of means. For the data in Table 7.1, we can look at the loadings plot in Figure 7.5. At first glance, it may appear that C and G are most appropriate to use, but this is not so. The problem is that the most diagnostic wavelengths for one compound may correspond to zero or very low intensity for the other one. This would mean that there will be regions of the chromatogram where one number is close to 0 or even negative (because of noise), leading to very large or even negative ratios. Measurements that are characteristic of both compounds but exhibit distinctly different features in each case are more appropriate. Figure 7.21(a) is of the ratio of intensities of variables D to F. Initially, this plot looks slightly discouraging because there are noise regions where almost any ratio could be obtained. Cutting the region down to times 5–18 (within this range, the intensities of both variables are positive) improves the situation. It is also helpful to consider the best way to display the graph because a ratio of 2:1 is no more significant to a ratio of 1:2, yet using a linear scale, there is an arbitrary asymmetry. To overcome this, either use a logarithmic scale (Figure 7.21(b)) or take the smaller of the ratio of D:F and F:D (Figure 7.21(c)).

The peak ratio plots would suggest that there is a composition 2 region starting between times 9 and 10 and finishing between times 14 and 15. There is some uncertainty about the exact start and end, largely because noise is imposed

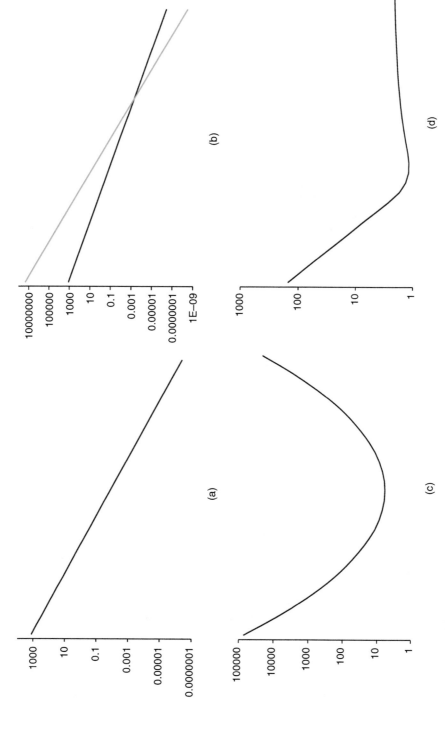

Figure 7.19 Ratios of peak intensities for the case studies (a)–(d) assuming ideal peak shapes and peaks detectable over an indefinite region.

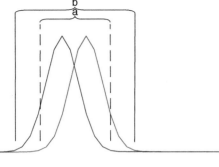

Figure 7.20 Regions of chromatogram (a) in Figure 7.18. Region *a* is where the ratio of the two components is between 50:1 and 1:50 and region *b* where the overall intensity is more than 1% of the maximum intensity.

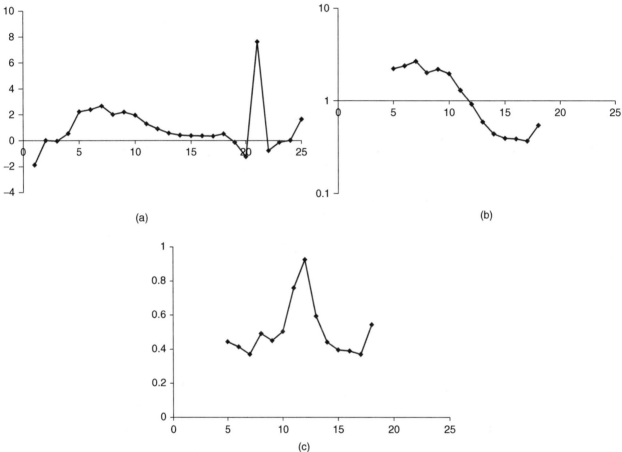

Figure 7.21 Ratio of intensity of measurements D to F for data set A. (a) Raw information, (b) logarithmic scale between points 5 and 18 and (c) the minimum of the ratio of intensity D:F and F:D between points 5 and 18.

upon the data. In some cases, peak ratio plots are very helpful, but they do depend on having adequate signal-to-noise ratios and finding suitable variables. If a spectrum is monitored over 200 wavelengths, this may not be so easy, and multivariate approaches that use all the wavelengths may be more successful. In addition, good diagnostic measurements are required, noise regions have to be eliminated and also the graphs can become quite complicated if there are several compounds in a portion of the chromatogram. An ideal situation would be to calculate several peak ratios simultaneously, but this then suggests that multivariate methods, as described below, have an important role to play.

Another simple trick is to sum the data to constant total at each point in time, as described above, so as to obtain values of

$$^{const}x_{ij} = \frac{x_{ij}}{\displaystyle\sum_{j=1}^{J} x_{ij}}$$

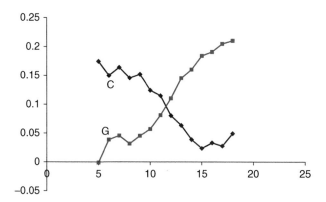

Figure 7.22 Intensities for wavelengths C and G using data of data set A summing the measurements at each successive point to constant total of 1.

Providing that noise regions are discarded, the relative intensities of diagnostic wavelengths should change according to the composition of the data; thus, we plot the value of x_{ij} for a diagnostic value of j against i. Unlike using ratio plots, we are able to choose strongly associated wavelengths such as C and G, as an intensity of 0 (or even a small negative number) will not unduly influence the appearance of the graph, given in Figure 7.22. The regions of composition 1 are somewhat flatter but influenced by noise, where the relative intensity changes most is in the composition 2 region.

This approach is not ideal, but the graphs of Figures 7.21 and 7.22 are intuitively easy for the practicing chromatographer (or spectroscopist) and result in the creation of a form of purity curve. The appearance of the curves can be enhanced by selecting variables that are least noisy and then calculating the relative intensity as a function of time for several (rather than just two) variables. Some will increase and others decrease according to whether the variable is most associated with the fastest or slowest eluting compound. Reversing the curve for one set of variables results in several superimposed purity curves, which can be averaged to give a good picture of changes over the chromatogram.

These methods can be extended to cases of embedded peaks, in which the purest point for the embedded peak does not correspond to a selective region; a weakness of using this method of ratios is that it is not always possible to determine whether a maximum (or minimum) in the purity curve is genuinely a consequence of a composition 1 region or simply the portion of the chromatogram where the concentration of one analyte is most.

Such simple approaches can become rather messy when there are several compounds in a cluster especially if the spectra are similar, but in favourable cases are very effective. Most chemometric experts advocate multivariate approaches, but traditional univariate ratio plots are still useful and often easy to understand.

1. Univariate approaches usually involve first selecting single marker variables for each compound in a mixture.
   (a) True
   (b) False

7.3.3 Correlation- and Similarity-Based Methods

Another set of methods are based on correlation coefficients. The principle is that the correlation coefficient between two successive points in time, defined by

$$r_{i-1,i} = \sum_{j=1}^{J} \frac{(x_{ij} - \overline{x}_i)(x_{i-1j} - \overline{x}_{i-1})}{s_i s_{i-1}}$$

where $\overline{x}_i$ is the mean measurement at time i and s_i the corresponding standard deviation, will have the following characteristics.

- It will be close to 1 in regions of composition 1.
- It will be close to 0 in noise regions.
- It will be substantially below 1 in regions of composition 2.

Table 7.5 Correlation coefficients for data set A between successive points (left-hand column) and between point 15 (right-hand column).

Time	$r_{i,i-1}$	$r_{i,15}$
1		−0.045
2	−0.480	0.227
3	−0.084	−0.515
4	0.579	−0.632
5	0.372	−0.651
6	0.802	−0.728
7	0.927	−0.714
8	0.939	−0.780
9	0.973	−0.696
10	0.968	−0.643
11	0.817	−0.123
12	0.489	0.783
13	0.858	0.974
14	0.976	0.990
15	0.990	1.000
16	0.991	0.991
17	0.968	0.967
18	0.942	0.950
19	0.708	0.809
20	0.472	0.633
21	0.332	0.326
22	−0.123	0.360
23	−0.170	0.072
24	−0.070	0.276
25	0.380	0.015

Table 7.5 is of the correlation coefficients between successive points in time for the data in Table 7.1. This is presented in Figure 7.23 and would suggest that

- points 7–9 are composition 1,
- points 10–13 are composition 2 and
- points 14–18 are composition 1.

Notice that because the correlation coefficient is between two successive points, the three-point dip in Figure 7.23 actually suggests there are four composition two points.

The principles can be further extended to finding the points in time corresponding to the purest regions of the data. This is sometimes useful, for example, to obtain the spectrum of each compound that has a composition 1 region. The highest correlation is between points 15 and 16; hence, one of these is the purest point. The correlation between points 14 and 15 is higher than between points 16 and 17; hence, point 15 is chosen as the elution time best representative of slowest eluting compound.

The next step is to calculate the correlation coefficient $r_{15,i}$ between this selected point in time and all other points. The lowest correlation is likely to belong to the second component. The data are presented in Table 7.5 and Figure 7.24. The trends are now very clear. The most negative correlation occurs at point 8, being the purity maximum for the fastest compound. The composition 2 region is somewhat smaller than estimated in the previous section, but probably more accurate. One reason why these graphs are an improvement on the univariate ones is that they take all the data into account rather than single variables. Where there are a hundred or more spectral frequencies, these approaches can

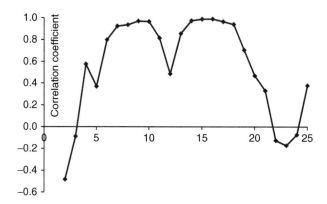

Figure 7.23 Graph of correlation between successive points in the data of data set A.

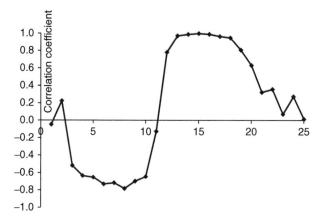

Figure 7.24 Correlation between point 15 and the data of data set A.

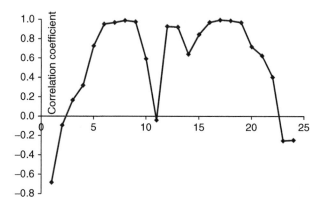

Figure 7.25 Graph corresponding to that of Figure 7.23 for data set B in Table 7.2.

have significant advantages, but it is important to ensure that most of the variables are meaningful. In the case of NMR or MS, 95% or more of the variables may simply arise from noise; thus, careful choice of variables using the methods of Section 7.2.5 is a prior necessity.

This method can be extended to quite complex peak clusters, as presented in Figure 7.25, for the data in Table 7.2. Ignoring the noise at the beginning, it is fairly clear that there are three components in the data as there are three plateaus. Notice that the central component eluting approximately between times 10 and 15 does not have a true composition 1 region because the correlation coefficient only reaches approximately 0.9, whereas the other two compounds have well-established selective areas. It is possible to determine the purest point for each component in the mixture successively by extending the approach illustrated above.

Another related aspect involves using these graphs to select pure variables; this is often useful in spectroscopy, where certain masses or frequencies are most diagnostic of different components in a mixture, finding these helps the later stages of the analysis.

It is not essential to use correlation coefficients as a method for determining similarity and other measures, even of dissimilarity, can be proposed, for example, Euclidean distances between normalised spectra (scaled to unit length rather than summed to 1), the larger the distance, the less the similarity. It is also possible to start with an average or most intense spectrum in the chromatogram and determine how the spectrum at each point in time differs from this. The variations on this theme are endless and most papers or software packages contain some differences reflecting the authors' favourite approaches. When using correlation- or similarity-based methods, it is important to work carefully through the details and not accept the results as a black box. There is insufficient space in this text to itemise each published method, but the general principles should be clear for any user.

1. If two successive points are composition 1, the correlation coefficient of their successive spectra should be

    (a) Close to 0
    (b) Close to 1
    (c) Cannot say

2. A region of a chromatogram consists of a large dominant peak enveloping a smaller embedded peak in the centre. The correlation coefficients are calculated between successive spectra.

    (a) There will be one region with correlation coefficients close to 1.
    (b) There will be two regions with correlation coefficients close to 1.
    (c) There will be three regions with correlation coefficients close to 1.

7.3.4 Eigenvalue-Based Methods

Many chemometricians like as multivariate approaches mostly based on PCA. A large number of methods, such as evolving factor analysis (EFA), fixed sized window factor analysis (WFA) or heuristic evolving latent projections (HELP), among many, are available in the literature, which contain one step that involves calculating eigenvalues to determine the composition of regions of evolutionary two-way data. The principle is that the more the components in a particular region of a chromatogram, the more the number of significant eigenvalues.

There are two fundamental groups of approaches. The first involves performing PCA on an expanding window which we will call EFA.

There are several variants on this theme, but a popular one is indicated below.

Perform uncentred PCA on the first few data points of the series, for example, points 1–4. In the case of Table 7.1.

- This will involve starting with a 4×12 matrix.
- Record the first few eigenvalues of this matrix, which should be more than the number of components expected in the mixture and cannot be more than the smallest dimension of the starting matrix. We will keep four eigenvalues.
- Extend the matrix by an extra point in time to a matrix of points 1–5 in this example and repeat PCA, keeping the same number of eigenvalues as above.
- Continue until the entire data matrix is covered; thus, the final step involves performing PCA on a data matrix of dimensions 25×12 and keeping 4 eigenvalues.
- Produce a table of the eigenvalues against matrix size. In this example, there will be 22 ($=25 - 4 + 1$) rows and 4 columns. This procedure is called *forward expanding factor analysis.*
- Next, take a matrix at the opposite end of the data set, from points 21 to 25, to give another 4×12 matrix, and calculate the eigenvalues of this matrix.
- Expand the matrix backwards; thus, the second calculation is from points 20 to 25, the third from points 19 to 25 and so on.
- Produce a table similar to that for forward expanding factor analysis. This procedure is called *backward expanding factor analysis.*

Table 7.6 Results of forward and backward EFA for the data set A.

Time	Forward				Backwards			
1	n/a	n/a	n/a	n/a	86.886	12.168	0.178	0.154
2	n/a	n/a	n/a	n/a	86.885	12.168	0.167	0.147
3	n/a	n/a	n/a	n/a	86.879	12.166	0.165	0.128
4	0.231	0.062	0.024	0.006	86.873	12.160	0.164	0.128
5	0.843	0.088	0.036	0.012	86.729	12.139	0.161	0.111
6	3.240	0.101	0.054	0.015	86.174	12.057	0.161	0.100
7	9.708	0.118	0.055	0.026	84.045	11.802	0.156	0.067
8	22.078	0.118	0.104	0.047	78.347	11.064	0.107	0.057
9	37.444	0.132	0.106	0.057	67.895	9.094	0.101	0.056
10	51.219	0.182	0.131	0.105	55.350	6.278	0.069	0.056
11	61.501	0.678	0.141	0.105	44.609	3.113	0.069	0.051
12	69.141	2.129	0.141	0.109	35.796	1.123	0.069	0.046
13	75.034	4.700	0.146	0.119	27.559	0.272	0.067	0.044
14	80.223	7.734	0.148	0.119	19.260	0.112	0.050	0.042
15	83.989	10.184	0.148	0.119	11.080	0.057	0.049	0.038
16	85.974	11.453	0.148	0.123	4.878	0.051	0.045	0.038
17	86.611	11.924	0.155	0.127	1.634	0.051	0.039	0.032
18	86.855	12.067	0.156	0.127	0.514	0.050	0.035	0.025
19	86.879	12.150	0.160	0.134	0.155	0.037	0.025	0.025
20	86.881	12.162	0.163	0.139	0.046	0.034	0.025	0.018
21	86.881	12.164	0.176	0.139	0.037	0.030	0.019	0.010
22	86.882	12.166	0.177	0.139	0.032	0.024	0.012	0.009
23	86.882	12.166	0.178	0.139	n/a	n/a	n/a	n/a
24	86.886	12.167	0.178	0.154	n/a	n/a	n/a	n/a
25	86.886	12.168	0.178	0.154	n/a	n/a	n/a	n/a

The results are given in Table 7.6. Notice that the first columns of the forward and backward eigenvalues should be properly aligned. Note also that the relative size of each eigenvalue is not proportional to the relative amount of each compound: as we will see below, this information requires a further step of rotation or transformation. Normally, for this reason, the eigenvalues are plotted on a logarithmic scale. The results for the first three eigenvalues (the fourth is omitted in order not to complicate the graph) are illustrated in Figure 7.26. What can we tell from this?

- Although the third eigenvalue increases slightly, this is largely because of the increase in size of the data matrix and it does not indicate a third component.
- In the forward plot, it is clear that the fastest eluting component has started to become significant in the matrix by time 4, so the elution window starts at time 4.
- In the forward plot, the slowest component starts to become significant by time 10.
- In the backward plot, the slowest eluting component starts to become significant at time 19.
- In the backward plot, the fastest eluting component starts to become significant at time 13.

Hence,

- there are two significant components in the mixture,
- the elution region of the fastest is between times 4 and 13 and
- the elution region of the slowest between times 10 and 19;
- hence, between times 10 and 13, the chromatogram is composition 2, consisting of overlapping elution.

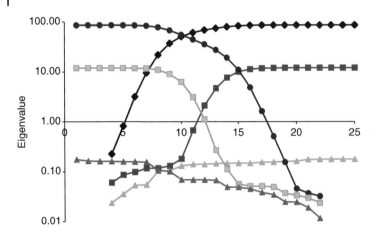

Figure 7.26 Forward and backward EFA plots of the first three eigenvalues from data set A.

We have interpreted the graphs visually, but, of course, some people like to use statistical methods, which can be useful for automation but rely on noise behaving in a specific manner. It is normally better to produce a graph to see that the conclusions are sensible.

The second approach involves using a fixed sized window (FSW) and is called *window factor analysis*. It is implemented as follows.

- Choose a window size, usually a small odd number such as 3 or 5. This window should be at least the maximum composition expected in the chromatogram, preferably one point wider. It does not need to be as large as the number of components expected in the system, only the maximum overlap anticipated. We will use a window size of 3.
- Perform uncentred PCA on the first points of the chromatogram corresponding to the window size, in this case points 1–3, resulting in a matrix of size 3×12.
- Record the first few eigenvalues of this matrix, which should be no more than the highest composition expected in the mixture and cannot be more than the smallest dimension of the starting matrix. We will keep three eigenvalues.
- Move the window successively along the chromatogram, so that the next window will consist of points 2–4, and the final one of points 23–25. In most implementations, the window is not changed in size.
- Produce a table of the eigenvalues against matrix centre. In this example, there will be 23 rows and three columns.

The results for the data in Table 7.1 are given in Table 7.7, with the graph, again presented on a logarithmic axis presented in Figure 7.27 for a window size of 3. What can we tell from this?

- It appears fairly clear that there are no regions where more than two components elute.
- The second eigenvalue appears to become significant between points 10 and 14. However, as a three-point window has been employed, this would suggest that the chromatogram is composition 2 between points 9 and 15. This is a slightly larger region than that found by expanding factor analysis. One problem about using a FSW in this case is that the data set is rather small, each matrix having a size of size 3×12, and so can be sensitive to noise. If more measurements are not available, a solution is to use a larger window size, but then the accuracy in time may be less. However, it is often not possible to predict elution windows to within 1 point in time, and the overall conclusions of the two methods are fairly similar in nature.
- The first eigenvalue mainly reflects the overall intensity of the chromatogram, see Figure 7.27.

The regions of elution for each component can be similarly defined as for EFA described above. There are numerous variations on fixed sized WFA, such as changing the window size across the chromatogram, and the results can change quite dramatically using different forms of data scaling. However, this is quite a simple visual technique that is popular. The region where the second eigenvalue is significant in Figure 7.27 can be compared with the dip in Figure 7.23, the ascent in Figure 7.24, the peak in Figure 7.21(c) and various features of the scores plots. In most cases, similar regions are predicted within a data point.

Eigenvalue-based methods are effective in many cases but may break down for unusual peak shapes. They normally depend on peak shapes being symmetrical and roughly equal widths for each compound in a mixture. The interpretation of eigenvalue plots for tailing peak shapes (Section 3.2.1.3) is quite hard. They also depend on a suitable selection of variables. If, as in the case of raw mass spectral data, the majority of variables are noise or consist mainly

Table 7.7 Fixed sized window factor analysis applied to data set A using a three-point window.

Centre			
1	n/a	n/a	n/a
2	0.063	0.027	0.014
3	0.231	0.036	0.013
4	0.838	0.057	0.012
5	3.228	0.058	0.035
6	9.530	0.054	0.025
7	21.260	0.098	0.045
8	34.248	0.094	0.033
9	41.607	0.133	0.051
10	39.718	0.391	0.046
11	32.917	0.912	0.018
12	27.375	1.028	0.024
13	25.241	0.587	0.028
14	22.788	0.169	0.022
15	17.680	0.049	0.014
16	10.584	0.031	0.013
17	4.770	0.024	0.013
18	1.617	0.047	0.013
19	0.505	0.046	0.018
20	0.147	0.031	0.016
21	0.036	0.027	0.015
22	0.031	0.021	0.009
23	0.030	0.021	0.010
24	0.032	0.014	0.009
25	n/a	n/a	n/a

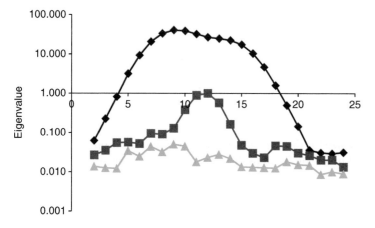

Figure 7.27 Three-point FSW graph for data set A.

of baseline; hence, they will not always give clear answers; however, reducing these will improve the appearance of the eigenvalue plots significantly.

1. The ratio of first to second eigenvalue approximates to the ratio of intensities of the two most significant compounds in a window.
    (a) True
    (b) False

2. A chromatogram consists of 40 points in time, and a FSW of width 3. How many points in time will be represented in the graph of eigenvalues versus time?
    (a) 34
    (b) 36
    (c) 38
    (d) 40

7.3.5 Derivatives

Finally, there are a number of approaches based on calculating derivatives. The principle is that a spectrum will not change significantly in nature during a selective or composition 1 region. Derivatives measure change, hence can exploit this.

There are a large number of approaches to incorporating information about derivatives into methods for determining composition; however, the method below, illustrated with reference to the data set of Table 7.1, is effective.

- Scale the spectra at each point in time, as described in Section 7.2.4.1 so all spectra are a constant total. For pure regions, the scaled spectra should not change in appearance.
- Calculate the first derivative at each wavelength and each point in time. Normally the Savitzky–Golay method, described in Section 3.3.1.2 can be employed. The simplest case is a 5 point quadratic first derivative, see Table 3.6, so that

$$\Delta_{ij} = \frac{dx_{ij}}{di} \approx (-2x_{(i-2)j} - x_{(i-1)j} + x_{(i+1)j} + 2x_{(i+2)j})/10$$

- Note that it is not always appropriate to choose a 5 point window, this depends very much on the nature of the raw data.
- The closer the magnitude of this is to 0, the more likely the point represents a pure region, hence, it is best to use the absolute value of the derivative, $|\Delta_{ij}|$. Sometimes there are points in time at the end of a region that represent noise and will dominate the overall average derivative calculation, these points may be discarded, often this can simply be done by removing points whose average intensity is below a given threshold.
- If the variables are fairly different in magnitude, it is first useful to scale each variable to a similar magnitude, but setting the sum of each column (or variable) to a constant total

$$^{const}\Delta_{ij} = |\Delta_{ij}|/ \sum_{i=w-1}^{I-w+1} |\Delta_{ij}|$$

where | indicates an absolute value, and the window size for the derivatives equals w. This step is optional.
- The final step involves averaging the values obtained in one of the steps discussed above (as appropriate) as follows:

$$d_i = \sum_{j=1}^{J} {}^{const}\Delta_{ij}/J \quad \text{or} \quad d_i = \sum_{j=1}^{J} |\Delta_{ij}|/J$$

The calculation illustrated for the data in Table 7.8(a) is of the data where each row is summed to a constant total. Notice that the first and last rows contain some large numbers because the absolute intensity is low, with several negative as well as positive numbers that are similar in magnitude. Table 7.8(b) is of the absolute value of the first derivatives. Notice the very large and not very meaningful numerical values at times 3 and 23. Row 22 also contains a fairly large number and thus is not very diagnostic. In Table 7.8(c), points 3, 22 and 23 have been discarded, and the

Table 7.8 Derivative calculation for determining purity of regions in data set A.

(a) Scaling the rows to constant total

1	−4.066	2.561	3.269	−2.295	0.292	1.240	−1.911	1.694	1.181	−2.380	−0.642	2.059
2	−0.176	0.000	0.183	−0.005	0.260	−0.136	0.045	0.498	−0.012	−0.005	0.128	0.220
3	0.145	0.111	0.404	−0.004	0.155	0.212	−0.246	−0.104	0.118	−0.099	0.114	0.193
4	0.157	0.117	0.136	0.080	0.014	0.143	0.037	−0.044	0.050	0.095	0.102	0.113
5	0.070	0.130	0.175	0.143	0.097	0.064	−0.001	0.056	0.042	0.046	0.090	0.088
6	0.101	0.126	0.150	0.143	0.066	0.060	0.039	0.027	0.069	0.087	0.086	0.045
7	0.084	0.139	0.164	0.126	0.075	0.047	0.046	0.041	0.066	0.068	0.074	0.069
8	0.093	0.135	0.146	0.123	0.097	0.061	0.032	0.030	0.047	0.074	0.084	0.078
9	0.087	0.123	0.152	0.126	0.093	0.057	0.046	0.045	0.056	0.068	0.075	0.071
10	0.081	0.120	0.124	0.127	0.097	0.064	0.057	0.043	0.062	0.075	0.079	0.069
11	0.060	0.102	0.115	0.113	0.095	0.086	0.081	0.071	0.070	0.071	0.076	0.061
12	0.046	0.075	0.081	0.104	0.108	0.112	0.110	0.101	0.077	0.062	0.065	0.058
13	0.034	0.042	0.064	0.080	0.105	0.134	0.145	0.134	0.104	0.061	0.048	0.048
14	0.010	0.025	0.039	0.068	0.114	0.154	0.160	0.143	0.111	0.078	0.051	0.046
15	0.008	0.020	0.024	0.065	0.110	0.163	0.184	0.158	0.109	0.063	0.045	0.051
16	0.008	−0.004	0.033	0.066	0.108	0.169	0.191	0.157	0.118	0.066	0.044	0.043
17	0.032	−0.003	0.028	0.061	0.097	0.163	0.205	0.200	0.114	0.054	0.013	0.036
18	0.004	−0.014	0.050	0.096	0.113	0.175	0.211	0.172	0.076	0.029	0.031	0.058
19	0.023	−0.091	−0.108	−0.038	−0.018	0.393	0.275	0.168	0.252	0.021	0.073	0.049
20	0.008	−0.027	−0.230	−0.358	0.439	0.297	0.135	0.750	0.096	−0.217	−0.089	0.195
21	1.664	−0.575	−0.659	0.719	1.151	0.094	−0.387	−1.732	0.394	0.245	0.115	−0.030
22	−0.057	0.235	0.597	−0.985	1.238	1.323	0.885	−0.708	0.070	−0.344	−0.048	−1.207
23	−0.422	−0.045	0.191	0.057	0.217	−0.553	0.287	0.373	0.079	0.264	0.631	−0.079
24	−0.221	0.037	−0.094	0.023	0.281	0.466	−0.009	0.090	−0.181	−0.116	0.305	0.420
25	−12.974	3.211	2.434	5.197	−3.829	3.105	−1.803	−3.461	2.053	0.395	3.118	3.553

(b) Absolute value of first derivative using a five-point Savitsky–Golay quadratic smoothing function

1												
2												
3	0.8605	0.4745	0.6236	0.4962	0.0636	0.2073	0.3814	0.3818	0.2216	0.4953	0.1439	0.4049
4	0.0480	0.0270	0.0296	0.0444	0.0448	0.0244	0.0235	0.0783	0.0087	0.0330	0.0108	0.0455
5	0.0178	0.0065	0.0466	0.0323	0.0108	0.0414	0.0586	0.0362	0.0085	0.0326	0.0095	0.0316
6	0.0114	0.0046	0.0009	0.0070	0.0144	0.0180	0.0036	0.0133	0.0017	0.0020	0.0052	0.0089
7	0.0026	0.0006	0.0049	0.0055	0.0023	0.0013	0.0085	0.0018	0.0007	0.0031	0.0032	0.0000
8	0.0038	0.0028	0.0064	0.0032	0.0081	0.0020	0.0036	0.0036	0.0024	0.0024	0.0013	0.0049
9	0.0059	0.0090	0.0120	0.0023	0.0041	0.0081	0.0096	0.0072	0.0025	0.0006	0.0002	0.0027
10	0.0119	0.0142	0.0168	0.0053	0.0025	0.0131	0.0193	0.0169	0.0075	0.0021	0.0039	0.0051
11	0.0140	0.0207	0.0221	0.0116	0.0035	0.0202	0.0253	0.0235	0.0111	0.0026	0.0070	0.0058
12	0.0167	0.0250	0.0222	0.0151	0.0043	0.0228	0.0271	0.0263	0.0131	0.0004	0.0085	0.0058
13	0.0141	0.0213	0.0224	0.0131	0.0035	0.0197	0.0256	0.0217	0.0111	0.0001	0.0075	0.0031
14	0.0103	0.0180	0.0134	0.0090	0.0004	0.0143	0.0200	0.0135	0.0087	0.0009	0.0044	0.0026
15	0.0007	0.0120	0.0077	0.0041	0.0023	0.0073	0.0150	0.0146	0.0027	0.0025	0.0076	0.0026
16	0.0011	0.0101	0.0025	0.0050	0.0015	0.0042	0.0121	0.0100	0.0064	0.0106	0.0072	0.0008
17	0.0027	0.0231	0.0247	0.0177	0.0251	0.0465	0.0202	0.0035	0.0245	0.0122	0.0043	0.0010
18	0.0008	0.0134	0.0661	0.0947	0.0548	0.0485	0.0042	0.1153	0.0095	0.0600	0.0206	0.0317
19	0.3269	0.1158	0.1653	0.0864	0.2434	0.0017	0.1260	0.3287	0.0581	0.0135	0.0085	0.0005
20	0.1518	0.0013	0.0543	0.1404	0.3419	0.1998	0.0687	0.3660	0.0130	0.0521	0.0115	0.2609
21	0.0956	0.0354	0.1423	0.0438	0.1269	0.0865	0.0773	0.1047	0.0374	0.0360	0.1157	0.1658
22	0.2543	0.0660	0.1121	0.0098	0.1250	0.0309	0.0386	0.0785	0.0871	0.0220	0.1303	0.0401
23	2.9440	0.7375	0.5495	0.9964	1.0917	0.5164	0.3725	0.2660	0.3065	0.0527	0.6359	0.8792
24												
25												

(Continued)

Table 7.8 (Continued)

(c) Rejecting points 3, 22 and 23 and putting the measurements on a common scale

1												
2												
3												
4	0.065	0.075	0.045	0.082	0.050	0.042	0.043	0.066	0.038	0.124	0.046	0.079
5	0.024	0.018	0.071	0.060	0.012	0.071	0.107	0.031	0.037	0.122	0.040	0.055
6	0.015	0.013	0.001	0.013	0.016	0.031	0.007	0.011	0.008	0.007	0.022	0.015
7	0.003	0.002	0.007	0.010	0.003	0.002	0.016	0.002	0.003	0.012	0.013	0.000
8	0.005	0.008	0.010	0.006	0.009	0.003	0.007	0.003	0.010	0.009	0.005	0.008
9	0.008	0.025	0.018	0.004	0.005	0.014	0.018	0.006	0.011	0.002	0.001	0.005
10	0.016	0.039	0.025	0.010	0.003	0.023	0.035	0.014	0.033	0.008	0.016	0.009
11	0.019	0.057	0.033	0.021	0.004	0.035	0.046	0.020	0.049	0.010	0.029	0.010
12	0.023	0.069	0.034	0.028	0.005	0.039	0.049	0.022	0.058	0.001	0.036	0.010
13	0.019	0.059	0.034	0.024	0.004	0.034	0.047	0.018	0.049	0.001	0.032	0.005
14	0.014	0.050	0.020	0.017	0.000	0.025	0.037	0.011	0.038	0.003	0.019	0.005
15	0.001	0.033	0.012	0.008	0.003	0.013	0.027	0.012	0.012	0.009	0.032	0.005
16	0.001	0.028	0.004	0.009	0.002	0.007	0.022	0.008	0.028	0.040	0.030	0.001
17	0.004	0.064	0.037	0.033	0.028	0.080	0.037	0.003	0.108	0.046	0.018	0.002
18	0.001	0.037	0.100	0.175	0.061	0.084	0.008	0.097	0.042	0.225	0.087	0.055
19	0.444	0.321	0.250	0.160	0.272	0.003	0.230	0.277	0.256	0.051	0.036	0.001
20	0.206	0.004	0.082	0.260	0.382	0.345	0.125	0.309	0.057	0.195	0.049	0.450
21	0.130	0.098	0.216	0.081	0.142	0.149	0.141	0.088	0.164	0.135	0.489	0.286
22												
23												
24												
25												

(d) Calculating the final consensus derivative

I	d_i
1	
2	
3	
4	0.063
5	0.054
6	0.013
7	0.006
8	0.007
9	0.010
10	0.019
11	0.028
12	0.031
13	0.027
14	0.020
15	0.014
16	0.015
17	0.038
18	0.081
19	0.192
20	0.205
21	0.177
22	
23	
24	
25	

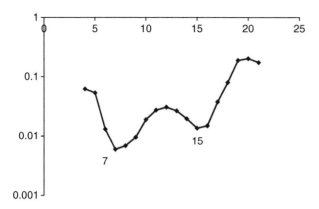

Figure 7.28 Derivative purity plot for data set A with purest points indicated.

columns have now been set to a constant total of 1. Finally, the consensus absolute value of the derivative is presented in Table 7.8(d).

The resultant value of d_i is best presented on a logarithmic scale, as shown in Figure 7.28. The regions of highest purity can be quite well pinpointed as minima at times 7 and 15. These graphs are most useful for determining the purest points in time rather than regions of differing composition, but the visual display is often very informative and can cope well with unusual peak shapes.

Of course, the derivative method is not restricted to coupled chromatography and can be used for any evolutionary process.

1. Which statement is true?
   (a) If the successive spectra in time do not change significantly in appearance, then if scaled to the same total of 1, their derivatives over time at each wavelength will be close to 0.
   (b) If successive spectra in time do not change significantly in appearance, then if scaled to the same total of 1, their derivatives over time at each wavelength will be close to 1.

7.4 Resolution

Resolution or deconvolution of two-way chromatograms or mixture spectra involves converting a cluster of peaks into its constituent parts, each ideally representing a component of the signal from a single compound. The number of named methods in the literature is enormous, and it would be out of the scope of this text to discuss each approach in detail. In areas such as chemical pattern recognition or calibration, certain generic approaches are accepted as part of an overall strategy, and the data pre-processing, variable selection and so on are regarded as extra steps. In the field of resolution of evolutionary multivariate data, there is a fondness for packaging a series of steps into a named method; hence, there are probably 50 or more named methods and maybe many unnamed approaches reported in the literature. Yet, most are based on a number of generic principles, which are described in this chapter.

There are several aims for resolution.

- Obtaining the profiles for each resolved compound. These might be the elution profiles (in chromatography) or the concentration distribution in a series of compounds (in spectroscopy of mixtures) or the pH profiles of different chemical species.
- Obtaining the spectra of each pure compound. This allows identification or library searching. In some cases, this procedure merely uses the multivariate signals to improve the quality of the individual spectra that may be noisy, but in other cases, such as an embedded peak, genuinely difficult information can be gleaned. This is particularly useful in impurity monitoring.
- Obtaining quantitative information. This involves using the resolved two-way data to provide concentrations (or relative concentrations when pure standards are not available).

- Automation. Complex chromatograms may consist of 50 or more peaks, some of which will be noisy and overlapping. Speeding up procedures, for example, using rapid chromatography in a matter of minutes resulting in considerable overlap, rather than taking half an hour per chromatogram, also results in embedded peaks. Chemometrics can ideally pull out the constituents' spectra and profiles.

The methods in this chapter differ from those of Chapter 6 in that pure standards are not required for the model.

Although some data sets can be very complicated, it is normal to divide the data into small regions where there are signals from only a few components. Even in spectroscopy of mixtures, in many cases such as MIR or NMR, it is normally easy to find regions of the spectra where only two or three compounds at the most absorb; thus, this process of finding windows rather than analysing an entire data set in one go is quite normal. Hence, we will limit the discussion to three peak clusters in this section. Naturally, the methods of Section 7.3 would usually first be applied to the entire data set to identify these regions. We will illustrate the discussion below primarily in the context of coupled chromatography, although it is also very well established in other areas such as equilibrium profiling.

There are several cases of increasing complexity.

7.4.1 Selectivity for All Components

The simplest problem is that all compounds (or chemical components) have some region of selectivity. These methods involve first finding some pure or selective (composition 1) region in the chromatogram or selective spectral measurement such as a m/z value for each compound in a mixture.

7.4.1.1 Pure Spectra and Selective Variables

The most straightforward situation is when each compound has an identifiable composition 1 region (often called *selective regions* in chromatography). The simplest approach is to estimate the pure spectrum in such a region. There are several methods.

- Take the spectrum at the point of maximum purity for each compound.
- Average the spectra for each compound over each composition 1 region.
- Perform PCA over each composition 1 region separately (so if there are three compounds, perform three independent PCA calculations) and then take the loadings of the first PC as an estimate of the pure spectrum. PCA is used as a smoothing technique, the idea being that the noise is banished to later PCs.

Some rather elaborate multivariate methods that use the profiles are also available, instead of using the spectra in the composition 1 regions. In the case of data set A in Table 7.1, we might guess that the fastest eluting compound A has a composition 1 region between points 4 and 8 and the slowest eluting compound B between points 15 and 19. Hence, we could divide the chromatogram as follows.

- Points 1–3: no compounds elute.
- Points 4–8: compound A elutes selectively.
- Points 9–14: co-elution.
- Points 15–19: compound B elutes selectively.
- Points 20–25: no compounds elute.

As discussed above, there can be slight variations on this theme. This is represented in Figure 7.29. Chemometrics is used to fill in the remaining pieces of the jigsaw. The only unknowns are the elution profiles in the composition 2 regions. The profiles in the composition 1 regions can be estimated either by using the summed profiles in these regions or by performing PCA in these regions and taking the scores of the first PC.

An alternative is to find pure variables rather than composition 1 regions. These methods are popular when using various types of spectroscopy such as in LC-MS or in the MIR of mixtures. Wavelengths, frequencies or masses belonging single compounds can often be identified. In the case of data set C in Table 7.3, we suspect that variables C and F belong to the two compounds (see Figure 7.15), and their profiles are presented in Figure 7.30. Note that these profiles are somewhat noisy. This is quite common in techniques such as mass spectrometry. It is possible to improve the quality of the profiles by using methods for smoothing, as described in Chapter 3, or to average profiles from several pure variables. The latter technique is useful in NMR or IR, where a peak might be defined by several data points or where there could be a number of selective regions in the spectrum.

The result of this section will be to produce either a first guess of all or part of the concentration profiles, represented by the matrix $\hat{C}$ or of the spectra $\hat{S}$, where the '^' or hat sign implies estimated. Although conceptually the easiest

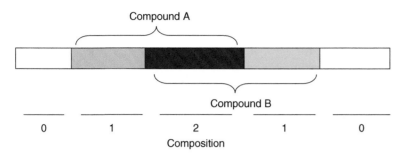

Figure 7.29 Composition of regions in chromatogram deriving from data set A.

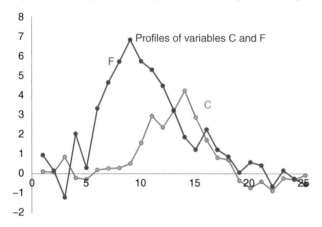

Figure 7.30 Profiles of variables *C* and *F* in Table 7.3.

approach to resolution is univariate, we see below that by using more variables, we can obtain better estimates using multiple linear regression (MLR).

1. The scores of the first principal component over a composition 1 region are an approximation of the profile of a pure compound.
   (a) True
   (b) False

7.4.1.2 Multiple Linear Regression

If pure profiles can be obtained from all components in a mixture, the next step in deconvolution is quite straightforward.

In the case of Table 7.1, we can guess the pure spectrum for *A* as the average of the data between times 4 and 8 and for *B* as the average of the data between times 15 and 19. These make up a 2×12 data matrix $\widehat{S}$. As

$$X \approx \widehat{C}\widehat{S}$$

therefore

$$\widehat{C} = X\widehat{S}'(\widehat{S}\,\widehat{S}')^{-1}$$

as discussed in Section 6.3. The estimated spectra are tabulated in Table 7.9 and the resultant profiles are presented in Figure 7.31. Notice that the vertical scale in fact has no direct physical meaning: intensity data can only be reconstructed by multiplying the profiles by the spectra. However, MLR has provided a very satisfactory estimate, and provided pure regions are available for each significant component, it is probably entirely adequate as a tool in many cases.

If pure variables such as spectral frequencies or *m/z* values can be determined, even if there are embedded peaks, it is also possible to use these to obtain first estimates of elution profiles, $\widehat{C}$, then the spectra can be obtained using all (or a great proportion of) the variables by

$$\widehat{S} = (\widehat{C}'\widehat{C})^{-1}\widehat{C}'X$$

Table 7.9 Estimated spectra obtained from the composition 1 regions in data set A.

A	B	C	D	E	F	G	H	I	J	K	L
0.519	0.746	0.862	0.713	0.454	0.341	0.194	0.176	0.312	0.410	0.465	0.404
0.041	0.006	0.087	0.221	0.356	0.603	0.676	0.575	0.395	0.199	0.136	0.162

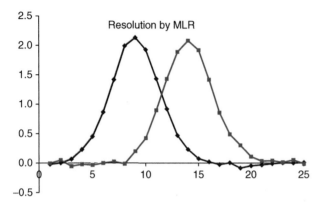

Figure 7.31 Reconstructed profiles for data set A using MLR.

The concentration profiles can be improved by increasing the number of variables; hence, for example, the first guess might involve using 1 variable per compound, the next 20 significant variables and the final 100 or more. This approach is also useful in spectroscopy of mixtures, if selective wavelengths can be identified for each compound. Using these for initial estimates of the concentrations of each compound in each spectrum, the full spectra can be reconstructed even when there are overlapping regions. Such approaches are useful in MIR, but not so valuable in NIR or UV/vis spectroscopy where it is often hard to find selective wavelengths and the effectiveness depends on the type of spectroscopy employed.

1. If a mixture consists of three components all of which have composition regions and is measured over 50 wavelengths, what are the dimensions of $\widehat{S}$?
   (a) 3×50
   (b) 50×3
   (c) We would need to know the dimensions of the selective regions.

2. Every component has to have at least one composition 1 region for MLR to be effective.
   (a) True
   (b) False

7.4.1.3 Principal Component Regression

Principal component regression (PCR) is an alternative to MLR (Section 5.4) and can be used in signal analysis just as in calibration. There are a number of ways of employing PCA, but a simple approach is to note that the scores and loadings can be related to the concentration profile and spectra by

$$X \approx \widehat{C}\,\widehat{S} = TRR^{-1}P$$

hence,

$$\widehat{C} = TR$$

and

$$\widehat{S} = R^{-1}P$$

Table 7.10 Estimation of profiles using PCA for the data in Table 7.9.

Loadings

0.215	0.321	0.375	0.372	0.333	0.305	0.288	0.255	0.248	0.237	0.236	0.214
−0.254	−0.377	−0.381	−0.191	0.085	0.369	0.479	0.422	0.203	−0.015	−0.113	−0.080

Matrix R^{-1}

1.667	−0.573
0.982	0.781

Matrix R

0.419	0.307
−0.527	0.894

Scores / **TR**

Scores		TR	
−0.019	0.007	−0.012	0.001
0.079	0.040	0.012	0.060
0.077	−0.076	0.072	−0.044
0.380	−0.147	0.237	−0.015
0.746	−0.284	0.462	−0.024
1.464	−0.493	0.873	0.009
2.412	−0.795	1.429	0.031
3.332	−1.147	2.000	−0.001
3.775	−1.066	2.143	0.208
3.646	−0.770	1.933	0.432
3.286	−0.116	1.438	0.906
2.954	0.593	0.925	1.438
2.650	1.210	0.473	1.896
2.442	1.504	0.231	2.095
2.020	1.461	0.077	1.927
1.432	1.098	0.022	1.421
0.803	0.682	−0.022	0.856
0.495	0.377	0.009	0.489
0.158	0.288	−0.086	0.306
0.037	0.112	−0.043	0.111
−0.013	0.046	−0.029	0.037
0.026	0.039	−0.010	0.043
0.026	0.009	0.006	0.016
0.057	0.041	0.003	0.054
0.011	−0.012	0.011	−0.007

If we perform PCA on the data set, keeping K components equal to the number of compounds in the mixture, and can estimate the pure spectra, it is possible to find the matrix R^{-1} simply by regression as

$$R^{-1} = \widehat{S}P'$$

(because the loadings are orthonormal (Section 4.3.1), this latter equation is quite simple). Matrix R is called a *rotation*, *transformation*, or *regression matrix* according to the author. It is then easy to obtain $\widehat{C}$. This procedure is illustrated in Table 7.10 using the spectra obtained from Table 7.9. The profiles are very similar to those presented in Figure 7.31; in this case, so are not presented graphically for brevity.

PCR can be employed in more elaborate ways using the known profiles in the composition 1 (and sometimes composition 0) regions for each compound. These methods were the basis of some of the earliest approaches to resolution of two-way chromatographic data. There are several variants and one is as follows.

Choose only those regions where one component elutes. In our example in Table 7.1:

- We will use the regions between times 4–8 and 15–19 inclusive, which involves 10 points.
- For each compound, use either the estimated profiles if the region is composition 1 or 0 if another compound elutes in this region (so a composition 1 region for compound A will be composition 0 for compound B). A matrix is obtained of size $Z \times 2$ whose columns correspond to each component where Z equals the total number of composition 1 data points. In our example, the matrix is of size 10×2, half of the values being 0 and half consisting of the profile in the composition 1 region. Call this matrix Z.
- Perform PCA on the overall matrix.
- Find a matrix R such that $Z \approx TR$ using the known profiles obtained in step 2, simply by using regression so that $R = (T'T)^{-1}T'Z$ but including the scores *only* of the composition 1 region.
- Knowing R, it is a simple matter to reconstruct the concentration profiles by including the scores over the entire data matrix as discussed above, and similarly the spectra.

The key steps in the calculation are presented in Table 7.11. Notice that the magnitude of the numbers in the matrix R differs from that presented in Table 7.10. This is simply because the magnitude of the estimates of the spectra and profiles is different and has no physical significance. The resultant profiles obtained by the multiplication $\hat{C} = TR$ on the entire data set are illustrated in Figure 7.32.

In straightforward cases, PCR is unnecessary and, if not carefully controlled, may provide worse results to MLR. However, for more complex systems, it can be very useful.

1. There are three compounds in a mixture, all with composition 1 regions. Fifty samples are recorded at 100 wavelengths. The matrix R has the following dimensions.

   (a) 50×3
   (b) 3×100
   (c) 3×3

7.4.2 Partial Selectivity

Somewhat harder situations occur when only some components exhibit selectivity. A common example is a completely embedded peak in HPLC-DAD. In the case of LC-MS or LC-NMR, this problem is often solved by finding pure variables; however, as UV/vis spectra are often completely overlapping, it is not always possible to treat data in this manner.

Fortunately, PCA comes to the rescue. In Chapter 6, we have discussed the different applicability of PCR and MLR. Using the former method, we stated that it was not necessary to have information about the concentration of every component in the mixture, simply a good idea of how many significant components are there. Thus, in the case of resolution of two-way data, these approaches can easily be extended. We will illustrate this using the data in Table 7.2, which correspond to three peaks, the middle one being completely embedded in the others. There are several different ways of exploiting this.

One approach uses the idea of a zero concentration window. The first step is to identify compounds that we know have selective regions and determine where they do not elute. This information may be obtained from a variety of approaches such as eigenvalue or PC plots. In this region, we expect the intensity of the data to be zero; hence, it is possible to find a vector r for each component so that

$$0 = T_0 r$$

where 0 is a vector of zeros, and T_0 denotes the portion of the scores in this region; normally, one excludes the region where no peaks elute and then finds the zero component region for each component. For example, if we record a cluster over 50 data points and we suspect that there are three peaks eluting between points 10 and 25, 20 and 35, and 30 and 45, then there are three T_0 matrices; for the fastest eluting component, this is between points 26 and 45. Notice that the number of PCs should be made equal to the number of significant compounds in the overall mixture. There is one small problem in that one value of the vector r must be set to an arbitrary number, usually the first coefficient is set

Table 7.11 Key steps in the calculation of rotation matrix for data set A using scores in composition 1 regions.

Time	Z		T	
	Compound A	Compound B	PC1	PC2
1				
2				
3				
4	1.341	0.000	0.380	−0.147
5	2.462	0.000	0.746	−0.284
6	4.910	0.000	1.464	−0.493
7	8.059	0.000	2.412	−0.795
8	11.202	0.000	3.332	−1.147
9				
10				
11				
12				
13				
14				
15	0.000	7.104	2.020	1.461
16	0.000	5.031	1.432	1.098
17	0.000	2.838	0.803	0.682
18	0.000	1.696	0.495	0.377
19	0.000	0.625	0.158	0.288
20				
21				
22				
23				
24				
25				

Matrix *R*

2.311	1.094
−3.050	3.197

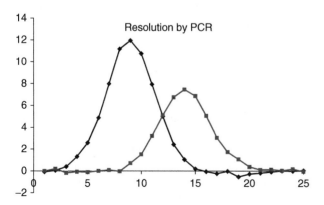

Figure 7.32 Profiles obtained as described in Section 7.4.1.3.

to 1, but this does not have a serious effect on the algorithm. The equation can then be solved as follows. Separate out the contribution from the first PC to that from all the others so, setting r_1 to 1

$$T_{0(2:K)}r_{2:k} \approx -t_{01}$$

so that

$$r_{2:K} = (T'_{0(2:K)}T_{0(2:K)})^{-1}T'_{0(2:K)}t_{01}$$

where $r_{2:K}$ is a column vector of length $K-1$, where there are K PCs, t_{01} is the scores of the first PC over the zero concentration window and $T_{0(2:K)}$ the scores of the remaining PCs. It is important to ensure that K equals the number of components suspected to elute within the cluster of peaks. It is possible to perform this operation on any embedded peaks because these also exhibit zero composition regions.

The profiles of all the compounds can now be obtained over the entire region by $\hat{C} = TR$ and the spectra by $\hat{S} = R^{-1}P$.

From inspecting Table 7.2, we might conclude that compound A elutes between times 4 and 13, B between times 9 and 17 and C between times 13 and 22. This information could be obtained by a variety of methods, as discussed in Section 7.3. Hence, the zero composition regions are as follows.

- Compound A: points 14–22.
- Compound B: points 4–8 and 18–22.
- Compound C: points 4–12.

Obviously, different approaches may identify slightly different regions. The calculation is presented in Table 7.12, for the data in Table 7.2, and the resultant profiles are presented in Figure 7.33. Of course, spectra could also be estimated this way.

Many a paper and thesis have been written about this problem, and there are a large number of modifications to this approach; however, in this chapter, we illustrate just using one of the best established approaches.

1. Compound A elutes between points 5 and 20, compound B between 15 and 35 and compound C between 30 and 45; data are recorded between points 1–50.

   (a) The zero concentration region for compound A is points 1–4 and 21–50.
   (b) The zero concentration region for compound A is points 21–45.
   (c) The zero concentration region for compound A is points 1–4 and 46–50.

7.4.3 Incorporating Constraints: ITTFA, ALS and MCR

Finally, it is important to mention another class of methods. In many cases, it is not possible to obtain a unique mathematical solution to the multivariate resolution of complex mixtures, and the problem of embedded components without selectivity, which may, for example, occur in impurity monitoring, causes difficulties while using many conventional approaches.

The methods above are non-iterative; that is, they have a single step that comes to a single solution. However, over the years, chemometricians have refined these approaches to develop iterative methods that are based on many of the principles mentioned above but do not stop at the first answer. These methods are particularly useful when there is partial selectivity, not only in chromatography but also in areas such as reaction monitoring or spectroscopy.

Common to all iterative methods is to start with a guess about some characteristic of each component. The methods described above are based on having quite good information about each component, whether it is a spectrum or composition 0 region, and if this information is poor, the methods will fall over, as there is only one chance to get it right, based on the starting information. In contrast, iterative methods are not so restrictive as to what the first guess is. If perfect knowledge is obtained about each component at the start, then this is useful, but iterative methods may then not be necessary.

There are, however, many other approaches for the estimation of the properties of each component in a mixture. One simple method is called a *needle* search. For each component in a mixture, the data point of maximum purity and the spectra at these points are taken as the starting guesses for each component, even though they may not necessarily correspond to pure variables. Statistical methods such as varimax rotation can also be used to provide first guesses,

Table 7.12 Determining spectrum and elution profiles of an embedded peak.

(a) Choosing matrices T_0: the composition 0 regions are used to identify the portions of the overall scores matrix for compounds A, B and C

Time	T			Composition 0 regions		
				A	B	C
1	0.011	−0.006	−0.052			
2	−0.049	0.036	0.035			
3	−0.059	−0.002	0.084			
4	0.120	−0.099	−0.033		0	0
5	0.439	−0.018	0.129		0	0
6	1.029	−0.205	0.476		0	0
7	2.025	−0.379	0.808		0	0
8	2.962	−0.348	1.133		0	0
9	3.505	−0.351	1.287			0
10	3.501	0.088	1.108			0
11	3.213	0.704	0.353			0
12	2.774	1.417	−0.224			0
13	2.683	1.451	−0.646			
14	2.710	1.091	−0.885	0		
15	2.735	0.178	−0.918	0		
16	2.923	−0.718	−0.950	0		
17	2.742	−1.265	−0.816	0		
18	2.359	−1.256	−0.761	0	0	
19	1.578	−0.995	−0.495	0	0	
20	0.768	−0.493	−0.231	0	0	
21	0.428	−0.195	−0.065	0	0	
22	0.156	−0.066	−0.016	0	0	
23	−0.031	−0.049	0.005			
24	−0.110	−0.007	−0.095			
25	0.052	−0.057	0.074			

(b) Determining a matrix R

A	B	C
1	1	1
0.078	2.603	−2.476
3.076	−1.560	−3.333

(Continued)

Table 7.12 (Continued)

(c) Determining the concentration profiles using $\widehat{C} = TR$

	A	B	C
1	−0.150	0.079	0.200
2	0.062	−0.009	−0.255
3	0.200	−0.196	−0.335
4	0.009	−0.085	0.475
5	0.835	0.192	0.052
6	2.476	−0.245	−0.050
7	4.480	−0.222	0.272
8	6.419	0.288	0.049
9	7.436	0.584	0.086
10	6.916	2.001	−0.410
11	4.354	4.494	0.295
12	2.194	6.813	0.012
13	0.809	7.466	1.244
14	0.072	6.931	2.959
15	−0.075	4.630	5.353
16	−0.055	2.536	7.869
17	0.133	0.722	8.596
18	−0.079	0.277	8.004
19	−0.022	−0.239	5.691
20	0.020	−0.155	2.757
21	0.213	0.022	1.126
22	0.100	0.009	0.374
23	−0.021	−0.167	0.076
24	−0.401	0.019	0.223
25	0.274	−0.211	−0.053

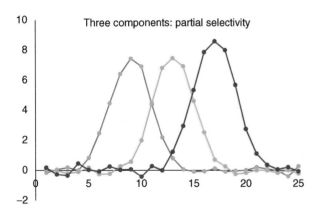

Figure 7.33 Profiles of three peaks obtained as in Section 7.4.2.

the loadings of the first varimax-rotated components corresponding to the first guesses of the spectra (in coupled chromatography). OPA and SIMPLISMA methods can also be employed, each pure variable being the starting guess of each component. A large variety of other methods have also been employed. None of these approaches are likely to provide pure estimates of spectra. For further details, readers are recommended to consult the specialist literature, the main principle being to find a guess of the characteristics of each component even though we know it will not be perfect, and there may not be a composition 1 region.

ITTFA (iterative target transform factor analysis) can then be used to improve these guesses. The principle is quite straightforward: PCA is performed first on the data set and a rotation or transformation matrix R is found to relate the loadings to the estimated spectra, just as described above. This then allows the concentration profiles to be predicted in turn, as discussed elsewhere in this chapter.

However, the next stage is the tricky part. The predicted concentration matrix $\hat{C}$ is examined. This matrix will, in the initial stages, usually contain elements that are not physically meaningful, for example, negative concentrations. These negative regions are changed, normally by the simple procedure of substituting the negative numbers by zero, to provide a new and second guess of the concentration profiles. This, in turn, results in fresh estimates of the spectra, which can then be corrected for negativity and so on. This iteration continues until $\hat{C}$ shows very little change. Several alternative criteria have also been proposed for convergence.

ITTFA is dependent on the quality of the initial guesses, but can be an effective algorithm. Non-negativity is the most common condition, but other constraints such as unimodality, if appropriate, can be added.

It is not necessary to start with estimates of spectra; concentration estimates are also legitimate; as this is an iterative method, it can be implemented either way round.

ALS (alternating least squares) rests on similar principles, but the criterion for convergence is somewhat different and looks at the error between the reconstructed data from the product of the estimated elution profiles and spectra ($\hat{C}\hat{S}$) and the observed data. In addition, as implemented, guesses of either the initial spectra or elution profiles can be used to the start of the iterations, whereas for ITTFA, it is normal in the case of chromatography to start with the elution profiles when applied to coupled chromatographic data (however, there is no real theoretical requirement for this restriction of course). The ALS procedure is often considered somewhat more general. ITTFA was developed very much with analytical applications in spectroscopy coupled to chromatography or another technique in mind, whereas the idea of ALS was borrowed from general statistics. In practice, neither method is radically different, the main distinction being the stopping criterion.

MCR (multivariate curve resolution) is in practical terms the same as ALS, but the terminology is often used for a packaged approach rather than an algorithm in its own right.

The original applications were quite simple, but ALS has since been applied to quite sophisticated problems, for example, from multi-way data sets. The MCR software is readily available in Matlab for downloading and because of this the method is very widespread. As compared with the methods discussed earlier in this section, MCR-ALS represents a sophisticated approach to find R, allowing the user to incorporate a wide variety of constraints and to estimate the solution when there is no unique theoretical answer.

1. ALS is usually applied to problems where there is no single unique mathematical solution.
    (a) True
    (b) False

2. It is possible to incorporate unimodality as a constraint using ALS but not ITTFA.
    (a) True
    (b) False

Problems

7.1   Determining of Purity within a Two-Component Cluster: Derivatives, Correlation Coefficients and PC Plots. Section 7.3.3 Section 7.3.5 Section 7.2.2

206	213	220	227	234	242	249	256	263	270	277	284	291	298	305	312	320	327	334	341	348	355	361	368	375	382	390
0.0204	0.0153	0.0127	0.0073	0.0051	0.0026	0.0016	−0.0020	0.0002	−0.0004	0.0111	0.0092	0.0006	0.0015	−0.0028	−0.0004	0.0008	−0.0011	0.0021	0.0026	0.0042	0.0035	−0.0045	0.0052	0.0026	0.0019	0.0062
0.0258	0.0211	0.0197	0.0141	0.0093	0.0078	0.0088	0.0058	0.0070	0.0081	0.0258	0.0260	0.0052	0.0032	−0.0012	0.0008	0.0022	0.0009	0.0048	0.0040	0.0038	0.0044	−0.0026	0.0053	0.0022	0.0033	0.0058
0.0361	0.0327	0.0347	0.0273	0.0169	0.0183	0.0228	0.0209	0.0214	0.0234	0.0545	0.0605	0.0136	0.0063	0.0011	0.0032	0.0048	0.0036	0.0077	0.0071	0.0050	0.0056	−0.0006	0.0081	0.0003	0.0050	0.0057
0.0522	0.0518	0.0620	0.0513	0.0294	0.0356	0.0468	0.0469	0.0463	0.0501	0.1046	0.1207	0.0277	0.0119	0.0036	0.0068	0.0087	0.0077	0.0111	0.0121	0.0086	0.0077	0.0015	0.0096	0.0003	0.0050	0.0056
0.0769	0.0812	0.1040	0.0883	0.0487	0.0611	0.0833	0.0871	0.0851	0.0922	0.1811	0.2120	0.0488	0.0209	0.0068	0.0117	0.0134	0.0145	0.0168	0.0198	0.0136	0.0116	0.0038	0.0088	0.0020	0.0049	0.0050
0.1098	0.1194	0.1589	0.1371	0.0744	0.0944	0.1313	0.1402	0.1371	0.1480	0.2813	0.3323	0.0756	0.0335	0.0114	0.0184	0.0199	0.0242	0.0254	0.0303	0.0203	0.0177	0.0071	0.0084	0.0049	0.0064	0.0044
0.1465	0.1617	0.2194	0.1910	0.1033	0.1316	0.1844	0.1984	0.1957	0.2103	0.3921	0.4654	0.1051	0.0477	0.0173	0.0265	0.0281	0.0351	0.0351	0.0405	0.0285	0.0250	0.0106	0.0130	0.0062	0.0079	0.0047
0.1795	0.1997	0.2742	0.2389	0.1291	0.1653	0.2323	0.2506	0.2485	0.2653	0.4917	0.5843	0.1315	0.0589	0.0233	0.0334	0.0359	0.0443	0.0432	0.0494	0.0360	0.0327	0.0150	0.0194	0.0065	0.0086	0.0065
0.1994	0.2243	0.3102	0.2697	0.1456	0.1872	0.2638	0.2850	0.2830	0.3015	0.5574	0.6613	0.1491	0.0660	0.0275	0.0371	0.0404	0.0506	0.0478	0.0553	0.0414	0.0381	0.0182	0.0215	0.0062	0.0091	0.0070
0.2030	0.2307	0.3194	0.2771	0.1498	0.1926	0.2724	0.2945	0.2919	0.3115	0.5748	0.6810	0.1540	0.0684	0.0292	0.0379	0.0404	0.0528	0.0490	0.0579	0.0423	0.0404	0.0195	0.0202	0.0051	0.0108	0.0064
0.1925	0.2193	0.3027	0.2622	0.1420	0.1822	0.2579	0.2797	0.2771	0.2956	0.5437	0.6451	0.1458	0.0656	0.0278	0.0365	0.0377	0.0505	0.0472	0.0552	0.0392	0.0393	0.0192	0.0197	0.0040	0.0108	0.0061
0.1724	0.1954	0.2688	0.2321	0.1261	0.1608	0.2277	0.2482	0.2469	0.2615	0.4789	0.5683	0.1295	0.0585	0.0247	0.0325	0.0337	0.0441	0.0424	0.0481	0.0348	0.0352	0.0166	0.0191	0.0028	0.0100	0.0064
0.1484	0.1672	0.2283	0.1961	0.1071	0.1356	0.1917	0.2114	0.2123	0.2207	0.4005	0.4732	0.1094	0.0492	0.0214	0.0271	0.0299	0.0355	0.0360	0.0395	0.0294	0.0292	0.0120	0.0174	0.0025	0.0093	0.0069
0.1253	0.1414	0.1903	0.1615	0.0893	0.1124	0.1589	0.1793	0.1837	0.1836	0.3249	0.3806	0.0899	0.0409	0.0190	0.0225	0.0269	0.0278	0.0292	0.0317	0.0241	0.0238	0.0066	0.0148	0.0026	0.0089	0.0069
0.1067	0.1213	0.1601	0.1326	0.0752	0.0944	0.1338	0.1577	0.1666	0.1550	0.2611	0.3032	0.0730	0.0349	0.0178	0.0201	0.0242	0.0215	0.0237	0.0253	0.0188	0.0198	0.0041	0.0120	0.0028	0.0086	0.0064
0.0931	0.1068	0.1382	0.1110	0.0647	0.0814	0.1164	0.1470	0.1612	0.1354	0.2104	0.2429	0.0603	0.0305	0.0176	0.0188	0.0222	0.0163	0.0194	0.0201	0.0145	0.0158	0.0041	0.0093	0.0026	0.0062	0.0064
0.0829	0.0963	0.1225	0.0950	0.0574	0.0718	0.1046	0.1430	0.1628	0.1227	0.1710	0.1964	0.0512	0.0272	0.0175	0.0181	0.0212	0.0126	0.0160	0.0167	0.0117	0.0121	0.0044	0.0081	0.0027	0.0060	0.0067
0.0750	0.0888	0.1105	0.0825	0.0519	0.0639	0.0954	0.1406	0.1650	0.1133	0.1389	0.1586	0.0440	0.0244	0.0169	0.0177	0.0212	0.0099	0.0142	0.0135	0.0094	0.0091	0.0039	0.0075	0.0026	0.0072	0.0069
0.0681	0.0815	0.0998	0.0710	0.0463	0.0561	0.0862	0.1354	0.1620	0.1032	0.1108	0.1261	0.0377	0.0219	0.0165	0.0169	0.0206	0.0076	0.0126	0.0104	0.0083	0.0071	0.0033	0.0066	0.0028	0.0079	0.0068
0.0609	0.0729	0.0882	0.0599	0.0400	0.0479	0.0756	0.1247	0.1515	0.0904	0.0866	0.0977	0.0319	0.0192	0.0161	0.0150	0.0190	0.0056	0.0116	0.0082	0.0075	0.0057	0.0013	0.0072	0.0034	0.0069	0.0069
0.0529	0.0630	0.0750	0.0488	0.0334	0.0397	0.0636	0.1083	0.1345	0.0754	0.0666	0.0735	0.0263	0.0158	0.0149	0.0119	0.0169	0.0037	0.0098	0.0059	0.0063	0.0051	−0.0004	0.0088	0.0034	0.0041	0.0073
0.0449	0.0531	0.0613	0.0387	0.0275	0.0319	0.0515	0.0887	0.1132	0.0601	0.0501	0.0540	0.0207	0.0122	0.0126	0.0087	0.0148	0.0025	0.0079	0.0039	0.0049	0.0048	−0.0015	0.0097	0.0033	0.0035	0.0073
0.0384	0.0447	0.0488	0.0308	0.0223	0.0252	0.0400	0.0699	0.0908	0.0462	0.0372	0.0387	0.0154	0.0094	0.0101	0.0061	0.0121	0.0017	0.0062	0.0023	0.0036	0.0050	−0.0024	0.0084	0.0021	0.0037	0.0074
0.0336	0.0375	0.0387	0.0247	0.0179	0.0198	0.0300	0.0546	0.0700	0.0345	0.0272	0.0273	0.0104	0.0074	0.0075	0.0050	0.0087	0.0008	0.0049	0.0013	0.0033	0.0046	−0.0032	0.0087	0.0018	0.0041	0.0069
0.0301	0.0316	0.0312	0.0195	0.0145	0.0157	0.0227	0.0425	0.0531	0.0252	0.0204	0.0201	0.0068	0.0060	0.0051	0.0048	0.0064	0.0005	0.0043	0.0008	0.0031	0.0033	−0.0044	0.0089	0.0020	0.0044	0.0061
0.0270	0.0272	0.0258	0.0149	0.0117	0.0124	0.0176	0.0324	0.0405	0.0176	0.0157	0.0160	0.0044	0.0045	0.0030	0.0043	0.0052	−0.0001	0.0033	−0.0002	0.0026	0.0017	−0.0050	0.0078	0.0023	0.0041	0.0055
0.0245	0.0232	0.0219	0.0115	0.0093	0.0094	0.0136	0.0238	0.0312	0.0118	0.0119	0.0131	0.0030	0.0028	0.0014	0.0033	0.0042	0.0000	0.0023	−0.0004	0.0026	0.0013	−0.0057	0.0069	0.0019	0.0029	0.0049
0.0232	0.0199	0.0188	0.0095	0.0073	0.0067	0.0096	0.0167	0.0239	0.0077	0.0097	0.0105	0.0017	0.0010	0.0005	0.0021	0.0026	−0.0003	0.0020	−0.0008	0.0027	0.0016	−0.0060	0.0079	0.0001	0.0012	0.0054
0.0228	0.0172	0.0164	0.0085	0.0059	0.0046	0.0061	0.0114	0.0172	0.0048	0.0085	0.0080	0.0006	−0.0002	−0.0005	0.0008	0.0006	−0.0009	0.0025	−0.0012	0.0025	0.0018	−0.0051	0.0092	−0.0009	0.0006	0.0070
0.0215																										

The table above represents a diode array HPLC chromatogram recorded at 27 wavelengths (the low digital resolution is used for illustrative purposes) and 30 points in time. The wavelengths in nanometre are presented in the top row.

1. Calculate the 29 correlation coefficients between successive points in time and plot a graph of these. Remove the first correlation coefficients and replot the graph. Comment on these graphs. How might you improve the graph still further?
2. Use first derivatives to look at purity as follows.
   (a) Sum the spectrum (to a total of 1) at each point in time.
   (b) At each wavelength and points 3–28 in time, calculate the absolute value of the 5-point quadratic Savitzky–Golay derivative (see Table 3.6). You should produce a matrix of size 26×27.
   (c) Average these overall wavelengths and plot this graph against time.
   (d) Improve this graph by using a logarithmic scale for the parameter calculated in step c.
   Comment on what you observe.
3. Perform PCA on the raw uncentred data, retaining the first two PCs. Plot a graph of the scores of PC2 versus PC1, labelling the points. What do you observe from this graph and how does it compare with the plots in questions 1 and 2?

7.2 Evolutionary and Window Factor Analysis in the Detection of an Embedded Peak.
Section 7.3.4 Section 7.4.1.3 Section 7.4.2
The following small data set represents an evolutionary process consisting to two peaks, one embedded, recorded at 16 points in time and over 6 variables.

0.156	0.187	0.131	0.119	0.073	0.028
0.217	0.275	0.229	0.157	0.096	0.047
0.378	0.4256	0.385	0.215	0.121	0.024
0.522	0.667	0.517	0.266	0.178	0.065
0.690	0.792	0.705	0.424	0.186	0.060
0.792	0.981	0.824	0.541	0.291	0.147
0.841	1.078	0.901	0.689	0.400	0.242
0.832	1.144	0.992	0.779	0.568	0.308
0.776	1.029	0.969	0.800	0.650	0.345
0.552	0.797	0.749	0.644	0.489	0.291
0.377	0.567	0.522	0.375	0.292	0.156
0.259	0.330	0.305	0.202	0.158	0.068
0.132	0.163	0.179	0.101	0.043	0.029
0.081	0.066	0.028	0.047	0.006	0.019
0.009	0.054	0.056	0.013	−0.042	−0.031
0.042	−0.005	0.038	−0.029	−0.013	−0.057

1. Perform EFA on the data (using uncentred PCA) as follows.
   (a) For forward EFA, perform PCA on the 3×6 matrix consisting of the first three spectra and retain the three eigenvalues.
   (b) Then, perform PCA on the 4×6 matrix consisting of the first four spectra retaining the first three eigenvalues.
   (c) Continue this procedure, increasing the matrix by one row at a time, until a 14×3 matrix is obtained whose rows correspond to the ends of each window (from 3 to 16) and columns to the eigenvalues.
   (d) Repeat the same process, but for backward EFA, the first matrix consisting of the three last spectra (14–16) to give another 14×3 matrix.
   If you are able to program in Matlab or VBA, it is easiest to automate this, but it is possible to simply repeatedly use the PCA add-in for each calculation.
2. Produce EFA plots, first converting the eigenvalues to a logarithmic scale, superimposing six graphs, always plot the eigenvalues against the extreme rather than middle or starting value of each window. Comment.

3. Perform WFA, again using an uncentred data matrix, with a window size of 3. To do this, simply perform PCA on spectra 1–3 and retain the first three eigenvalues. Repeat this for spectra 2–4, 3–5 and so on. Plot the logarithms of the first three eigenvalues against window centre and comment.
4. There are clearly two components in this mixture. Show how you could distinguish the situation of an embedded peak from that of two peaks with a central region of co-elution and demonstrate that we are dealing with an embedded peak in this situation.
5. From the EFA plot, it is possible to identify the composition 1 regions for the main peak. What are these? Calculate the average spectrum over these regions and use this as an estimate of the spectrum of the main component.

7.3 Variable Selection and PC Plots in LC-MS.
Section 7.2.2   Section 7.2.4.2   Section 7.2.4   Section 7.2.4.1

m/z																									
41	0	334	905	1286	1917	1620	1831	1694	1423	1860	1678	1770	1668	1148	1492	1571	1402	1181	1207	1173	1146	919	527	1146	1002
42	530	578	386	245	229	119	69	164	162	176	34	286	283	106	72	67	315	239	284	335	191	288	0	638	236
44	242	567	353	188	178	0	158	328	341	455	325	517	162	262	265	180	137	110	102	91	141	158	157	354	219
50	66	111	60	73	39	0	3	39	52	83	88	59	75	63	7	73	91	68	46	8	24	34	64	76	67
54	165	83	107	88	11	110	30	35	45	0	13	98	69	99	132	134	77	158	110	107	65	135	208	128	104
59	1111	1143	593	706	43	558	19	139	499	39	379	222	0	240	537	299	234	803	375	558	41	299	658	623	739
61	360	278	515	375	14	131	0	162	206	130	180	363	224	322	225	252	98	310	70	106	179	194	364	173	463
68	0	430	770	919	1626	1530	1238	1260	722	661	550	523	648	594	749	771	600	632	494	324	612	364	479	568	650
69	594	562	440	367	550	259	233	124	0	108	57	131	292	125	258	234	438	263	425	514	430	436	320	403	385
72	287	135	427	153	314	278	16	118	109	104	0	124	105	160	268	196	72	84	180	212	152	199	103	265	163
76	962	1256	1114	589	498	453	0	130	398	405	173	325	56	506	574	128	608	1136	580	882	584	726	132	1028	754
77	1225	1150	622	1094	0	54	41	289	758	294	592	323	301	906	665	449	473	759	639	811	125	340	901	123	925
78	35	256	73	0	271	671	1713	2733	2633	4479	3830	4564	4476	4396	3225	4688	5054	4824	4836	4122	3714	3703	4279	2996	3313
82	246	123	98	120	107	126	81	47	40	0	36	46	15	111	110	128	102	134	135	139	139	158	100	194	
86	131	59	89	94	68	48	72	33	2	5	0	15	5	9	61	42	92	50	143	86	121	131	115	66	96
94	188	108	139	46	40	92	45	23	5	25	0	17	61	104	115	121	117	59	125	173	198	187	40	38	92
95	0	721	2172	3216	3545	5536	4551	5060	4589	4578	4817	5014	4648	4550	4515	4203	5025	5535	3103	4273	2273	3468	2416	3073	2926
96	0	2246	5742	9371	13004	13709	11352	14185	12855	12223	9457	13212	12832	12833	11816	11459	11006	9069	9768	11437	7258	7432	7375	10484	8036
97	0	631	377	1123	1489	1309	1170	1437	1515	1613	1507	1426	1717	1528	1766	1624	1191	1201	1250	1322	965	828	704	1439	872
104	132	113	121	112	83	104	0	56	57	59	46	41	46	85	81	69	102	84	130	117	132	116	99	116	92
108	221	289	198	97	70	105	67	47	18	29	36	138	96	7	43	69	6	0	49	85	76	10	76	128	112
109	470	109	448	212	0	184	203	407	380	396	687	318	333	317	496	335	282	160	337	280	89	233	170	443	249
110	274	212	247	108	191	90	114	70	0	61	61	54	129	91	133	127	117	131	187	205	149	130	125	184	166
118	527	284	295	339	122	231	214	122	86	0	91	43	98	231	254	238	263	292	250	310	164	300	350	192	300
119	484	855	357	206	0	110	213	416	50	294	552	348	296	430	302	340	345	28	431	291	124	149	163	525	467
126	263	177	168	155	166	134	115	117	0	16	18	35	44	127	109	199	161	269	172	211	149	299	235	261	281
127	663	861	675	669	793	538	766	906	710	884	507	868	770	673	645	740	849	914	766	331	1050	758	0	880	1113
128	217	223	102	135	57	85	25	53	36	11	0	9	98	83	9	122	149	184	138	89	147	135	154	168	160
131	174	76	91	44	82	90	43	40	5	27	0	48	65	57	89	40	93	43	69	120	116	104	103	83	113
135	441	598	555	249	235	103	0	359	436	524	422	372	512	548	377	504	209	296	491	301	254	228	325	712	524
136	2452	3865	4081	0	1256	1492	485	3372	892	2608	1527	1531	2492	2470	1717	3113	2322	1753	2687	4391	3904	3323	2472	4982	2813
137	1118	234	222	68	123	72	557	412	485	190	606	352	0	151	278	74	681	518	437	574	458	303	768	268	783
138	216	152	15	184	71	57	145	22	21	11	0	27	54	111	42	66	176	133	166	146	172	184	143	87	221
140	286	262	149	189	237	139	151	74	40	14	0	32	47	90	85	57	138	86	128	148	185	177	213	186	219
141	283	295	194	107	45	13	435	9	96	66	71	57	28	91	13	10	50	76	27	27	19	39	25	0	108
142	553	304	187	332	256	243	269	38	51	95	0	151	53	178	110	123	217	343	320	304	351	363	262	245	378
144	360	190	191	124	158	138	104	62	0	4	13	11	28	65	81	52	100	110	149	197	179	196	164	157	173
149	204	80	133	104	73	104	119	27	33	0	19	35	43	86	60	104	129	112	103	106	87	196	133	141	115
154	0	497	1225	1782	2516	2625	2111	2027	1806	1737	1678	1478	1444	1358	1566	1200	1224	1046	907	1044	796	779	967	765	874
155	0	423	1179	1823	1670	1904	1900	1988	2013	2259	1929	2232	1886	2031	1819	2036	1909	1720	1708	1427	1408	1375	1250	1434	1388
158	101	19	45	54	0	41	27	5	46	28	43	57	42	74	63	52	69	91	61	50	54	83	51	11	88
167	547	396	231	196	286	282	160	77	75	0	29	89	180	112	147	155	172	314	238	301	377	390	320	337	259
168	269	146	132	207	163	93	85	73	21	68	0	35	58	19	98	124	137	184	161	101	220	189	112	129	147
169	334	207	206	229	92	132	43	141	148	0	107	105	123	179	139	173	202	218	183	253	180	180	140	163	206
172	58	0	137	345	633	876	1141	1261	1321	1575	1567	1328	1390	1657	1467	1447	1183	1310	1122	1005	815	838	920	679	773
195	380	187	184	195	164	159	0	37	20	6	51	78	64	95	180	221	98	221	204	302	206	318	223	217	305
196	200	415	160	125	82	28	71	202	172	145	115	124	15	47	11	63	52	71	96	0	45	57	88	11	54
197	109	178	250	115	77	50	0	151	236	206	257	211	140	184	246	146	147	150	166	166	20	140	220	249	159
198	92	97	54	39	44	33	63	26	5	45	0	17	8	35	46	122	48	70	37	29	64	38	25	29	49

The table above represents the intensity of 49 masses in LC-MS of a peak cluster recorded at 25 points in time. The aim of this exercise is to look at variable selection and the influence on PC plots. The data have been transposed to fit on a page, with the first column representing the mass numbers. Some pre-processing has already been performed with the original masses reduced a little and the ion current at each mass is set to a minimum of 0. You will probably wish to transpose the matrix so that the columns represent mass numbers.

1. Plot the total ion current (using the masses above) against time. This is done by summing the intensity over all masses at each point in time.
2. Perform PCA on the data set, but standardise the intensities at each mass and retain two PCs. Present the scores plot of PC2 versus PC1, labelling all the points in time, starting from 1 the lowest to 25 the highest. Produce a similar loadings plot, also labelling the points and comment on the correspondence between these graphs.
3. Repeat this but sum the intensities at each point in time to 1 before standardising and performing PCA and produce scores plot of PC2 versus PC1 and comment. Why might it be desirable to remove points in time 1–3? Repeat the procedure, this time using only points 4–25 in time. Produce PC2 versus PC1 scores and loadings plots and comment.
4. A very simple approach to variable selection involves sorting according to standard deviation. Take the standard deviation of the 49 masses using the raw data and list the 10 masses with highest standard deviations.
5. Perform PCA, standardised, again on the reduced 25 × 10 data set consisting of the best 10 masses according to the criterion in question 4 and present the labelled scores and loadings plots. Comment. Can you assign m/z values to the components in the mixture?

7.4 Use of Derivatives, MLR and PCR in Signal Analysis.
Section 7.3.5   Section 7.4.1.2   Section 7.4.1.3
The following data represent HPLC data recorded at 30 points in time and 10 wavelengths.

0.042	0.076	0.043	0.089	0.105	−0.004	0.014	0.030	0.059	0.112
0.009	0.110	0.127	0.179	0.180	0.050	0.015	0.168	0.197	0.177
−0.019	0.118	0.182	0.264	0.362	0.048	0.147	0.222	0.375	0.403
0.176	0.222	0.329	0.426	0.537	0.115	0.210	0.328	0.436	0.598
0.118	0.304	0.494	0.639	0.750	0.185	0.267	0.512	0.590	0.774
0.182	0.364	0.554	0.825	0.910	0.138	0.343	0.610	0.810	0.935
0.189	0.405	0.580	0.807	1.005	0.209	0.404	0.623	0.811	1.019
0.193	0.358	0.550	0.779	0.945	0.258	0.392	0.531	0.716	0.964
0.156	0.302	0.440	0.677	0.715	0.234	0.331	0.456	0.662	0.806
0.106	0.368	0.485	0.452	0.666	0.189	0.220	0.521	0.470	0.603
0.058	0.262	0.346	0.444	0.493	0.188	0.184	0.336	0.367	0.437
0.159	0.281	0.431	0.192	0.488	0.335	0.196	0.404	0.356	0.265
0.076	0.341	0.629	0.294	0.507	0.442	0.252	0.592	0.352	0.196

0.138	0.581	0.883	0.351	0.771	0.714	0.366	0.805	0.548	0.220
0.223	0.794	1.198	0.543	0.968	0.993	0.494	1.239	0.766	0.216
0.367	0.865	1.439	0.562	1.118	1.130	0.578	1.488	0.837	0.220
0.310	0.995	1.505	0.572	1.188	1.222	0.558	1.550	0.958	0.276
0.355	0.895	1.413	0.509	1.113	1.108	0.664	1.423	0.914	0.308
0.284	0.723	1.255	0.501	0.957	0.951	0.520	1.194	0.778	0.219
0.350	0.593	0.948	0.478	0.738	0.793	0.459	0.904	0.648	0.177
0.383	0.409	0.674	0.454	0.555	0.629	0.469	0.684	0.573	0.126
0.488	0.220	0.620	0.509	0.494	0.554	0.580	0.528	0.574	0.165
0.695	0.200	0.492	0.551	0.346	0.454	0.695	0.426	0.584	0.177
0.877	0.220	0.569	0.565	0.477	0.582	0.747	0.346	0.685	0.168
0.785	0.230	0.486	0.724	0.346	0.601	0.810	0.370	0.748	0.147
0.773	0.204	0.435	0.544	0.321	0.442	0.764	0.239	0.587	0.152
0.604	0.141	0.417	0.504	0.373	0.458	0.540	0.183	0.504	0.073
0.493	0.083	0.302	0.359	0.151	0.246	0.449	0.218	0.392	0.110
0.291	0.050	0.096	0.257	0.034	0.199	0.238	0.142	0.271	0.018
0.204	0.034	0.126	0.097	0.092	0.095	0.215	0.050	0.145	0.034

The aim of this problem is to explore different approaches to signal resolution using a variety of common chemometric methods.

1. Plot a graph of the sum of intensities at each point in time. Verify that it looks as if there are three peaks in the data.
2. Calculate the derivative of the spectrum, scaled at each point in time to a constant sum and at each wavelength as follows.
   (a) Re-scale the spectrum at each point in time by dividing the total intensity at that point in time so that the total intensity at each point in time equals 1.
   (b) Then, calculate the smoothed 5-point quadratic Savitsky–Golay first derivatives, as presented in Table 3.6, independently for each of the 10 wavelengths. A table consisting of derivatives at 26 times and 10 wavelengths should be obtained.
   (c) Superimpose the 10 graphs of derivatives at each wavelength.
3. Summarise the change in derivative with time by calculating the mean of the absolute value of the derivative over all 10 wavelengths at each point in time. Plot a graph of this and explain why a value close to zero indicates a good pure or composition 1 point in time. Show that this suggests points 6, 17 and 26 are good estimates of pure spectra for each component.
4. The concentration profiles of each component can be estimated using MLR as follows.
   (a) Obtain estimates of the spectra of each pure component at the three points of highest purity, to give an estimated spectral matrix $\widehat{S}$.
   (b) Using MLR, calculate $\widehat{C} = X\widehat{S}'(\widehat{S}\widehat{S}')^{-1}$
   (c) Plot a graph of the predicted concentration profiles.
5. An alternative method is PCR. Perform *uncentred* PCA on the raw data matrix $\mathbf{X}$ and verify that there are approximately three components.
6. Using estimates of each pure component given in question 4(a), perform PCR as follows.
   (a) Using regression, find the matrix R for which $\widehat{S} = R^{-1}P$, where P is the loadings matrix obtained in question 5; keep three PCs only.
   (b) Estimate the elution profiles of all the three peaks as $\widehat{C} \approx TR$.
   (c) Plot these graphically.

7.5 Titration of Three Spectroscopically Active Compounds with pH.
Section 7.2.2   Section 7.3.3   Section 7.4.1.2
The following data represent the spectra of a mixture if three spectroscopically active species are recorded at 25 wavelengths over 36 different values of pH.

	2.15	2.24	2.44	2.68	3.00	3.25	3.47	3.72	4.04	4.40	4.77	5.06	5.40	5.68	5.98	6.25	6.49	6.85	7.00	7.47	7.75	7.96	8.12	8.51	8.82	9.11	9.38	9.61	9.89	10.38	10.57	10.74	11.01	11.27	11.47	11.64
336	0.000	0.000	0.000	0.000	0.001	0.001	0.001	0.002	0.002	0.002	0.002	0.002	0.002	0.002	0.002	0.002	0.002	0.003	0.003	0.003	0.004	0.004	0.004	0.005	0.004	0.005	0.005	0.005	0.006	0.006	0.006	0.006	0.007	0.007	0.008	0.008
332	0.000	0.000	0.001	0.001	0.001	0.002	0.002	0.002	0.002	0.001	0.001	0.002	0.002	0.002	0.002	0.002	0.001	0.003	0.003	0.003	0.003	0.004	0.004	0.005	0.005	0.004	0.005	0.007	0.006	0.006	0.006	0.007	0.007	0.008	0.008	0.008
328	0.000	0.001	0.001	0.001	0.001	0.003	0.003	0.002	0.002	0.002	0.002	0.002	0.002	0.003	0.003	0.003	0.003	0.003	0.004	0.004	0.004	0.004	0.005	0.005	0.005	0.005	0.006	0.005	0.006	0.006	0.006	0.006	0.007	0.007	0.007	0.008
324	0.001	0.001	0.001	0.002	0.001	0.003	0.003	0.003	0.003	0.003	0.002	0.002	0.003	0.003	0.003	0.003	0.003	0.003	0.004	0.004	0.006	0.005	0.005	0.006	0.006	0.006	0.006	0.007	0.007	0.007	0.007	0.008	0.009	0.008	0.008	0.009
320	0.002	0.002	0.002	0.003	0.002	0.005	0.005	0.004	0.003	0.003	0.003	0.003	0.003	0.004	0.004	0.003	0.004	0.004	0.004	0.005	0.005	0.006	0.005	0.005	0.006	0.006	0.007	0.007	0.007	0.008	0.008	0.008	0.009	0.009	0.009	0.009
316	0.005	0.005	0.004	0.005	0.004	0.005	0.005	0.005	0.004	0.004	0.004	0.004	0.004	0.005	0.005	0.004	0.004	0.005	0.005	0.005	0.006	0.006	0.006	0.006	0.006	0.006	0.007	0.007	0.007	0.008	0.008	0.008	0.008	0.008	0.009	0.010
312	0.006	0.006	0.005	0.006	0.006	0.007	0.007	0.007	0.005	0.005	0.005	0.005	0.005	0.004	0.004	0.004	0.005	0.006	0.005	0.006	0.006	0.006	0.007	0.007	0.007	0.007	0.008	0.010	0.008	0.010	0.009	0.009	0.009	0.009	0.012	0.010
308	0.013	0.012	0.011	0.012	0.011	0.011	0.011	0.010	0.009	0.008	0.007	0.007	0.007	0.007	0.007	0.006	0.007	0.008	0.007	0.010	0.008	0.008	0.008	0.009	0.009	0.009	0.010	0.011	0.010	0.011	0.010	0.012	0.012	0.012	0.013	0.012
304	0.026	0.025	0.023	0.023	0.021	0.021	0.019	0.018	0.015	0.013	0.010	0.009	0.009	0.009	0.008	0.009	0.008	0.008	0.009	0.013	0.010	0.010	0.010	0.011	0.011	0.011	0.012	0.013	0.012	0.013	0.013	0.013	0.013	0.014	0.015	0.015
300	0.056	0.054	0.051	0.048	0.045	0.042	0.039	0.034	0.028	0.023	0.018	0.016	0.014	0.013	0.013	0.012	0.013	0.013	0.013	0.015	0.014	0.015	0.014	0.015	0.015	0.016	0.017	0.016	0.016	0.015	0.016	0.015	0.016	0.017	0.016	0.017
296	0.104	0.102	0.098	0.091	0.084	0.079	0.072	0.064	0.053	0.044	0.037	0.032	0.028	0.026	0.026	0.025	0.025	0.026	0.025	0.026	0.027	0.027	0.027	0.028	0.027	0.027	0.028	0.026	0.025	0.021	0.020	0.021	0.020	0.020	0.021	0.022
292	0.166	0.162	0.156	0.147	0.137	0.129	0.119	0.107	0.092	0.079	0.069	0.063	0.058	0.056	0.055	0.055	0.054	0.055	0.055	0.058	0.056	0.057	0.057	0.056	0.055	0.055	0.054	0.050	0.047	0.040	0.037	0.037	0.035	0.035	0.036	0.036
288	0.234	0.229	0.223	0.214	0.202	0.192	0.180	0.167	0.149	0.135	0.124	0.116	0.110	0.109	0.109	0.107	0.107	0.108	0.108	0.108	0.109	0.110	0.110	0.110	0.108	0.107	0.104	0.100	0.094	0.085	0.081	0.081	0.079	0.078	0.077	0.079
284	0.306	0.302	0.296	0.287	0.274	0.263	0.253	0.238	0.218	0.203	0.192	0.185	0.180	0.179	0.178	0.177	0.178	0.178	0.179	0.179	0.181	0.181	0.183	0.181	0.179	0.178	0.175	0.172	0.169	0.163	0.160	0.158	0.157	0.156	0.156	0.157
280	0.379	0.376	0.372	0.364	0.352	0.343	0.331	0.317	0.300	0.285	0.274	0.269	0.264	0.262	0.263	0.262	0.263	0.264	0.264	0.266	0.268	0.269	0.270	0.270	0.270	0.270	0.269	0.270	0.270	0.270	0.270	0.270	0.271	0.270	0.270	0.271
276	0.449	0.448	0.444	0.437	0.428	0.421	0.410	0.397	0.382	0.371	0.361	0.356	0.354	0.353	0.352	0.353	0.354	0.355	0.356	0.358	0.361	0.362	0.362	0.364	0.366	0.368	0.371	0.375	0.380	0.390	0.392	0.395	0.396	0.399	0.399	0.399
272	0.520	0.518	0.516	0.508	0.501	0.496	0.487	0.477	0.464	0.457	0.449	0.447	0.445	0.446	0.446	0.445	0.447	0.449	0.450	0.451	0.456	0.456	0.457	0.458	0.462	0.469	0.471	0.478	0.487	0.503	0.508	0.510	0.514	0.516	0.517	0.517
268	0.584	0.582	0.578	0.574	0.567	0.566	0.559	0.550	0.542	0.537	0.531	0.532	0.531	0.532	0.533	0.534	0.536	0.538	0.538	0.539	0.545	0.546	0.547	0.548	0.553	0.557	0.563	0.570	0.580	0.598	0.604	0.608	0.612	0.614	0.614	0.615
264	0.632	0.630	0.626	0.623	0.616	0.617	0.613	0.607	0.602	0.601	0.599	0.601	0.601	0.602	0.605	0.606	0.608	0.611	0.611	0.613	0.618	0.619	0.620	0.622	0.628	0.627	0.633	0.640	0.649	0.664	0.670	0.674	0.678	0.679	0.680	0.680
260	0.663	0.661	0.657	0.655	0.650	0.653	0.652	0.650	0.648	0.650	0.649	0.651	0.654	0.656	0.658	0.660	0.662	0.665	0.667	0.668	0.673	0.676	0.674	0.673	0.677	0.677	0.680	0.684	0.688	0.699	0.701	0.704	0.707	0.708	0.709	0.708
256	0.673	0.672	0.671	0.669	0.668	0.674	0.673	0.670	0.672	0.675	0.676	0.680	0.683	0.685	0.687	0.690	0.692	0.696	0.697	0.698	0.703	0.704	0.705	0.702	0.702	0.702	0.701	0.702	0.702	0.707	0.708	0.709	0.710	0.712	0.712	0.710
252	0.638	0.639	0.640	0.642	0.643	0.651	0.652	0.651	0.653	0.658	0.661	0.666	0.670	0.671	0.675	0.677	0.680	0.683	0.684	0.685	0.691	0.692	0.692	0.689	0.688	0.685	0.684	0.682	0.680	0.680	0.681	0.680	0.680	0.680	0.681	0.680
248	0.571	0.573	0.576	0.582	0.586	0.597	0.599	0.599	0.604	0.611	0.615	0.623	0.626	0.628	0.631	0.635	0.636	0.639	0.641	0.642	0.647	0.648	0.649	0.645	0.644	0.642	0.639	0.638	0.634	0.632	0.631	0.631	0.631	0.631	0.631	0.631
244	0.479	0.482	0.488	0.496	0.503	0.515	0.519	0.522	0.528	0.537	0.544	0.550	0.555	0.559	0.562	0.566	0.566	0.570	0.571	0.572	0.578	0.580	0.580	0.579	0.578	0.578	0.576	0.575	0.573	0.573	0.573	0.573	0.573	0.573	0.575	0.576
240	0.382	0.386	0.391	0.402	0.409	0.424	0.430	0.435	0.444	0.454	0.462	0.470	0.476	0.480	0.483	0.484	0.487	0.490	0.492	0.494	0.500	0.501	0.503	0.503	0.506	0.509	0.511	0.515	0.516	0.520	0.522	0.523	0.524	0.526	0.527	0.529

1. Perform PCA on the raw uncentred data and obtain the scores and loadings for the first three PCs.
2. Plot a graph of the loadings of the first PC and superimpose this on the graph of the average spectrum over all the observed pHs, scaling the two graphs so that they are of approximately similar size. Comment why the first PC is not very useful for discriminating between the compounds.
3. Calculate the log of the correlation coefficient between each successive spectrum and plot this against pH (there will be 35 numbers, plot the log of the correlation between the spectra at pH 2.15 and 2.24 against the lower pH). Show how this is consistent with three different spectroscopic species in the mixture. On the basis of three components, are there pure regions for each component and over which pH ranges are these?
4. Centre the data and produce three scores plots, PC2 versus PC1, PC3 versus PC1 and PC3 versus PC2. Label each point with pH (Excel users will have to adapt the macro provided). Comment on these plots, especially in light of the correlation graph in question 3.
5. Normalise the scores of the first two PCs obtained in question 4 by dividing the square root of the sum of squares at each pH. Plot the graph of the normalised scores of PC2 versus PC1, labelling each point as in question 4, and comment.
6. Using the information given above, choose one pH that best represents the spectra for each of the three compounds (there may be several answers to this, but they should not differ by a great deal). Plot the spectra of each pure compound, superimposed on one another.
7. Using the guesses of the spectra for each compound in question 7 given above, perform MLR to obtain estimated profiles for each species by $\hat{C} = XS'(SS')^{-1}$. Plot a graph of the pH profiles of each species.

7.6 Resolution of Mid-Infrared Spectra of a Three-Component Mixture.
Section 7.2.2   Section 7.2.4.1   Section 7.4.1.2

	2000	1984	1968	1952	1936	1920	1904	1888	1872	1856	1840	1824	1808	1792	1776	1760	1744	1728	1712	1696	1680	1664	1648	1632	1616
1	211	214	256	381	422	531	456	354	382	421	447	312	398	426	425	657	617	472	647	359	343	378	488	840	5859
2	143	145	167	241	262	301	259	246	247	224	237	197	218	222	215	581	514	318	521	252	221	224	271	688	6503
3	164	169	258	489	484	374	285	265	424	412	368	272	472	495	336	475	487	379	461	303	285	265	325	738	5062
4	184	189	252	427	440	416	339	307	395	384	371	282	409	428	341	618	588	418	582	330	304	300	370	834	6563
5	211	217	310	552	574	578	469	340	495	566	549	355	590	639	524	459	502	485	513	368	374	399	525	756	3027
6	272	277	336	508	556	673	576	458	503	542	570	404	521	556	539	866	812	611	844	468	444	481	617	1110	7995
7	190	197	305	587	581	455	347	304	504	508	455	324	579	612	421	491	524	440	497	349	337	319	399	811	4825

	1600	1584	1568	1552	1536	1520	1504	1488	1472	1456	1440	1424	1408	1392	1376	1360	1344	1328	1312	1296	1280	1264	1248	1232	1216
1	4251	1936	1507	1617	1956	2318	3339	5298	5282	4086	2429	1090	1153	1810	1453	771	585	457	434	382	356	397	432	339	363
2	4053	1407	1282	1412	1739	2051	2599	3195	2844	2635	1515	720	853	1245	1117	605	432	294	272	248	214	223	225	181	201
3	3587	1355	1163	1281	1662	2006	3724	4358	3345	2862	1755	879	866	1377	1141	600	432	332	326	284	248	267	319	270	316
4	4408	1657	1433	1573	1976	2356	3669	4573	3874	3343	1992	961	1025	1575	1337	713	516	380	363	322	283	304	338	277	314
5	2983	1707	1173	1237	1543	1869	3785	6011	5666	4036	2505	1143	1042	1771	1299	669	524	461	456	387	371	426	507	406	448
6	5693	2501	1985	2139	2602	3085	4468	6806	6642	5229	3106	1409	1497	2338	1898	1008	759	586	557	490	454	503	546	431	465
7	3678	1520	1236	1352	1759	2134	4243	5233	4140	3354	2080	1029	971	1586	1268	661	484	390	387	333	298	326	397	333	387

	1200	1184	1168	1152	1136	1120	1104	1088	1072	1056	1040	1024	1008	992	976	960	944	928	912	896	880	864	848	832	816
1	372	352	384	356	329	390	1091	1462	972	839	988	887	936	959	718	410	343	553	420	382	290	274	1556	2919	684
2	218	194	216	213	207	223	434	579	506	650	751	507	595	558	353	215	228	466	316	286	186	186	1802	3472	663
3	314	270	290	271	258	334	696	1151	994	830	1143	982	585	578	447	300	289	468	362	351	260	229	1180	2182	502
4	322	285	310	295	282	338	732	1094	904	865	1099	885	725	707	507	320	310	546	401	375	267	248	1691	3198	674
5	437	407	435	389	353	468	1371	2007	1355	873	1174	1235	881	952	808	479	367	468	408	391	331	287	582	951	409
6	478	448	488	455	424	500	1349	1828	1255	1112	1323	1159	1190	1209	898	519	444	732	551	503	376	355	2120	3987	910
7	381	331	354	326	306	406	927	1509	1232	938	1313	1196	686	698	568	374	339	506	408	399	308	266	1049	1891	495

	800	784	768	752	736	720	704	688	672	656	640	624	608	592	576	560	544	528
1	296	1238	4468	3372	2742	2709	2229	2593	1091	264	240	230	258	287	347	554	625	598
2	171	452	1404	1130	1084	1089	1071	2363	972	169	155	134	163	149	210	265	322	366
3	220	625	1924	2592	4946	5042	3565	3207	1376	252	222	216	239	288	357	451	538	658
4	240	725	2336	2396	3520	3568	2710	3175	1340	251	225	211	240	264	338	444	527	606
5	327	1482	5384	4846	5896	5921	4201	2737	1200	313	277	287	304	395	447	713	797	798
6	377	1510	5384	4180	3686	3659	2982	3550	1493	344	312	297	335	370	453	701	797	784
7	267	867	2821	3492	6235	6342	4432	3533	1529	299	263	262	285	357	430	571	670	791

The table given above is of seven spectra consisting of different mixtures of three compounds, 1,2,3-trimethylbenzene, 1,3,5-trimethylbenzene and toluene, whose mid-infrared spectra have been recorded at $16 \, \text{cm}^{-1}$ intervals between 528 and 2000 nm, which you will need to reorganise as a matrix of dimensions 7×93.

1. Scale the data so that the sum of the spectral intensities at each wavelength equals 1 (note that this differs from the usual method, which is along the rows, and is a way of putting equal weight on each wavelength). Perform PCA, without further pre-processing, and produce a plot of the loadings of PC2 versus PC1.

2. Many wavelengths are not very useful if they are of low intensity. Identify those wavelengths for which the sum of overall seven spectra is greater than 10% of the wavelength that has the maximum sum and label these in the graph of question 1.

3. Comment on the appearance of the graph from question 2 and suggest three wavelengths that are typical of each of the compounds.

4. Using the three wavelengths selected in question 3, obtain a 7×3 matrix of relative concentrations in each of the spectra and call this $\hat{C}$.

5. Calling the original data X, obtain the estimated spectra for each compound by $S = (\hat{C}'\hat{C})^{-1}\hat{C}'X$ and plot these graphically.

Appendix

A.1 Vectors and Matrices

A.1.1 Notation and Definitions

A single number is often called a *scalar* and is represented by italics, for example, x.

A vector consists of a row or column of numbers and is represented by bold lower case italics, for example, $\boldsymbol{x}$. For example, $\boldsymbol{x} = \begin{bmatrix} 3 & -11 & 9 & 0 \end{bmatrix}$ is a row vector and $\boldsymbol{y} = \begin{bmatrix} 5.6 \\ 2.8 \\ 1.9 \end{bmatrix}$ is a column vector.

A matrix is a two-dimensional array of numbers and is represented by bold upper case italics, for example, X. For example, $X = \begin{bmatrix} 12 & 3 & 8 \\ -2 & 14 & 1 \end{bmatrix}$ is a matrix.

The dimensions of a matrix are normally presented with the number of rows first and the number of columns second, and vectors can be considered as matrices with one dimension equal to 1, so that $\boldsymbol{x}$ above has dimensions 1×4 and X has dimensions 2×3.

A square matrix is one where the number of columns equals the number of rows. For example, $Y = \begin{bmatrix} -7 & 4 & -1 \\ 11 & -3 & 6 \\ 2 & 4 & -12 \end{bmatrix}$ is a square matrix.

An identity matrix is a square matrix whose elements are equal to 1 in the diagonal and 0 elsewhere and is often denoted by I. For example, $I = \begin{bmatrix} 1 & 0 \\ 0 & 1 \end{bmatrix}$ is an identity matrix.

Each number in a matrix is called an *element* and is often referenced as a scalar, with subscripts referring to the row and column; hence, in the matrix above, $y_{21} = 11$, which is the element in row 2 and column 1. Optionally, a comma can be placed between the subscripts for clarity; this is useful if one of the dimensions exceeds 9.

A.1.2 Matrix and Vector Operations

A.1.2.1 Addition and Subtraction

Addition and subtraction are the most straightforward operations. Each matrix (or vector) must have the same dimensions and simply involves performing the operation element by element. Hence,

$$\begin{bmatrix} 9 & 7 \\ -8 & 4 \\ -2 & 4 \end{bmatrix} + \begin{bmatrix} 0 & -7 \\ -11 & 3 \\ 5 & 6 \end{bmatrix} = \begin{bmatrix} 9 & 0 \\ -19 & 7 \\ 3 & 10 \end{bmatrix}$$

A.1.2.2 Transpose

Transposing a matrix involves swapping the columns and rows around and may be denoted by a right-hand side superscript "'". For example, if $Z = \begin{bmatrix} 3.1 & 0.2 & 6.1 & 4.8 \\ 9.2 & 3.8 & 2.0 & 5.1 \end{bmatrix}$, then $Z' = \begin{bmatrix} 3.1 & 9.2 \\ 0.2 & 3.8 \\ 6.1 & 2.0 \\ 4.8 & 5.1 \end{bmatrix}$. Some authors used a superscripted 'T' instead.

Chemometrics: Data Driven Extraction for Science, Second Edition. Richard G. Brereton.
© 2018 John Wiley & Sons Ltd. Published 2018 by John Wiley & Sons Ltd.
Companion website: http://booksupport.wiley.com

A.1.2.3 Multiplication

Matrix and vector multiplication using the 'dot' product is sometimes denoted by the symbol '·' between matrices, to distinguish from the cross product sometimes encountered in physics but rarely in chemometrics. However, authors often omit the '·' as we will do in this text.

It is only possible to multiply two matrices together if the number of columns of the first matrix equals the number of rows of the second matrix. The number of rows of the product will equal the number of rows of the first matrix, and the number of columns equals the number of columns of the second matrix. Hence, a 3×2 matrix when multiplied by a 2×4 matrix will give a 3×4 matrix.

Multiplication of matrices is not commutative, that is generally $AB \neq BA$ even if the second product is allowed. Matrix multiplication can be expressed in the form of summations. For arrays with more than two dimensions (e.g. tensors), conventional symbolism can be quite awkward and it is probably easier to think in terms of summations.

If matrix A has dimensions $I \times J$ and matrix B has dimensions $J \times K$, then the product C of dimensions $I \times K$ has elements defined by

$$c_{ik} = \sum_{j=1}^{J} a_{ij} b_{jk}$$

Hence,

$$\begin{bmatrix} 1 & 7 \\ 9 & 3 \\ 2 & 5 \end{bmatrix} \begin{bmatrix} 6 & 10 & 11 & 3 \\ 0 & 1 & 8 & 5 \end{bmatrix} = \begin{bmatrix} 6 & 17 & 67 & 38 \\ 54 & 93 & 123 & 42 \\ 12 & 25 & 62 & 31 \end{bmatrix}$$

To illustrate this, the element of the first row and second column of the product is given by

$$17 = 1 \times 10 + 7 \times 1$$

When several matrices are multiplied together, it is normal to take any two neighbouring matrices, multiply them together and then multiply this product with another neighbouring matrix. It does not matter what order this is done in; hence, $ABC = (AB)C = A(BC)$. Hence, matrix multiplication is associative. Matrix multiplication is also distributive, that is $A(B + C) = AB + AC$.

A.1.2.4 Inverse

Most square matrices have inverses, defined by the matrix which when multiplied with the original matrix gives the identity matrix and is represented by a superscript '−1' as a right-hand side superscript, so that $DD^{-1} = I$. Note that some square matrices do not have inverses: this is caused by there being correlations in the original matrix which results in the determinant being zero: such matrices are called *singular* matrices.

A.1.2.5 Pseudo-Inverse

In several sections of this text, we use the idea of a pseudo-inverse. If matrices are not square, it is not possible to calculate an inverse, but the analogous concept of a pseudo-inverse exists and is employed in regression analysis.

If $A = BC$ then $B'A = B'BC$ so $(B'B)^{-1}B'A = C$ and $(B'B)^{-1}B'$ is said to be the left pseudo-inverse of B.

Equivalently, $AC' = BCC'$ so $AC'(CC')^{-1} = B$ and $C'(CC')^{-1}$ is said to be the right pseudo-inverse of C.

Sometimes, the pseudo-inverse is denoted by a superscript '+'. Hence, B^+ is the pseudo-inverse of B. There is not usually a distinction in notation between right and left, but the reader should be able to work this out from context. Note that if a matrix is not square (in which case there is no need to calculate a pseudo-inverse), only one of the two possible pseudo-inverses will exist: if the number of rows is less than the columns, it is the right pseudo-inverse, otherwise the left one.

In regression, the equation $A \approx BC$ is an approximation; for example, A may represent a series of spectra that are approximately equal to the product of two matrices such as scores and loadings matrices, and the pseudo-inverse can be used to obtain the best fit model (or estimate) of C knowing A and B or of B knowing A and C.

A.1.2.6 Trace and Determinant

Other properties of square matrices sometimes encountered are the trace, which is the sum of the diagonal elements and the determinant which relates to the size of the matrix. A determinant of 0 indicates a matrix without an inverse. A determinant close to 0 often suggests that the data are quite correlated or a poor experimental design resulting in fairly unreliable predictions or that there are many redundant noisy variables. If the dimensions of matrices are large and the magnitudes of the elements are small (e.g. 10^{-3}), it is sometimes possible to obtain a determinant close to zero even though the matrix has an inverse; a solution to this problem is to multiply each element by a number such as 10^3 and then remember to readjust the magnitude of the numbers in resultant computation to take into account of this later.

A.1.2.7   Vector Magnitude

An interesting property chemometricians sometimes use is that the product of the transpose of a column vector with itself equals the sum of square of elements of the vector, so that $x'x = \Sigma x^2$. The magnitude, or length, of a vector is given by $\sqrt{(x'x)} = \sqrt{\sum x^2}$ or the square root of the sum of its elements. This can be visualised in geometry as the length of the line from the origin to the point in space indicated by the vector co-ordinates. Note that there is some confusion in terminology, as the length of a vector can also denote the number of elements in a vector, so the term magnitude is less ambiguous.

A.2   Algorithms

There are many different descriptions of the various algorithms in the literature. This Appendix describes one algorithm for each of four regression methods.

A.2.1   Principal Components Analysis

NIPALS is a common, iterative algorithm often used for principal components analysis (PCA). Some authors use another method called *singular value decomposition* (SVD). The main difference is that NIPALS extracts components one at a time and can be stopped after the desired number of PCs is obtained. In the case of large data sets with, for example, 200 variables (e.g. in spectroscopy), this can be very useful and reduce the amount of effort required. In this text, we will restrict to NIPALS for simplicity.

The steps are as follows.

Initialisation

1. Take a matrix Z and if required pre-process (e.g. mean centre or standardise) to give the matrix X, which is used for PCA.

New Principal Component

2. Take a column of this matrix (often the column with greatest sum of squares) as the first guess of the scores first principal component, call it $^{initial}\hat{t}$.

Iteration for Each Principal Component

3. Calculate

$$^{unnorm}\hat{p} = \frac{^{initial}\hat{t}' X}{\sum \hat{t}^2}$$

4. Normalise the guess of the loadings, so

$$\hat{p} = \frac{^{unnorm}\hat{p}}{\sqrt{\sum {}^{unnorm}\hat{p}^2}}$$

5. Now calculate a new guess of the scores

$$^{new}\hat{t} = X\hat{p}'$$

Check for Convergence

6. Check if this new guess differs from the first guess, a simple approach is to look at the size of the sum of square difference in the old and new scores, that is, $\sum (^{initial}\hat{t} - {}^{new}\hat{t})^2$. If this is small, the PC has been extracted, set the PC scores (t) and loadings (p) for the current PC to $\hat{t}$ and $\hat{p}$. Otherwise, return to step 3, substituting the initial scores by the new scores.

Compute the Component and Calculate Residuals

7. Subtract the effect of the new PC from the data matrix to get a residual data matrix

$$^{resid}X = X - tp$$

Further PCs

8. If it is desired to compute further PCs, substitute the residual data matrix for X and go to step 2.

The scores of a new sample that was not part of the original training set (e.g. test set) can be estimated by

$$\hat{t} = xP'$$

where $\hat{t}$ is a row vector of dimensions $1 \times A$ with A the number of components.

A.2.2   PLS1

There are several implementations, the one below is non-iterative.

Initialisation

1. Take a matrix $I \times J$ and if required pre-process (e.g. mean centre or standardise) to give the matrix X which is used for PLS.
2. Take the concentration (or c) vector and pre-process it to give the vector c, which is used for PLS. Note that if the data matrix Z is centred down the columns, c must also be centred. Generally, centring is the only form of pre-processing useful for PLS1. Start with an estimate of $\hat{c}$ that is a vector of 0s (equal to its mean if the vector is already centred).

New PLS Component

3. Calculate the $J \times 1$ vector h

$$h = X'c$$

4. Normalise this vector to give w, which is called the *weights vector*

$$w = \frac{h}{\sqrt{\sum h^2}}$$

5. Calculate the $I \times 1$ scores vector, which is simply given by

$$t = Xw$$

6. Calculate the $1 \times Jx$ loadings by

$$p = \frac{t'X}{\sum t^2}$$

7. Calculate the c loading (a scalar) by

$$q = \frac{c't}{\sum t^2}$$

Compute the Component and Calculate Residuals

8. Subtract the effect of the new PLS component from the data matrix to get a residual data matrix

$$^{resid}X = X - tp$$

9. Determine the new estimate of c by

$$^{new}\hat{c} = {}^{initial}\hat{c} + tq$$

and sum the contribution of all components calculated to give an estimated $\hat{c}$. Note that the initial estimate is 0 (or the mean) before the first component has been computed. Calculate

$$^{resid}c = {}^{true}c - {}^{new}\hat{c}$$

where ^{true}c is, like all values of c, after the data have been pre-processed (such as centring).

Further PLS Components

10. If further components are required, replace both X and c by the residuals and return to step 3.

Note that in the implementation used in this text, the PLS loadings are neither normalised nor orthogonal. However, the weights vectors are both orthogonal and normal. Remember to add back the mean when comparing the estimates with the observed data.

There are several different PLS1 algorithms; hence, it is useful to check exactly what method a particular package uses, although the resultant concentration estimates should be identical for each method (unless there is a problem with convergence in iterative approaches).

A separate step involves prediction of test set or unknown samples. Unlike PCA, PLS does not result in a best least squares fit to the original X matrix, so prediction is more complicated than for PCA. There are several ways of expressing this, of which the following is an easy one for A components.

1. Calculate $b = W \, (PW)^{-1} q$, where W is the $J \times A$ weights matrix, each column representing a component. P the $A \times J x$ loadings matrix, and q an $A \times 1$ vector of the c loadings.
2. As appropriate, calculate $b_0 = \bar{c} - \bar{x} b$ if it is required to transform the predictions to the original data.
3. Estimate $\hat{c} = b_0 + xb$.

Similar estimations could be made for the x block but the reader is referred to the specialist literature.

A.2.3 PLS2

This is a straightforward, iterative extension of PLS1. Only small variations are required. Instead of c being a vector, it is now a $I \times N$ matrix C and instead of q being a scalar, it is now a $1 \times N$ vector q, where there are N columns in the c matrix (e.g. concentrations of analytes).

Initialisation
 1. Take the data matrix and if required pre-process (e.g. mean centre or standardise) to give the matrix X, which is used for PLS.
 2. Take the concentration matrix and pre-process it to give the matrix C, which is used for PLS. Note that if the data matrix is centred down the columns, the vector matrix must also be centred. Generally, centring is the only form of pre-processing useful for PLS2. Start with an estimate of $\hat{C}$ that is a matrix of 0s (equal to the mean value of c if already centred).

New PLS Component
 3. An extra step is required to identify a vector u, which can be a guess (as in PCA), but can be chosen as one of the columns in the initial pre-processed concentration matrix, C.
 4. Calculate the vector

$$h = X' u$$

 5. Normalise to get the weights by

$$w = \frac{h}{\sqrt{\sum h^2}}$$

 6. Calculate the guessed scores by

$$^{new}\hat{t} = Xw$$

 7. Calculate the guessed x loadings by

$$\hat{p} = \frac{\hat{t}' X}{\sum \hat{t}^2}$$

 8. Calculate the c loadings (a $1 \times N$ vector rather than scalar in PLS2) by

$$\hat{q} = \frac{C' \hat{t}}{\sum \hat{t}^2}$$

 9. If this is the first iteration, remember the scores and call them $^{initial}t$, then produce a new vector u by

$$u = \frac{C \hat{q}}{\sum q^2}$$

and return to step 4.

Check for Convergence

10. If this is the second time, round compare the new and old scores vectors, for example, by looking at the size of the sum of square difference in the old and new scores, that is, $\sum (^{initial}\hat{t} - ^{new}\hat{t})^2$. If this is small, the PLS component has been adequately modelled, set the PLS scores (t) and both types of loadings (p and c) for the current PC to $\hat{t}$ and $\hat{p}$, and $\hat{q}$. Otherwise, calculate a new value of u as in step 8 and return to step 4.

Compute the Component and Calculate the Residuals

11. Subtract the effect of the new PLS component from the data matrix to get a residual data matrix

$$^{resid}X = X - tp$$

12. Determine the new c matrix estimate by

$$^{new}\hat{C} = ^{initial}\hat{C} + tq$$

and sum the contribution of all components calculated to give an estimated $\hat{c}$. Calculate

$$^{resid}C = ^{true}C - \hat{C}$$

Further PLS Components

13. If further components are required, replace both X and C by the residuals and return to step 3.

Prediction of new samples that are not part of the training set involves similar principles to PLS1, except we are predicting a C matrix rather than vector.

A.2.4   Tri-Linear PLS1

The algorithm below is closely based on PLS1 and is suitable when there is only one column in the c vector.

Initialisation

1. Take a three-way tensor $\underline{Z}$ and if required pre-process (e.g. mean centre or standardise) to give the tensor $\underline{X}$, which is used for PLS. Perform all pre-processing on this tensor. The tensor has dimensions $I \times J \times K$.
2. Pre-process the concentrations if appropriate to give a vector c.

New PLS Component

3. From the original tensor, create a new matrix H with dimensions $J \times K$, which is the sum of each of the I matrices for each of the samples multiplied by the concentration of the analyte for the relevant sample, that is,

$$H = X_1 c_1 + X_2 c_2 + \cdots + X_I c_I$$

or, as a summation

$$h_{jk} = \sum_{i=1}^{I} c_i x_{ijk}$$

4. Perform PCA on H to obtain the scores and loadings, $^h t$ and $^h p$, for the first PC of H. Note that only the first PC is retained, and for each new PLS component, a fresh H matrix is obtained.
5. Calculate the two x loadings for the current PLS component of the overall data set by normalising the scores and loadings of H, that is,

$$^j p = \frac{^h t'}{\sqrt{\sum {^h t^2}}}$$

$$^k p = \frac{^h p}{\sqrt{\sum {^h p^2}}}$$

(This step is generally not necessary for most PCA algorithms.)

6. Calculate the overall scores by

$$t_i = \sum_{j=1}^{J} \sum_{k=1}^{K} x_{ijk} {^j p_j} {^k p_k}$$

7. Calculate the c loadings vector

$$q = (T'T)^{-1}T'c$$

where T is the scores matrix each column consisting of one component (a vector for the first PLS component).

Compute the Component and Calculate Residuals

8. Subtract the effect of the new PLS component from the original data matrix to get a residual data matrix (for each sample i)

$$^{resid}X_i = X_i - t_i^j p^k p$$

9. Determine the new concentration estimates by

$$\hat{c} = Tq$$

Calculate

$$^{resid}c = {}^{true}c - \hat{c}$$

Further PLS Components

10. If further components are required, replace both X and c by the residuals and return to step 3.

A.3 Basic Statistical Concepts

There are numerous texts on basic statistics, some of them oriented towards chemists. It is not the aim of this section to provide a comprehensive background, but simply to provide the main definitions and tables that are helpful for using this text.

A.3.1 Descriptive Statistics

A.3.1.1 Mean
The mean of a series of measurements is defined by

$$\bar{x} = \sum_{i=1}^{I} x_i / I$$

Conventionally, a bar is placed above the letter. Sometimes, the letter m is used, but in this text, we will avoid this, as m is often used to denote an index. Hence, the mean of the measurement

$$4 \quad 8 \quad 5 \quad -6 \quad 2 \quad -5 \quad 6 \quad 0$$

$$\bar{x} = (4 + 8 + 5 - 6 + 2 - 5 + 6 + 0)/8 = 1.75$$

Statistically, this sample mean is often considered an estimate of the true population mean, sometimes denoted by μ. The population involves all possible samples of a given provenance, whereas only a selection is observed. In some cases, in chemometrics, this distinction is not so clear; for example, the mean intensity at a given wavelength over a chromatogram is an experimentally observed variable and it is not clear whether this mean is an estimate of all possible spectra of a given provenance or just something the experimenter obtained.

A.3.1.2 Variance and Standard Deviation
The estimated or sample variance of a series of measurements is defined by

$$v = \sum_{i=1}^{I} (x_i - \bar{x})^2 / (I - 1)$$

which can also be calculated using the equation

$$v = \sum_{i=1}^{I} x_i^2 / (I - 1) - \bar{x}^2 \times I / (I - 1)$$

So the estimated (sample) variance of the data in Section A.3.1.1 is

$$v = (4^2 + 8^2 + 5^2 + 6^2 + 2^2 + 5^2 + 6^2 + 0^2)/7 - 1.75^2 \times 8/7 = 25.928$$

This above equation is useful when it is required to estimate the population variance from a sample. However, the population variance is defined by

$$v = \sum_{i=1}^{I} (x_i - \bar{x})^2/I = \sum_{i=1}^{I} x_i^2/I - \bar{x}^2$$

One reason why there is a factor of $I - 1$ when using measurements in a number of samples to estimate population statistics is because one degree of freedom is lost when determining variance experimentally. For example, if we record one sample, the sum of squares $\sum_{i=1}^{I} (x_i - \bar{x})^2$ must be equal to 0, but this does not imply that the variance of the parent population is 0. As the number of samples increases, this small correction is not very important, and sometimes ignored. Statisticians might say that if we take, for example, 100 people from a town and calculate the variance of their heights, we are in practice estimating the variance of all people in the town and using our 100 measurements as a sample of maybe the full population of 1000. However, after we measure more than 10 heights, for example, it is unlikely that the difference between the population and sample means can be accurately estimated, indeed sampling would have to be very representative to obtain a detectable and reproducible difference.

The standard deviation, s, is simply the square root of the variance. The population standard deviation is sometimes denoted by σ.

In chemometrics, it is usual to use the population and not the sample standard deviation for standardising a data matrix. The reason is that we are not trying to estimate parameters in this case, but just to put different variables on a similar scale, when using Excel or Matlab, please ensure you use the right type.

A.3.1.3 Covariance and Correlation Coefficient

The covariance between two variables is a method for determining how closely they follow similar trends. It will never exceed in magnitude the geometric mean of the variance of the two variables, the lower the value, the less close the trends. Both variables must be measured on an identical number of samples, I in this case. The sample or estimated covariance between variables x and y is defined by

$$\text{cov}_{xy} = \sum_{i=1}^{I} (x_i - \bar{x})(y_i - \bar{y})/(I - 1)$$

whereas the population statistic is given by

$$\text{cov}_{xy} = \sum_{i=1}^{I} (x_i - \bar{x})(y_i - \bar{y})/I$$

Unlike the variance, it is perfectly possible for a covariance to take on negative values.

Many chemometricians prefer to use the Pearson's correlation coefficient, given by

$$r_{xy} = \frac{\text{cov}_{xy}}{s_x \cdot s_y} = \frac{\sum_{i=1}^{I} (x_i - \bar{x})(y_i - \bar{y})}{\sqrt{\sum_{i=1}^{I} (x_i - \bar{x})^2 \sum_{i=1}^{I} (y_i - \bar{y})^2}}$$

Notice that the definition of the correlation coefficient is identical both for samples and populations.

The correlation coefficient has a value between -1 and $+1$. If close to $+1$, the two variables are perfectly correlated. In many applications, correlation coefficients of -1 also indicate a perfect relationship. Under such circumstances, the value of y can be exactly predicted if we know x. The closer the correlation coefficients are to 0, the harder it is to use one variable to predict another. Some people prefer to use the square of the correlation coefficient, which varies between 0 and 1.

If two columns of a matrix have a correlation coefficient of ± 1, the matrix is said to be rank deficient and has a determinant of 0, and so no inverse, this has consequences both in experimental design and in regression. There are various ways around this such as by removing selected variables or performing PCA.

In some areas of chemometrics, we used a *variance–covariance* matrix. This is a square matrix, whose dimensions usually equal the number of variables in a data set; for example, if there are 20 variables, the matrix has dimensions

20×20. The diagonal elements equal the variance of each variable and the off-diagonal elements the covariances. This matrix is symmetric about the diagonal. It is usual to employ population rather than sample statistics for this calculation as discussed above.

A.3.2 Normal Distribution

The normal distribution is an important statistical concept. There are many ways of introducing such distributions. Many texts use a probability density function

$$f(x) = \frac{1}{\sigma\sqrt{2\pi}} \exp\left(-\frac{1}{2}\left(\frac{x-\mu}{\sigma}\right)^2\right)$$

This rather complicated equation can be interpreted as follows. The function $f(x)$ is proportional to the probability that a measurement has a value x for a normally distributed population of mean μ and standard deviation σ. The function is scaled so that the area under the normal distribution curve is 1.

Most tables deal with the *standardised* normal distribution. This involves first standardising the raw data to give a new value z, the equation simplifies to

$$f(z) = \frac{1}{\sqrt{2\pi}} \exp\left(-\frac{z^2}{2}\right)$$

Instead of calculating $f(z)$, most people look at the area under the normal distribution curve. This is proportional to the probability that a measurement is between certain limits. For example, the probability that a measurement is between 1 and 2 standard deviations can be calculated by taking the proportion of the overall area for which $1 \le z \le 2$.

These numbers can be obtained using simple functions, for example, in a spreadsheet, but are often conventionally presented in a tabular form. There are a surprisingly large number of types of tables, but Table A.1 allows the reader to calculate relevant information. This table is of the cumulative normal distribution and represents the area to the left of the curve for a specified number of standard deviations from the mean. The number of standard deviations equals the sum of the left-hand column and the top row, so, for example, the area for 1.17 standard deviations equals 0.87900.

Using this table, it is then possible to determine the probability of a measurement between any specific limits.

- The probability that a measurement is above 1 standard deviation from the mean is equal to $1 - 0.84134 = 0.15866$.
- The probability that a measurement is more than 1 standard deviation from the mean will be twice this because both positive and negative deviations are possible and the curve is symmetrical and is equal to 0.31732. Put another way, around a third of all measurements will fall outside 1 standard deviation from the mean.
- The probability that a measurement falls between -2 and $+1$ standard deviations from the mean can be calculated as follows.
  - The probability that a measurement falls between 0 and -2 standard deviations is the same as the probability it falls between 0 and $+2$ standard deviations and is equal to $0.97725 - 0.5 = 0.47725$.
  - The probability that a measurement falls between 0 and $+1$ standard deviations is equal to $0.84134 - 0.5 = 0.34134$.
  - Therefore, the total probability is $0.47725 + 0.34134 = 0.81859$.

The normal distribution curve is not only a probability distribution but is also used to describe peak shapes in spectroscopy and chromatography.

Instead of using tables, there are various functions in almost all common packages such as Excel and Matlab to calculate the normal probability distribution. In this section, for brevity, we just describe the use of traditional statistical tables for brevity, for the normal distribution and related distributions. This is expanded in the sections on Excel and Matlab.

A.3.3 χ^2-Distribution

When there is one variable measured, the χ^2 distribution can be very simply related to the normal distribution. χ^2 is simply the square of the distance to the mean. If a series of samples are normally distributed, we can describe the distribution of their squares by χ^2. Therefore, for example, if data are standardised, the proportion of samples less than 4 units of χ^2 from the mean is the same as the proportion of samples within 2 normalised units either side of the mean (i.e. having a value of $z \le \pm 2$).

The χ^2 distribution becomes important when more than one variable is measured, that is, multivariate data. For the normal distribution, we can look at how far a sample is from the mean in terms of the standard deviation, this may be

Table A.1 Cumulative standardised normal distribution.

	0.00	0.01	0.02	0.03	0.04	0.05	0.06	0.07	0.08	0.09
0.0	0.50000	0.50399	0.50798	0.51197	0.51595	0.51994	0.52392	0.52790	0.53188	0.53586
0.1	0.53983	0.54380	0.54776	0.55172	0.55567	0.55962	0.56356	0.56749	0.57142	0.57535
0.2	0.57926	0.58317	0.58706	0.59095	0.59483	0.59871	0.60257	0.60642	0.61026	0.61409
0.3	0.61791	0.62172	0.62552	0.62930	0.63307	0.63683	0.64058	0.64431	0.64803	0.65173
0.4	0.65542	0.65910	0.66276	0.66640	0.67003	0.67364	0.67724	0.68082	0.68439	0.68793
0.5	0.69146	0.69497	0.69847	0.70194	0.70540	0.70884	0.71226	0.71566	0.71904	0.72240
0.6	0.72575	0.72907	0.73237	0.73565	0.73891	0.74215	0.74537	0.74857	0.75175	0.75490
0.7	0.75804	0.76115	0.76424	0.76730	0.77035	0.77337	0.77637	0.77935	0.78230	0.78524
0.8	0.78814	0.79103	0.79389	0.79673	0.79955	0.80234	0.80511	0.80785	0.81057	0.81327
0.9	0.81594	0.81859	0.82121	0.82381	0.82639	0.82894	0.83147	0.83398	0.83646	0.83891
1.0	0.84134	0.84375	0.84614	0.84849	0.85083	0.85314	0.85543	0.85769	0.85993	0.86214
1.1	0.86433	0.86650	0.86864	0.87076	0.87286	0.87493	0.87698	0.87900	0.88100	0.88298
1.2	0.88493	0.88686	0.88877	0.89065	0.89251	0.89435	0.89617	0.89796	0.89973	0.90147
1.3	0.90320	0.90490	0.90658	0.90824	0.90988	0.91149	0.91308	0.91466	0.91621	0.91774
1.4	0.91924	0.92073	0.92220	0.92364	0.92507	0.92647	0.92785	0.92922	0.93056	0.93189
1.5	0.93319	0.93448	0.93574	0.93699	0.93822	0.93943	0.94062	0.94179	0.94295	0.94408
1.6	0.94520	0.94630	0.94738	0.94845	0.94950	0.95053	0.95154	0.95254	0.95352	0.95449
1.7	0.95543	0.95637	0.95728	0.95818	0.95907	0.95994	0.96080	0.96164	0.96246	0.96327
1.8	0.96407	0.96485	0.96562	0.96638	0.96712	0.96784	0.96856	0.96926	0.96995	0.97062
1.9	0.97128	0.97193	0.97257	0.97320	0.97381	0.97441	0.97500	0.97558	0.97615	0.97670
2.0	0.97725	0.97778	0.97831	0.97882	0.97932	0.97982	0.98030	0.98077	0.98124	0.98169
2.1	0.98214	0.98257	0.98300	0.98341	0.98382	0.98422	0.98461	0.98500	0.98537	0.98574
2.2	0.98610	0.98645	0.98679	0.98713	0.98745	0.98778	0.98809	0.98840	0.98870	0.98899
2.3	0.98928	0.98956	0.98983	0.99010	0.99036	0.99061	0.99086	0.99111	0.99134	0.99158
2.4	0.99180	0.99202	0.99224	0.99245	0.99266	0.99286	0.99305	0.99324	0.99343	0.99361
2.5	0.99379	0.99396	0.99413	0.99430	0.99446	0.99461	0.99477	0.99492	0.99506	0.99520
2.6	0.99534	0.99547	0.99560	0.99573	0.99585	0.99598	0.99609	0.99621	0.99632	0.99643
2.7	0.99653	0.99664	0.99674	0.99683	0.99693	0.99702	0.99711	0.99720	0.99728	0.99736
2.8	0.99744	0.99752	0.99760	0.99767	0.99774	0.99781	0.99788	0.99795	0.99801	0.99807
2.9	0.99813	0.99819	0.99825	0.99831	0.99836	0.99841	0.99846	0.99851	0.99856	0.99861

	0.0	0.1	0.2	0.3	0.4	0.5	0.6	0.7	0.8	0.9
3.0	0.99865	0.99903	0.99931	0.99952	0.99966	0.99977	0.99984	0.99989	0.99993	0.99995
4.0	0.999968	0.999979	0.999987	0.999991	0.999995	0.999997	0.999998	0.999999	0.999999	1.000000

used to determine whether it is an outlier or whether it belongs to a predefined group. If the underlying distribution is assumed to be normal, a probability can be calculated, as discussed in Section A.3.2, that a sample is between or greater than any specified distance(s) from the mean. We would like to perform a similar calculation in cases where there is more than one variable. Once the number of variables exceeds 1, there is no specific positive or negative direction to the mean or centre of the data, hence instead a squared distance (the Mahalanobis distance) must be calculated. We can see that when one variable is recorded, if it is distributed normally, its square is distributed according to χ^2. This statistic has a different distribution according to the number of variables or degrees of freedom.

Table A.2 Critical values of χ^2.

	0.99	0.975	0.95	0.9	0.1	0.05	0.025	0.01
1	0.00016	0.00098	0.00393	0.01579	2.706	3.841	5.024	6.635
2	0.0201	0.0506	0.1026	0.2107	4.605	5.991	7.378	9.210
3	0.115	0.216	0.352	0.584	6.251	7.815	9.348	11.345
4	0.297	0.484	0.711	1.064	7.779	9.488	11.143	13.277
5	0.554	0.831	1.145	1.610	9.236	11.07	12.832	15.086
6	0.872	1.237	1.635	2.204	10.645	12.592	14.449	16.812
7	1.239	1.690	2.167	2.833	12.017	14.067	16.013	18.475
8	1.647	2.180	2.733	3.490	13.362	15.507	17.535	20.09
9	2.088	2.700	3.325	4.168	14.684	16.919	19.023	21.666
10	2.558	3.247	3.940	4.865	15.987	18.307	20.483	23.209
11	3.053	3.816	4.575	5.578	17.275	19.675	21.920	24.725
12	3.571	4.404	5.226	6.304	18.549	21.026	23.337	26.217
13	4.107	5.009	5.892	7.041	19.812	22.362	24.736	27.688
14	4.660	5.629	6.571	7.790	21.064	23.685	26.119	29.141
15	5.229	6.262	7.261	8.547	22.307	24.996	27.488	30.578
20	8.260	9.591	10.851	12.443	28.412	31.410	34.170	37.566
25	11.524	13.12	14.611	16.473	34.382	37.652	40.646	44.314
30	14.953	16.791	18.493	20.599	40.256	43.773	46.979	50.892
40	22.164	24.433	26.509	29.051	51.805	55.758	59.342	63.691
50	29.707	32.357	34.764	37.689	63.167	67.505	71.420	76.154
100	70.065	74.222	77.929	82.358	118.498	124.342	129.561	135.807

Usually, the χ^2 distribution is presented in a table of critical values. A typical table is Table A.2. Along the columns, we have critical levels, the numbers in the table representing critical values at a given p. The rows are for different degrees of freedom. A critical level of 0.10 represents the value of χ^2 above which we expect 10% of the samples to lie, also called a p value of 0.10. As χ^2 can be computed for any number of variables or degrees of freedom (denoted v), we could have a table analogous to Table A.1 for any value of v, but this would result in hundreds of tables, so presenting just critical values simplifies the information: usually, we ask whether χ^2 is above the specified p value.

We can see that for one degree of freedom, χ^2 is the same as the normal distribution but using squared distances.

- The critical value of χ^2 using one degree of freedom and $p = 0.10$ is 2.706 from Table A.1.
- That means, we expect 10% of samples to have a χ^2 greater than 2.706.
- If we look at Table A.1, we find the value of z below which the 0.95 of the samples should lie.
- Because this corresponds to a p value of 0.05 which is 0.10/2.
- χ^2 contains only positive numbers, as it is squared. Hence, a χ^2 of 4 represents samples having a z value of either +2 or −2.
- We find that this value for one degree of freedom is approximately 1.645; the table has to be interpolated.
- Squaring 1.645 gives us 2.706.

There is no advantage of using χ^2 over z if there is only one degree of freedom. However, if more than 1, there is no positive or negative direction; hence, we use the squared value. If the underlying data are normally distributed, the distribution of the squared Mahalanobis distance should follow χ^2. Hence, we could answer, for example, if there are 10 variables, what is the p value corresponding to a squared Mahalanobis distance of 19. Looking at Table A.2, we find that this is somewhere between 0.05 and 0.025. If we were to ask whether a sample exceeded a critical value of $p = 0.01$, we would answer it does not.

There are several uses of χ^2, but in our cases, it is most useful when looking at multivariate distributions, to see whether samples at specified distance from the mean can be regarded as members of the parent distribution or not.

A.3.4 *t*-Distribution

When sample sizes are relatively small, we find that ideal distributions can be distorted. Many early statistical studies used limited sample sizes; thus, it was necessary to develop approaches, not just for estimating means and standard deviations but to deduce more information about the underlying population. Even in more modern studies, some applications such as process control and metabolomics often suffer from modest sample sizes.

The *t*-distribution is similar in nature to the normal distribution but is distorted when the number of degrees of freedom is small. In this context, the number of degrees of freedom (v) equals the number of samples −1. Once v reaches about 10, the normal distribution and *t*-distribution are very similar and cannot easily be distinguished. The *t*-distribution can only be used when there is one variable.

Most tables of *t*-distributions present the critical value of the *t*-statistic for different degrees of freedom. Table A.3 is of the two-tailed *t*-distribution, which we employ in this text, and asks whether a parameter differs significantly from another. If there are 10 degrees of freedom and the *t*-statistic equals 2.32, then the probability is quite high, slightly above 95% ($p = 0.05$), that it is, significant (or more accurately a low probability of rejecting the null hypothesis).

From Table A.3, we can see that as the number of degrees of freedom becomes very large, the *t*-distribution converges to the normal distribution.

- The critical value at $p = 0.01$ for ∞ degrees of freedom is 2.576.
- As this is two tailed, it represents all data above 0.995 and below 0.995 standard deviations from the mean.
- From Table A.1, we do see that this is indeed between 2.57 and 2.58 standard deviations from the mean, so the two tables agree.

Note that there are several ways in which the critical values of t can be presented, and some tables are for the one-tailed critical levels, simply double the p values while using the two-tailed critical levels.

The one-tailed test is used to see whether a parameter is significantly larger than another; for example, does the mean of a series of samples significantly exceed that of a series of reference samples? In this book, we are mainly concerned with using the *t*-statistic to determine the significance of coefficients when analysing the results of designed experiments; in such cases, both a negative and positive coefficient are equally significant, so a two-tailed test is most appropriate.

A.3.5 *F*-Distribution

The *t*-distribution cannot be used when there is more than one variable, analogous to the normal distribution. The *F*-distribution can be regarded as the equivalent extension of the *t*-distribution when there is more than one variable but small sample sizes, or analogous to χ^2, but when there are small sample sizes. The *F*-statistic is always positive and represents a squared number.

There are numerous ways of introducing this distribution in the literature, which are widely employed in many diverse areas. The *F*-distribution is often introduced in the context of ANOVA, but can also be used in a modified form (called Hotelling's T^2) to determine the significance of the Mahalanobis distance from the centre of a group in multivariate space.

- The *F*-distribution is characterised by two different types of degrees of freedom.
- It is often written as $F(v_1, v_2)$.
- In the context of multivariate matrices, if we consider a data set consisting of I samples and J variables, then v_1 represents the number of variables and v_2 the number of observations minus the number of variables ($I - J$). The first degree of freedom represents the number of columns in a data matrix, which is the reverse of the way dimensions are denoted; thus, a multivariate matrix has dimensions $v_2 \times (v_1 + v_2)$.
- The *F*-distribution is often used in a wide number of contexts.
- Note that $F(v_1, v_2) \neq F(v_2, v_1)$.

The *F*-test is also commonly used to compare two variances or errors and ask whether either one variance is significantly greater than the other (one-tailed) or whether it differs significantly (two-tailed). In this book, we only use the one-tailed *F*-test, mainly to see whether one error (e.g. lack-of-fit) is significantly greater than a second one (e.g. experimental or analytical).

Table A.3 Critical values of two-tailed *t*-distribution.

Df	10%	5%	1%	0.1%
1	6.314	12.706	63.656	636.578
2	2.920	4.303	9.925	31.600
3	2.353	3.182	5.841	12.924
4	2.132	2.776	4.604	8.610
5	2.015	2.571	4.032	6.869
6	1.943	2.447	3.707	5.959
7	1.895	2.365	3.499	5.408
8	1.860	2.306	3.355	5.041
9	1.833	2.262	3.250	4.781
10	1.812	2.228	3.169	4.587
11	1.796	2.201	3.106	4.437
12	1.782	2.179	3.055	4.318
13	1.771	2.160	3.012	4.221
14	1.761	2.145	2.977	4.140
15	1.753	2.131	2.947	4.073
16	1.746	2.120	2.921	4.015
17	1.740	2.110	2.898	3.965
18	1.734	2.101	2.878	3.922
19	1.729	2.093	2.861	3.883
20	1.725	2.086	2.845	3.850
25	1.708	2.060	2.787	3.725
30	1.697	2.042	2.750	3.646
35	1.690	2.030	2.724	3.591
40	1.684	2.021	2.704	3.551
45	1.679	2.014	2.690	3.520
50	1.676	2.009	2.678	3.496
100	1.660	1.984	2.626	3.390
∞	1.645	1.960	2.576	3.291

The F-statistic is the ratio between these two variances, normally presented as a number greater than 1, that is the largest over the smallest. The F-distribution depends on the number of degrees of freedom of each variable, so, if the highest variance is obtained from a data set consisting of 10 samples, and the lowest of 7 samples, the two variances have 9 and 6 degrees of freedom, respectively. The F-distribution differs according to the number of degrees of freedom, and it would be theoretically possible to produce an F-distribution table for every possible combination of degrees of freedom, rather similar to the normal distribution table. However, this would mean an enormous number of tables, and it is more usual simply to calculate the F-statistic at certain well-defined critical p levels.

A one-tailed F-statistic at 1% probability level is the value of the F-ratio above which only 1% of measurements would fall if the two variances were significantly different. If the F-statistic exceeds this value, then we have more than 99% confidence that there is a significant difference in the two variances (in other words, reject the null hypothesis that there is not difference at $p = 0.01$), for example, that the lack-of-fit error really is bigger than the analytical error.

We present in this chapter the 1% and 5% one-tailed F-critical levels (Tables A.4 and A.5). The number of degrees of freedom belonging to the data with the highest variance is always along the top, and the degrees of freedom belonging to the data with the lowest variance down the side. It is easy to use Excel or most statistically based packages to calculate

Table A.4 One-tailed critical values of the F-distribution at 1% level.

	1	2	3	4	5	6	7	8	9	10	15	20	25	30	50	100	∞
1	4052.18	4999.34	5403.53	5624.26	5763.96	5858.95	5928.33	5980.95	6022.40	6055.93	6156.97	6208.66	6239.86	6260.35	6302.26	6333.92	6365.59
2	98.5019	99.0003	99.1640	99.2513	99.3023	99.3314	99.3568	99.3750	99.3896	99.3969	99.4332	99.4478	99.4587	99.4660	99.4769	99.4914	99.4987
3	34.1161	30.8164	29.4567	28.7100	28.2371	27.9106	27.6714	27.4895	27.3449	27.2285	26.8719	26.6900	26.5791	26.5045	26.3544	26.2407	26.1252
4	21.1976	17.9998	16.6942	15.9771	15.5219	15.2068	14.9757	14.7988	14.6592	14.5460	14.1981	14.0194	13.9107	13.8375	13.6897	13.5769	13.4633
5	16.2581	13.2741	12.0599	11.3919	10.9671	10.6722	10.4556	10.2893	10.1577	10.0511	9.7223	9.5527	9.4492	9.3794	9.2377	9.1300	9.0204
6	13.7452	10.9249	9.7796	9.1484	8.7459	8.4660	8.2600	8.1017	7.9760	7.8742	7.5590	7.3958	7.2960	7.2286	7.0914	6.9867	6.8801
7	12.2463	9.5465	8.4513	7.8467	7.4604	7.1914	6.9929	6.8401	6.7188	6.6201	6.3144	6.1555	6.0579	5.9920	5.8577	5.7546	5.6496
8	11.2586	8.6491	7.5910	7.0061	6.6318	6.3707	6.1776	6.0288	5.9106	5.8143	5.5152	5.3591	5.2631	5.1981	5.0654	4.9633	4.8588
9	10.5615	8.0215	6.9920	6.4221	6.0569	5.8018	5.6128	5.4671	5.3511	5.2565	4.9621	4.8080	4.7130	4.6486	4.5167	4.4150	4.3106
10	10.0442	7.5595	6.5523	5.9944	5.6364	5.3858	5.2001	5.0567	4.9424	4.8491	4.5582	4.4054	4.3111	4.2469	4.1155	4.0137	3.9090
11	9.6461	7.2057	6.2167	5.6683	5.3160	5.0692	4.8860	4.7445	4.6315	4.5393	4.2509	4.0990	4.0051	3.9411	3.8097	3.7077	3.6025
12	9.3303	6.9266	5.9525	5.4119	5.0644	4.8205	4.6395	4.4994	4.3875	4.2961	4.0096	3.8584	3.7647	3.7008	3.5692	3.4668	3.3608
13	9.0738	6.7009	5.7394	5.2053	4.8616	4.6203	4.4410	4.3021	4.1911	4.1003	3.8154	3.6646	3.5710	3.5070	3.3752	3.2723	3.1654
14	8.8617	6.5149	5.5639	5.0354	4.6950	4.4558	4.2779	4.1400	4.0297	3.9394	3.6557	3.5052	3.4116	3.3476	3.2153	3.1118	3.0040
15	8.6832	6.3588	5.4170	4.8932	4.5556	4.3183	4.1416	4.0044	3.8948	3.8049	3.5222	3.3719	3.2782	3.2141	3.0814	2.9772	2.8684
16	8.5309	6.2263	5.2922	4.7726	4.4374	4.2016	4.0259	3.8896	3.7804	3.6909	3.4090	3.2587	3.1650	3.1007	2.9675	2.8627	2.7528
17	8.3998	6.1121	5.1850	4.6689	4.3360	4.1015	3.9267	3.7909	3.6823	3.5931	3.3117	3.1615	3.0676	3.0032	2.8694	2.7639	2.6531
18	8.2855	6.0129	5.0919	4.5790	4.2479	4.0146	3.8406	3.7054	3.5971	3.5081	3.2273	3.0771	2.9831	2.9185	2.7841	2.6779	2.5660
19	8.1850	5.9259	5.0103	4.5002	4.1708	3.9386	3.7653	3.6305	3.5225	3.4338	3.1533	3.0031	2.9089	2.8442	2.7092	2.6023	2.4893
20	8.0960	5.8490	4.9382	4.4307	4.1027	3.8714	3.6987	3.5644	3.4567	3.3682	3.0880	2.9377	2.8434	2.7785	2.6430	2.5353	2.4212
25	7.7698	5.5680	4.6755	4.1774	3.8550	3.6272	3.4568	3.3239	3.2172	3.1294	2.8502	2.6993	2.6041	2.5383	2.3999	2.2888	2.1694
30	7.5624	5.3903	4.5097	4.0179	3.6990	3.4735	3.3045	3.1726	3.0665	2.9791	2.7002	2.5487	2.4526	2.3860	2.2450	2.1307	2.0062
35	7.4191	5.2679	4.3958	3.9082	3.5919	3.3679	3.1999	3.0687	2.9630	2.8758	2.5970	2.4448	2.3480	2.2806	2.1374	2.0202	1.8910
40	7.3142	5.1785	4.3126	3.8283	3.5138	3.2910	3.1238	2.9930	2.8876	2.8005	2.5216	2.3689	2.2714	2.2034	2.0581	1.9383	1.8047
45	7.2339	5.1103	4.2492	3.7674	3.4544	3.2325	3.0658	2.9353	2.8301	2.7432	2.4642	2.3109	2.2129	2.1443	1.9972	1.8751	1.7374
50	7.1706	5.0566	4.1994	3.7195	3.4077	3.1864	3.0202	2.8900	2.7850	2.6981	2.4190	2.2652	2.1667	2.0976	1.9490	1.8248	1.6831
100	6.8953	4.8239	3.9837	3.5127	3.2059	2.9877	2.8233	2.6943	2.5898	2.5033	2.2230	2.0666	1.9651	1.8933	1.7353	1.5977	1.4273
∞	6.6349	4.6052	3.7816	3.3192	3.0172	2.8020	2.6393	2.5113	2.4073	2.3209	2.0385	1.8783	1.7726	1.6964	1.5231	1.3581	1.0000

Table A.5 One-tailed critical values of the *F*-distribution at 5% level.

	1	2	3	4	5	6	7	8	9	10	15	20	25	30	50	100	∞
1	161.45	199.50	215.71	224.58	230.16	233.99	236.77	238.88	240.54	241.88	245.95	248.02	249.26	250.10	251.77	253.04	254.32
2	18.5128	19.0000	19.1642	19.2467	19.2963	19.3295	19.3531	19.3709	19.3847	19.3959	19.4291	19.4457	19.4557	19.4625	19.4757	19.4857	19.4957
3	10.1280	9.5521	9.2766	9.1172	9.0134	8.9407	8.8867	8.8452	8.8123	8.7855	8.7028	8.6602	8.6341	8.6166	8.5810	8.5539	8.5265
4	7.7086	6.9443	6.5914	6.3882	6.2561	6.1631	6.0942	6.0410	5.9988	5.9644	5.8578	5.8025	5.7687	5.7459	5.6995	5.6640	5.6281
5	6.6079	5.7861	5.4094	5.1922	5.0503	4.9503	4.8759	4.8183	4.7725	4.7351	4.6188	4.5581	4.5209	4.4957	4.4444	4.4051	4.3650
6	5.9874	5.1432	4.7571	4.5337	4.3874	4.2839	4.2067	4.1468	4.0990	4.0600	3.9381	3.8742	3.8348	3.8082	3.7537	3.7117	3.6689
7	5.5915	4.7374	4.3468	4.1203	3.9715	3.8660	3.7871	3.7257	3.6767	3.6365	3.5107	3.4445	3.4036	3.3758	3.3189	3.2749	3.2298
8	5.3176	4.4590	4.0662	3.8379	3.6875	3.5806	3.5005	3.4381	3.3881	3.3472	3.2184	3.1503	3.1081	3.0794	3.0204	2.9747	2.9276
9	5.1174	4.2565	3.8625	3.6331	3.4817	3.3738	3.2927	3.2296	3.1789	3.1373	3.0061	2.9365	2.8932	2.8637	2.8028	2.7556	2.7067
10	4.9646	4.1028	3.7083	3.4780	3.3258	3.2172	3.1355	3.0717	3.0204	2.9782	2.8450	2.7740	2.7298	2.6996	2.6371	2.5884	2.5379
11	4.8443	3.9823	3.5874	3.3567	3.2039	3.0946	3.0123	2.9480	2.8962	2.8536	2.7186	2.6464	2.6014	2.5705	2.5066	2.4566	2.4045
12	4.7472	3.8853	3.4903	3.2592	3.1059	2.9961	2.9134	2.8486	2.7964	2.7534	2.6169	2.5436	2.4977	2.4663	2.4010	2.3498	2.2962
13	4.6672	3.8056	3.4105	3.1791	3.0254	2.9153	2.8321	2.7669	2.7144	2.6710	2.5331	2.4589	2.4123	2.3803	2.3138	2.2614	2.2064
14	4.6001	3.7389	3.3439	3.1122	2.9582	2.8477	2.7642	2.6987	2.6458	2.6022	2.4630	2.3879	2.3407	2.3082	2.2405	2.1870	2.1307
15	4.5431	3.6823	3.2874	3.0556	2.9013	2.7905	2.7066	2.6408	2.5876	2.5437	2.4034	2.3275	2.2797	2.2468	2.1780	2.1234	2.0659
16	4.4940	3.6337	3.2389	3.0069	2.8524	2.7413	2.6572	2.5911	2.5377	2.4935	2.3522	2.2756	2.2272	2.1938	2.1240	2.0685	2.0096
17	4.4513	3.5915	3.1968	2.9647	2.8100	2.6987	2.6143	2.5480	2.4943	2.4499	2.3077	2.2304	2.1815	2.1477	2.0769	2.0204	1.9604
18	4.4139	3.5546	3.1599	2.9277	2.7729	2.6613	2.5767	2.5102	2.4563	2.4117	2.2686	2.1906	2.1413	2.1071	2.0354	1.9780	1.9168
19	4.3808	3.5219	3.1274	2.8951	2.7401	2.6283	2.5435	2.4768	2.4227	2.3779	2.2341	2.1555	2.1057	2.0712	1.9986	1.9403	1.8780
20	4.3513	3.4928	3.0984	2.8661	2.7109	2.5990	2.5140	2.4471	2.3928	2.3479	2.2033	2.1242	2.0739	2.0391	1.9656	1.9066	1.8432
25	4.2417	3.3852	2.9912	2.7587	2.6030	2.4904	2.4047	2.3371	2.2821	2.2365	2.0889	2.0075	1.9554	1.9192	1.8421	1.7794	1.7110
30	4.1709	3.3158	2.9223	2.6896	2.5336	2.4205	2.3343	2.2662	2.2107	2.1646	2.0148	1.9317	1.8782	1.8409	1.7609	1.6950	1.6223
35	4.1213	3.2674	2.8742	2.6415	2.4851	2.3718	2.2852	2.2167	2.1608	2.1143	1.9629	1.8784	1.8239	1.7856	1.7032	1.6347	1.5580
40	4.0847	3.2317	2.8387	2.6060	2.4495	2.3359	2.2490	2.1802	2.1240	2.0773	1.9245	1.8389	1.7835	1.7444	1.6600	1.5892	1.5089
45	4.0566	3.2043	2.8115	2.5787	2.4221	2.3083	2.2212	2.1521	2.0958	2.0487	1.8949	1.8084	1.7522	1.7126	1.6264	1.5536	1.4700
50	4.0343	3.1826	2.7900	2.5572	2.4004	2.2864	2.1992	2.1299	2.0733	2.0261	1.8714	1.7841	1.7273	1.6872	1.5995	1.5249	1.4383
100	3.9362	3.0873	2.6955	2.4626	2.3053	2.1906	2.1025	2.0323	1.9748	1.9267	1.7675	1.6764	1.6163	1.5733	1.4772	1.3917	1.2832
∞	3.8414	2.9957	2.6049	2.3719	2.2141	2.0986	2.0096	1.9384	1.8799	1.8307	1.6664	1.5705	1.5061	1.4591	1.3501	1.2434	1.0000

critical F values for any probability and combination of degrees of freedom, but it is still worth being able to understand the use of tables.

If one error (e.g. the lack-of-fit) is measured using eight degrees of freedom and another error (e.g. replicate) is measured using six degrees of freedom, then if the F-ratio between the mean lack-of-fit and replicate errors is 7.89, is it significant? The critical F-statistic at 1% is 8.1017 and at 5% is 4.1468. Hence, the F-ratio is significant at almost 99% ($=100-1$)% level because 7.89 is almost equal to 8.1017. Thus, we are 99% certain that the lack-of-fit is really significant (or $p < 0.01$).

Some texts also present tables for a two-tailed F-test, but for brevity, we omit this. However, a two-tailed F-statistic at 10% significance is the same as a one-tailed F-statistic at 5% significance, and so on.

The F-distribution can be used in many other contexts, for example, to determine how likely a sample is to belong to a parent distribution if it is a specified Mahalanobis distance from the centre of the distribution, according to the number of samples and variables characterising a training set: however, if the sample size is large, it is also possible to use the χ^2 distribution, and it is unlikely that the difference could be observed experimentally. Note that when dealing with multivariate distances, a scaling factor is usually necessary, resulting in Hotelling's T^2, which has the same shape as F. These measures all do depend on the underlying data being normally distributed.

A.4 Excel for Chemometrics

There are many excellent books on Excel in general, and the package in itself is associated with an extensive help system. It is not the purpose of this text to duplicate these books, which in themselves are regularly updated as new versions of Excel become available, but to primarily indicate the features that the user of advanced data analysis might find useful. The examples in this book are illustrated using Office 365 using Excel 2016 under Windows 10, but most features are applicable to most versions of Excel and are also likely to be upwards compatible. It is recommended to use 32 bit rather than 64 Excel as some macros are not compatible with the 64 bit version. To check whether your version of Excel is 32 bit, open the 'File' tab, then the 'Account' menu item and then the 'About Excel' button. It is assumed that the reader has already some experience in Excel, and the aim of this section is to indicate some features that will be useful to the scientific user of Excel, especially the chemometrician. This section should be regarded primarily as one of tips and hints and the best way forward is by practice. Comprehensive help facilities and websites are also available for the specialist user of Excel. The specific chemometric add-ins available to accompany this text are also described.

A.4.1 Names and Addresses

There are a surprisingly large number of methods for naming cells and portions of spreadsheets in Excel, and it is important to be aware of all the possibilities.

A.4.1.1 Alphanumeric Format

The default naming convention is alphanumeric, each cell's address involving one or two letters referring to the column followed by a number referring to the row. Thus, cell *C5* is the address of the fifth row of the third column. The alphanumeric method is a historic feature of this and most other spreadsheets but is at odds with the normal scientific convention of quoting rows before columns. After the letters A–Z are exhausted (the first 26 columns), two-letter names are used starting at AA, AB, AC up to AZ and then BA, BB, then after ZZ, AAA, AAB and so on.

A.4.1.2 Maximum Size

The size of a worksheet is limited, to a maximum of 16 384 ($=2^{14}$) columns and 1 048 576 ($=2^{20}$) rows, which means the highest address is *XFD*1048576. There is no difference in size between 32 bit and 64 bit Excel in modern implementations providing Windows is 64 bit. 64 bit Excel is faster when data sets are very large, but still has disadvantages, in that macros and add-ins are not upward compatible.

A.4.1.3 Numeric Format

Some people prefer using a numeric format for addressing cells. The columns are numbered rather than labelled with letters. To change from the default, select the 'File' tab, choose 'Options' menu item, then 'Formulas' menu item and select *R1C1* checkbox (Figure A.1). Deselecting this returns to the alphanumeric system. When using the RC notation, the rows are cited first, as is normal for matrices, and the columns are cited second, which is the opposite to the default convention. Cell *B5* is now called cell *R5C2*. Below we employ the alphanumeric notation unless specifically indicated.

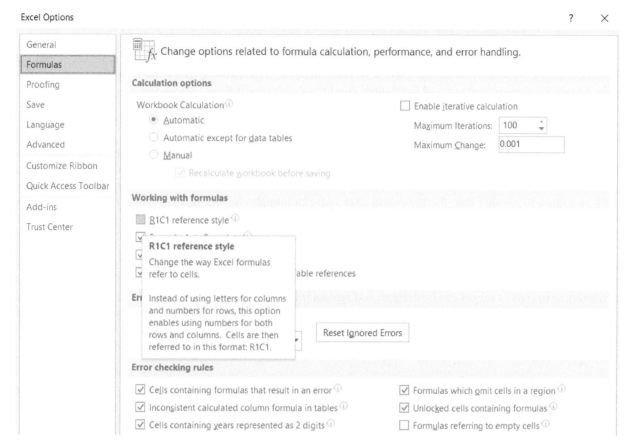

Figure A.1 Changing to numeric cell addresses.

A.4.1.4 Worksheets

Each spreadsheet file can contain several worksheets, each of which either contain data or graphs and which can also be named. The default is *Sheet1*, *Sheet2* and so on, but it is possible to call the sheets by almost any name, such as *Data*, *Results* and *Spectra*, and it is also possible to contain spaces, for example, *Sample 1A*. A cell specific to one sheet has the address of the sheet included, separated by a !; hence, the cell *Data!E3* is the address of the third row and fifth column on a worksheet *Data*. This address can be used in any worksheet, thus allows results of calculations to be placed in different worksheets. If the name contains a space, it is cited in quotes, for example, *'Spectrum 1'!B12*.

New sheets can be inserted using the Home tab, and cells ribbon, select 'Insert Sheet', so any number of sheets can be inserted this way and named as appropriate.

A.4.1.5 Spreadsheets

In addition, it is possible to reference data in another spreadsheet file, for example *[First.xlsx]Result!A3* refers to the address of cell *A3* in worksheet *Result* and file *First.xlsx*. This might be potentially useful, for example, if the file *First* consists of a series of spectra or a chromatogram, whereas the current file consists of the results of processing the data, such as graphs or statistics. This flexibility comes with a price as all files have to be available simultaneously and is somewhat elaborate especially in cases where files are regularly reorganised and backed up.

Note that there are several file extensions for Excel, the commonest being *.xls* (old versions of Excel), *.xlsx* (default) and *.xlsm* (contains macros). If you get into a muddle, always check the file extension, which unless specified by you will be a default or previously used extension. While referring to other spreadsheets, you must include the full extension to the filename.

A.4.1.6 Invariant Addresses

When copying an equation or a cell address within a spreadsheet, it is worth remembering another convention. The $ sign means that the row or column is invariant. For example, if the equation '=A1' is placed in cell *C2*, and then this is copied to cell *D4*, the relative references to the rows and columns are also moved; hence, cell *D4* will actually contain the contents of the original cell *B3*. This is often quite useful because it allows operations on entire rows and columns, but sometimes we need to avoid this. Placing the equation '=$A1' in cell *C2* has the effect of fixing the column, so that the contents of cell *D4* will now be equal to those of cell *A3*. Placing '=A$1' in cell *C2* makes the contents of cell *D4* equal to *B1*, and placing '=$A$1' in cell *C2* makes cell *D4* equal to *A1*. The naming conventions can be combined, for example, *Data!B$3*. Experienced users of Excel often combine a variety of different tricks and it is best to learn these ways by practice rather than reading lots of books.

A.4.1.7 Ranges

A range is a set of cells, often organised as a matrix. Hence, the range *A2:C3* consists of six numbers organised into two rows and three columns, as illustrated in Figure A.2. It is possible to calculate the function of a range, for example =*AVERAGE(A2:C3)*, these will be described in more detail in Section A.4.2.4. If this function is placed in cell *D4*, then if it is copied to another cell, the range will alter correspondingly; for example, moving from cell *D4* to cell *E6* would change the range, automatically, to *B4:D5*. There are various ways to overcome this, the simplest being to use the $ convention as described above. Note that ! and [] can also be used in the address of a range, so that =*AVERAGE([first.xls]Data!$B$3:$C$10)* is an entirely legitimate statement.

It is not necessary for a range to be a single contiguous matrix. The statement =*AVERAGE(A1:B5,C8,B9:D11)* will involve $10 + 1 + 9 = 20$ different cells as indicated in Figure A.3.

The references to cells or ranges can be dragged and remain invariant. If the middle of a cell is 'hit', the cursor turns into an arrow. Dragging this drags the cell, but the original reference remains unchanged, see Figure A.4, where the

	A	B	C	D
1				
2	9	7	3	
3	0	-4	6	
4				
5				

Figure A.2 The range A2:C3.

A14			f_x	=AVERAGE(A1:B5,C8,B9:D11)			
	A	B	C	D	E	F	G
1	1.6	2.7					
2	-3.5	8.2					
3	-4.0	6.1					
4	-2.2	0.8					
5	7.0	1.1					
6							
7							
8			3.9				
9		5.6	1.5	2.8			
10		-2.7	-1.9	-3.4			
11		6.0	0.0	2.4			
12							
13							
14	1.6						
15							

Figure A.3 The operation =AVERAGE(A1:B5,C8,B9:D11).

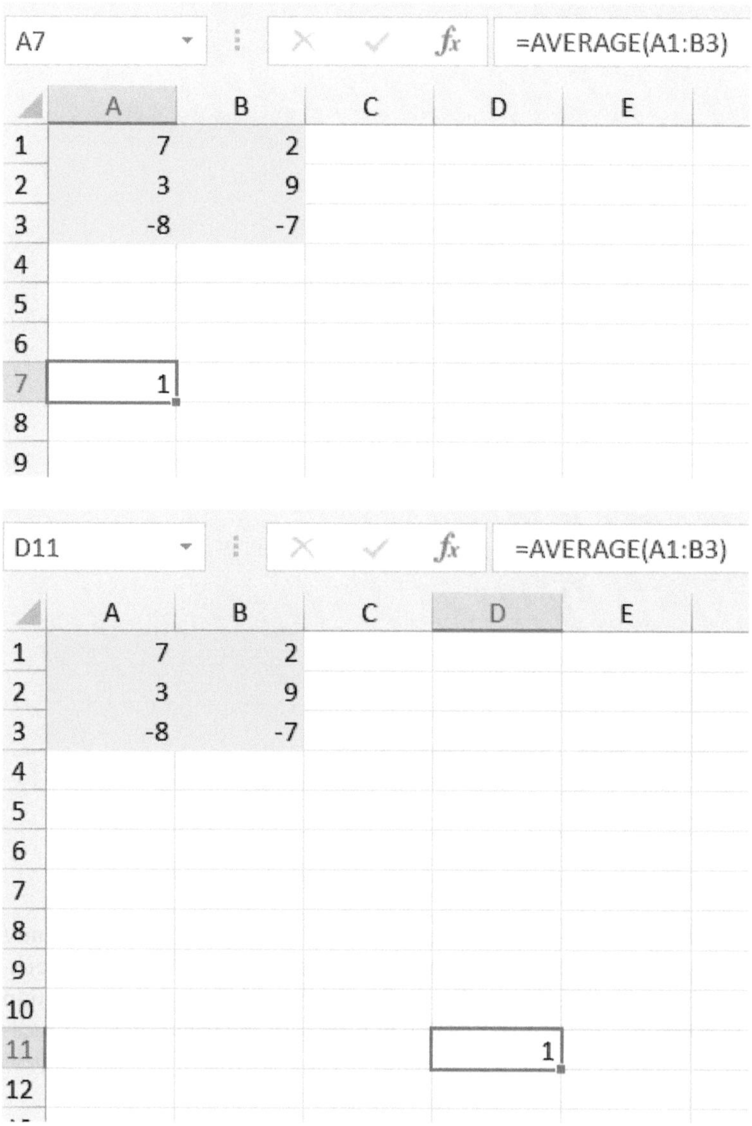

Figure A.4 Dragging a cell so that the reference is invariant.

equation *=AVERAGE(A1:B3)* had been placed originally in cell *A7*. If a cell is copied, however, the relative references change unless the $ sign has been used as above.

A.4.1.8 Naming Matrices

It is particularly useful to be able to name matrices or vectors. This is quite straightforward. Using the 'Formulas' ribbon and choosing 'Name Manager', it is possible to select a portion of a worksheet and give it a name. Figure A.5 shows how to call the range A1–B3 'X', creating a 3×2 matrix. It is then possible to perform operations on these matrices, so that *=SUMSQ(X)* gives the sum of squares of the elements of X and is an alternative to *=SUMSQ(A1:B3)*. It is possible to perform matrix operations; for example, if both X and Y are 4×2 matrices, select a third 4×2 region and place the command *=X + Y* in that region (as discussed in Section A.4.2.2, it is necessary to end all matrix operations by simultaneously pressing the <SHIFT> <CTRL> <ENTER> keys). Quite elaborate commands can then be nested; for example, *=3*(X + Y) − Z* is entirely acceptable where all the three arrays are named matrices, notice that the '3*' is a scalar multiplication: more about this will be discussed below.

The name of a matrix is common to an entire spreadsheet, rather than any individual worksheet. This means that it is possible to store a matrix called 'Data' on one worksheet and then perform operations on it in a separate worksheet.

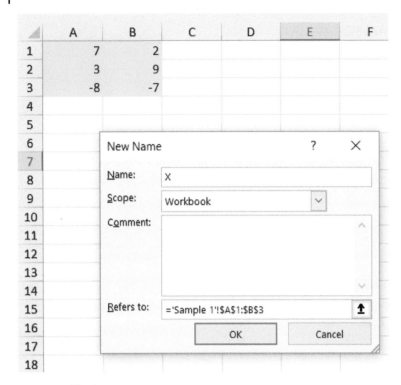

Figure A.5 Naming a range.

A.4.2 Equations and Functions

A.4.2.1 Scalar Operations

The operations are straightforward and can be performed on cells, numbers and functions. The operations +, −, * and / indicate addition, subtraction, multiplication and division as in most environments; powers are indicated by ^. Brackets can be used and there is no practicable limit to the size of the expression or the amount of nesting. A destination cell must be chosen. Hence, the operation $=3*(2-5)\wedge2/(8-1)+6$ gives a value of 9.857. All the usual rules of precedence are obeyed. Operations can mix cells, functions and numbers, for example $=4*A1 + SUM(E1:F3) - SUMSQ(Y)$ involves adding

- four times the value of cell *A1* to
- the sum of the values of the six cells in the range *E1–F3* and
- subtracting the sum of the squares of the matrix **Y**.

Equations can be copied around the spreadsheet, the $, ! and [] conventions being used if appropriate.

A.4.2.2 Matrix Operations

There are a number of matrix operations, of especial use in chemometrics. These operations must be terminated by simultaneously pressing the <SHIFT>, <CTRL> and <ENTER> keys. It is first necessary to select a portion of the spreadsheet where the destination matrix will be displayed. The most useful functions are as follows.

- *MMULT* multiplies two matrices together. The inner dimensions must be identical. If not, there is an error message. The destination (third) matrix must be selected; if its dimensions are wrong, a result is still given, but some numbers may be missing or duplicated. The syntax is $= MMULT(A,B)$ and Figure A.6 illustrates the result of multiplying a 4×3 matrix with a 3×2 matrix to give a 4×2 matrix.
- *TRANSPOSE* gives the transpose of a matrix. Select the destination of the correct shape and use the syntax $=TRANSPOSE(A)$ as illustrated in Figure A.7.
- *MINVERSE* is an operation that can only be performed on square matrices and gives the inverse. An error message is presented if the matrix has no inverse or is not square. The syntax is $= MINVERSE(A)$ as illustrated in Figure A.8.

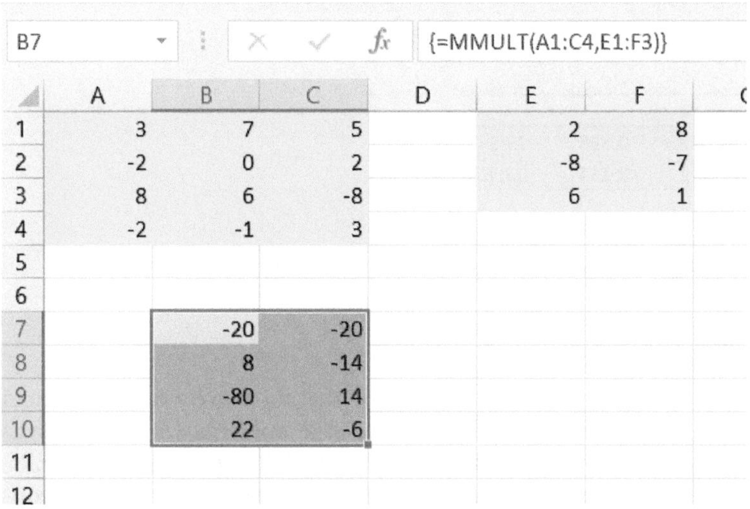

Figure A.6 Matrix multiplication in Excel.

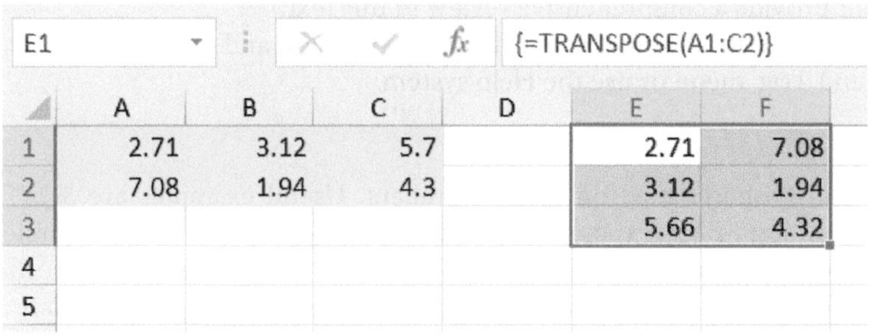

Figure A.7 Matrix transpose in Excel.

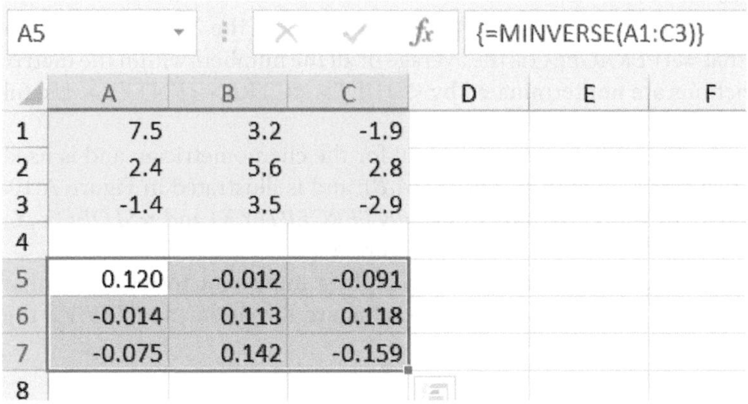

Figure A.8 Matrix inverse in Excel.

It is not necessary to give a matrix a name, so the expression $=TRANSPOSE(C11:G17)$ is acceptable, and it is entirely possible to mix terminology, for example, $=MMULT(X,B6:D9)$ will work, providing the relevant dimensions are correct.

It is a very useful facility to be able to combine matrix operations. This means that more complex expressions can be performed. A common example is to calculate the pseudo-inverse $(Y'Y)^{-1}Y'$, in Excel, as follows: $=MMULT(MINVERSE(MMULT(TRANSPOSE(Y),Y)),TRANSPOSE(Y))$. Figure A.9 illustrates this, where Y is a 4×2 matrix, and its pseudo-inverse, a 2×4 matrix. Of course, each intermediate step of the calculation could be displayed separately if required. In addition, via macros, we can automate this to save typing long equations each time. However, for learning the basics of chemometrics, it is useful, in the first instant, to present the equation in full.

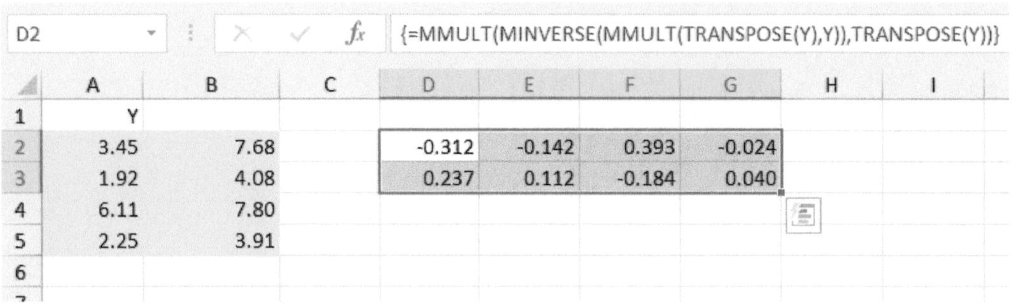

D2		▾	⋮	×	✓	*fx*	{=MMULT(MINVERSE(MMULT(TRANSPOSE(Y),Y)),TRANSPOSE(Y))}		

◢	A	B	C	D	E	F	G	H	I
1	Y								
2	3.45	7.68		-0.312	-0.142	0.393	-0.024		
3	1.92	4.08		0.237	0.112	-0.184	0.040		
4	6.11	7.80							
5	2.25	3.91							
6									
7									

Figure A.9 Pseudo-inverse of a matrix.

It is possible to add and subtract matrices using + and −, but remember to ensure that the two (or more) matrices have the same dimensions as has the destination. It is also possible to mix matrix and scalar operations, so that the syntax $=2*MMULT(X,Y)$ is acceptable. Remember always to complete the statement by <SHIFT> <CTRL> <ENTER> if performing arithmetic on matrices rather than scalars. Furthermore, it is possible to add (or subtract) matrices consisting of a constant number, for example, $=Y+2$ would add 2 to each element of Y. Other conventions for mixing matrix and scalar variables and operations can be deduced by practice, although in most cases, the result is what we would logically expect, and for brevity, we will not provide a comprehensive review in this text.

There are a number of other matrix functions in Excel, for example, to calculate determinant and trace of a matrix, to use these select the 'Formulas' tab and 'Math and Trig' menu or use the Help system.

A.4.2.3 Arithmetic Functions of Scalars

There are numerous arithmetic functions that can be performed on single numbers. Useful examples are *SQRT* (square root), *LOG* (log to the base 10), *LN* (natural log), *EXP* (exponential) and *ABS* (absolute value), for example $=SQRT(A1+2*B1)$. A few functions have no number to operate on, such as *ROW()* which is the row number of a cell, *COLUMN()* the column number of a cell and *PI()* the number π. Trigonometric functions operate on angles in radians; thus, be sure to convert if your original numbers are in degrees or cycles, for example $=COS(PI())$ gives a value of −1.

A.4.2.4 Arithmetic Functions of Ranges and Matrices

It is often useful to calculate the function of a range; for example, $=SUM(A1:C9)$ is the sum of the 27 numbers within the range. It is possible to use matrix notation so that $=AVERAGE(X)$ is the average of all the numbers within the matrix X. Notice that as the answer is a scalar, these functions are not terminated by <SHIFT> <CTRL> <ENTER>. Useful functions include *SUM*, *AVERAGE*, *SUMSQ* (sum of squares) and *MEDIAN*.

Some functions require more than one range. The *CORREL* function is useful for the chemometrician and is used to compute the correlation coefficient between two arrays, syntax $=CORREL(A,B)$, and is illustrated in Figure A.10. Another couple of useful functions involve linear regression. The functions $=INTERCEPT(Y,X)$ and $=SLOPE(Y,X)$ provide the parameters b_0 and b_1 in the equation $y = b_0 + b_1 x$, as illustrated in Figure A.11.

Standard deviations, variances and covariances are useful common functions. It is important to recognise that there are both population and sample functions, so that *STDEV.S* is the sample standard deviation and *STDEV.P* the

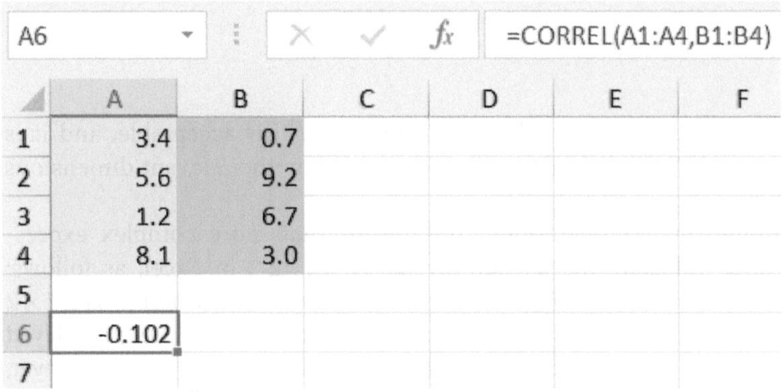

A6		▾	⋮	×	✓	*fx*	=CORREL(A1:A4,B1:B4)	

◢	A	B	C	D	E	F
1	3.4	0.7				
2	5.6	9.2				
3	1.2	6.7				
4	8.1	3.0				
5						
6	-0.102					
7						

Figure A.10 Correlation between two ranges.

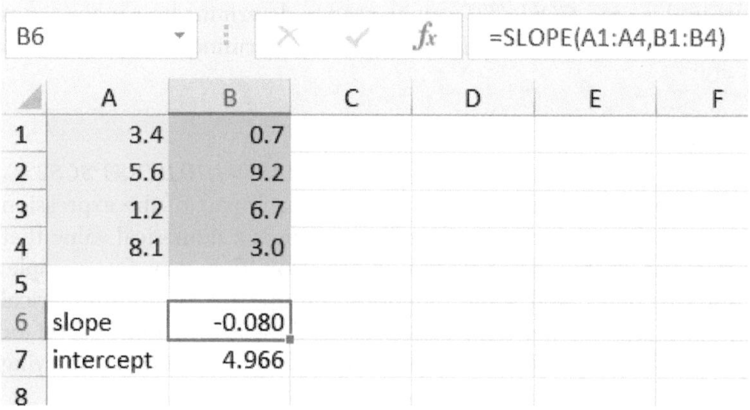

Figure A.11 Finding the slope and intercept when fitting a linear model to two ranges.

equivalent population standard deviation. Note that for standardising matrices, it is a normal convention to use the population standard deviation. Similar comments apply to *VAR.S* and *VAR.P*. For the covariance, use *COVARIANCE.S* and *COVARIANCE.P*. There are legacy functions from previous versions of Excel, which are not so logical, be careful if choosing these functions.

A.4.2.5 Statistical Functions

Surprisingly, a large number of common statistical functions are available in Excel. Conventionally, many such functions are presented in tabular format as in this book, for completeness, but most information can be easily obtained from Excel.

The inverse normal distribution is quite useful and allows a determination of the number of standard deviations from the mean to give a defined probability; for example, *=NORM.INV(0.9,0,1)* is the value within which 90% (0.9) of the readings will fall if the mean is 0 and standard deviation is 1, and equals 1.282, which can be verified using Table A.1. The function *NORM.DIST* returns the probability of lying within a particular value; for example, *=NORM.DIST(1.5,0,1,TRUE)* is the probability of a value that is less than 1.5, for a mean of 0 and a standard deviation of 1, using the cumulative normal distribution (*=TRUE*), and equals 0.99319 (see Table A.1). Similar functions such as *T.DIST*, *T.INV*, *CHISQ.DIST*, *CHISQ.INV*, *F.DIST* and *F.INV* can be employed if required, eliminating the need for tables, although these do not allow a standard deviation or mean to be specified, but require the relevant degrees of freedom to be specified. It is important to check whether the functions are one tailed or two tailed if appropriate, or right side/left side of the distribution as required and there are several alternatives: the reader is advised to read the description of the functions in Excel when first using them. There are also a likewise number of legacy functions to be compatible with early versions of Excel.

A.4.2.6 Logical Functions

There are several useful logical functions in Excel. *IF* is a common one. Figure A.12 is of the function *=IF(A1<B1,A1,B1)* and places the lower of the values of columns A and B in column D. Notice that this has been copied down the column

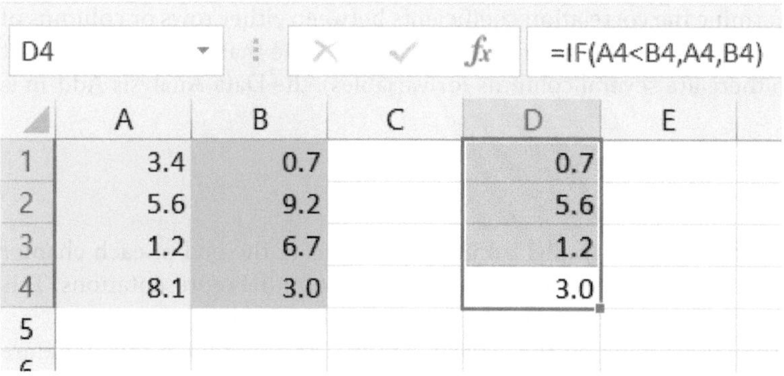

Figure A.12 Use of IF in Excel.

and also that there are no $ signs in the arguments in this case. *COUNTIF* can be used to determine how many times an expression is valid within a region of a worksheet. This is quite useful, for example, to determine how many values of a matrix are above a threshold or equal zero.

A.4.2.7 Nesting and Combining Functions and Equations

It is possible to nest and combine functions and equations. The expression =$C6+IF(A$7>1,10,IF(B$3*$C$2>5, 15,0))^2-2*SUMSQ(X+Y) is entirely legitimate, although it is important to ensure that each part of the expression results in compatible type of information (in this case, the result of using the *IF* function is a numerical value that is squared). Note that spreadsheets are not restricted to containing numerical information, they may, for example, also contain names (characters) or logical variables or dates, and some functions will operate on non-numerical information. In this section, we have concentrated primarily on numerical functions, as these are the most useful for the chemometrician, but it is important to recognise that nonsensical results would be obtained, for example, if trying to add a character to a numerical expression to a date.

A.4.3 Add-Ins

A very important capability of Excel consists of add-ins. In this section, we will describe only those add-ins that are part of the standard Excel package. It is possible to write one's own add-ins or download a number of useful add-ins from the web. This book is associated with an add-ins specifically for chemometrics, as will be described in Section A.4.5.2.

If properly installed, there should be a 'Data Analysis' item in the 'Data' ribbon. If this does not appear, you should select the 'File' menu, then select 'Options' and the 'Add-ins' and the find 'Manage Excel Add-ins', as illustrated in Figure A.13(a). Select 'Go' and then you will be presented with a list of add-ins, which will depend on your installation, and select the 'Analysis Toolpak' typically illustrated in Figure A.13(b). One occasional problem is that some institutes use Excel over a network. The problem with this is that it is not always possible to install these facilities on an individual computer, but this must be performed by the Network administrator, dependent on your configuration.

Once the menu item is selected, the dialog box of Figure A.14 should appear. There are several useful facilities, but probably, the most important for the purpose of chemometrics is the 'Regression' feature. The default notation in Excel differs from that in this book. A multiple linear model is formed between a single response y and any number of x variables. Figure A.15 illustrates the result of performing regression on one x variable to give the best fit model $y \approx b_0 + b_1 x_1 + b_2 x_2$. There are quite a number of statistics produced. Notice in the dialog box one selects 'constant is zero' if one does not want to have a b_0 term, this is equivalent to forcing the intercept to be equal to 0. The answer, in the case illustrated, is $y \approx -0.15762 + 0.29993 \, x_1$, see cells *B30–B31*. Notice that this answer could also have been performed using matrix manipulations with the pseudo-inverse, after first adding a column of 1s to the X matrix, as described in Section A.1.2.5 and elsewhere. There are quite a few options in the regression function, which the interested reader can gain experience of, if required.

A second facility that is sometimes useful is the random number generator function. There are several possible distributions, but the most usual is the normal distribution. It is necessary to specify a mean and standard deviation. If one wants to be able to return to the distribution later, also specify a seed, which must be an integer number. Figure A.16 illustrates the generation of 10 random numbers coming from a distribution of mean 6 and standard deviation 2 placed in cells *A1–A10* (note that the standard deviation is of the parent population and will not be exactly the same for a sample). This facility is very helpful in simulations and can be employed to study the effect of noise on a data set.

The 'Correlation' facility that allows one to determine the correlation coefficients between either rows or columns of a matrix is also useful in chemometrics, for example, as the first step in cluster analysis. Note that for two columns, it is better to use the *CORREL* function, but when there are several columns (or variables), the Data Analysis Add-in is easier.

A.4.4 Charts

Most graphs in this text have been produced in Excel, and all graphs from the problems at the end of each chapter can be produced either in Excel or in Matlab. The graphics facilities are quite good except for 3D representations. This section will briefly outline some of the main features of the chart tool useful for applications in this text.

Graphs can be produced either by selecting the 'Insert' tab and the 'Charts' group. Most graphs in this book are produced using an *xy* or Scatter plot, allowing the value of one parameter (e.g. the score of PC2) to be plotted against

Excel Options ? ×

General	View and manage Microsoft Office Add-ins.
Formulas	
Proofing	**Add-ins**
Save	
Language	
Advanced	
Customize Ribbon	
Quick Access Toolbar	
Add-ins	
Trust Center	

Name ▲	Location	Type
Active Application Add-ins		
Analysis ToolPak	C:\...t\Office16\Library\Analysis\ANALYS32.XLL	Excel Add-in
Multivariate Analysis	C:\...Microsoft\AddIns\MultivariateAnalysis.xla	Excel Add-in
Variance/Covariance Matrix Add-in	C:\...r 5 second edition\cov-matrix-2007.xlam	Excel Add-in
Inactive Application Add-ins		
Analysis ToolPak - VBA	C:\...ffice16\Library\Analysis\ATPVBAEN.XLAM	Excel Add-in
Date (XML)	C:\...s\Microsoft Shared\Smart Tag\MOFL.DLL	Action
Euro Currency Tools	C:\...e\root\Office16\Library\EUROTOOL.XLAM	Excel Add-in
Microsoft Actions Pane 3		XML Expansion Pack
Microsoft Power Map for Excel	C:\...ap Excel Add-in\EXCELPLUGINSHELL.DLL	COM Add-in
Solver Add-in	C:\...t\Office16\Library\SOLVER\SOLVER.XLAM	Excel Add-in
Document Related Add-ins		
No Document Related Add-ins		
Disabled Application Add-ins		
No Disabled Application Add-ins		

Add-in: Analysis ToolPak
Publisher: Microsoft Corporation
Compatibility: No compatibility information available
Location: C:\Program Files (x86)\Microsoft Office\root\Office16\Library\Analysis\ANALYS32.XLL

Description: Provides data analysis tools for statistical and engineering analysis

Manage: Excel Add-ins ▼ Go...

(a)

Add-ins ? ×

Add-ins available:

☑ Analysis ToolPak   OK
☐ Analysis ToolPak - VBA
☐ Euro Currency Tools   Cancel
☑ Multivariate Analysis
☐ Solver Add-in   Browse...
☑ Variance/Covariance Matrix Add-in
                            Automation...

Analysis ToolPak

Provides data analysis tools for statistical and engineering
analysis

(b)

Figure A.13 Finding the Analysis Toolpak.

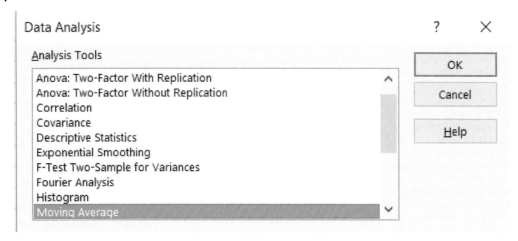

Figure A.14 Data Analysis Add-in dialog box.

another (e.g. the score of PC1). Various enhancements to the appearance of graphs can be learnt by experience and we will not list all these in detail for brevity.

It is often desirable to use different symbols for groups of parameters or classes of compounds. This can be done by superimposing several graphs, each being represented by a separate series. This is illustrated in Figure A.17 in which cells *A1–B10* represent Series 1 and *A11–B20* represent Series 2, each set of measurements having a different symbol. To achieve this, once a chart has been developed for Series 1, go to the 'Chart Tools' tab and then select the 'Design' tab. You will find a 'Select Data' item, and this allows you to add as many additional series as you like. Each will be default and have different symbols, although, of courses, you can change the symbols for each series as you wish.

The default graphics options are not necessarily the most appropriate and are designed primarily for display on a screen. Typically, we often remove gridlines and remove the box around the graph. In addition, often it is a good idea to label the axes and add a title and increase the font size. These facilities are available and a case of personal preference. The default appearance of Excel Charts has improved considerably over the years.

It is also possible to place the chart on a sheet of its own. In the 'Design Tab', choose 'Move Chart' and select 'Location'. A typical finalised chart is illustrated in Figure A.18.

There are numerous resources on the web to learn more about how to create charts in Excel if desired. One difficulty involves attaching a label to each point in a chart; this may be the name of an object or a variable. With this text, we produce a downloadable macro that can be edited to permit this facility, as described in Section A.4.5 There are, of course, also numerous facilities available on the web as alternatives.

A.4.5 Downloadable Macros

To facilitate the use of charts and also common multivariate methods, we provide some downloadable macros in the companion website. Macros are programs written in a language called VBA that can be run in Excel to enhance the normal facilities. For more details, read the Excel literature or search the web.

A.4.5.1 Labelling Macro
A macro that allows points to be added to be labelled in a scatterplot is available.

The simplest approach is to download the file 'label', which is in the old Excel format to be compatible with previous versions and open it in the same directory as your current file that contains your data. In the file containing your data, you should have a 'Developer' tab, which can be opened, to reveal a 'Macros' item, which if clicked open will reveal some macros that have been pre-written. If there is no 'Developer' tab, go to the 'File' tab, then 'Options' menu item and then select 'Customise Ribbon' and ensure you can add the 'Developer' tab. You should see the AddChartLabels macro that allows labels to be added to points in a graph.

First, produce a scatterplot in Excel using the Chart Wizard. Make sure that there are two columns to the left of the data in Excel. The first should contain the labels you want for each point. The second (on the far left) should contain an 'x' for each point you want to label. This allows you to select points on the graph. If you want to label all the points, then put at 'x' all the way down the column, see Figure A.19. Select the graph. Then, simply run the macro, you will be

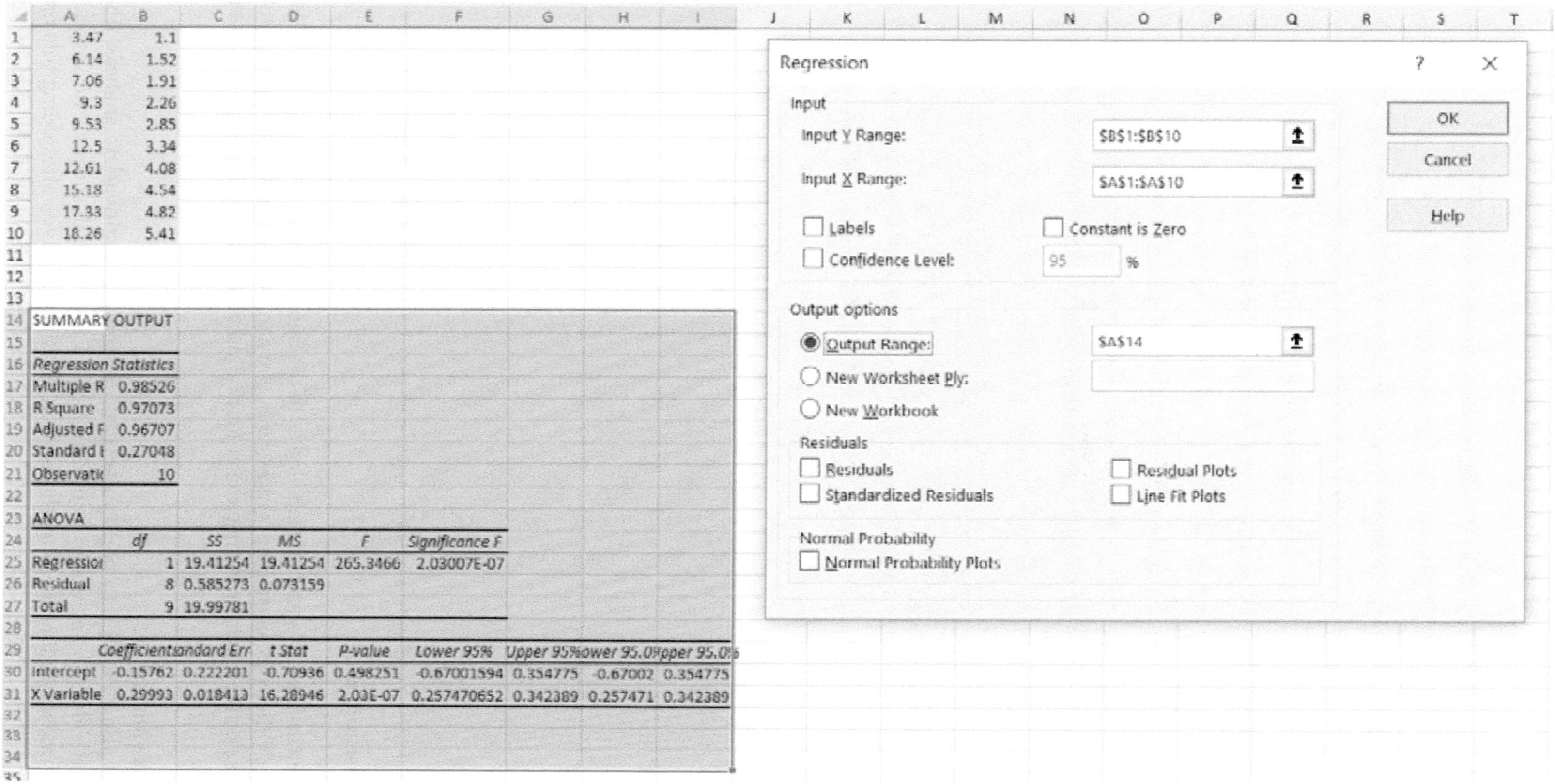

A	B
3.47	1.1
6.14	1.52
7.06	1.91
9.3	2.26
9.53	2.85
12.5	3.34
12.61	4.08
15.18	4.54
17.33	4.82
18.26	5.41

SUMMARY OUTPUT

Regression Statistics

Multiple R	0.98526
R Square	0.97073
Adjusted R	0.96707
Standard E	0.27048
Observatic	10

ANOVA

	df	SS	MS	F	Significance F
Regression	1	19.41254	19.41254	265.3466	2.03007E-07
Residual	8	0.585273	0.073159		
Total	9	19.99781			

	Coefficients	Standard Err	t Stat	P-value	Lower 95%	Upper 95%	Lower 95.0%	Upper 95.0%
Intercept	-0.15762	0.222201	-0.70936	0.498251	-0.670011594	0.354775	-0.67002	0.354775
X Variable	0.29993	0.018413	16.28946	2.03E-07	0.257470652	0.342389	0.257471	0.342389

Regression

Input

Input Y Range: $B$1:$B$10

Input X Range: $A$1:$A$10

☐ Labels ☐ Constant is Zero
☐ Confidence Level: 95 %

Output options
◉ Output Range: $A$14
○ New Worksheet Ply:
○ New Workbook

Residuals
☐ Residuals ☐ Residual Plots
☐ Standardized Residuals ☐ Line Fit Plots

Normal Probability
☐ Normal Probability Plots

OK Cancel Help

Figure A.15 Linear regression using the Excel Data Analysis Add-in.

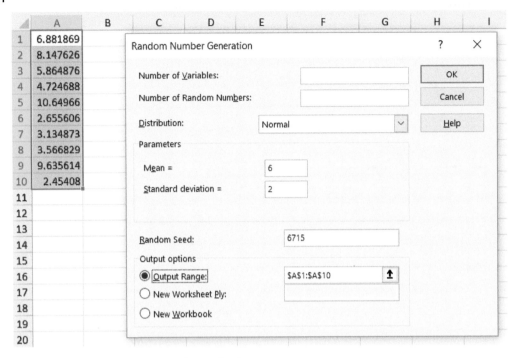

Figure A.16 Generating random numbers in Excel.

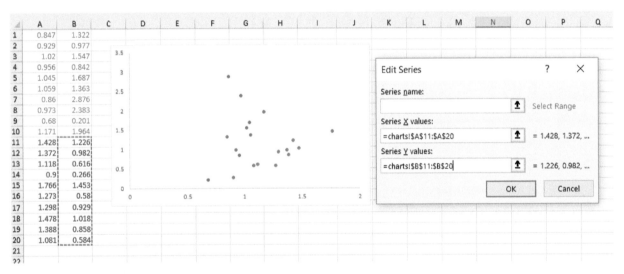

Figure A.17 Adding an extra series in Excel.

asked the font size and then each point should be labelled as in the figure. If some of the labels overlap after the macro has been run, for example, if there are close points in a graph, you can manually select each label and move it around the graph, or even delete selective labels or change the colour of individual labels, for example, to emphasize specific data points.

A.4.5.2 Multivariate Analysis Add-In

Accompanying the text is also an add-in to perform several methods for multivariate analysis. The reader is urged first to understand the methods by using matrix commands in Excel or Matlab scripts, and several examples in this book guide the reader to understanding these methods from scratch. However, after doing this once, it is probably unnecessary to repeat the full calculations from scratch and convenient to have available add-ins in Excel. Although the performance has been tested on computers of a variety of configurations, the software was originally developed for Office 2000 and Windows 98, using 64 MB memory, which almost all present-day computers exceed. There may

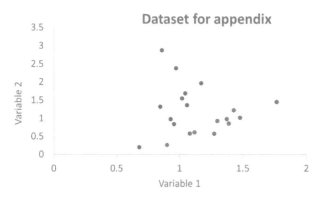

Figure A.18 Finalised chart from Excel.

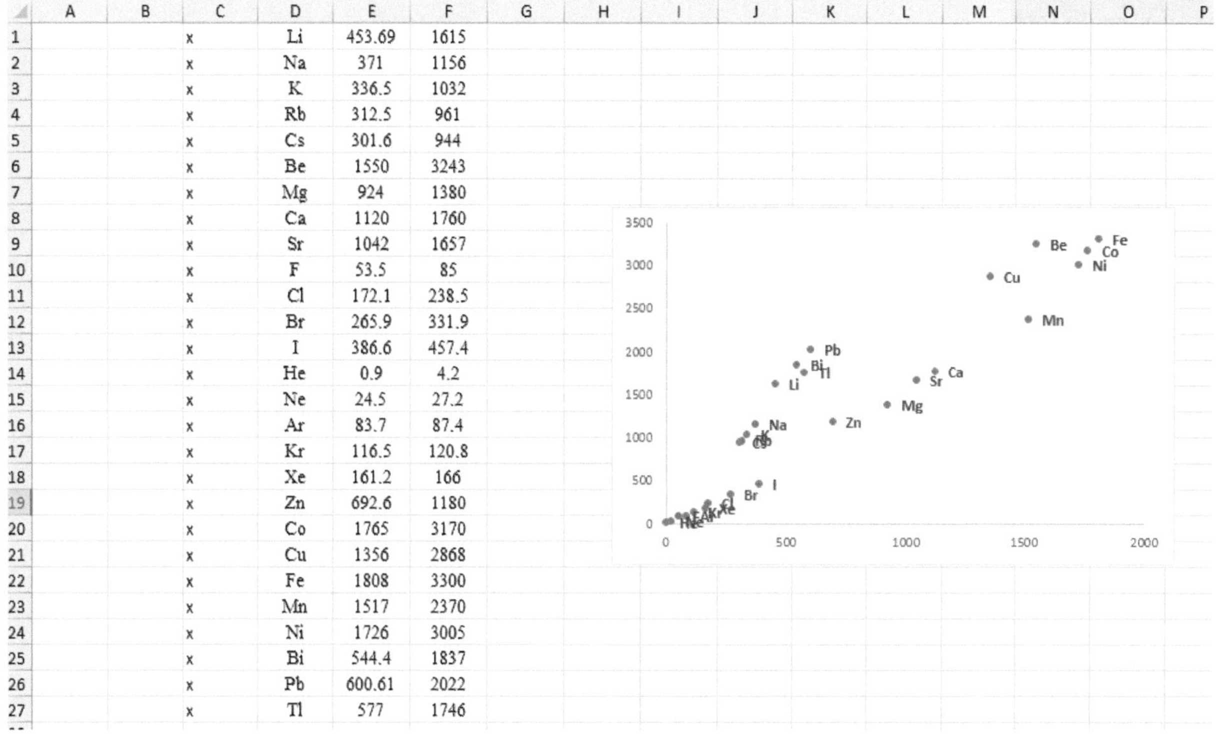

	A	B	C	D	E	F	G	H	I	J	K	L	M	N	O	P
1			x	Li	453.69	1615										
2			x	Na	371	1156										
3			x	K	336.5	1032										
4			x	Rb	312.5	961										
5			x	Cs	301.6	944										
6			x	Be	1550	3243										
7			x	Mg	924	1380										
8			x	Ca	1120	1760										
9			x	Sr	1042	1657										
10			x	F	53.5	85										
11			x	Cl	172.1	238.5										
12			x	Br	265.9	331.9										
13			x	I	386.6	457.4										
14			x	He	0.9	4.2										
15			x	Ne	24.5	27.2										
16			x	Ar	83.7	87.4										
17			x	Kr	116.5	120.8										
18			x	Xe	161.2	166										
19			x	Zn	692.6	1180										
20			x	Co	1765	3170										
21			x	Cu	1356	2868										
22			x	Fe	1808	3300										
23			x	Mn	1517	2370										
24			x	Ni	1726	3005										
25			x	Bi	544.4	1837										
26			x	Pb	600.61	2022										
27			x	Tl	577	1746										

Figure A.19 Labelling a graph in Excel.

be problems with lower configurations, but it is upward compatible, for example, using Office 2016 and Windows 10; however, you must use 32 bit and not 64 bit Excel; this is the only significant limitation as is common for many macros. The VBA software was written by Tom Thurston and the associated C DLLs by Les Erskine.

You need to download the add-ins from the publisher's website. You will obtain a set-up file, click this to obtain the screen in Figure A.20, and follow the instructions. If in doubt, please contact whoever is responsible for maintaining computing facilities within your department or office. Please note that sometimes there can be problems with networks; under such circumstances, you may be required to consult the systems manager. If all is well (remember to ensure that you are using a 32 bit version of Excel), the add-in should be visible in the 'Add-ins' tab. If it is not, go to the 'File' tab, then select 'Options' and 'Add-ins' just as for the Analysis Toolpak described in Section A.4.3 to eventually reach the screen presented in Figure A.21 to add the facility.

Once all is sorted, select the 'Multivariate Analysis' Add-in, and the dialog box of Figure A.22 should appear, allowing four options that will be described below.

The PCA dialog box is illustrated in Figure A.23. It is first necessary to select the data range, and the number of PCs to be calculated. By default, the objects are along the rows and the variables down the columns, but it is possible to

Figure A.20 Setup screen for the Excel chemometrics add-in.

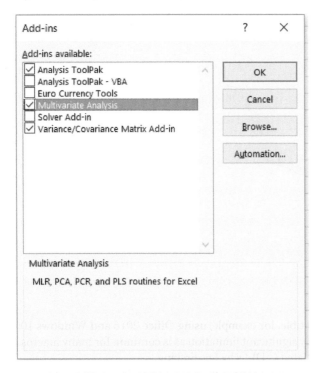

Figure A.21 Selecting the Multivariate Analysis Add-in

transpose the data matrix, in PCA and all other options. The data may be mean centred in the direction of variables, or standardised (this uses the population rather than sample standard deviation as recommended in this book).

It is possible to cross-validate the PCs by leaving one sample out at a time approach (see Section 4.3.2.2); this option is useful if one wants guidance as to how many PCs are relevant to the model. You are also asked to select the number of PCs required.

An output range must be chosen; it is only necessary to select the top left-hand cell of this range, but be careful that it does not overwrite the existing data. For normal PCA, choose which of eigenvalues, scores and loadings you wish to display. If you select eigenvalues, you will also be given the total sum of squares of the pre-processed (rather than raw) data together with the percentage variance of each eigenvalue.

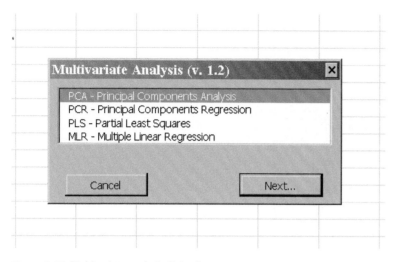

Figure A.22 Multivariate analysis dialog box.

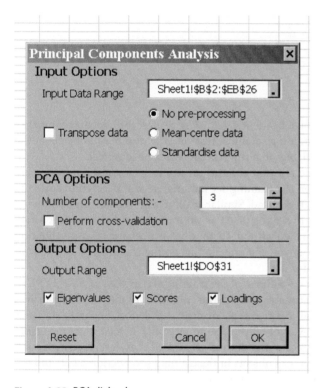

Figure A.23 PCA dialog box.

Although cross-validation is always performed on the pre-processed data, the RSS and PRESS values are always calculated on the x block in the original units, as discussed in Section 4.3.2.2. The reason for this relates to rather complex problems that occur when standardising a column after one sample has been removed. There are, of course, many other possible approaches. When performing cross-validation, the only output available involves error analysis.

The PCR dialog box, illustrated in Figure A.24, is considerably more complicated. It is always necessary to have a training set consisting of an x block and a c block. The latter may consist of more than one column. For PCR, unlike PLS, all columns are treated independently; hence, there is no analogy to PLS2. You can choose three options. (a) 'Training set only' is primarily for building and validating models. It only uses the training set. You need only to specify a x and c block training set. The number of objects in both sets must be identical. (b) 'Predict concentrations' is used to predict concentrations from an unknown series of samples. It is necessary to have a x and c block training set as well as a x block for the unknowns. A model will be built from the training set and applied to the unknowns. There can be any number

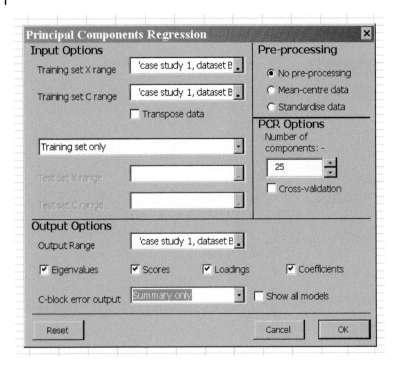

Figure A.24 PCR dialog box.

of unknowns, but the number of variables in the two *x* blocks must be identical. (c) 'Use test set (predict and compare)' allows two sets of blocks where concentrations are known, a training set and a test set. The number of objects in the training and test set will normally differ, but the number of variables in both data sets must be identical.

There are three methods for data scaling, as in PCA, but the relevant column means and standard deviations are always obtained from the training set. If there is a test set, then the training set parameters will be used to scale the test set so that the test set is unlikely to be mean centred or standardised. Similar scaling is performed on both the '*c*' and '*x*' block simultaneously. If you want to apply other forms of scaling (such as summing rows to a constant total), this can be performed manually in Excel and PCA can be performed without further pre-processing. Cross-validation is performed only on the '*c*' block; if you choose cross-validation, you can only do this on the training set. If you want to perform cross-validation on the '*x*' block, use the PCA facility.

There are a number of types of output. Eigenvalues, scores and loadings (of the training set) are the same as in PCA, whereas the coefficients relate the PCs to the concentration estimates and are the columns of matrix $\boldsymbol{R}$ as described in Section 6.4.1. This information is available if requested in all cases except for cross-validation. Separate statistics can be obtained for the '*c*' block predictions. There are three levels of output. 'Summary only' involves just the errors including the training set error (adjusted by the number of degrees of freedom to give *RMSEC* as described in Section 6.6.1), the cross-validated error *RMSECV* (divided by the number of objects in the training set, Section 6.6.2) and the test set error *RMSEP* (Section 6.6.3), as appropriate to the relevant calculation. If the 'Predictions' option is selected, then the predicted concentrations are also displayed, and 'Predictions and Residuals' provides the residuals as well (if appropriate for the training and test sets), although these can also be calculated manually. If the 'How all models' option is selected, then predicted *c* values and the relevant errors (according to the information required) for 1, 2, 3 up to the chosen number of PCs are displayed. If this option is not selected, only information for the full model is provided.

The PLS dialog box, illustrated in Figure A.25, is very similar to PCR, except that there is an option to perform PLS1 ('One *c* variable at a time') (see Section 6.5.1) as well as PLS2 (Section 6.5.2). However, even when performing PLS1, it is possible to use several variables in the *c* block, each variable, however, is modelled independently. Instead of coefficients (in PCR), we have 'C-loadings' ($\boldsymbol{Q}$) for PLS, as well as the 'X-loadings' ($\boldsymbol{P}$), although there is only one scores matrix. Strictly speaking, there are no eigenvalues for PLS, but the size of each component is given by the magnitude, which is the product of the sum of squares of the scores and loadings for each PLS component. Note that the loadings in the method described in this text are neither normalised nor orthogonal. If one selects PLS2, there will be a single set of 'Scores' and 'X-loadings' matrices; however, many columns in the *c* block, but 'C-loadings', will be in the form of a matrix. If PLS1 is selected and there is more than one column in the '*c*' block, separate 'Scores' and 'X-loadings'

Partial Least Squares

Input Options

Training set X range   `'case study 1, dataset B`

Training set C range   `'case study 1, dataset B`

☐ Transpose data

`Training set only` ▾

Test set X range

Test set C range

Pre-processing

◉ No pre-processing

○ Mean-centre data

○ Standardise data

PLS Options

Number of components: -

`12`

☐ Cross-validation

☐ PLS 1

Output Options

Output Range   `'case study 1, dataset B`

☑ Magnitudes   ☑ Scores   ☑ P (X Loadings)   ☑ Q (C Loadings)

C-block error output   `Predictions` ▾   ☑ Show all models

Reset   Cancel   OK

Figure A.25 PLS dialog box.

matrices are generated for each compound variable, as well as an associated 'C-loadings' vector; thus, the output can become quite extensive unless one is careful to select the appropriate options.

For both PCR and PLS, it is, of course, possible to transpose data, but *both* the x block and the c block must be transposed. These facilities are not restricted to predicting concentrations in spectra of mixtures and can be used for any purpose, such as QSAR or sensory statistics.

The MLR dialog box, illustrated in Figure A.26, is somewhat simpler than the others and is mainly used if two out of X, C and S are known. The type of unknown matrix is chosen and then regions of the spreadsheet of the correct size

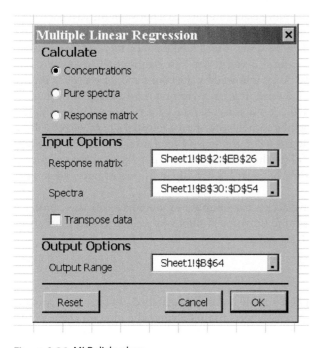

Multiple Linear Regression

Calculate

◉ Concentrations

○ Pure spectra

○ Response matrix

Input Options

Response matrix   `Sheet1!$B$2:$EB$26`

Spectra   `Sheet1!$B$30:$D$54`

☐ Transpose data

Output Options

Output Range   `Sheet1!$B$64`

Reset   Cancel   OK

Figure A.26 MLR dialog box.

must be selected. This facility also performs regression using the pseudo-inverse and is mainly provided for completion. Note that it is not necessary to restrict the data to spectra or concentrations. MLR can also be performed using standard matrix operations in Excel as described in Section A.4.2.2.

This add-in provides a basic functionality for many of the multivariate methods described in Chapters 4–7 and can be used when solving the problems at the end of each chapter in Excel.

A.5   Matlab for Chemometrics

Many chemometricians use Matlab. In order to appreciate the popularity of this approach, it is important to understand the vintage of chemometrics. The first applications of quantum chemistry, another type of computational chemistry, were developed in the 1960s and 1970s where Fortran was the main numerical programming environment. Hence, large libraries of routines were established over this period, and to this day, most quantum chemists still program in Fortran. Were the disciplines of quantum chemistry to start over again, probably Fortran would not be the main programming environment of choice, but tens of thousands (or more) man-years would need to be invested to rewrite the entire historic databases of programs. If we were developing an operating system that would be used by tens or hundreds of millions of people, that investment might be worthwhile, but the scientific market is much smaller; hence, once the environment is established, new researchers tend to stick to it as they can then exchange code and access libraries.

Although some early chemometrics code was developed in Fortran (the Arthur package of Kowalski) and Basic (Wold's early version of SIMCA) and commercial packages are mainly written in C, most public domain chemometrics code first became available in the 1980s where Matlab was an upcoming new environment. An advantage of Matlab is that it is very much oriented towards matrix operations and most chemometrics algorithms are best expressed this way. It can be awkward to write matrix-based programs in C, Basic or Fortran unless one has access to or develops specialised libraries. Matlab was originally a technical programming environment mainly for engineers and physical scientists, but over the years, the user base has expanded strongly and Matlab has kept pace with new technology including extensive graphics, interface to Excel, numerous toolboxes for specialist use and the ability to compile software. In this section, we will primarily concentrate on the basics required for chemometrics and also to solve the problems in this book; for the more experienced user, there are numerous other outstanding texts on Matlab, including the extensive documentation produced by the developer of the software, the MathWorks, which maintains an excellent website. In this book, you will be introduced to a number of main features to help you solve the problems, but as you gain experience, you will undoubtedly develop your own personal favourite approaches. Matlab can be used at many levels, and it is now possible to develop sophisticated packages with good graphics in this environment.

There are many versions of Matlab and of Windows, and for the more elaborate interface between the two packages, it is necessary to refer to technical manuals. We will illustrate this section with Matlab R2016a running under Windows 10, although some readers may have access to more up-to-date editions. Most facilities are forward compatible, although there are a few small deletions of commands, so be careful if you try to run elaborate 20-year-old code. There is quite a good online help facility in Matlab, you can type `help` followed by the command, or else click on the '?' icon on the top right of the screen to search documentation. However, it is useful to first have a grasp of the basics, which will be described below. We will primarily describe the main commands necessary for the sort of chemometric calculations described in this book and introduce methods in their simplest form: more experienced users will be able to expand on these basics, and this Appendix is not intended as a comprehensive reference, the user to get more familiar and to provide some tricks of the trade specifically useful for the chemometrics expert as an aid to producing graphs and performing calculations relevant to this book. It should be regarded as a dip in the water, once aided, next steps should be independent.

In addition to having access to core Matlab, it is useful to have access to a number of toolboxes. For this book, we will assume that the reader has access to the 'Statistics and Machine Learning' Toolbox as well. There are numerous toolboxes, but this one is the most important for the chemometrics expert.

A.5.1   Getting Started

To start Matlab, it is easiest to simply click the icon that should be available if properly installed, the first time you use Matlab, and a blank screen as in Figure A.27 will appear. It is probably wise to first change the directory you would like to work in: this will be the directory files are read from and saved to and can be done by navigating through the file icon on the top left corner. You can always change this directory at any stage and also set up a default start up

Figure A.27 Default Matlab window.

directory should you so wish. If in a script (or program) you want to change the directory name, use the `cd` command, but remember that a script is often designed to be used on different computers; hence, in many cases, this is unwise.

To enter commands, find the main Command window, where you can type code in. The easiest way to start is by typing commands in after the '≫' prompt. Each Matlab command is typed on a separate line, terminated by the <ENTER> key. If the <ENTER> key is preceded by a semicolon (;), there is no output from Matlab (unless you have made an error) and on the next line you type the next command and so on. Otherwise, you are given some output, for example the result of multiplying matrices together, which can be useful, but if the information contains several lines of numbers that fill up a screen and which may not be very interesting, it is best to suppress this. Hence, if writing a long program (or script in Matlab terminology), if you wish to supress output except when you choose, terminate all statements with a semicolon.

Matlab is case sensitive; hence, the variable x is different to X. Commands are all lower case.

A.5.2 File Types

There are several types of files that one may wish to create and use, but there are three main kinds that are useful for the beginner.

A.5.2.1 Mat Files

These files store the 'workspace' or variables created during a session. All matrices, vectors and scalars with unique names are saved. Many chemometricians exchange data in this format. The command `save` places all this information into a file called `matlab.mat` in the current working directory. If you wish to save into a named file, type a filename, so the command `save mydata` saves the workspace into a file called `mydata.mat` in the current directory. If you want a space in the filename, enclose in single quotes for example `'Tuesday file'`.

In order to access these data in Matlab from an existing file, for example, you may have done some calculations a few days ago, or stored some spectra, simply use the `load` command, remembering what directory you are in, for example, type `load mydata`. If only one or two arrays are required, typing the array names after the filename will just load these, for example, `load mydata X`, just loads array X.

Alternatively, on the left (using the default window) is a list of files in the current directory and below that, if a mat file is selected, a list of arrays. Clicking either the filename (to input all arrays) or an individual array name will read these data into Matlab. This is illustrated in Figure A.28.

A.5.2.2 m Files

Quite often, it is useful to create programs that can be run again. This is done via m files. The same rules about directories apply as discussed above.

These files are simple text files and may be created in a variety of ways. A simple way of creating an m file is to click on the Home menu item on the main Matlab window, then 'New' and 'Script'. You can then type commands into an Editor window, and when you are happy you can save it, and if you wish, close this window again.

There are various ways in which you can run the script. For the simplest implementation, make sure everything is in the same directory. Then in the Matlab Command window, you can just type the name of the m file, for example, `Simplecode`, as shown in Figure A.29.

Some m files are functions: to create such an m file, after 'New', select 'Function'. These differ slightly from straight scripts, in that they usually involve the input of one or more arrays and the output of one or more arrays, although functions without a specified input or output (the function could, example.g. create a graph) are legitimate. Consider a simple function `codeforsum`. This takes a matrix and returns the sum of all its elements. The syntax can be written

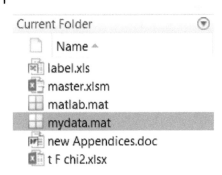

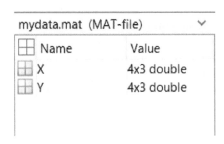

Figure A.28 File and array listing in Matlab.

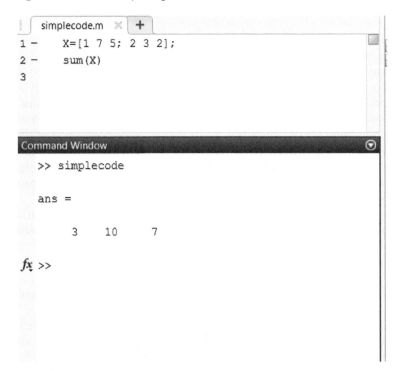

Figure A.29 Running an m file script in Matlab.

function [s] = codeforsum (X) where matrix X is the input and s the output, consisting the sum of all the elements of X. In the Command window, there is no need to use the same variable names as in the function; hence, we could have a matrix Y whose sum is p, and we can call the function by the statement p = codeforsum(Y). This is very convenient because we could compute the sum of several very different matrices and use the same function. The principle is illustrated in Figure A.30.

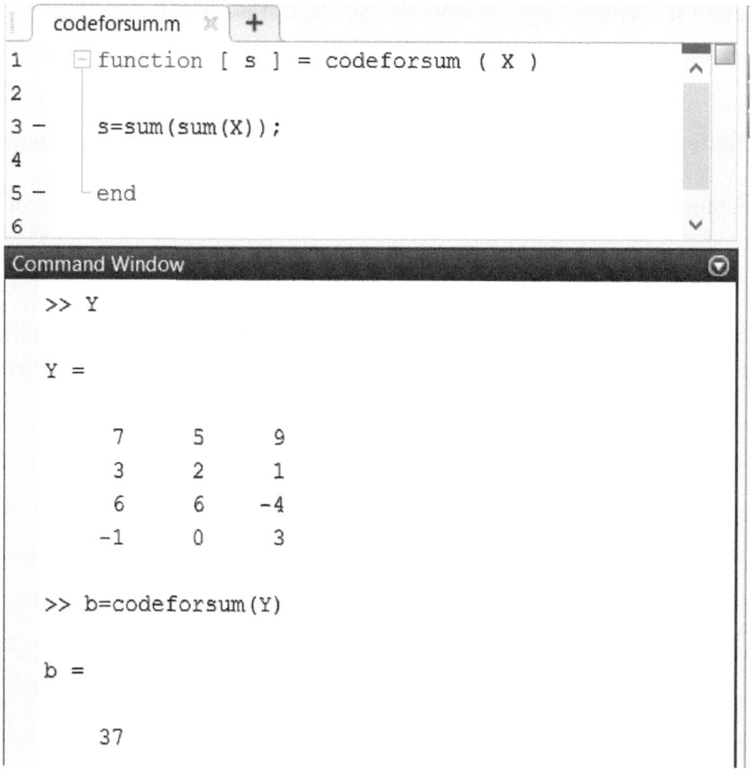

```
codeforsum.m   ×   +
1    □ function [ s ] = codeforsum ( X )
2
3 -      s=sum(sum(X));
4
5 -      └ end
6
```

```
Command Window
    >> Y

    Y =

           7      5      9
           3      2      1
           6      6     -4
          -1      0      3

    >> b=codeforsum(Y)

    b =

          37
```

Figure A.30 Running an m file function in Matlab.

The experienced Matlab user can establish libraries of m files and use these in different applications. Different programmers have different styles. Equally some programmers just like running all their code in the Command window. For most of the problems in this text, only a few statements are required, so the code can be either developed as one single file, even copied and pasted into the Command window, or as interactive Matlab commands. Obviously, if you decide to use Matlab for your personal research, or even to swap code with colleagues, you are likely to want to develop more elaborate habits.

A.5.2.3 Diary Files

These files keep a record of a session. The simplest approach is not to use diary files but just to copy and paste the text of a Matlab session, but diary files can be useful because one can selectively save just certain commands. In order to start a diary file, type `diary` (a default file called `diary` will be created in the current directory) or `diary filename` where `filename` is the name of the file. This automatically opens a file into which all subsequent commands used in a session, together with their results, are stored. To stop recording, simply type `diary off` and to start again (in the same file) type `diary on`.

The file can be viewed as a text file, in the Text Editor. Note that you must close the diary session before the information is saved.

A.5.3 Matrices

The key to Matlab is matrices. Understanding how Matlab copes with matrices is essential for the user of this environment.

A.5.3.1 Scalars, Vectors and Matrices

It is possible to handle scalars, vectors and matrices in Matlab. The package automatically determines the nature of a variable when first introduced. A scalar is simply a number so

```
X = 2
```

sets up a scalar $X = 2$. Notice that there is a distinction between upper and lower case, and it is entirely possible that another scalar x (lower case) coexists

```
x = 7
```

It is not necessary to restrict a name to a single letter, but all matrix names must start with an alphabetic rather than numeric character without spaces.

For one- and two-dimensional arrays, it is important to enclose the information within square brackets. A row vector can be defined by

```
Y = [2 8 7]
```

resulting in a 1×3 row vector. A column vector is treated rather differently as a matrix of three rows and one column. If a matrix or vector is typed on a single line, each new row starts a semicolon, so a 3×1 column vector may be defined by

```
Z = [1; 4; 7]
```

Alternatively, it is possible to place each row on a separate line, so

```
Z = [1
     4
     7]
```

has the same effect. Another trick is to enter as a row vector and then take the transpose (see Section A.5.3.3).

Matrices can be similarly defined, for example

```
W = [2 7 8; 0 1 6]
```

or

```
W = [2 7 8
     0 1 6]
```

are alternative ways, in the Matlab window, of setting up a 2×3 matrix.

One can specifically obtain the value of any element of a matrix; for example, $W(2,1)$ gives the element on the second row and first column of W, which equals 0 in this case. For vectors, only one dimension is needed, so $Z(2)$ equals 4 and $Y(3)$ equals 7.

It is also possible to extract single rows or columns from a matrix, by using a colon operator. The second row of matrix W is denoted by $W(2,:)$. This is exemplified in Figure A.31. It is possible to define any rectangular region of a matrix, using the colon operator. For example, if S is a matrix having dimensions 12×8, we may want a sub-matrix between rows 7–9 and columns 5–12, it is simply necessary to define $S(7:9, 5:12)$.

If you want to find out how many matrices are in memory, use the function who that lists all current matrices available to the program, or whos that contains details about their size. This is sometimes useful if you have had a long Matlab session or have imported a number of data sets.

There is a special notation for the identity matrix. The command eye (3) sets up a 3×3 identity matrix, the number enclosed in the brackets referring to the dimensions.

To delete all matrices from memory, use the command clear. If you only want to delete a few rather than all, specify the matrices, for example, clear X Y.

```
Command Window
  >> W = [2 7 8; 0 1 6];
  >> W(2,:)

  ans =

         0     1     6
```

Figure A.31 Obtaining vectors from matrices.

A.5.3.2 Basic Matrix Operations

The basic matrix operations +, − and * correspond to the normal matrix addition, subtraction and multiplication (using the dot product); for scalars, these are also defined in the usual way with the addition of the / symbol for division. For the first two operations, the two matrices should have the same dimensions, and for multiplication, the number of columns of the first matrix should equal the number of rows of the second matrix. It is possible to place the results in a target or else simply display them on the screen as a default variable called `ans`. Figure A.32 illustrates setting up three matrices, a 3×2 matrix X, a 2×3 matrix Y and a 3×3 matrix Z and calculating $X*Y+Z$.

There are quite a number of elaborations based on these basic operations, but the first time user is recommended to keep things simple. However, it is worth noting that it is possible to add scalars to matrices. An example involves adding the number 2 to each element of W, as defined above, either type `W + 2` or first define a scalar (e.g. `P = 2`) and then add this using the command `W + P`. Similarly, one can multiply, subtract or divide all elements of a matrix by a scalar. Notice that it is not possible to add a vector to a matrix even if the vector has one dimension identical to that of the matrix.

A.5.3.3 Matrix Functions

A significant advantage of Matlab is that there are several further very useful matrix operations. Most are in the form of functions; the arguments are enclosed in brackets. Three arguments that are important in chemometrics are as follows.

- Transpose is denoted by ' so that `W`' is the transpose of W.
- Inverse is a function `inv` so that `inv(Q)` is the inverse of a square matrix Q.
- The pseudo-inverse can simply be obtained by the function `pinv`, without any further commands, see Figure A.33, where we also verify that the product of a matrix with its pseudo-inverse is the identity matrix.

For a comprehensive list of functions, see the help files that come with Matlab; however, a few that are useful to the reader of this book are as follows. The `size` function gives the dimensions of a matrix, so `size(W)` will return a 2×1 vector with elements, in the example of Section A.5.3.1, of 2 and 3. It is possible to create a new vector, for example, `s = size(W)`, in such a situation `s(1)` will equal 2, or the number of rows. The element `W(s(1), s(2))` represents the last element in the matrix `W`. In addition, it is possible to use the functions `size(W,1)` and `size(W,2)`, which provide the number of rows and columns directly. These functions are very useful when writing simple programs as discussed below.

The `mean` function can be used in various ways. By default, this function produces the mean of each column in a matrix, so that `mean(W)` results in a 1×3 row vector containing the means. It is possible to specify which dimension one wishes to take the mean over, the default being the first one, so `mean(W,2)` is a 2×1 column vector. The overall mean of an entire matrix can be obtained using the mean function twice, that is, `mean(mean(W))`. Note that the mean of a vector is always a single number, whether the vector is a column or row vector. This function is illustrated in Figure A.34. The functions such as `min` and `max` compute minima and maxima of columns, for rows use a transpose, such as `min(X')`.

The functions `std` and `var` calculate the column standard deviations and variances: however, the default is the sample statistics, and in most calculations in chemometrics, we use the population statistics as our aim is to scale the

```
Command Window

>> X = [9 8; 11 4; 5 6];
>> Y = [2 7 1; 5 3 8];
>> Z=[2 3 5; 6 0 1; 11 4 8];
>> X*Y+Z


ans =

      60      90      78
      48      89      44
      51      57      61
```

Figure A.32 Simple matrix operations in Matlab.

```
Command Window                                    ⊙
  >> testmatrix=[2      7    8
0    1    6
];
  >> inverse=pinv(testmatrix)

inverse =

      0.0567    -0.0844
      0.1564    -0.2055
     -0.0261     0.2009

  >> testmatrix*inverse

ans =

      1.0000     0.0000
     -0.0000     1.0000
```

Figure A.33 Calculating a pseudo-inverse in Matlab.

```
Command Window                                    ⊙
  >> W = [2 7 8; 0 1 6];
  >> mean(W)

ans =

       1      4      7

  >> mean(W,2)

ans =

      5.6667
      2.3333
```

Figure A.34 Mean function in Matlab.

data not to estimate a population parameter from a sample. Under such circumstances, it is essential to use std(X,1) and similarly for the variance. This is illustrated in Figure A.35.

The norm function of a matrix is quite often useful and consists of the square root of the sum of squares, so in our example, norm(W) equals 12.0419. This can be useful when scaling data, especially for vectors. Note that if Y is a row vector, then sqrt(Y*Y') is the same as norm(Y).

It is useful to combine some of these functions; for example, min(s) would be the minimum dimension of matrix **W**, where *s* is as defined above. The enthusiasts can increase the number of variables within a function, an example being min([s 2 4]), which finds the minimum of all the numbers in vector s together with 2 and 4. This facility can be useful if it is desired to limit to number of principal components or eigenvalues displayed. If Spec is a spectral matrix, and we know that we will never have more than 10 significant components, then min([size(Spec)] 10) will choose a number that is the minimum of the two dimensions of Spec or equals 10 if this value is larger.

Some functions operate on individual elements rather than rows or columns. For example, sqrt(W) results in a new matrix of identical dimensions to **W** containing the square root of all the elements. In most cases whether a function returns a matrix, vector or scalar is common sense, but there are certain linguistic features, a few rather historic, so if in doubt test out the function first.

```
Command Window                                                    ⊙

  >> data=[9   3    7
  5    6    4
  8    4    0
  3    5    9
  1    6    1
  ];
  >> std(data)

  ans =

       3.3466      1.3038      3.8341

  >> std(data,1)

  ans =

       2.9933      1.1662      3.4293
```

Figure A.35 Calculating standard deviations in Matlab: the second calculation is preferred for most chemometric calculations where the aim is to scale a matrix.

A.5.3.4 Pre-Processing

Pre-processing is slightly awkward in Matlab. One way is to write a small program with loops as described in Section A.5.5. If you think in terms of vectors and matrices, however, it is quite easy to come up with a simple approach. If W is our original 2×3 matrix and we want to mean centre the columns, we can easily obtain a 1×3 vector $\overline{w}$, which corresponds to the means of each column, multiply this by a 2×1 vector $\mathbf{1}$ gives a 2×3 vector consisting of the means, and so our new mean centred matrix V can be calculated as $V = W - \mathbf{1}\overline{w}$. There is a special function in Matlab called ones that creates vectors or matrices that just consist of the number 1, an array ones (5,3) would create a matrix of dimensions 5×3 solely of 1s, so a 2×1 vector could be specified using the function ones(2,1). Hence, we can write V=W-ones(2,1)*mean(W) to create a new mean centred matrix V as illustrated in Figure A.36.

The experienced user of Matlab can build on this to perform other common methods for pre-processing such as standardisation.

A.5.3.5 Principal Components Analysis

There are several ways of performing PCA in Matlab. We will look at each method and how they relate to the results of NIPALS. We assume X is a $I \times J$ matrix.

SVD is performed by svd. A typical statement is [a b c] = svd(X).

- The matrix b is of dimensions $I \times J$. The values b(g,g) where g is the component number consists of the square root of the eigenvalue obtained by NIPALS. All other elements of this matrix are 0. This matrix is always positive.
- The matrix a is of dimensions $I \times I$. It corresponds to the scores matrix in NIPALS but is scaled so that the sum of squares of each column is 1, that is, sum(a(:,g).^2) = 1. To obtain the same answer as NIPALS, simply multiply each column by its eigenvalue, for example, b(g,g)*a(:,g). Each column represents a component.
- The matrix c is of dimensions $J \times J$. This corresponds to the loadings obtained in NIPALS. Each successive column corresponds to a successive component.

There are some important things to note. First of all, the sign after PCA cannot be controlled; thus, in some cases, using svd may give components opposite in sign to, for example, Excel. What is important is that the product of scores and loadings always has the same sign. The other is that if the two dimensions are different, only components up to the minimum of I and J are non-zero. The third is that the data are not automatically centred using svd, if you want to centre it, you must do it yourself.

```
Command Window                                    ⌄

  >> W

  W =

         2      7      8
         0      1      6

  >> V=W-ones(2,1)*mean(W)

  V =

         1      3      1
        -1     -3     -1
```

Figure A.36 Mean centring a matrix in Matlab.

The second method is `princomp`. This is due to be withdrawn in later versions of Matlab but is still widely employed. Unlike `svd`, by default the data are centred. Once centred, it is easy to relate the output to that obtained using NIPALS. Assume we calculate `[d e f] = princomp(X)`.

- The $J \times J$ matrix d corresponds exactly to the loadings matrix obtained via NIPALS, with each column corresponding to successive components.
- The $I \times I$ matrix e corresponds exactly to the scores matrix obtained via NIPALS, with each column corresponding to successive components. Note that using `princomp`, the first matrix is the loadings rather than scores matrix.
- f is a column vector rather than a matrix, unlike `svd`, each element corresponding to an eigenvalue. To get to the eigenvalue obtained using NIPALS, multiply each element by $I - 1$.

Remember that always there will be some zero components as in `svd`.

The replacement function `pca` is very similar to `princomp` but has more options. In particular, the calculation can be performed uncentred, so `[p q r] = pca(X,'Centred',false)` will result in the same scores as columns in matrix q and loadings as columns in matrix p as NIPALS on the raw uncentred data: the eigenvalues have to be multiplied by $I - 1$.

A.5.4 Importing and Exporting Data

In chemometrics, we want to perform operations on numerical data. There are many ways of getting information into Matlab generally straight into matrix format. Some of the simplest are as follows.

- Type the numerical information in the Command window as described above; for small data sets, this is probably easiest.
- If the information is available in a space delimited form with each row on a separate line, for example, as a text file, copy the data, type a command such as

`X = [`

but do NOT terminate this by the enter key, then paste the data into the Matlab window and finally terminate with

`]`

using a semicolon if you do not want to see the data displayed again (useful if the original data set is large such as a series of spectra).

- Information can be saved as mat files (Section A.5.2.1) and these can be imported into Matlab. Many public domain chemometrics data sets are stored in this format.

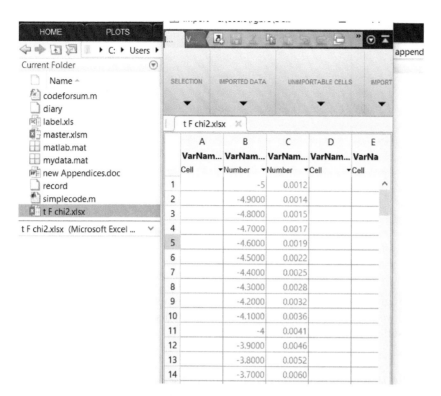

Figure A.37 Importing from Excel to Matlab.

- Information can be imported from Excel. The simplest way is to select a spreadsheet from the current directory. Clicking on this opens it, and one can select the data to import as illustrated in Figure A.37.

Data are best exported as mat files but can also be exported into a variety of formats. The `xlswrite` function is particularly useful when sending data to Excel.

Non-numerical data such as a character data, for example, names, can also be exported.

A.5.5 Introduction to Programming and Structure

For the enthusiasts, it is possible to write quite elaborate programs and develop very professional looking m files. The beginner is also advised to have a basic idea of a few of the main features of Matlab as a programming environment.

First and foremost is the ability to make comments (statements that are not executed), by starting a line with the % sign. Anything after this is simply ignored by Matlab but helps make large m files comprehensible.

Loops commence with the `for` statement, which has a variety of different syntaxes, the simplest being `for i = a : b`, which increments the variable i from the number a (which must be a scalar) to b by 1. An increment (which can be negative and does not need to be an integer) can be specified using the syntax `for i = a : n : b`, notice how, unlike many programming languages, this is the middle value of the three variables. Loops finish with the `end` statement. As an example, the operation of mean centring (Section A.5.3.4) is written in the form of a loop, see Figure A.38, the interested reader should be able to interpret the commands using the information given above. Obviously, for this quite small operation, a loop is not strictly necessary, but for more elaborate programs, it is important to be able to use loops, and there is a lot of flexibility about addressing matrices which make this facility very useful.

`If` and `while` facilities are also useful commands.

Many programmers like to organise their work into functions. In this introductory text, we will not delve too far into this, but a library of m files that consist of different functions can be easily set up as described in Section A.5.2.1.

This text does not aim to be a comprehensive manual about Matlab programming. The best way to learn is from hands-on experience, with a few tips to start up. Every programmer has their own style. If learning for chemometrics, a good starting point is to program in the algorithms of Section A.2, as practice, but also as possible building blocks for more elaborate computations. Most professional chemometrics have some programming experience, the majority in Matlab but also some other environments such as R or Python. Once one environment is mastered, it is normally easy to pick another.

```
Command Window                                    ⊙

  >> W=[2  7  8  ;  0  1  6];
  Y=mean(W);
  for  i=1:size(W,1);
  V(i,:)  =  W(i,:)  -  Y;
  end
  V


  V  =


        1        3        1
       -1       -3       -1
```

Figure A.38 A simple loop used for mean centring.

A.5.6   Graphics

There are a large number of different types of graph available in Matlab. Below we discuss a few methods that can be used to produce diagrams of the type employed in this text. The enthusiast will soon discover further approaches. Matlab is a very powerful tool for data visualisation.

A.5.6.1   Creating Figures

There are several ways to create new graphs. The simplest is by plotting a command as discussed in the next sections. A new window consisting of a figure is created. Unless otherwise indicated, each time a graphics command is executed, the graph in the figure window is overwritten.

In order to organise the figures better, it is preferable to use the `figure` command. Each time this is typed in the Matlab command window, a new blank figure, as illustrated in Figure A.39, is produced, so typing this three times in succession results in three blank figures each of which is able to contain a graph. The figures are automatically

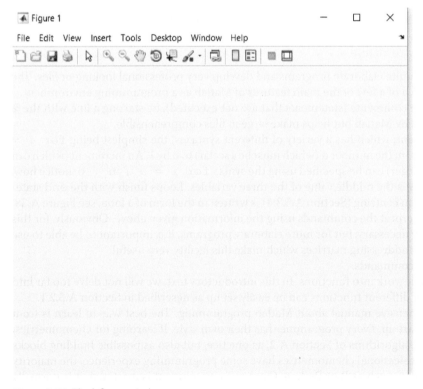

Figure A.39 Blank figure window.

numbered from 1 onwards. In order to return the second figure (number 2), simply type `figure(2)`. All plotting commands apply to the currently open figure. If you wish to produce a graph in the most recently opened window, it is not necessary to specify a number. Therefore, if you were to type the command `figure` three times, unless specified, the current graph will be displayed in Figure 3. The figures can be accessed either as small icons or through the `Window` menu item. It is possible to skip figure numbers; thus, the command `figure(10)` will create a figure number 10, even if no other figures have been created.

Each time you plot a graph if you do not open a new figure, the new graph will overwrite the old one. So if you want `figure; plot(x(:,1),x(:,2)); plot (y(:,1), y(:2))` where a semicolon is used to separate commands, you will only see the graph of `y(:,2)` versus `y(:,1)`. To avoid this, use the very useful `hold on` facility. This allows the user to superimpose several graphs. The statements `figure; plot(x(:,1),x(:,2)); hold on ; plot (y(:,1), y(:2))` will plot both on the same figure. This is illustrated in Figure A.40.

If you want to produce several small graphs on one figure, use the `subplot` command. This has the syntax `subplot(n,m,i)`. It divides the figure into $n \times m$ small graphs and puts the current plot into the ith position, where the first row is numbered from 1 to m, the second from $m + 1$ to $2m$ and so on. Figure A.41 illustrates the case where the commands `subplot(2,2,1)` and `subplot(2,2,2)` have been used to divide the window into a 2×2 grid, capable of holding up to four graphs, and figures have been inserted into positions 1 (top left) and 2 (top right). Further figures can be inserted into the grid in the vacant positions, or the current figures can be replaced and overwritten.

Once the figure is complete, you can copy it using the `Copy Figure` menu item and then place it in documents. In this section, we will illustrate the figures by screen snapshots showing the grey background of the Matlab screen. Alternatively, the figures can be saved in Matlab format, using the menu item under the current directory, as a `fig` file, which can then be opened and edited in Matlab in the future.

If you have several figures and want to remove them from Matlab (they may have been saved or the program may generate more figures than you need), the statement `close all` starts from afresh. Using just `close` only removes the most current figure.

A.5.6.2 Line Graphs

The simplest type of graph is a line graph. If Y is a vector, then `plot(Y)` will simply produce a graph of each element against row number. Often, we want to plot a row or column of a matrix against element number, for example, if each successive point corresponds to a point in time or a spectral wavelength. This is quite is to do, the command `plot(X(:,2))` plots the second column of X. Plotting a subset is also possible, for example, `plot(X(11:20,2))` produces a graph of column 2 rows 11–20, in practice allowing an expansion of the interesting region. Plotting more than one row against another can be done by specifying two arguments; hence, for example, `plot(X(:,1),X(:,2))` plots the first column on the horizontal scale and the second column on the vertical scale.

Once you have produced a line graph, it is possible to change its appearance. There are numerous ways of doing this, which can be learnt by experience, but there are two approaches.

The first approach is to expand the `plot` statement using specifiers. If writing a program to be used regularly so that plots of a certain type are always generated, this is the preferred approach. Common reasons are to change the line colour or whether there are markers (symbols). The Matlab help files provide a comprehensive list, but this can be specified using quotes, within the plot command. Common colours include blue ('b'), red ('r'), green ('g') and black ('b'). Common markers include circles ('o'), squares ('s') and diamonds ('d'). The line type can be continuous ('−'), none – just markers ('·'), or dotted (':'). These specifiers are combined together. For example, `plot(x(:,1),x(:,2),' ro')` will plot a graph with red circular markers and no line, as illustrated in Figure A.42.

Almost everything you want can be specified within a program, such as axis legends, chart titles and limits to axes, which the reader will learn by experience. One very useful statement is the `axis square` statement. This makes the axes square. So long as the axes are scaled correctly, this ensures that the angles in the graph reflect the angles in the original data space. Figure A.43 illustrates the importance. In the left-hand graph, the two lines do not appear to be at right angles, whereas in the right-hand graph they are. In certain applications, it is useful to see the true angles between vectors.

The second approach is interactive. Rather than type in statements, it is possible to change properties interactively. On the figure, click the white arrow on the top menu, and then any part of the graph you like. This will bring up the property editor as illustrated in Figure A.44. This is very flexible and allows the appearance of graphs to be customised. Particularly useful is to be able to label the axes and change the axis limits.

Almost all specifications can be changed both using the Property Editor and within code, and sometimes a combination is quickest.

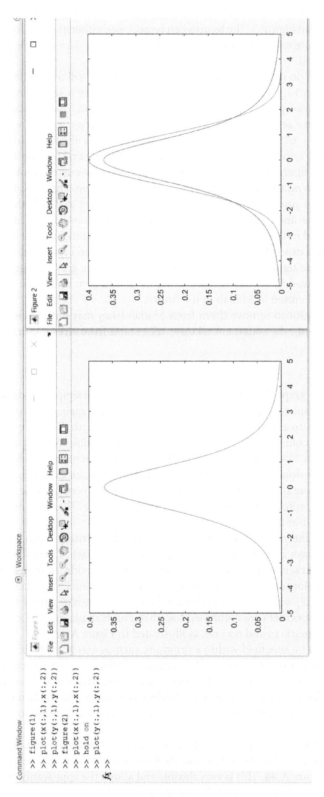

Figure A.40 Use of hold on.

Command Window

```
>> figure
>> subplot(2,2,1)
>> plot(x(:,1),x(:,2))
>> subplot(2,2,2)
>> plot(y(:,1),y(:,2))
fx >>
```

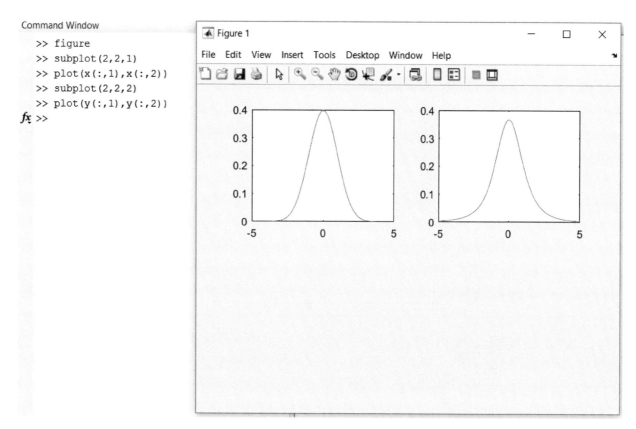

Figure A.41 Use of multiple plot facility.

Command Window

```
>> plot(x(:,1),x(:,2),' ro')
fx >>
```

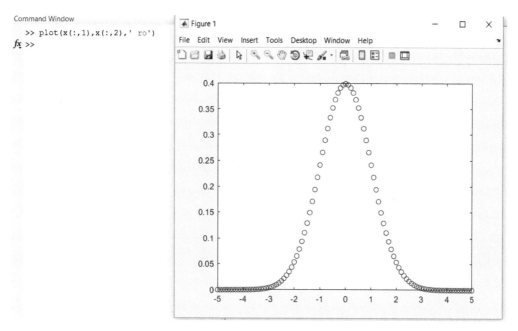

Figure A.42 Use of specifiers to change the properties of a graph in Matlab.

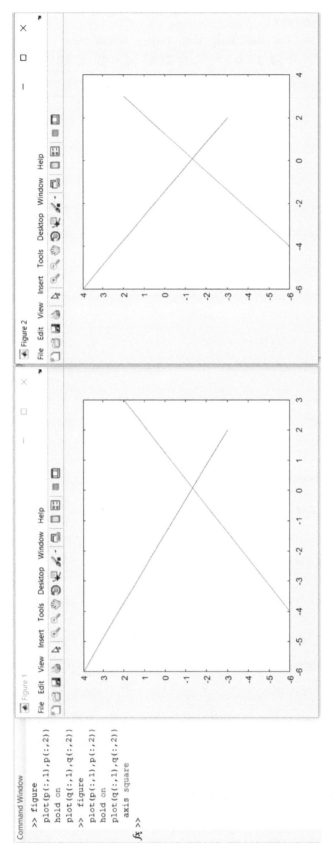

Figure A.43 Use of axis square statement to view correct angles between vectors.

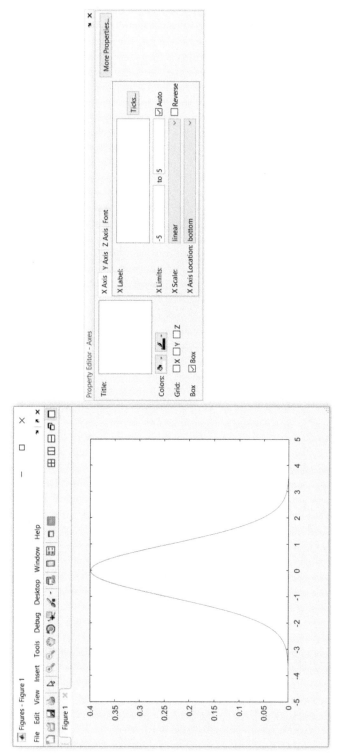

Figure A.44 Matlab Property Editor.

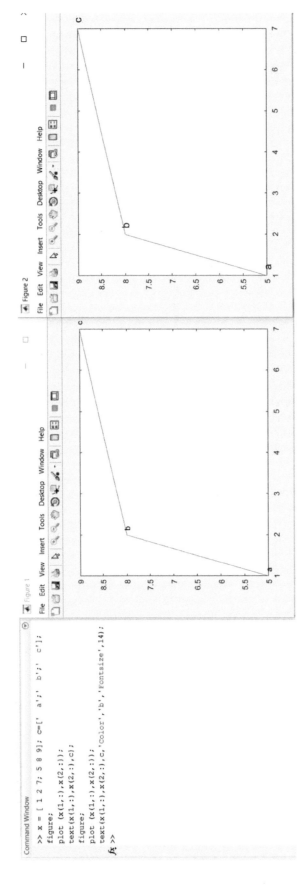

Figure A.45 Use of text command in Matlab.

In addition to line graphs obtained using the `plot` statement, there are several other common types of graphs. For brevity, we do not describe these, except for 3D graphics in Section A.5.6.4.

A.5.6.3 Labelling Points

Points in a graph can be labelled using the `text` command. The basic syntax is `text (A,B,name)`, where the A and B are arrays with the same number of elements, and it is recommended that `name` is an array of names or characters likewise with the identical number of elements. Matlab is rather awkward in handling of string (or character) variables. There are various ways telling Matlab that a variable is a string rather than numerical variable. Any data surrounding by single quotes is treated as a string, so the array `c = ['a' ; 'b' ; 'c']` will be treated by Matlab as a 3×1 character array. Figure A.45 illustrates the use of this method. Note that in order to prevent the labels overlapping with the points in the graph, leaving one or two spaces before the actual text helps. It is possible to move the labels, in the interactive graph editor if there is still some overlap, or else code this in. The appearance of the labelling can be altered by adding a specification; for example, `text(x(1,:),x(2,:),c,'Color','b','Fontsize',14)` specifies that the text is blue and with a font size of 14. The effect is illustrated in Figure A.45. The positions of the labels can be changed manually, or in order to prevent overlap with axes, the axis limits can also be changed.

Sometimes, the labels are originally in a numerical format; for example, they may consist of numerical values of points in time or wavelengths. For Matlab to recognise this, the numbers can be converted to strings using the `num2str` function, so that the numbers are changed to text and can be used for labelling points, for example, important wavelengths or key samples.

A.5.6.4 Three-Dimensional Graphics

One of the most flexible and important aspects of Matlab are the comprehensive 3D graphics facilities. We have used these in Chapter 7 when we illustrate the 3D PC plots. Matlab has facilities for contour plots, mesh plots and so on, but in this section, we focus only on the 3D plot statement.

Consider a scores matrix of dimensions 25×3 ($\boldsymbol{T}$) and a loadings matrix of dimensions 12×3 ($\boldsymbol{P}$); note that the default in Matlab is to represent the loadings as column rather than row vectors, see the discussion about PCA – whereas in this book, we normally represent loadings as row vectors. The statement `plot3 (T(:,1),T(:,2),T(:,3))` produces a graph of all the three columns of the scores against one another, see Figure A.46. Often, the default orientation is not the most informative for our purposes, and we may well wish to change this. There are a huge number of commands in Matlab to do this, which is a big bonus for the enthusiast, but for the first-time user, the easiest is to select the rotation icon, and interactively change the view (see Figure A.47). If that is the desired view, leave go of the icon.

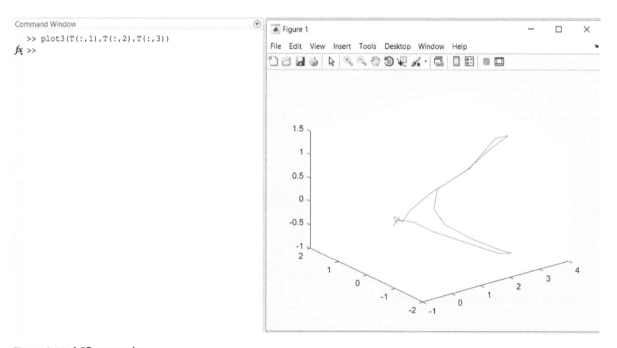

Figure A.46 A 3D scores plot.

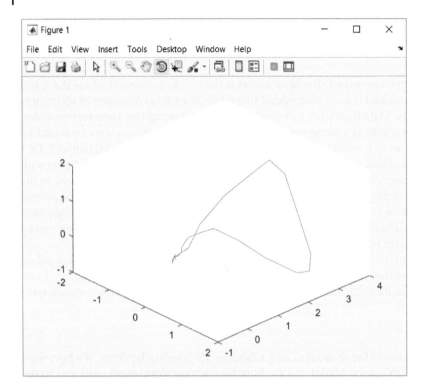

Figure A.47 Using the rotation icon to obtain a better view.

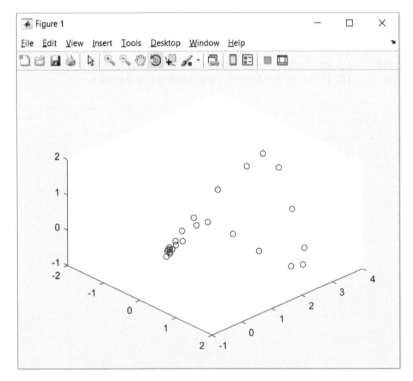

Figure A.48 Changing the appearance of the 3D plot.

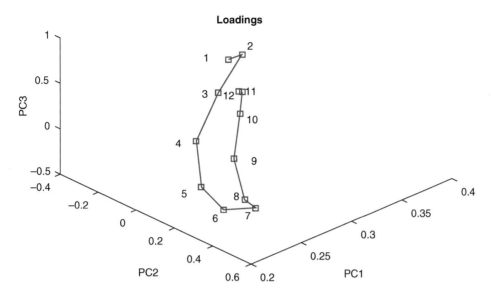

Figure A.49 Loadings plot with identical orientation to the scores plot, labelled and copied into Word.

Quite often, we want to return to the view, and a way of keeping the same perspective is via the `view` command. Typing `[A B]` = `view` will save this information. The enthusiasts will be able to interpret these in fundamental terms, but it is not necessary to understand this when first using 3D graphics in Matlab. However, in chemometrics, we often wish to look simultaneously at 3D scores and loadings plots and it is important that both have identical orientations. Thus, the way to do this is to ensure that the loadings have the same orientation as the scores (Figure A.48). The commands

```
figure(2)
    plot3(P(:,1),P(:,2),P(:,3))
    view(A B)
```

should place a loadings plot with the same orientation as the scores. This plot with the axes labelled, and a title as finalised and copied into Word is presented in Figure A.49.

The experienced user can improve these graphs just as 2D graphs, but with many additional facilities. Matlab is at its most powerful when visualising in 3D, which is a weakness of Excel.

Answers to the Multiple Choice Questions

2.2.1
1 a   2 b

2.2.2
1 a   2 b   3 b

2.2.3.1
1 c

2.2.3.2
1 b   2 b

2.2.3.3
1 b   2 b

2.2.3.4
1 b

2.2.4.1
1 b   2 b

2.2.4.2
1 b

2.2.4.3
1 b   2 b   3 a

2.2.4.4
1 b   2 b

2.2.4.5
1 a

2.2.4.6
1 b

2.2.4.7
1 a

2.2.5
1 b   2 b   3 a

2.3.1
1 b   2 c   3 b

2.3.2
1 b   2 b

2.3.3
1 a   2 a   3 c

2.3.4
1 a   2 b   3 b

2.4.1
1 d   2 b

2.4.2
1 a

2.4.3
1 a   2 b

2.4.4
1 a

2.4.5
1 a

2.5.1
1 b

2.5.2.1
1 c

2.5.2.2
1 b

2.5.2.3
1 a   2 b

2.5.3
1 b

2.5.4
1 d   2 a

2.5.5
1 b

2.6.1
1 d   2 a

2.6.2
1 a

2.6.3
1 c

2.6.4
1 b

3.2.1
1 b

3.2.1.1
1 b   2 c

3.2.1.2
1 c   2 a

3.2.1.3
1 b

3.2.1.4
1 b

3.2.2
1 b

3.2.3.1
1 b

3.2.3.2
1 a

3.2.3.3
1 a

3.2.4
1 a

3.3.1.1
1 a

3.3.1.2
1 a

3.3.1.3
1 c

3.3.1.4
1 a

3.3.2
1 b

3.3.3
1 a

3.4.1
1 a   2 c

3.4.2
1 b

3.4.3
1 a

3.5.1.1
1 a

3.5.1.2
1 b   2 b

3.5.1.3
1 a

3.5.1.4
1 a

3.5.1.5
1 a

3.5.2.1
1 b

3.5.2.2
1 b

3.5.2.3
1 a

3.5.3
1 a

3.6.1
1 b   2 c

3.6.2
1 b   2 b

3.6.3
1 c

3.6.4.1
1 b   2 a

3.6.4.2
1 b

4.2.1
1 b

4.2.2
1 a

4.2.3.1
1 c

4.2.3.2
1 c

4.2.3.3
1 b

4.2.4
1 b

4.3.1
1 b   2 b   3 c

4.3.2.1
1 a   2 b

4.3.2.2
1 a

4.4
1 a

4.5.1
1 c   2 a

4.5.2
1 a   2 b

4.6.1
1 b   2 b

4.6.2
1 b

4.6.3
1 b

4.6.4
1 a   2 c

4.6.5
1 b

4.7.1
1 b

4.7.2
1 a

4.8.1
1 a   2 b

4.8.2
1 c

4.8.3
1 b

4.8.4
1 b

4.9.1
1 b

4.9.2
1 a

4.9.3
1 a

5.1.2
1 b

5.2
1 a

5.2.1
1 c

5.2.1.1
1 d

5.2.1.2
1 b   2 a

5.2.1.3
1 c   2 a

5.2.1.4
1 b   2 b

5.2.2
1 b

5.2.3
1 b

5.3
1 c

5.3.1
1 d   2 c

5.3.2
1 c

5.4
1 b

5.5
1 b

5.5.1.1
1 a

5.5.1.2
1 a

5.5.1.3
1 b

5.5.1.4
1 d

5.5.2
1 b

5.6
1 b

5.6.1
1 d

5.6.2
1 b

5.6.3
1 a

6.1.1
1 b

6.1.2
1 a

6.2.1
1 a

6.2.2
1 a

6.2.3
1 c

6.3.1
1 c

6.3.2
1 a

6.3.3
1 b   2 b

6.4.1
1 c

6.4.2
1 c   2 b

6.5.1
1 b   2 a

6.5.2
1 b

6.5.3.1
1 d

6.5.3.2
1 b

6.6.1
1 a

6.6.2
1 c

6.6.3
1 a

7.2.1
1 a

7.2.2
1 b

7.2.3
1 c   2 b

7.2.4.1
1 b

7.2.4.2
1 a

7.2.5
1 a

7.3.1
1 c

7.3.2
1 a

7.3.3
1 b   2 b

7.3.4
1 b   2 c

7.3.5
1 a

7.4.1.1
1 a

7.4.1.2
1 a   2 a

7.4.1.3
1 c

7.4.2
1 b

7.4.3
1 a   2 b

Index